MODERN COSMOLOGY

MODERN COSMOLOGY

Scott Dodelson

Fermi National Accelerator Laboratory
University of Chicago

ACADEMIC PRESS

An Imprint of Elsevier

Amsterdam   Boston   London   New York   Oxford   Paris   San Diego
San Francisco   Singapore   Sydney   Tokyo

Senior Publishing Editor	Jeremy Hayhurst
Senior Project Manager	Julio Esperas
Editorial Coordinator	Nora Donaghy
Cover Design	Dick Hannus
Copyeditor	Kirstin Landon

This book is printed on acid-free paper.

On the cover: The distribution of galaxies in the universe as recorded by the Two Degree Field
Galaxy Survey. Courtesy http://msowww.anu.edu.au/2dFGRS/

ACADEMIC PRESS
An Imprint of Elsevier
525 B Street, Suite 1900, San Diego, California 92101-4495, USA
http://www.academicpress.com

Academic Press
84 Theobald's Road, London WC1X 8RR, UK
http://www.academicpress.com

Academic Press
200 Wheeler Road, Burlington, Massachusetts 01803, USA
www.academicpressbooks.com

Library of Congress Control Number: 2002117793

ISBN-13: 978-0-12-219141-1 ISBN-10: 0-12-219141-2
Transferred to Digital Printing 2011

CONTENTS

PREFACE

The Men of the Great Assembly had three sayings: "Be patient before reaching a decision; Enable many students to stand on their own; Make a fence around your teaching."

Ethics of the Fathers 1:1

There are two aspects of cosmology today that make it more alluring than ever. First, there is an enormous amount of data. To give just one example of how rapidly our knowledge of the structure of the universe is advancing, consider galaxy surveys which map the sky. In 1985, the state-of-the-art survey was the one carried out by the Center for Astrophysics; it consisted of the positions of 1100 galaxies. Today, the Sloan Digital Sky Survey and the Two Degree Field between them have recorded the 3D positions of half a million galaxies.

The other aspect of modern cosmology which distinguishes it from previous efforts to understand the universe is that we have developed a consistent theoretical framework which agrees *quantitatively* with the data. These two features are the secret of the excitement in modern cosmology: we have a theory which makes predictions, and these predictions can be tested by observations.

Understanding what the theory is and what predictions it makes is not trivial. First, many of the predictions are statistical. We don't predict that there should be a hot spot in the cosmic microwave background (CMB) at $RA = 15h$, dec= 27°. Rather, predictions are about the distribution and magnitude of hot and cold spots. Second, these predictions, and the theory on which they are based, involve lots of steps, many arguments drawn from a broad range of physics. For example, we will see that the distribution of hot and cold spots in the CMB depends on quantum mechanics, general relativity, fluid dynamics, and the interaction of light with matter. So we will indeed follow the first dictum of the Men of the Great Assembly and be patient before coming to judgment. Indeed, the fundamental measures of structure in the universe — the power spectra of matter and radiation — agree extraordinary well with the current cosmological theory, but we won't have the tools to understand this agreement completely until Chapters 7 and 8. Sober minds have always knows that it pays to be patient before pronouncing judgment on ideas as lofty as those necessary to understand our Universe. The modern twist on this "Be patient" theme is that we need to set up the framework (in this case Chapters

1–6) before we can appreciate the success of the current cosmological model.

Pick a random page in the book, and you will see that I have tried very hard to fulfill the second part of the aphorism. The hand-waving and qualitative arguments that facilitate understanding are here, but the main purpose of the book is to give you the tools to get in the game, to do calculations yourself, and follow cosmological arguments from first principles. Once you have mastered these tools, you will be prepared for any changes in the basic theoretical model. For example, much of the book is predicated on the notion that inflation seeded the structure we see today. If this turns out to be incorrect, the tools developed to study perturbations in Chapters 4 and 5 and the observations and analysis techniques described in the last half of the book will still be very relevant. As a more exotic example: all of the book assumes that there are three spatial dimensions in the universe. This seems like a plausible assumption (to me), but many theoretical physicists are now exploring the possibility that extra dimensions may have played a role in the early universe. If extra dimensions do turn out to be important, perturbations still need to be evolved and measured on our 3D brane. The tools developed here will still be useful.

The final part of the quote above is particularly relevant today since cosmology is such a broad subject. Many important papers, discoveries, and even subbranches of cosmology must be left ouside the fence. I think I have built the fence in a natural place. Enclosed within is the smooth expanding universe, with linear perturbations generated by inflation and then evolved with the Boltzmann–Einstein equations. The fence thus encloses not just the classical pillars of the Big Bang — the CMB, the expansion of the universe, and the production of the light elements — but also the modern pillars: the peaks and troughs in the CMB anisotropy spectrum; clustering of matter on large scales at just the right level; dark matter production and evolution; dark energy; inflation; the abundance of galaxy clusters; and velocity surveys. It also leaves room for important future developments such as weak lensing and polarization.

Outside the fence are some topics that will stay there forever, such as the steady state universe and similar alternatives. Other topics — notably cosmic strings and other topological defects — have been relegated beyond the fence only recently. Indeed, given the exciting research still being carried out to understand the cosmological implication of defects, it was a difficult decision to omit them entirely. Still other topics are crucial to an understanding of the universe and are the subject of active research, but are either too difficult or too unsettled. The most important of these is the study of nonlinearities. It would have been impossible to do justice to the advances over the last decade made in the study of nonlinear evolution. However, the linear theory presented here is a necessary prerequisite to understanding the growth of nonlinearities and their observational implications. A hint of the way in which our understanding of linear perturbations informs the nonlinear discussions is given in Section 9.5, where I discuss the attempts to predict the abundance of galaxy clusters (very nonlinear beasts) using linear theory.

Who is this book for? Researchers in one branch of cosmology wishing to learn about another should benefit. For example, inflation model builders who wish to understand the impact of their models on the CMB and large scale structure can learn the basics here. Experimentalists striving to understand the theoretical implications of their measurements can learn where those theory curves come from. People with no previous experience with statistics can use Chapter 11 to get up to date on the latest techniques. Even theorists who have heretofore worked only in one field, say large scale structure, can learn about new theoretical topics such as the CMB, weak lensing, and polarization. I have tried to emphasize the common origin of all these phenomena (small perturbations around a smooth background). More generally, researchers in other fields of physics who wish to understand the recent advances in cosmology can learn about them, and the physics on which they depend, here.

My main goal though is that the book should be accessible to beginning graduate students in physics and astronomy and to advanced undergrads looking to get an early start in cosmology. The only math needed is ordinary calculus and differential equations. As mentioned above, quite a bit of physics impacts on cosmology; however, you needn't have taken classes in all these fields to learn cosmology. General relativity is an essential tool, so a course in GR would be helpful, but I have tried to introduce the features we will need when we need them. For example, while we won't derive the Einstein equations, we will use them, and using them is pretty easy as long as one is comfortable with indices. Similarly, although inflation in Chapter 6 is based on field theory, you certainly do not need to have taken a course in field theory to understand the minimal amount needed for inflation. It can be easily understood if you understand the quantum mechanics of the harmonic oscillator.

To make the book easy to use, I have included summaries at the ends of some of the chapters. The idea is that you may not be interested in how the Boltzmann equations are derived, but you still need to know what they are to obtain the main cosmology results in Chapters 7–10. In that case, you can skip the bulk of Chapter 4 and simply skim the summary.

Writing the book has been almost pure pleasure in no small part because it forced me to read carefully papers I had previously been only dimly aware of. Thus a big acknowledgment to the many people who have pushed cosmology into the 21st century with all of their hard work. In the *"Suggested Reading"* sections at the end of each chapter, I have pointed to other books that should be useful, but also to the papers that influenced me most while working to understand the material in the chapter. These references, and others sprinkled throughout the text, are far from complete: they simply offer one entry into a vast literature which has grown dramatically in the last decade.

Many thanks to people who looked over early versions of the book and provided helpful comments, especially Mauricio Calvao, Douglas Scott and Uros Seljak. Kev Abazajian, Jeremy Bernstein, Pawel Dyk, Marc Kamionkowski, Manoj Kaplinghat, Eugene Lim, Zhaoming Ma, Angela Olinto, Eduardo Rozo, Ryan Scranton, Tristan Smith, and Iro Tasitsiomi also offered useful suggestions. Jeremy Berstein, Sanghamitra Deb, James Dimech, Jim Fry, Donghui Jeong, Bob Klauber, Chung-

Pei Ma, Olga Mena, Aravind Natarajan, Mark Alan Peot, Eduardo Rozo, Suharyo Sumowidagdo, and Tong-Jie Zhang found mistakes in earlier printings and graciously let me know about them. Thanks also to Andy Albrecht who introduced me to Susan Rabiner, and to Susan who was very supportive throughout. Thanks to Nora Donaghy, Julio Esperas, Jeremy Hayhurst, and Lakshmi Sadasiv, my contacts at Academic Press. I was supported by a grant from Academic Press, by NASA Grant NAG5-10842, by the DOE, and by NSF Grant PHY-0079251. Finally, I am most grateful to Marcia, Matthew, Ilana, David, and Coby for their support and love.

1

THE STANDARD MODEL AND BEYOND

Einstein's discovery of general relativity in the last century enabled us for the first time in history to come up with a compelling, testable theory of the universe. The realization that the universe is expanding and was once much hotter and denser allows us to modernize the deep age-old questions "Why are we here?" and "How did we get here?" The updated versions are now "How did the elements form?", "Why is the universe so smooth?", and "How did galaxies form from this smooth origin?" Remarkably, these questions and many like them have quantitative answers, answers that can be found only by combining our knowledge of fundamental physics with our understanding of the conditions in the early universe. Even more remarkable, these answers can be tested against astronomical observations.

This chapter describes the idea of an expanding universe, without using the equations of general relativity. The success of the Big Bang rests on three observational pillars: the Hubble diagram exhibiting expansion; light element abundances which are in accord with Big Bang nucleosynthesis; and the blackbody radiation left over from the first few hundred thousand years, the cosmic microwave background. After introducing these pieces of evidence, I move beyond the *Standard Model* embodied by the three pillars. Developments in the last two decades of the 20th century — both theoretical and observational — point to

- the existence of dark matter and perhaps even dark energy
- the need to understand the evolution of perturbations around the zero order, smooth universe
- inflation, the generator of these perturbations

The emergent picture of the early universe is summarized in the time line of Figure 1.15.

1.1 THE EXPANDING UNIVERSE

We have good evidence that the universe is expanding. This means that early in its history the distance between us and distant galaxies was smaller than it is

1

today. It is convenient to describe this effect by introducing the scale factor a, whose present value is set to one. At earlier times a was smaller than it is today. We can picture space as a grid as in Figure 1.1 which expands uniformly as time evolves. Points on the grid maintain their coordinates, so the *comoving distance* between two points — which just measures the difference between coordinates — remains constant. However, the physical distance is proportional to the scale factor, and the physical distance does evolve with time.

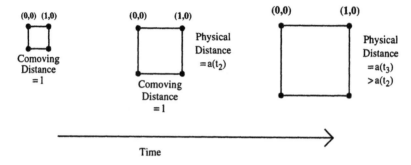

Figure 1.1. Expansion of the universe. The comoving distance between points on a hypothetical grid remains constant as the universe expands. The physical distance is proportional to the comoving distance times the scale factor, so it gets larger as time evolves.

In addition to the scale factor and its evolution, the smooth universe is characterized by one other parameter, its geometry. There are three possibilities: flat, open, or closed universes. These different possibilities are best understood by considering two freely traveling particles which start their journeys moving parallel to each other. A *flat* universe is Euclidean: the particles remain parallel as long as they travel freely. General relativity connects geometry to energy. Accordingly, a flat universe is one in which the energy density is equal to a critical value, which we will soon see is approximately 10^{-29} g cm^{-3}. If the density is higher than this value, then the universe is *closed*: gradually the initially parallel particles converge, just as all lines of constant longitude meet at the North and South Poles. The analogy of a closed universe to the surface of a sphere runs even deeper: both are said to have *positive curvature*, the former in three spatial dimensions and the latter in two. Finally, a low-density universe is *open*, so that the initially parallel paths diverge, as would two marbles rolling off a saddle.

To understand the history of the universe, we must determine the evolution of the scale factor a with cosmic time t. Again, general relativity provides the connection between this evolution and the energy in the universe. Figure 1.2 shows how the scale factor increases as the universe ages. Note that the dependence of a on t varies as the universe evolves. At early times, $a \propto t^{1/2}$ while at later times the dependence switches to $a \propto t^{2/3}$. How the scale factor varies with time is determined by the energy density in the universe. At early times, one form of energy, radiation, dominates, while at later times, nonrelativistic matter accounts for most of the energy density. In fact, one way to explore the energy content of the universe is to measure changes in the scale factor. We will see that, partly as a result of such exploration, we now believe that, very recently, a has stopped growing as $t^{2/3}$, a signal that a new form of energy has come to dominate the cosmological landscape.

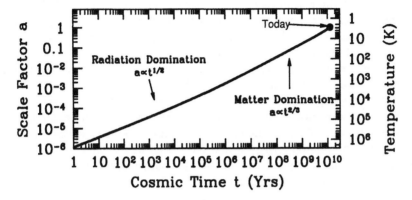

Figure 1.2. Evolution of the scale factor of the universe with cosmic time. When the universe was very young, radiation was the dominant component, and the scale factor increased as $t^{1/2}$. At later times, when matter came to dominate, this dependence switched to $t^{2/3}$. The right axis shows the corresponding temperature, today equal to 3K.

To quantify the change in the scale factor and its relation to the energy, it is first useful to define the Hubble rate

$$H(t) \equiv \frac{da/dt}{a} \qquad (1.1)$$

which measures how rapidly the scale factor changes. For example, if the universe is flat and matter-dominated, so that $a \propto t^{2/3}$, then $H = (2/3)t^{-1}$. Thus a powerful test of this cosmology is to measure separately the Hubble rate today, H_0, and the age of the universe today. Here and throughout, subscript 0 denotes the value of a quantity today. In a flat, matter-dominated universe, the product $H_0 t_0$ should equal 2/3.

More generally, the evolution of the scale factor is determined by the Friedmann equation

$$H^2(t) = \frac{8\pi G}{3}\left[\rho(t) + \frac{\rho_{\mathrm{cr}} - \rho_0}{a^2(t)}\right] \qquad (1.2)$$

where $\rho(t)$ is the energy density in the universe as a function of time with ρ_0 the present value. The *critical density*

$$\rho_{\mathrm{cr}} \equiv \frac{3H_0^2}{8\pi G} \qquad (1.3)$$

where G is Newton's constant.

To use Einstein's equation, we must know how the energy density evolves with time. This turns out to be a complicated question because ρ in Eq. (1.2) is the sum of several different components, each of which scale differently with time. Consider first nonrelativistic matter. The energy of one such particle is equal to its rest mass energy, which remains constant with time. The energy density of many of these is therefore equal to the rest mass energy times the number density. When the scale factor was smaller, the densities were necessarily larger. Since number density is inversely proportional to volume, it should be proportional to a^{-3}. Therefore the energy density of matter scales as a^{-3}.

The photons which make up the cosmic microwave background (CMB) today have a well-measured temperature $T_0 = 2.725 \pm 0.002K$ (Mather *et al.*, 1999). A photon with an energy $k_B T_0$ today has a wavelength $\hbar c / k_B T_0$. Early on, when the scale factor was smaller than it is today, this wavelength would have been correspondingly smaller. Since the energy of a photon is inversely proportional to its wavelength, the photon energy would have been larger than today by a factor of $1/a$. This argument applied to the thermal bath of photons implies that the temperature of the plasma as a function of time is

$$T(t) = T_0/a(t). \tag{1.4}$$

At early times, then, the temperature was higher than it is today, as indicated in Figure 1.2. The energy density of radiation, the product of number density times average energy per particle, therefore scales as a^{-4}.

Evidence from distant supernovae (Chapter 2; Riess *et al.*, 1998; Perlmutter *et al.*, 1999) suggests that there may well be energy, *dark energy*, besides ordinary matter and radiation. One possibility is that this new form of energy remains constant with time, i.e., acts as a *cosmological constant*, a possibility first introduced (and later abandoned) by Einstein. Cosmologists have explored other forms though, many of which behave very differently from the cosmological constant. We will see more of this in later chapters.

Equation (1.2) allows for the possibility that the universe is not flat: if it were flat, the sum of all the energy densities today would equal the critical density, and the last term in Eq. (1.2) would vanish. If the universe is not flat, the *curvature* energy scales as $1/a^2$. In most of this book we will work within the context of a flat universe. In such a universe, the evolution of perturbations is much easier to calculate than in open or closed universes. Further, there are several persuasive arguments, both theoretical and more recently observational, which strongly support the flatness of the universe. More on this in Chapters 2 and 8.

Figure 1.3 illustrates how the different terms in Eq. (1.2) vary with the scale factor. While today matter, and possibly a cosmological constant dominate the landscape, early on, because of the a^{-4} scaling, radiation was the dominant constituent of the universe.

Let's introduce some numbers. The expansion rate is a measure of how fast the universe is expanding, determined (Section 1.2) by measuring the velocities of distant galaxies and dividing by their distance from us. So the expansion is often

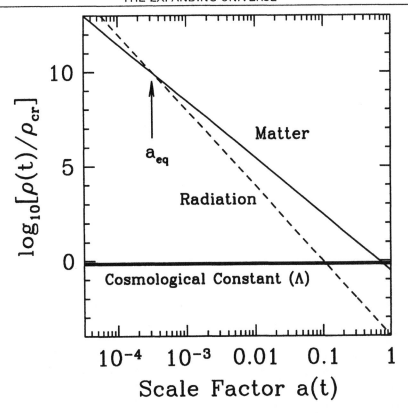

Figure 1.3. Energy density vs scale factor for different constituents of a flat universe. Shown are nonrelativistic matter, radiation, and a cosmological constant. All are in units of the critical density today. Even though matter and cosmological constant dominate today, at early times, the radiation density was largest. The epoch at which matter and radiation are equal is a_{eq}.

written in units of velocity per distance. Present measures of the Hubble rate are parameterized by h defined via

$$H_0 = 100h \ \text{km sec}^{-1} \, \text{Mpc}^{-1}$$

$$= \frac{h}{0.98 \times 10^{10} \ \text{years}} = 2.133 \times 10^{-33} h \ \text{eV}/\hbar \qquad (1.5)$$

where h has nothing to do with Planck's constant $\hbar$. The astronomical length scale of a megaparsec (Mpc) is equal to 3.0856×10^{24} cm. Current measurements set $h = 0.72 \pm 0.08$ (Freedman *et al.*, 2001).

    The predicted age for a flat, matter-dominated universe, $(2/3)H_0^{-1}$, is then of order 8 to 10 Gyr. The current best estimate for the age of the universe is 12.6 Gyr, with a 95% confidence level lower limit of 10.4 Gyr (Krauss and Chaboyer, 2001), so this test suggests that a flat, matter-dominated universe is barely viable. You will show in Exercise 2 that the age of the universe with a cosmological constant is larger (for fixed h); in fact one of the original arguments in favor of this excess energy was to make the universe older.

Newton's constant in Eq. (1.3) is equal to $6.67 \times 10^{-8}\mathrm{cm}^3\mathrm{g}^{-1}\mathrm{sec}^{-2}$. This, together with Eq. (1.5), enables us to get a numerical value for the critical density:

$$\rho_{\mathrm{cr}} = 1.88h^2 \times 10^{-29}\mathrm{g\ cm}^{-3}. \qquad (1.6)$$

An important ramification of the higher densities in the past is that the rates for particles to interact with each other, which scale as the density, were also much higher early on. Figure 1.4 shows some important rates as a function of the scale

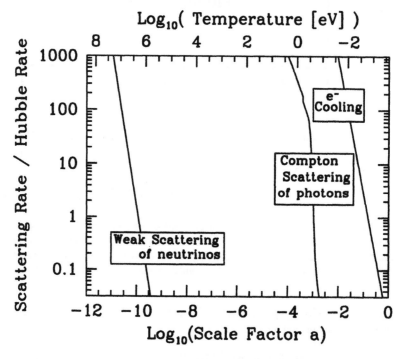

Figure 1.4. Rates as a function of the scale factor. When a given rate becomes smaller than the expansion rate H, that reaction falls out of equilibrium. Top scale gives (k_B times) the temperature of the universe, an indication of the typical kinetic energy per particle.

factor. For example, when the temperature of the universe was greater than several MeV$/k_B$, the rate for electrons and neutrinos to scatter was larger than the expansion rate. Thus, before the universe could double in size, a neutrino scattered many times off background electrons. All these scatterings brought the neutrinos into equilibrium with the rest of the cosmic plasma. This is but one example of a very general, profound fact: if a particle scatters with a rate greater than the expansion rate, that particle stays in equilibrium. Since rates were typically quite large, the early universe was a relatively simple environment: not only was it very smooth, but many of its constituents were in equilibrium. Chapter 2 explores some manifestations of the equilibrium conditions, while Chapter 3 touches on several cases where equilibrium could not be maintained because the reaction rates dropped beneath the expansion rate.

1.2 THE HUBBLE DIAGRAM

If the universe is expanding as depicted in Figure 1.1, then galaxies should be moving away from each other. We should therefore see galaxies receding from us. Recall that the wavelength of light or sound emitted from a receding object is stretched out so that the observed wavelength is larger than the emitted one. It is convenient to define this stretching factor as the redshift z:

$$1 + z \equiv \frac{\lambda_{\text{obs}}}{\lambda_{\text{emit}}} = \frac{1}{a}. \tag{1.7}$$

For low redshifts, the standard Doppler formula applies and $z \simeq \frac{v}{c}$. So a measurement of the amount by which absorption and/or emission lines are redshifted is a direct measure of how fast the structures in which they reside are receding from us.

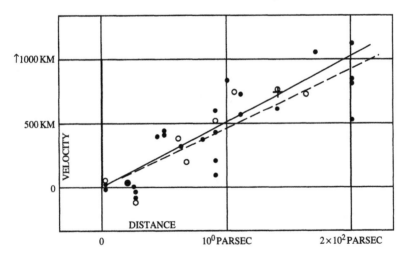

Figure 1.5. The original Hubble diagram (Hubble, 1929). Velocities of distant galaxies (units should be km sec^{-1}) are plotted vs distance (units should be Mpc). Solid (dashed) line is the best fit to the filled (open) points which are corrected (uncorrected) for the sun's motion.

Hubble (1929) first found that distant galaxies are in fact receding from us. He also noticed the trend that the velocity increases with distance. This is exactly what we expect in an expanding universe, for the physical distance between two galaxies is $d = ax$ where x is the comoving distance. In the absence of any comoving motion ($\dot{x} = 0$, no *peculiar* velocity) the relative velocity $v = \dot{d}$ is therefore equal to $\dot{a}x = Hd$. Therefore, velocity should increase linearly with distance (at least at low redshift) with a slope given by H, the Hubble constant. Hubble's Hubble constant can be easily extracted from Figure 1.5. It is simply $H = 1000/2$ km sec^{-1} Mpc^{-1}, almost a factor of 10 higher than current estimates. Also notice that Hubble's data went out to redshift $z = 1000$ km sec$^{-1}/c \simeq 0.003$.

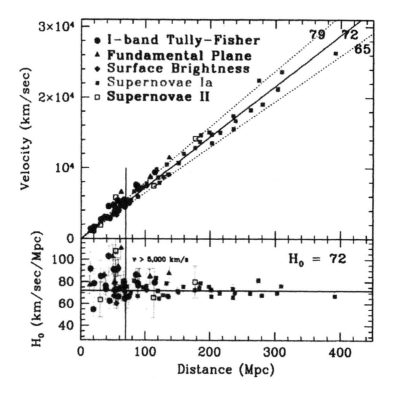

Figure 1.6. Hubble diagram from the Hubble Space Telescope Key Project (Freedman *et al.*, 2001) using five different measures of distance. Bottom panel shows H_0 vs distance with the horizontal line equal to the best fit value of 72 km sec^{-1} Mpc^{-1}.

The Hubble diagram is still the most direct evidence we have that the universe is expanding. Current incarnations use the same principle as the original: find the distance and the redshift of distant objects. Measuring redshifts is straightforward; the hard part is determining distances for objects of unknown intrinsic brightness. One of the most popular techniques is to try to find a *standard candle*, a class of objects which have the same intrinsic brightness. Any difference between the apparent brightness of two such objects then is a result of their different distances from us. This method is typically generalized to find a correlation between an observable and intrinsic brightness. For example, Cepheid variables are stars for which intrinsic brightness is tightly related to period. The Hubble Space Telescope measured the periods of thousands of Cepheid variables in galaxies as far away as 20 Mpc. With distances to these galaxies fixed, five different distance measures were used to go much further, as far away as 400 Mpc. Figure 1.6 shows that all of these five indicators agree with one another and have converged on $H_0 = 72$ km sec^{-1} Mpc^{-1} with 10% errors.

As shown in Figure 1.6 the standard candle that can be seen at largest distances is a Type Ia supernova. Since they are so bright, supernovae can be used to extend the Hubble diagram out to very large redshifts (the current record is of order $z \simeq 1.7$), a regime where the simple Doppler law ceases to work. Figure 1.7 shows a recent Hubble diagram using very these very distant objects. In the next chapter, we will derive the correct expression for the distance (in this case the *luminosity* distance) as a function of redshift. For now, I simply point out that this expression depends on the energy content of the universe. The three curves in Figure 1.7 depict three different possibilities: flat matter dominated; open; and flat with a cosmological constant (Λ). The high-redshift data are now good enough to distinguish among these possibilities, strongly disfavoring the previously favored flat, matter-dominated universe. The current best fit is a universe with about 70% of the energy in the form of a cosmological constant, or some other form of dark energy. More on this in Chapter 2.

1.3 BIG BANG NUCLEOSYNTHESIS

When the universe was much hotter and denser, when the temperature of order an MeV/k_B, there were no neutral atoms or even bound nuclei. The vast amounts of radiation in such a hot environment ensured that any atom or nucleus produced would be immediately destroyed by a high energy photon. As the universe cooled well below the binding energies of typical nuclei, light elements began to form. Knowing the conditions of the early universe and the relevant nuclear cross-sections, we can calculate the expected primordial abundances of all the elements (Chapter 3).

Figure 1.8 shows the predictions of Big Bang Nucleosynthesis (BBN) for the light element abundances[1]. The boxes and arrows in Figure 1.8 show the current estimates for the light element abundances. These are consistent with the predictions, and this consistency test provides yet another ringing confirmation of the Big Bang. The measurements do even more though. The theoretical predictions, which we will explore in detail in Chapter 3, depend on the density of protons and neutrons at the time of nucleosynthesis. The combined proton plus neutron density

---

[1] Recall nuclear notation: The 4 in ^{4}He refers to the total number of nucleons (protons and neutrons). So ^{4}He has two neutrons and two protons, while ^{3}He has two protons and one neutron. See the box on page 63 for more details.

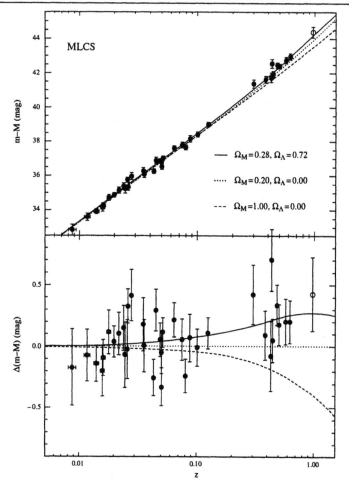

Figure 1.7. Hubble diagram from distant Type Ia supernovae. Top panel shows apparent magnitude (an indicator of the distance) vs redshift. Lines show the predictions for different energy contents in the universe, with Ω_M the ratio of energy density today in matter compared to the critical density and Ω_Λ the ratio of energy density in a cosmological constant to the critical density. Bottom panel plots the residuals, making it clear that the high-redshift supernovae favor a Λ-dominated universe over a matter-dominated one.

is called the *baryon* density since both protons and neutrons have baryon number one and these are the only baryons around at the time. Thus, BBN gives us a way of measuring the baryon density in the universe. Since we know how those densities scale as the universe evolves (they fall as a^{-3}), we can turn the measurements of light element abundances into measures of the baryon density today.

In particular, the measurement of primordial deuterium pins down the baryon density extremely accurately to only a few percent of the critical density. Ordi-

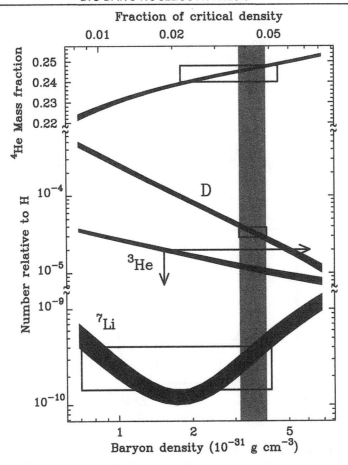

Figure 1.8. Constraint on the baryon density from Big Bang Nucleosynthesis (Burles, Nollett, and Turner, 1999). Predictions are shown for four light elements — ^{4}He, deuterium, ^{3}He, and lithium — spanning a range of 10 orders of magnitude. The solid vertical band is fixed by measurements of primordial deuterium. The boxes are the observations; there is only an upper limit on the primordial abundance of ^{3}He.

nary matter (baryons) contributes at most 5% of the critical density. Since the total matter density today is almost certainly larger than this — direct estimates give values of order 20–30% — nucleosynthesis provides a compelling argument for nonbaryonic dark matter.

The deuterium measurements (Burles and Tytler, 1998) are the new developments in the field. These measurements are so exciting because they explore the deuterium abundance at redshifts of order 3–4, well before much processing could have altered the primordial abundances. Figure 1.9 shows one such detection. The basic idea is that light from distant QSOs is absorbed by intervening neutral hydrogen systems. The key absorption feature arises from transition from the ($n = 1$)

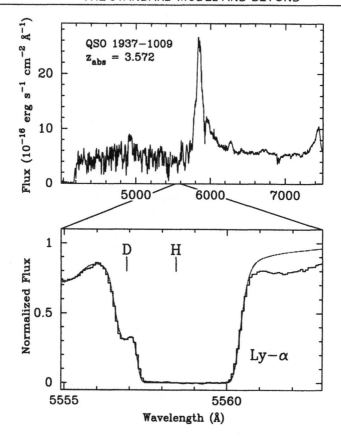

Figure 1.9. Spectrum from a distant QSO (Burles, Nollett, and Turner, 1999). Absorption of photons with rest wavelength 1216 Å corresponding to the $n = 1$ to $n = 2$ state of hydrogen is redshifted up to $1216\,(1 + 3.572)$ Å. Bottom panel provides details of the spectrum in this range, with the the presence of deuterium clearly evident.

ground state of hydrogen to the first excited state ($n = 2$), requiring a photon with wavelength $\lambda = 1215.7$ Å. Since photons are absorbed when exciting hydrogen in this fashion, there is a trough in the spectrum at $\lambda = 1215.7$ Å, redshifted by a factor of $1 + z$. The corresponding line from deuterium should be (i) shifted over by $0.33\,(1 + z)$ Å(see Exercise 3) and (ii) much less damped since there is much less deuterium. Figure 1.9 shows just such a system; there are now half a dozen with detections precisely in the neighborhood shown in Figure 1.8. Note that the steep decline in deuterium as a function of baryon density helps here: even relatively large errors in D measurements translate into small errors on the baryon density.

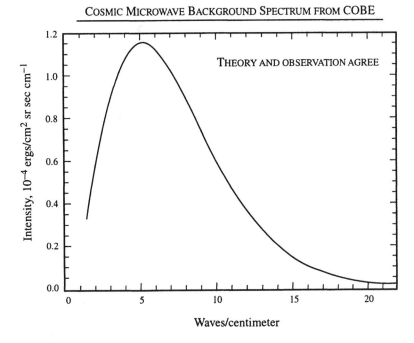

Figure 1.10. Intensity of cosmic microwave radiation as a function of wavenumber from Far InfraRed Absolute Spectrophotometer (FIRAS) (Mather *et al.*, 1994), an instrument on the COBE satellite. Hidden in the theoretical blackbody curve are dozens of measured points, all of which have uncertainties smaller than the thickness of the curve!

1.4 THE COSMIC MICROWAVE BACKGROUND

The CMB offers us a look at the universe when it was only $300,000$ years old. The photons in the cosmic microwave background last scattered off electrons at redshift 1100; since then they have traveled freely through space. When we observe them today, they literally come from the earliest moments of time. They are therefore the most powerful probes of the early universe. We will spend an inordinate amount of time in this book working through the details of what happened before the epoch of last scattering and also developing the mathematics of the freestreaming process since then. A crucial fact about this history, though, is that the collisions with electrons before last scattering ensured that the photons were in equilibrium. That is, they should have a blackbody spectrum.

The specific intensity of a gas of photons with a blackbody spectrum is

$$I_\nu = \frac{4\pi\hbar\nu^3/c^2}{\exp\{2\pi\hbar\nu/k_B T\} - 1}. \tag{1.8}$$

Figure 1.10 shows the remarkable agreement between this prediction (see Exercise 4) of Big Bang cosmology and the observations by the FIRAS instrument aboard the

COBE spacecraft. We have been told[2] that detection of the 3K background by Penzias and Wilson in the mid-1960s was sufficient evidence to decide the controversy in favor of the Big Bang over the Steady State universe. Penzias and Wilson, though, measured the radiation at just one wavelength. If even their one-wavelength result was enough to tip the scales, the current data depicted in Figure 1.10 should send skeptics from the pages of physics journals to the far reaches of radical Internet chat groups.

The most important fact we learned from our first 25 years of surveying the CMB was that the early universe was very smooth. No anisotropies were detected in the CMB. This period, while undoubtedly frustrating for observers searching for anisotropies, solidified the view of a smooth Big Bang. We are now moving on. We have discovered anisotropies in the CMB, indicating that the early universe was not completely smooth. There were small perturbations in the cosmic plasma. To understand these, we must go beyond the Standard Model.

1.5 BEYOND THE STANDARD MODEL

While the three pillars put the Big Bang model on firm footing, other observations cry out for more details. I hinted above at one of these, the notion that there must be nonbaryonic matter in the universe. *Dark matter* is a familiar concept to astronomers; the first suggestion was put forth by Zwicky in 1933(!). Figure 1.11 illustrates the way dark matter can be found in galaxies, with the use of rotation curves probing the gravitational field. Indeed, a mismatch between the matter inferred from gravity and that we can see exists on almost all observable scales.

Because of the limits inferred from Big Bang nucleosynthesis, the dark matter, or at least an appreciable fraction of it, must be nonbaryonic. What is this new form of matter? And how did it form in the early universe? The most popular idea currently is that the dark matter consists of elementary particles produced in the earliest moments of the Big Bang. In Chapter 3, we will explore this possibility in detail, arguing that dark matter was likely produced when the temperature of the universe was of order hundreds of GeV/k_B. As we will see, the hypothesis that dark matter consists of fundamental relics from the early universe may soon be tested experimentally.

The last decades of the 20th century saw a number of large surveys of galaxies designed to measure structure in the universe. These culminated in two large surveys, the Sloan Digital Sky Survey and the Two Degree Field Galaxy (Figure 1.12) Redshift Survey, which between them will compile the redshifts of, and hence the distances to, a million galaxies. Galaxies in Figure 1.12 are clearly not distributed randomly: the universe has structure on large scales. To understand this structure, we must go beyond the Standard Model not only by including dark matter, but also by allowing for deviations from smoothness. We must develop the tools to study

---

[2]For a fascinating first-hand account of the history of the discovery of the CMB, see Chapter 1 of Partridge (1995).

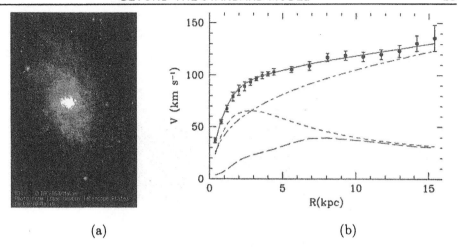

(a)   (b)

Figure 1.11. (a) Image of spiral galaxy M33. The inner brightest region has a radius of several kpc. (b) Rotation curve for M33 (Corbelli and Salucci, 2000). Points with error bars come from the 21-cm line of neutral hydrogen. Solid line is a model fitting the data. Different contributions to the total rotation curve are: dark matter halo (dot-dashed line), stellar disk (short dashed line), and gas (long dashed line). At large radii, dark matter dominates.

perturbations around the smooth background of the Standard Model. We will see in Chapters 4 and 5 that this is straightforward in theory, as long as the perturbations remain small.

The best ways to learn about the evolution of structure and to compare theory with observations are to look at anisotropies in the CMB and at how matter is distributed on large scales. To compare theory with observations, we must at first try to avoid scales dominated by nonlinearities. As an extreme example, we can never hope to understand cosmology by carefully examining rock formations on Earth. The intermediate steps — collapse of matter into a galaxy; molecular cooling; star formation; planetary formation; etc. — are much too complicated to allow comparison between linear theory and observations. While perturbations to the matter on small scales (less than about 10 Mpc) have grown nonlinear, large-scale perturbations are still small. So they have been processed much less than the corresponding small-scale structure. Similarly, anisotropies in the CMB have remained small because the photons that make up the CMB do not clump.

Identifying large-scale structure and the CMB as the two most promising areas of study solves just one issue. Another very important challenge is to understand how to characterize these distributions so that theory can be compared to experiment. It is one thing to look at a map and quite another to quantitative tests of cosmological models. To make such tests, it is often useful to take the Fourier transform of the distribution in question; as we will see, working in Fourier space makes it easier to separate large from small scales. The most important statistic in the cases of

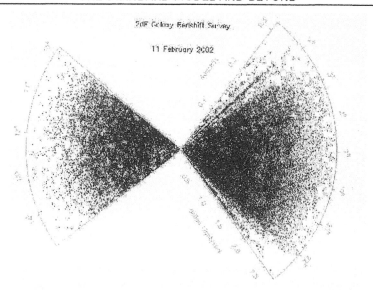

Figure 1.12. Distribution of galaxies in the Two Degree Field Galaxy Redshift Survey (2dF) (Colless *et al.*, 2001). By the end of the survey, redshifts for 250,000 galaxies will have been obtained. As shown here, they probe structure in the universe out to $z = 0.3$, corresponding to distances up to $1000\,h^{-1}$ Mpc away from us (we are located at the center). See color Plate 1.12.

both the CMB and large-scale structure is the *two-point function*, called the *power spectrum* in Fourier space. If the mean density of the galaxies is $\bar{n}$, then we can characterize the inhomogeneities with $\delta(\vec{x}) = (n(\vec{x}) - \bar{n})/\bar{n}$, or its Fourier transform $\tilde{\delta}(\vec{k})$. The power spectrum $P(k)$ is defined via

$$\langle \tilde{\delta}(\vec{k}) \tilde{\delta}(\vec{k}') \rangle = (2\pi)^3 P(k) \delta^3(\vec{k} - \vec{k}'). \tag{1.9}$$

Here the angular brackets denote an average over the whole distribution, and $\delta^3()$ is the Dirac delta function which constrains $\vec{k} = \vec{k}'$. The details aside, Eq. (1.9) indicates that the power spectrum is the spread, or the variance, in the distribution. If there are lots of very under- and overdense regions, the power spectrum will be large, whereas it is small if the distribution is smooth. Figure 1.13 shows the power spectrum of the galaxy distribution. Since the power spectrum has dimensions of k^{-3} or (length)^3, Figure 1.13 shows the combination $k^3 P(k)/2\pi^2$, a dimensionless number which is a good indication of the clumpiness on scale k.

The best measure of anisotropies in the CMB is also the two-point function of the temperature distribution. There is a subtle technical difference between the two power spectra which are used to measure the galaxy distribution and the CMB, though. The difference arises because the CMB temperature is a two-dimensional field, measured everywhere on the sky (i.e., with two angular coordinates). Instead of Fourier transforming the CMB temperature, then, one typically expands it in

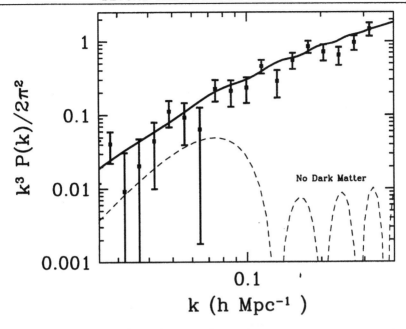

Figure 1.13. The variance $\Delta^2 \equiv k^3 P(k)/2\pi^2$ of the Fourier transform of the galaxy distribution as a function of scale. On large scales, the variance is smaller than unity, so the distribution is smooth. The solid line is the theoretical prediction from a model in which the universe contains dark matter, a cosmological constant, with perturbations generated by inflation. The dashed line is a theory with only baryons and no dark matter. Data come from the PSCz survey (Saunders *et al.*, 2000) as analyzed by Hamilton and Tegmark (2001).

spherical harmonics, a basis more appropriate for a 2D field on the surface of a sphere. Therefore the two-point function of the CMB is a function of multipole moment l, not wave number k. Figure 1.14 shows the measurements of dozens of groups since 1992, when COBE first discovered large-angle (low l in the plot) anisotropies.

Figures 1.13 and 1.14 both have theoretical curves in them which appear to agree well with the data. The main goal of much of this book is to develop a first-principles understanding of these theoretical predictions. Indeed, understanding the development of structure in the universe has become a major goal of most cosmologists today. Note that this second aspect of cosmology beyond the Standard Model reinforces the first: i.e., observations of structure in the universe lead to the conclusion that there must be dark matter. In particular, the dashed curve in Figure 1.13 is the prediction of a model with baryons only, with no dark matter. The inhomogeneities expected in this model (when normalized to the CMB observations) are far too small. In Chapter 7, we will come to understand the reason why a baryon-only universe would be so smooth. For now, though, the message is clear: Dark matter is needed not only to explain rotation curves of galaxies but to explain

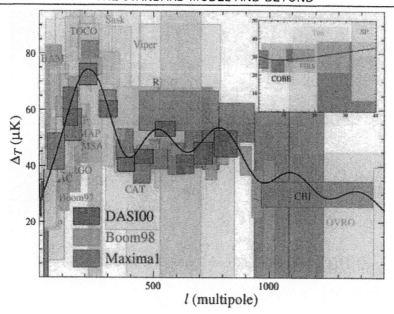

Figure 1.14. Anisotropies in the CMB predicted by the theory of inflation compared with observations. x-axis is multipole moment (e.g., $l = 1$ is the dipole, $l = 2$ the quadrupole) so that large angular scales correspond to low l; y-axis is the root mean square anisotropy (the square root of the two-point function) as a function of scale. The characteristic signature of inflation is the series of peaks and troughs, a signature which has been verified by experiment. See color Plate 1.14.

structure in the universe at large!

While trying to understand the evolution of structure in the universe, we will be forced to confront the question of what generated the initial conditions, the primordial perturbations that were the seeds for this structure. This will lead us to a third important aspect of cosmology beyond the Standard Model: the theory of inflation. Chapter 6 introduces this fascinating proposal, that the universe expanded exponentially fast when it was but 10^{-35} sec old. Until recently, there was little evidence for inflation. It survived as a viable theory mainly because of its aesthetic appeal. The discoveries of the past several years have changed this. They have by and large confirmed some of the basic predictions of inflation. Most notably, this theory makes concrete predictions for the initial conditions, predictions that have observable consequences today. For me, the most profound and exciting discovery in cosmology has been the observation of anisotropies in the CMB, with a characteristic pattern predicted by inflation.

The theory encompassing all these Beyond the Standard Model ingredients — dark matter plus evolution of structure plus inflation — is called *Cold Dark Matter*, or CDM. The "Cold" part of this moniker comes from requiring the dark matter particles to be able to clump efficiently in the early universe. If they are *hot* instead, i.e., have large pressure, structure will not form at the appropriate levels.

1.6 SUMMARY

By way of summarizing the features of an expanding universe that I have outlined above and that we will explore in great detail in the coming chapters, let's construct a time line. We can characterize any epoch in the universe by the time since the Big Bang; by the value of the scale factor at that time; or by the temperature of the cosmic plasma. For example, today, $a = 1$; $t \simeq 14$ billion years; and $T = 2.725\text{K} = 2.35 \times 10^{-4}$ eV$/k_B$. Figure 1.15 shows a time line of the universe using both time and temperature as markers. The milestones indicated on the time line range from those about which we are quite certain (e.g., nucleosynthesis and the CMB) to those that are more speculative (e.g., dark matter production, inflation, and dark energy today).

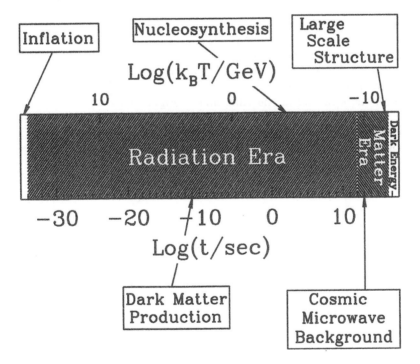

Figure 1.15. A history of the universe. Any epoch can be associated with either temperature (top scale) or time (bottom scale).

The time line in Figure 1.15 shows the dominant component of the universe at various times. Early on, most of the energy in the universe was in the form of radiation. Eventually, since the energy of a relativistic particle falls as $1/a$ while that of nonrelativistic matter remains constant at m, matter overtook radiation.

At relatively recent times, the universe appears to have become dominated not by matter, but by some dark energy, whose density remains relatively constant with time. The evidence for this unexplained form of energy is new and certainly not conclusive, but it is very suggestive.

The classical results in cosmology can be understood in the context of a smooth universe. Light elements formed when the universe was several minutes old, and the CMB decoupled from matter at a temperature of order $k_B T \sim 1/4$ eV. Heavy elementary particles may make up the dark matter in the universe; if they do, their abundance was fixed at very high temperatures of order $k_B T \sim 100$ GeV.

We will be mostly interested in this book in the perturbations around the smooth universe. The early end of the time line allows for a brief period of inflation, during which primordial perturbations were produced. These small perturbations began to grow when the universe became dominated by matter. The dark matter grew more and more clumpy, simply because of the attractive nature of gravity. An overdensity of dark matter of 1 part in 1000 when the temperature was 1 eV grew to 1 part in 100 by the time the temperature dropped to 0.1 eV. Eventually, at relatively recent times, perturbations in the matter ceased to be small; they became the nonlinear structure we see today. Anisotropies in the CMB today tell us what the universe looked like when it was several hundred thousand years old, so they are wonderful probes of the perturbations.

Some of the elements in the time line in Figure 1.15 may well be incorrect. However, since most of these ideas are testable, the data which will be taken during the coming decade will tell us which parts of the time line are correct and which need to be discarded. This in itself seems a sufficient reason to study the CMB and large-scale structure.

SUGGESTED READING

There are many good textbooks covering the homogeneous Big Bang. I am most familiar with *The Early Universe* (Kolb and Turner), which has especially good discussions on nucleosynthesis and inflation. Peacock's *Cosmological Physics* is the most up-to-date and perhaps the broadest of the standard cosmology texts, with more of an emphasis on extragalactic astronomy than either *The Early Universe* or this book. A popular account which still captures the essentials of the homogeneous Big Bang (testifying to the success of the model: it hasn't changed that much in 25 years) is *The First Three Minutes* (Weinberg). More recently, three books of note are: *The Whole Shebang* (Ferris), *The Little Book of the Big Bang* (Hogan), and *A Short History of the Universe* (Silk).

A nice article summarizing the evidence for an expanding universe and some methods to quantify it is Freedman (1998). Two of the pioneers in the field of Big Bang nucleosynthesis, Schramm and Turner, wrote a very clear review article (1998) right before a tragic accident took the life of the first author. An excellent account of the evidence for dark matter in spiral galaxies is Vera Rubin's 1983 article in Scientific American.

I have not attempted to record the history of the discovery of the Big Bang. Three books I am familiar with which treat this history in detail are *Blind Watchers of the Sky* (Kolb), *3K: The Cosmic Microwave Background* (Partridge), and *Three Degrees Above Zero: Bell Labs in the Information Age* (Bernstein). An article which sheds light on this history is Alpher and Herman (1988).

EXERCISES

Exercise 1. Suppose (incorrectly) that H scales as temperature squared all the way back until the time when the temperature of the universe was 10^{19} GeV$/k_B$ (i.e., suppose the universe was radiation dominated all the way back to the *Planck time*). Also suppose that today the dark energy is in the form of a cosmological constant Λ, such that ρ_Λ today is equal to $0.7\rho_{cr}$ and ρ_Λ remains constant throughout the history of the universe. What was $\rho_\Lambda/(3H^2/8\pi G)$ back then?

Exercise 2. Assume the universe today is flat with both matter and a cosmological constant, the latter with energy density which remains constant with time. Integrate Eq. (1.2) to find the present age of the universe. That is, rewrite Eq. (1.2) as

$$dt = H_0^{-1} \frac{da}{a} \left[\Omega_\Lambda + \frac{1 - \Omega_\Lambda}{a^3} \right]^{-1/2} \tag{1.10}$$

where Ω_Λ is the ratio of energy density in the cosmological constant to the critical density. Integrate from $a = 0$ (when $t = 0$) until today at $a = 1$ to get the age of the universe today. In both cases below the integral can be done analytically.
(a) First do the integral in the case when $\Omega_\Lambda = 0$.
(b) Now do the integral in the case when $\Omega_\Lambda = 0.7$. For fixed H_0, which universe is older?

Exercise 3. Using the fact that the reduced mass of the electron–nucleus in the D atom is larger than in hydrogen, and the fact that the Lyman α ($n = 1 \rightarrow n = 2$) transition in H has a wavelength 1215.67Å, find the wavelength of the photon emitted in the corresponding transition in D. Astronomers often define

$$v \equiv c\frac{\Delta\lambda}{\lambda} \qquad\qquad (1.11)$$

to characterize the splitting of two nearby lines. What is v for the H–D pair?

Exercise 4. Convert the specific intensity in Eq. (1.8) into an expression for what is plotted in Figure 1.10, the energy per square centimeter per steradian per second. Note that the x-axis is $1/\lambda$, the inverse wavelength of the photons. Show that the peak of a 2.73K blackbody spectrum does lie at $1/\lambda = 5\,\mathrm{cm}^{-1}$.

2

THE SMOOTH, EXPANDING UNIVERSE

Just as the early navigators of the great oceans required sophisticated tools to help them find their way, we will need modern technology to help us work through the ramifications of an expanding universe. In this chapter I introduce two of the necessary tools, general relativity and statistical mechanics. We will use them to derive some of the basic results laid down in Chapter 1: the expansion law of Eq. (1.2), the dependence of different components of energy density on the scale factor which governs expansion, the epoch of equality a_{eq} shown in Figure 1.3, and the luminosity distance needed to understand the implications of the supernovae diagram in Figure 1.7. Indeed, with general relativity and statistical mechanics, we can go a long way toward performing a cosmic inventory, identifying those components of the universe that dominate the energy budget at various epochs.

Implicit in this discussion will be the notion that the universe is smooth (none of the densities vary in space) and in equilibrium (the consequences of which will be explored in Section 2.3). In succeeding chapters, we will see that the deviations from equilibrium and smoothness are the source of much of the richness in the universe. Nonetheless, if only in order to understand the framework in which these deviations occur, a basic knowledge of the "zero order" universe is a must for any cosmologist.

In this chapter, I begin using units in which

$$\hbar = c = k_B = 1. \tag{2.1}$$

Many papers employ these units, so it is important to get accustomed to them. Please work through Exercise 1 if you are uncomfortable with the idea of setting the speed of light to 1.

2.1 GENERAL RELATIVITY

Most of cosmology can be learned with only a passing knowledge of general relativity. One must be familiar with the concept of a metric, understand geodesics, and be

23

able to apply the Einstein equations to the Friedmann–Robertson–Walker (FRW) metric thereby relating the parameters in the metric to the density in the universe. Eq. (1.2) is the result of applying the Einstein equations to the zero order universe. We will derive it in this section. Chapters 4 and 5 apply them to the perturbed universe. With the experience we gain in this section, there will be nothing difficult about these subsequent chapters. The principles are identical; only the algebra will be a touch harder.

2.1.1 The Metric

Figure 1.1 from Chapter 1 highlights the fact that even if one knows the components of a vector, say the difference between two grid points there, the physical distance associated with this vector requires additional information. In the case of a smooth expanding universe, the scale factor connects the coordinate distance with the physical distance. More generally, the *metric* turns coordinate distance into physical distance and so will be an essential tool in our quest to make quantitative predictions in an expanding universe.

We are familiar with the metric for the Cartesian coordinate system which says that the square of the physical distance between two points separated by dx and dy in a 2D plane is $(dx)^2 + (dy)^2$. However, were we to use polar coordinates instead, the square of the physical distance would no longer be the sum of the square of the two coordinate differences. Rather, if the differences dr and $d\theta$ are small, the square of the distance between two points is $(dr)^2 + r^2(d\theta)^2 \neq (dr)^2 + (d\theta)^2$. This distance is *invariant*: an observer using Cartesian coordinates to find it would get the same result as one using polar coordinates. Thus another way of stating what a metric does is this: it turns observer-dependent coordinates into invariants. Mathematically, in the 2D plane, the invariant distance squared $dl^2 = \sum_{i,j=1,2} g_{ij} dx^i dx^j$. The metric g_{ij} in this 2D example is a 2×2 symmetric matrix. In Cartesian coordinates the metric is diagonal with each element equal to 1. In polar coordinates (taking $x^1 = r$ and $x^2 = \theta$) it is also diagonal with $g_{11} = 1$, but g_{22} which multiplies $(d\theta)^2$ is equal to r^2.

There is yet another way of thinking about a metric, using pictures. When handed a vector, we immediately think of a line with an arrow attached, the length of the line corresponding to the length of the vector and the arrow to its direction. In fact, this notion is rooted too firmly in Euclidean space. In actuality, the length of the vector depends on the metric. An intuitive way of understanding this is to consider the contour map in Figure 2.1. The number of lines crossed by a vector is a measure of the vertical distance traveled by a hiker. Vectors of the same apparent 2D length — corresponding to identical coordinate distances — can correspond to significantly different physical distances. Mathematically the surface of the Earth can be parametrized by two coordinates, say θ and ϕ. Then the metric is a very nontrivial function of θ and ϕ which accounts for all the elevation changes on the surface.

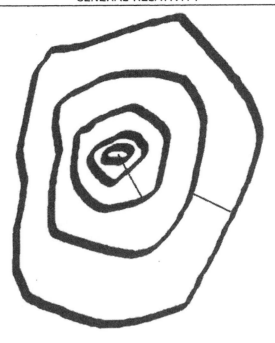

Figure 2.1. Contour map of a mountain. The closely spaced contours near the center correspond to rapid elevation gain. The two thin lines correspond to hikes of significantly different difficulty even though they appear to be of the same length. Similarly, the true length of a vector requires knowledge of the metric.

The great advantage of the metric is that it incorporates gravity. Instead of thinking of gravity as an external force and talking of particles moving in a gravitational field, we can include gravity in the metric and talk of particles moving freely in a distorted or curved space-time, one in which the metric cannot be converted everywhere into Euclidean form.

In four space-time dimensions the invariant includes time intervals as well, so

$$ds^2 = \sum_{\mu,\nu=0}^{3} g_{\mu\nu} dx^\mu dx^\nu \tag{2.2}$$

where the indices μ and ν range from 0 to 3 (see the box on page 27), with the first one reserved for the time-like coordinate ($dx^0 = dt$) and the last three for spatial coordinates. Here I have explicitly written down the summation sign, but from now on we will use the convention that repeated indices are summed over. The metric $g_{\mu\nu}$ is necessarily symmetric, so in principle has four diagonal and six off-diagonal components. It provides the connection between values of the coordinates and the more physical measure of the interval ds^2 (sometimes called *proper time*). Special relativity is described by Minkowski space-time with the metric: $g_{\mu\nu} = \eta_{\mu\nu}$, with

$$\eta_{\mu\nu} = \begin{pmatrix} -1 & 0 & 0 & 0 \\ 0 & 1 & 0 & 0 \\ 0 & 0 & 1 & 0 \\ 0 & 0 & 0 & 1 \end{pmatrix}. \tag{2.3}$$

What is the metric which describes the expanding universe? Let us return to the grid depicted in Figure 1.1. We said earlier that two grid points move away from each other, so that the distance between the two points is always proportional to the scale factor. If the comoving distance today is x_0, the physical distance between the two points at some earlier time t was $a(t)x_0$. At least in a flat (as opposed to open or closed) universe, the metric then is almost identical to the Minkowski metric, except that distance must be multiplied by the scale factor. This suggests that the metric in an expanding, flat universe is

$$g_{\mu\nu} = \begin{pmatrix} -1 & 0 & 0 & 0 \\ 0 & a^2(t) & 0 & 0 \\ 0 & 0 & a^2(t) & 0 \\ 0 & 0 & 0 & a^2(t) \end{pmatrix}. \tag{2.4}$$

This is called the Friedmann–Robertson–Walker (FRW) metric.

As noted in Eq. (1.2), which we will shortly derive, the evolution of the scale factor depends on the density in the universe. When perturbations are introduced, the metric will become more complicated, and the perturbed part of the metric will be determined by the inhomogeneities in the matter and radiation.

Indices

In three dimensions, a vector $\vec{A}$ has three components, which we refer to as A^i, superscript i taking the values $1, 2$, or 3. The dot product of two vectors is then

$$\vec{A} \cdot \vec{B} = \sum_{i=1}^{3} A^i B^i \equiv A^i B^i \qquad (2.5)$$

where I have introduced the Einstein summation convention of not explicitly writing the $\sum$ sign when an index (in this case i) appears twice. Similarly, matrices can be written in component notation. For example, the product of two matrices $\mathbf{M}$ and $\mathbf{N}$ is

$$(\mathbf{MN})_{ij} = M_{ik} N_{kj} \qquad (2.6)$$

again with implicit summation over k.

In relativity, two generalizations must be made. First, in relativity a vector has a fourth component, the time component. Since the spatial indices run from 1 to 3, it is conventional to use 0 for the time component. Greek letters are used to represent all four components, so $A^\mu = (A^0, A^i)$. The second, more subtle, feature of relativity is the distinction between upper and lower indices, the former associated with vectors and the latter with 1-forms. One goes back and forth with the metric tensor, so

$$A_\mu = g_{\mu\nu} A^\nu \qquad ; \qquad A^\mu = g^{\mu\nu} A_\nu. \qquad (2.7)$$

A vector and a 1-form can be contracted to produce an invariant, a scalar. For example, the statement that the four-momentum squared of a massless particle must vanish is

$$P^2 \equiv P_\mu P^\mu = g_{\mu\nu} P^\mu P^\nu = 0. \qquad (2.8)$$

This contraction is the equivalent of counting the contours crossed by a vector, as depicted in Figure 2.1.

Just as the metric can turn an upper index on a vector into a lower index, the metric can be used to raise and lower indices on tensors with an arbitrary number of indices. For example, raising the indices on the metric tensor itself leads to

$$g^{\mu\nu} = g^{\mu\alpha} g^{\nu\beta} g_{\alpha\beta}. \qquad (2.9)$$

If the index $\alpha = \nu$, then the first term on the right is equal to the term on the left, so if the combination of the last two terms on the right force α to be equal to ν, then the equation is satisfied. Therefore,

$$g^{\nu\beta} g_{\alpha\beta} = \delta^\nu{}_\alpha, \qquad (2.10)$$

where $\delta^\nu{}_\alpha$ is the Kronecker delta equal to zero unless $\nu = \alpha$ in which case it is 1.

2.1.2 The Geodesic Equation

In Minkowski space, particles travel in straight lines unless they are acted on by a force. Not surprisingly, the paths of particles in more general space-times are more complicated. The notion of a straight line gets generalized to a *geodesic*, the path followed by a particle in the absence of any forces. To express this in equations, we must generalize Newton's law with no forces, $d^2\vec{x}/dt^2 = 0$, to the expanding universe.

The machinery necessary to generalize $d^2\vec{x}/dt^2 = 0$ is perhaps best introduced by starting with a simple example: particle motion in a Euclidean 2D plane. In that case, the equations of motion in Cartesian coordinates $x^i = (x, y)$ are

$$\frac{d^2x^i}{dt^2} = 0. \tag{2.11}$$

However, if we use polar coordinates $x'^i = (r, \theta)$ instead, the equations for a free particle look significantly different. The fundamental difference between the two coordinate systems is that the basis vectors for polar coordinates $\hat{r}, \hat{\theta}$ vary in the plane. Therefore, $d^2\vec{x}'/dt^2 = 0$ does *not* imply that each coordinate r and θ satisfies $d^2x'^i/dt^2 = 0$.

To determine the equation satisfied by the polar coordinates, we can start from the Cartesian equation and then transform. In particular,

$$\frac{dx^i}{dt} = \frac{\partial x^i}{\partial x'^j}\frac{dx'^j}{dt}. \tag{2.12}$$

$\partial x^i/\partial x'^j$ is called the *transformation matrix* going from one basis to another. In the case of Cartesian to polar coordinates in 2D, $x^1 = x'^1 \cos x'^2$ and $x^2 = x'^1 \sin x'^2$, so the transformation matrix is

$$\frac{\partial x^i}{\partial x'^j} = \begin{pmatrix} \cos x'^2 & -x'^1 \sin x'^2 \\ \sin x'^2 & x'^1 \cos x'^2 \end{pmatrix}. \tag{2.13}$$

Therefore, the geodesic equation becomes

$$\frac{d}{dt}\left[\frac{dx^i}{dt}\right] = \frac{d}{dt}\left[\frac{\partial x^i}{\partial x'^j}\frac{dx'^j}{dt}\right] = 0. \tag{2.14}$$

The derivative with respect to time acts on both terms inside the brackets. If the transformation from the Cartesian basis to the new basis was linear, then the derivative acting on the transformation matrix would vanish, and the geodesic equation in the new basis would still be $d^2x'^i/dt^2 = 0$. In the case of polar coordinates, though, the transformation is not linear, and we need the fact that

$$\frac{d}{dt}\left(\frac{\partial x^i}{\partial x'^j}\right) = \frac{\partial}{\partial x'^j}\left(\frac{dx^i}{dt}\right)$$

$$= \frac{\partial^2 x^i}{\partial x'^j \partial x'^k} \frac{dx'^k}{dt} \tag{2.15}$$

where the first equality holds since derivatives commute and the second comes from inserting dx^i/dt from Eq. (2.12), changing dummy indices from $j \to k$. The geodesic equation in the new coordinates therefore becomes

$$\frac{d}{dt}\left[\frac{\partial x^i}{\partial x'^j}\frac{dx'^j}{dt}\right] = \frac{\partial x^i}{\partial x'^j}\frac{d^2 x'^j}{dt^2} + \frac{\partial^2 x^i}{\partial x'^j \partial x'^k}\frac{dx'^k}{dt}\frac{dx'^j}{dt} = 0. \tag{2.16}$$

To get this in a more recognizable form, note that the term multiplying the second time derivative is the transformation matrix. If we multiply the equation by the inverse of this transformation matrix, then the second time derivative will stand alone, leaving

$$\frac{d^2 x'^l}{dt^2} + \left[\left(\left\{\frac{\partial x}{\partial x'}\right\}^{-1}\right)^l_{\;i} \frac{\partial^2 x^i}{\partial x'^j \partial x'^k}\right]\frac{dx'^k}{dt}\frac{dx'^j}{dt} = 0. \tag{2.17}$$

You can check that this rather cumbersome expression does indeed give the correct equations of motion in polar coordinates. More importantly, by keeping things general, we have derived the geodesic equation in a non-Cartesian basis.

It is convenient to define the *Christoffel symbol*, $\Gamma^l_{\;jk}$, to be the coefficient of the $(dx'^k/dt)(dx'^j/dt)$ term in Eq. (2.17). Note that by definition it is symmetric in its lower indices j and k. In a Cartesian coordinate system, the Christoffel symbol vanishes and the geodesic equation is simply $d^2 x^i/dt^2 = 0$. But in general, the Christoffel symbol does not vanish; its presence describes geodesics in nontrivial coordinate systems. The reason why this generalized geodesic equation is so powerful is that in a nontrivial space-time such as the expanding universe it is not *possible* to find a fixed Cartesian coordinate system, so we need to know how particles travel in the more general case.

There are two small changes we need to make when importing the geodesic equation (2.17) into relativity. The first is trivial: allow the indices to range from 0 to 3 to include time and the three spatial dimensions. The second is also not surprising: since time is now one of our coordinates, it will not do to use it as the evolution parameter. Instead introduce a parameter λ which monotonically increases along the particle's path as in Figure 2.2. The geodesic equation then

Figure 2.2. A particle's path is parametrized by λ, which monotonically increases from its initial value λ_i to its final value λ_f.

becomes

$$\frac{d^2 x^\mu}{d\lambda^2} = -\Gamma^\mu{}_{\alpha\beta} \frac{dx^\alpha}{d\lambda} \frac{dx^\beta}{d\lambda}.$$ (2.18)

We derived this equation transforming from a Cartesian basis, so that the Christoffel symbol is given by the term in square brackets in Eq. (2.17). It is almost always more convenient, however, to obtain the Christoffel symbol from the metric directly. A convenient formula expressing this dependence is

$$\Gamma^\mu{}_{\alpha\beta} = \frac{g^{\mu\nu}}{2} \left[\frac{\partial g_{\alpha\nu}}{\partial x^\beta} + \frac{\partial g_{\beta\nu}}{\partial x^\alpha} - \frac{\partial g_{\alpha\beta}}{\partial x^\nu} \right].$$ (2.19)

Note that the raised indices on $g^{\mu\nu}$ are important: $g^{\mu\nu}$ is the inverse of $g_{\mu\nu}$ (see the box on page 27). So $g^{\mu\nu}$ in the flat, FRW metric is identical to $g_{\mu\nu}$ except that its spatial elements are $1/a^2$ instead of a^2.

Using the general expression in Eq. (2.19) and the FRW metric in Eq. (2.4), we can derive the Christoffel symbol in an expanding, homogeneous universe. First we compute the components with upper index equal to zero, $\Gamma^0{}_{\alpha\beta}$. Since the metric is diagonal, the factor of $g^{0\nu}$ vanishes unless $\nu = 0$ in which case it is -1. Therefore,

$$\Gamma^0{}_{\alpha\beta} = \frac{-1}{2} \left[\frac{\partial g_{\alpha 0}}{\partial x^\beta} + \frac{\partial g_{\beta 0}}{\partial x^\alpha} - \frac{\partial g_{\alpha\beta}}{\partial x^0} \right].$$ (2.20)

The first two terms here reduce to derivatives of g_{00}. Since the FRW metric has constant g_{00}, these terms vanish, and we are left with

$$\Gamma^0{}_{\alpha\beta} = \frac{1}{2} \frac{\partial g_{\alpha\beta}}{\partial x^0}.$$ (2.21)

The derivative is nonzero only if α and β are spatial indices, which will be identified with Roman letters i, j running from 1 to 3. Since $x^0 = t$, we have

$$\Gamma^0{}_{00} = 0$$

$$\Gamma^0{}_{0i} = \Gamma^0{}_{i0} = 0$$

$$\Gamma^0{}_{ij} = \delta_{ij} \dot{a} a$$ (2.22)

where overdots indicate derivatives with respect to time.[1] It is a straightforward, but useful, exercise to show that $\Gamma^i{}_{\alpha\beta}$ is nonzero only when one of its lower indices is zero and one is spatial, so that

$$\Gamma^i{}_{0j} = \Gamma^i{}_{j0} = \delta_{ij} \frac{\dot{a}}{a}$$ (2.23)

with all other $\Gamma^i{}_{\alpha\beta}$ zero.

---

[1] I will use this convention until Chapter 4. After that, overdots will denote derivatives with respect to conformal time.

This has been a long, rather formal subsection, opening with the generalization of the geodesic equation to curved space-time and then proceeding with a calculation of the Christoffel symbol in the expanding universe described by the FRW metric. Before completing our main task and using the Einstein equations to derive Eq. (1.2), let's take a break and apply the geodesic equation to a single particle. In particular let's see how a particle's energy changes as the universe expands. We'll do the calculation here for a massless particle; an almost identical problem for a massive particle is relegated to Exercise 4.

Start with the four-dimensional energy–momentum vector $P^\alpha = (E, \vec{P})$, whose time component is the energy. We use this four-vector to define the parameter λ in Eq. (2.18):

$$P^\alpha = \frac{dx^\alpha}{d\lambda}. \tag{2.24}$$

This is an implicit definition of λ. Fortunately, one never needs to find λ explicitly, for it can be directly eliminated by noting that

$$\frac{d}{d\lambda} = \frac{dx^0}{d\lambda} \frac{d}{dx^0}$$

$$= E\frac{d}{dt}. \tag{2.25}$$

The zeroth component of the geodesic equation (2.18) then becomes

$$E\frac{dE}{dt} = -\Gamma^0{}_{ij} P^i P^j \tag{2.26}$$

where the equality holds since only the spatial components of $\Gamma^0{}_{\alpha\beta}$ are nonzero. Inserting these components leads to a right-hand side equal to $-\delta_{ij} a\dot{a} P^i P^j$. A massless particle has energy–momentum[2] vector $(E, \vec{P})$ with zero magnitude:

$$g_{\mu\nu} P^\mu P^\nu = -E^2 + \delta_{ij} a^2 P^i P^j = 0 \tag{2.27}$$

which enables us to write the right hand side of Eq. (2.26) as $-(\dot{a}/a)E^2$. Therefore, the geodesic equation yields

$$\frac{dE}{dt} + \frac{\dot{a}}{a}E = 0, \tag{2.28}$$

the solution to which is

$$E \propto \frac{1}{a}. \tag{2.29}$$

This confirms our hand-waving argument in Chapter 1 that the energy of a massless particle should decrease as the universe expands since it is inversely proportional to its wavelength, which is being stretched along with the expansion. In Chapter 4 we will rederive this result in yet another way using the Boltzmann equation.

---

[2]Note that $\vec{P}$ measures motion on the comoving (nonexpanding) grid. The physical momentum which measures changes in physical distance is related to $\vec{P}$ by a factor of a. Hence the factor of a^2 in Eq. (2.27).

2.1.3 Einstein Equations

If you did a word search on the previous two subsections, you might be surprised to discover that the words "general relativity" never appeared. The concept of a metric and the realization that nontrivial metrics affect geodesics both exist completely independently of general relativity. The part of general relativity that is hidden above is that gravitation can be described by a metric, in our case by Eq. (2.4). There is a second aspect of general relativity, though: one which relates the metric to the matter and energy in the universe. This second part is contained in the Einstein equations, which relate the components of the Einstein tensor describing the geometry to the energy–momentum tensor describing the energy:

$$G_{\mu\nu} \equiv R_{\mu\nu} - \frac{1}{2} g_{\mu\nu} \mathcal{R} = 8\pi G T_{\mu\nu}. \tag{2.30}$$

Here $G_{\mu\nu}$ is the Einstein tensor; $R_{\mu\nu}$ is the *Ricci tensor*, which depends on the metric and its derivatives; $\mathcal{R}$, the *Ricci scalar*, is the contraction of the Ricci tensor ($\mathcal{R} \equiv g^{\mu\nu} R_{\mu\nu}$); G is Newton's constant; and $T_{\mu\nu}$ is the energy–momentum tensor. We will spend some time on the energy–momentum tensor in Section 2.3. For now, all we need to know is that it's a symmetric tensor describing the constituents of the universe. The left-hand side of Eq. (2.30) is a function of the metric, the right a function of the energy: the Einstein equations relate the two.

The Ricci tensor is most conveniently expressed in terms of the Christoffel symbol,

$$R_{\mu\nu} = \Gamma^{\alpha}{}_{\mu\nu,\alpha} - \Gamma^{\alpha}{}_{\mu\alpha,\nu} + \Gamma^{\alpha}{}_{\beta\alpha}\Gamma^{\beta}{}_{\mu\nu} - \Gamma^{\alpha}{}_{\beta\nu}\Gamma^{\beta}{}_{\mu\alpha}. \tag{2.31}$$

Here commas denote derivatives with respect to x. So, for example, $\Gamma^{\alpha}{}_{\mu\nu,\alpha} \equiv \partial\Gamma^{\alpha}{}_{\mu\nu}/\partial x^{\alpha}$. Although this expression looks formidable, we have already done the hard work by computing the Christoffel symbol in an FRW universe. It turns out that there are only two sets of nonvanishing components of the Ricci tensor: one with $\mu = \nu = 0$ and the other with $\mu = \nu = i$.

Consider

$$R_{00} = \Gamma^{\alpha}{}_{00,\alpha} - \Gamma^{\alpha}{}_{0\alpha,0} + \Gamma^{\alpha}{}_{\beta\alpha}\Gamma^{\beta}{}_{00} - \Gamma^{\alpha}{}_{\beta0}\Gamma^{\beta}{}_{0\alpha}. \tag{2.32}$$

Recall that the Christoffel symbol vanishes if its two lower indices are zero, so the first and third terms on the right vanish. Similarly, the indices α and β in the second and fourth terms must be spatial. We are left with

$$R_{00} = -\Gamma^{i}{}_{0i,0} - \Gamma^{i}{}_{j0}\Gamma^{j}{}_{0i}. \tag{2.33}$$

Using Eq. (2.23) leads directly to

$$R_{00} = -\delta_{ii} \frac{\partial}{\partial t} \left(\frac{\dot{a}}{a} \right) - \left(\frac{\dot{a}}{a} \right)^2 \delta_{ij}\delta_{ij}$$

$$= -3 \left[\frac{\ddot{a}}{a} - \frac{\dot{a}^2}{a^2} \right] - 3 \left(\frac{\dot{a}}{a} \right)^2$$

$$= -3\frac{\ddot{a}}{a}. \tag{2.34}$$

The factors of 3 on the second line arise since δ_{ii} means sum over all three spatial indices, counting one for each. I will leave the space-space component as an exercise; it is

$$R_{ij} = \delta_{ij}\left[2\dot{a}^2 + a\ddot{a}\right]. \tag{2.35}$$

The next ingredient in the Einstein equations is the Ricci scalar, which we can now compute since

$$\mathcal{R} \equiv g^{\mu\nu}R_{\mu\nu}$$

$$= -R_{00} + \frac{1}{a^2}R_{ii}. \tag{2.36}$$

Again the sum over i leads to a factor of 3, so

$$\mathcal{R} = 6\left[\frac{\ddot{a}}{a} + \left(\frac{\dot{a}}{a}\right)^2\right]. \tag{2.37}$$

To understand the evolution of the scale factor in a homogeneous universe, we need consider only the time-time component of the Einstein equations:

$$R_{00} - \frac{1}{2}g_{00}\mathcal{R} = 8\pi G T_{00}. \tag{2.38}$$

The terms on the left sum to $3\dot{a}^2/a^2$, and the time-time component of the energy–momentum tensor is simply the energy density ρ. So we finally have

$$\left(\frac{\dot{a}}{a}\right)^2 = \frac{8\pi G}{3}\rho. \tag{2.39}$$

To get this into the form of Eq. (1.2), recall that the left-hand side here is the square of the Hubble rate and that the critical density was defined as $\rho_{\rm cr} \equiv 3H_0^2/8\pi G$. So, dividing both sides by H_0^2 leads to

$$\frac{H^2(t)}{H_0^2} = \frac{\rho}{\rho_{\rm cr}}. \tag{2.40}$$

Here the energy density ρ counts the energy density from all species: matter, radiation, and the dark energy. In our derivation, we have assumed the universe is flat, so Eq. (2.40) does not contain a term corresponding to the curvature of the universe. I leave it as an exercise to derive the Einstein equation in an open universe.

2.2 DISTANCES

We can anticipate that measuring distance in an expanding universe will be a tricky business. Referring back to the expanding grid of Figure 1.1, we immediately see two possible ways to measure distance, the comoving distance which remains fixed as the universe expands or the physical distance which grows simply because of the expansion. Frequently, neither of these two measures accurately describes the process of interest. For example light leaving a distant QSO at redshift 3 starts its journey towards us when the scale factor was only a quarter of its present value and ends it today when the universe has expanded by a factor of 4. Which distance do we use in that case to relate, say, the luminosity of the QSO to the flux we see?

The fundamental distance measure, from which all others may be calculated, is the distance on the comoving grid. If the universe is flat, as we will assume through most of this book, then computing distances on the comoving grid is easy: the distance between two points $\vec{x}_1$ and $\vec{x}_2$ is equal to $[(x_1 - x_2)^2 + (y_1 - y_2)^2 + (z_1 - z_2)^2]^{1/2}$.

One very important comoving distance is the distance light could have traveled (in the absence of interactions) since $t = 0$. In a time dt, light travels a comoving distance $dx = dt/a$ (recall that we are setting $c = 1$), so the total comoving distance light could have traveled is

$$\eta \equiv \int_0^t \frac{dt'}{a(t')}. \tag{2.41}$$

The reason this distance is so important is that no information could have propagated further (again on the comoving grid) than η since the beginning of time. Therefore, regions separated by distance greater than η are not causally connected. If they appear similar, we should be suspicious! We can think of η then as the *comoving horizon*. We can also think of η, which is monotonically increasing, as a time variable and call it the *conformal time*. Just like the time t, the temperature T, the redshift z, and the scale factor a, η can be used to discuss the evolution of the universe. In fact, for most purposes η is the most convenient time variable, so when we begin to study the evolution of perturbations, we will use it instead of t. In some simple cases, η can be expressed analytically in terms of a (Exercise 11). For example, in a matter-dominated universe, $\eta \propto a^{1/2}$, while $\eta \propto a$ in a radiation-dominated universe.

Another important comoving distance is that between a distant emitter and us. In that case, the comoving distance out to an object at scale factor a (or redshift $z = 1/a - 1$) is

$$\chi(a) = \int_{t(a)}^{t_0} \frac{dt'}{a(t')} = \int_a^1 \frac{da'}{a'^2 H(a')}. \tag{2.42}$$

Here I have changed the integration over t' to one over a', which brings in the additional factor of $da/dt = aH$ in the denominator. Typically we can see objects out to $z \lesssim 6$; at these late times radiation can be ignored (recall Figure 1.3). If the universe is purely matter dominated at such times, then $H \propto a^{-3/2}$ and we can do the integral in Eq. (2.42) analytically,

$$\chi^{\text{Flat,MD}}(a) = \frac{2}{H_0} \left[1 - a^{1/2} \right]$$

$$= \frac{2}{H_0} \left[1 - \frac{1}{\sqrt{1+z}} \right]. \tag{2.43}$$

This comoving distance goes as z/H_0 for small z (verifying our hand-waving discussion of the small-z Hubble diagram in Section 1.2) and then asymptotes to $2/H_0$ as z gets very large.

A classic way to determine distances in astronomy is to measure the angle θ subtended by an object of known physical size l. The distance to that object (assuming the angle subtended is small) is then

$$d_A = \frac{l}{\theta}. \tag{2.44}$$

The subscript $_A$ here denotes *angular diameter distance*. To compute the angular diameter distance in an expanding universe, we first note that the comoving size of the object is l/a. The comoving distance out to the object is given by Eq. (2.42), so the angle subtended is $\theta = (l/a)/\chi(a)$. Comparing with Eq. (2.44), we see that the angular diameter distance is

$$d_A^{\text{flat}} = a\chi = \frac{\chi}{1+z}. \tag{2.45}$$

Note that the angular diameter distance is equal to the comoving distance at low redshift, but actually decreases at very large redshift. At least in a flat universe, objects at large redshift appear larger than they would at intermediate redshift! The superscript here is a warning that this result holds only in a flat universe. In an open or closed universe, the curvature density is defined as $\Omega_k = 1 - \Omega_0$ where Ω_0 is the ratio of total to critical density today, including contributions from matter, radiation, and any other form of energy such as a cosmological constant. If the curvature is nonzero, the angular diameter distance generalizes to

$$d_A = \frac{a}{H_0\sqrt{|\Omega_k|}} \begin{cases} \sinh\left[\sqrt{\Omega_k}H_0\chi\right] & \Omega_k > 0 \\ \sin\left[\sqrt{-\Omega_k}H_0\chi\right] & \Omega_k < 0 \end{cases}. \tag{2.46}$$

Note that both of these expressions reduce to the flat case in the limit that the curvature density Ω_k goes to zero. Figure 2.3 shows the angular diameter distance in a flat universe, both with and without a cosmological constant.

Another way of inferring distances in astronomy is to measure the flux from an object of known luminosity. Recall that (forgetting about expansion for the moment) the observed flux F a distance d from a source of known luminosity L is

$$F = \frac{L}{4\pi d^2} \tag{2.47}$$

since the total luminosity through a spherical shell with area $4\pi d^2$ is constant. How does this result generalize to an expanding universe? Again it is simplest to work

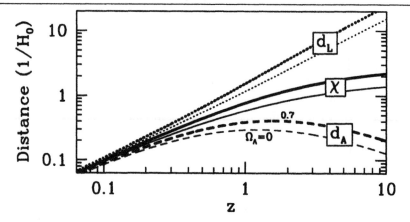

Figure 2.3. Three distance measures in a flat expanding universe. From top to bottom, the luminosity distance, the comoving distance, and the angular diameter distance. The pair of lines in each case is for a flat universe with matter only (light curves) and 70% cosmological constant Λ (heavy curves). In a Λ-dominated universe, distances out to fixed redshift are larger than in a matter-dominated universe.

on the comoving grid, this time with the source centered at the origin. The flux we observe is

$$F = \frac{L(\chi)}{4\pi\chi^2(a)} \tag{2.48}$$

where $L(\chi)$ is the luminosity through a (comoving) spherical shell with radius $\chi(a)$. To further simplify, let's assume that the photons are all emitted with the same energy. Then $L(\chi)$ is this energy multiplied by the number of photons passing through a (comoving) spherical shell per unit time. In a fixed time interval, photons travel farther on the comoving grid at early times than at late times since the associated physical distance at early times is smaller. Therefore, the number of photons crossing a shell in the fixed time interval will be smaller today than at emission, smaller by a factor of a. Similarly, the energy of the photons will be smaller today than at emission, because of expansion. Therefore, the energy per unit time passing through a comoving shell a distance $\chi(a)$ (i.e., our distance) from the source will be a factor of a^2 smaller than the luminosity at the source. The flux we observe therefore will be

$$F = \frac{La^2}{4\pi\chi^2(a)} \tag{2.49}$$

where L is the luminosity at the source. We can keep[3] Eq. (2.47) in an expanding universe as long as we define the *luminosity distance*

$$d_L \equiv \frac{\chi}{a}. \tag{2.50}$$

---

[3] Actually there is one more difference that needs to be accounted for: the observed luminosity is related to the emitted luminosity at a different wavelength. Here we have assumed a detector which counts all the photons.

The luminosity distance is shown in Figure 2.3.

All three distances are larger in a universe with a cosmological constant than in one without. This follows from the fact that the energy density, and therefore the expansion rate, is smaller in a Λ-dominated universe. The universe was therefore expanding more slowly early on, and light had more time to travel from distant objects to us. These distant objects will therefore appear fainter to us than if the universe was dominated by matter only.

2.3 EVOLUTION OF ENERGY

Let us return to the energy–momentum tensor on the right-hand side of the Einstein equations. We will eventually include perturbations to $T^{\mu}{}_{\nu}$, but in the spirit of this chapter, first consider the case of a perfect isotropic fluid. Then,

$$T^{\mu}{}_{\nu} = \begin{pmatrix} -\rho & 0 & 0 & 0 \\ 0 & \mathcal{P} & 0 & 0 \\ 0 & 0 & \mathcal{P} & 0 \\ 0 & 0 & 0 & \mathcal{P} \end{pmatrix} \tag{2.51}$$

where $\mathcal{P}$ is the pressure of the fluid.

How do the components of the energy–momentum tensor evolve with time? Consider first the case where there is no gravity and velocities are negligible. The pressure and energy density in that case evolve according to the continuity equation, $\partial \rho / \partial t = 0$, and the Euler equation, $\partial \mathcal{P} / \partial x^i = 0$. This can be promoted to a 4 component conservation equation for the energy–momentum tensor: $\partial T^{\mu}{}_{\nu} / \partial x^{\mu} = 0$. In an expanding universe, however, the conservation criterion must be modified. Instead, conservation implies the vanishing of the *covariant* derivative:

$$T^{\mu}{}_{\nu;\mu} \equiv \frac{\partial T^{\mu}{}_{\nu}}{\partial x^{\mu}} + \Gamma^{\mu}{}_{\alpha\mu} T^{\alpha}{}_{\nu} - \Gamma^{\alpha}{}_{\nu\mu} T^{\mu}{}_{\alpha}. \tag{2.52}$$

The vanishing of $T^{\mu}{}_{\nu;\mu}$ is four separate equations; let's consider the $\nu = 0$ component. This is

$$\frac{\partial T^{\mu}{}_{0}}{\partial x^{\mu}} + \Gamma^{\mu}{}_{\alpha\mu} T^{\alpha}{}_{0} - \Gamma^{\alpha}{}_{0\mu} T^{\mu}{}_{\alpha} = 0. \tag{2.53}$$

Since we are assuming isotropy, $T^{i}{}_{0}$ vanishes, so the dummy indices μ in the first term and α in the second must be equal to zero:

$$\frac{-\partial \rho}{\partial t} - \Gamma^{\mu}{}_{0\mu} \rho - \Gamma^{\alpha}{}_{0\mu} T^{\mu}{}_{\alpha} = 0. \tag{2.54}$$

From Eq. (2.23), $\Gamma^{\alpha}{}_{0\mu}$ vanishes unless α, μ are spatial indices equal to each other, in which case it is $\dot{a}/a$. So, the conservation law in an expanding universe reads

$$\frac{\partial \rho}{\partial t} + \frac{\dot{a}}{a} [3\rho + 3\mathcal{P}] = 0. \tag{2.55}$$

Rearranging terms, we have

$$a^{-3} \frac{\partial \left[\rho a^3 \right]}{\partial t} = -3 \frac{\dot{a}}{a} \mathcal{P}. \tag{2.56}$$

The conservation law can be applied immediately to glean information about the scaling of both matter and radiation with the expansion. Matter has effectively zero pressure, so

$$\frac{\partial \left[\rho_{\mathrm{m}} a^3 \right]}{\partial t} = 0 \tag{2.57}$$

implying that the energy density of matter $\rho_{\mathrm{m}} \propto a^{-3}$. We anticipated this result in Chapter 1 based on the simple notion that the mass remains constant, while the number density scales as the inverse volume. The application to radiation also offers no surprises. Radiation has $\mathcal{P} = \rho/3$ (Exercise 14), so working from Eq. (2.55),

$$\frac{\partial \rho_r}{\partial t} + \frac{\dot{a}}{a} 4 \rho_r = a^{-4} \frac{\partial \left[\rho_r a^4 \right]}{\partial t}$$

$$= 0. \tag{2.58}$$

Therefore, the energy density of radiation $\rho_r \propto a^{-4}$, accounting for the decrease in energy per particle as the universe expands.

Through most of the early universe, reactions proceeded rapidly enough to keep particles in equilibrium, different species sharing a common temperature. We will often want to express the energy density and pressure in terms of this temperature. For this reason, and many others which will emerge over the next few chapters, it is convenient to introduce the occupation number, or *distribution function*, of a species. This counts the number of particles in a given region in phase space around position $\vec{x}$ and momentum $\vec{p}$.[4] The energy of a species is then obtained by summing the energy over all of phase space elements: $\sum f(\vec{x}, \vec{p}) E(p)$ with $E(p) = \sqrt{p^2 + m^2}$. How many phase space elements are there in a region of "volume" $d^3x d^3p$? By Heisenberg's principle, no particle can be localized into a region of phase space smaller than $(2\pi\hbar)^3$, so this is the size of a fundamental element. Therefore, the number of phase space elements in $d^3x d^3p$ is $d^3x d^3p/(2\pi\hbar)^3$ (see Figure 2.4), and the energy density is

$$\rho_i = g_i \int \frac{d^3p}{(2\pi)^3} \, f_i(\vec{x}, \vec{p}) E(p) \tag{2.59}$$

where i labels different species, g_i is the degeneracy of the species (e.g., equal to 2 for the photon for its spin states), and I have gone back to $\hbar = 1$. In equilibrium at temperature T, bosons such as photons have Bose–Einstein distributions,

$$f_{\mathrm{BE}} = \frac{1}{e^{(E-\mu)/T} - 1}, \tag{2.60}$$

and fermions such as electrons have Fermi-Dirac distributions,

---

[4]By p here I mean not the comoving momentum defined in Eq. (2.24), but rather the proper momentum which decreases with the expansion. See Exercise 15 for a discussion.

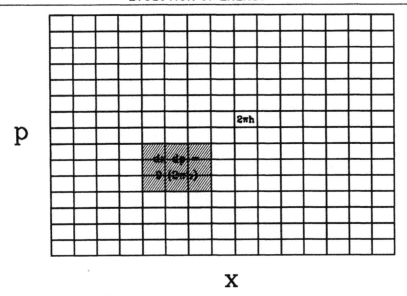

p

X

Figure 2.4. Phase space of position and momentum in one dimension. Volume of each cell is $2\pi\hbar$, the smallest region into which a particle can be confined because of Heisenberg's principle. Shaded region has infinitesmal volume $dxdp$. This covers nine cells. To count the appropriate number of cells, therefore, the phase space integral must be $\int dxdp/(2\pi\hbar)$.

$$f_{\rm FD} = \frac{1}{e^{(E-\mu)/T} + 1},\qquad (2.61)$$

with μ the chemical potential. It should be noted that these distributions do not depend on position $\vec{x}$ or on the direction of the momentum $\hat{p}$, simply on the magnitude p. This is a feature of the zero-order, smooth universe. When we come to consider inhomogenities and anisotropies, we will see that the distribution functions have small perturbations around these zero order values, and the perturbations do depend on position and on the direction of propagation.

The pressure can be similarly expressed as an integral over the distribution function,

$$\mathcal{P}_i = g_i \int \frac{d^3p}{(2\pi)^3}\ f_i(\vec{x}, \vec{p}) \frac{p^2}{3E(p)}.\qquad (2.62)$$

For almost all particles at almost all times in the universe, the chemical potential is much smaller than the temperature. To a good approximation, then, the distribution function depends only on E/T and the pressure satisfies (Exercise 14)

$$\frac{\partial \mathcal{P}_i}{\partial T} = \frac{\rho_i + \mathcal{P}_i}{T}.\qquad (2.63)$$

This relation can be used to show that the entropy density in the universe scales as a^{-3}. To see this, let's rewrite Eq. (2.56) as

$$a^{-3} \frac{\partial \left[(\rho + \mathcal{P}) a^3 \right]}{\partial t} - \frac{\partial \mathcal{P}}{\partial t} = 0. \qquad (2.64)$$

The derivative of the pressure with respect to time can be written as $(dT/dt)(\partial \mathcal{P}/\partial T)$ so

$$a^{-3} \frac{\partial \left[(\rho + \mathcal{P}) a^3 \right]}{\partial t} - \frac{dT}{dt} \frac{\rho + \mathcal{P}}{T} = a^{-3} T \frac{\partial}{\partial t} \left[\frac{(\rho + \mathcal{P}) a^3}{T} \right]$$

$$= 0. \qquad (2.65)$$

So the entropy density[5]

$$s \equiv \frac{\rho + \mathcal{P}}{T} \qquad (2.66)$$

scales as a^{-3}. Although we have framed the argument in terms of a single species, this scaling holds for the total entropy including all species in equilibrium. In fact, even if two species have different temperatures, the sum of their entropy densities still scales as a^{-3}. We will make use of this fact shortly when computing the relative temperatures of neutrinos and photons in the universe.

2.4 COSMIC INVENTORY

Armed with an expression for the energy density of a given species (Eq. (2.59)), and a knowledge of how it evolves in time (Eq. (2.56)), we can now tackle quantitatively the question of how much energy is contributed by the different components of the universe.

2.4.1 Photons

The temperature of the CMB photons has been measured extraordinarily precisely by the FIRAS instrument aboard the COBE satellite, $T = 2.725 \pm 0.002$K (Mather *et al.*, 1999). The energy density associated with this radiation is

$$\rho_\gamma = 2 \int \frac{d^3 p}{(2\pi)^3} \frac{1}{e^{p/T} - 1} p. \qquad (2.67)$$

The factor of 2 in front of Eq. (2.67) accounts for the two spin states of the photon. The energy of a given state is simply equal to p since the photon is massless. The chemical potential is zero; we expect this theoretically because early in the universe, photon number is not conserved (e.g., electrons and positrons can annihilate to produce photons). We also know it observationally because the spectrum of the CMB has been measured so accurately. The limits on a chemical potential are $\mu/T < 9 \times 10^{-5}$ (Fixsen *et al.*, 1996), so μ can be safely ignored. Since there is

---

[5]Technically, there is another term in the entropy density—proportional to the chemical potential—but, as mentioned above, this term is usually irrelevant in cosmology. Even with nonzero chemical potential, though, the entropy density scales as a^{-3}.

no angular dependence in the integrand of Eq. (2.67), the angular integral yields a factor of 4π and we are left with a one-dimensional integral. Define a dummy variable $x \equiv p/T$; then

$$\rho_\gamma = \frac{8\pi T^4}{(2\pi)^3} \int_0^\infty \frac{dx\, x^3}{e^x - 1}. \tag{2.68}$$

The integral can be expressed in terms of the Riemann ζ function; it is $6\zeta(4) = \pi^4/15$, so

$$\rho_\gamma = \frac{\pi^2}{15} T^4. \tag{2.69}$$

Since we derived (Eq. (2.58)) that the energy density of radiation scales as a^{-4}, the temperature of the CMB must scale as a^{-1}.

It will be useful to have all energy densities in the same units. The simplest way to do this is to divide all energy densities by the critical density today.[6] Thus,

$$\frac{\rho_\gamma}{\rho_{\rm cr}} = \frac{\pi^2}{15} \left(\frac{2.725 {\rm K}}{a} \right)^4 \frac{1}{8.098 \times 10^{-11} h^2 {\rm eV}^4}$$

$$= \frac{2.47 \times 10^{-5}}{h^2 a^4} \tag{2.70}$$

where to get the last line, it is useful to remember the conversion between kelvin and eV: 11605 K = 1 eV. To reiterate an important point, the photon energy density in Eq. (2.70) depends on time via the scale factor, but has no spatial dependence. This is because we have used the zero-order distribution function, the Bose–Einstein function, for the photons. In fact there are small perturbations around this zero-order distribution function. These do have a spatial dependence and correspond to the anisotropies in the CMB.

2.4.2 Baryons

Unlike the CMB, baryons[7] cannot be simply described as a gas with a temperature and zero chemical potential. Therefore, the baryon density must be measured directly, not via a temperature. There are now four established ways of measuring the baryon density, and these all seem to agree reasonably well (Fukugita, Hogan, and Peebles 1998). These are all measurements at different redshifts, and we know that the density scales as a^{-3}, so to facilitate comparison, one defines Ω_b via

$$\frac{\rho_b}{\rho_{\rm cr}} = \Omega_b a^{-3}. \tag{2.71}$$

That is, Ω_b is the ratio of the baryon density to the critical density today.

---

[6]The critical density — just like the Hubble rate which defines it — changes with time. However, it is common to define $\rho_{\rm cr}$ to be a constant, the critical density today, and I will follow this convention.

[7]I refer to all the nuclei and electrons in the universe as baryons. This is technically incorrect (electrons are *leptons*), but nuclei are so much more massive than electrons that virtually all the mass *is* in the baryons.

The simplest way is to observe the baryons today in galaxies. The greatest contribution to the density, though, comes not from stars in galaxies, but rather from gas in groups of galaxies. In these groups, Ω_b is about 0.02. The second way to count baryons is by looking at the spectra of distant quasars. The amount of light absorbed from these beacons is a measure of the intervening hydrogen, and hence the baryon density. These estimates (Rauch *et al.*, 1997) suggest $\Omega_b h^{1.5} \simeq 0.02$ with a fairly large uncertainty. A third method is to infer the baryon density by careful scrutiny of the anisotropies in the universe. As we will see in Chapter 8, these depend on the baryon density. Preliminary results (Pryke *et al.*, 2001; Netterfield *et al.*, 2001) give $\Omega_b h^2 = 0.024^{+0.004}_{-0.003}$ from the CMB. Finally, we will see in Chapter 3 that the light element abundances are sensitive to the baryon density, and that estimates of these abundances pin down $\Omega_b h^2 = 0.0205 \pm 0.0018$.

Remarkably, then, these estimates of the baryon density with very different techniques all agree.[8] They all place the baryon density at roughly 2–5% of the critical density. The total matter density in the universe is higher than this, so there must be matter in the universe that is nonbaryonic.

2.4.3 Matter

All of the methods of measuring the baryon density mentioned above involve the interaction of matter and radiation. For example, simply counting stars works at some level because we roughly know how much mass is required to output the light from a typical star. There are, however, methods of measuring the mass of matter that do not rely on the way light and matter interact. These classically have involved measuring the gravitational field in a given system, thereby inferring information about the mass responsible for that field. Figure 2.5 shows the inferred mass-to-light ratios of many systems, ranging from galaxies to superclusters. Historically this ratio was measured on small scales first, suggesting that the density in the universe was far less than critical. As more large-scale data were obtained, the steady increase in the mass/light ratio led some cosmologists to speculate that eventually we would find that the density was critical. Bahcall and collaborators (Bahcall, Lubin, and Dorman 1995; Bahcall *et al.*, 2000), however, have argued that mass-to-light ratios do not increase past $R \sim 1$Mpc; a leveling off occurs consistent with a matter density $\Omega_m \simeq 0.3$, where Ω_m is the ratio of the total matter density today to the critical density and

$$\rho_{\mathrm{m}} = \Omega_m \rho_{\mathrm{cr}} a^{-3}. \tag{2.72}$$

Recently a number of other techniques for inferring the matter density have emerged. We will see in Chapter 7 that the distribution of galaxies in the universe, in particular the power spectrum of this distribution, is very sensitive to $\Omega_m h$;

---

[8]Whether or not agreement holds is subject to debate. There have been claims (e.g., Persic and Salucci, 1992) that there is a *missing baryon problem* because the present-day abundance appears to be lower than that inferred from the light element abundances.

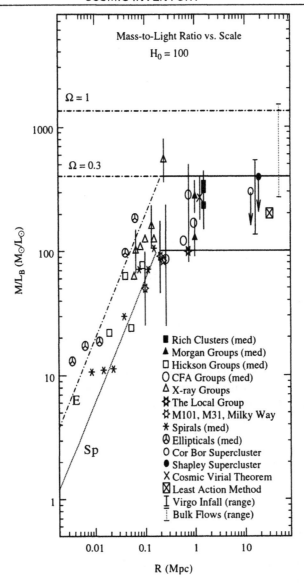

Figure 2.5. Mass vs. light ratio as a function of scale (Bahcall *et al.*, 2000). On the largest scales, the ratio flattens so that $\Omega_m \simeq 0.3$.

virtually all galaxy surveys[9] have inferred $\Omega_m h \simeq 0.2$. Another cosmological probe

---

[9]To mention three examples, the Automated Plate Measuring (APM) Survey, to be discussed further in Chapter 9, has been analyzed by Efstathiou and Moody (2001); the Two Degree Field (2DF) by Percival *et al.* (2001), and early data from the Sloan Digital Sky Survey by Dodelson *et al.* (2001). These groups found $\Omega_m h = 0.14^{+0.24}_{-0.09}, 0.20 \pm 0.03$, and $0.14^{+0.11}_{-0.06}$, respectively.

we will encounter in Chapter 9 is the cosmic velocity field (Strauss and Willick, 1995) and its relation to the observed galaxy distribution. These are related by the continuity equation, a relation sensitive to Ω_m. Again most of the measurements are clustered around $\Omega_m \simeq 0.3$.

Another way of measuring the total mass density is to pick out observations sensitive to Ω_b/Ω_m and use the apparent value of Ω_b to infer the matter density. For example, most of the baryonic mass in a galaxy cluster is in the form of hot gas. The ratio of the mass of gas in clusters of galaxies to the total mass can be measured either by looking for X-ray emission (White *et al.*, 1993) or by looking at the electron-heated CMB in the direction of the cluster (Grego *et al.*, 2001). If this ratio is characteristic of the universe as a whole — and it probably is, because clusters are so large — then the cosmic baryon to matter ratio is around 20%. Since baryons make up only about 5% of the critical density, the total matter density is inferred to be about 0.25. Another way of inferring the baryon/matter ratio is by looking for features in the power spectrum of galaxies; if the baryon fraction is truly of order 20%, then there will be wiggles in the spectrum (again Chapter 7). There are tentative hints of these wiggles in the early data from the Two Degree Field (2DF) survey (Percival *et al.*, 2001). These pin down $\Omega_b/\Omega_m = 0.15 \pm 0.07$, consistent with the cluster observations. Finally, the anisotropies in the CMB (Chapter 8) are sensitive to the matter density $\Omega_m h^2$. Recent determinations indicate $\Omega_m h^2 = 0.16 \pm 0.04$ (Pryke *et al.*, 2001; Netterfield *et al.*, 2001). Given the fact that current best estimates of the Hubble constant give $h = 0.72$, the CMB observations also are consistent with a matter density equal to 30% of the critical density.

We therefore have an enormous amount of evidence telling us that the baryon density is of order 5% the critical density, while the total matter density is some five times larger. Most of the matter in the universe must not be baryons. It must be some new form of matter: dark matter.

2.4.4 Neutrinos

The next component we need consider is the neutrino. Unlike photons and baryons, cosmic neutrinos have not been observed, so arguments about their contribution to the energy density are necessarily theoretical. However, these theoretical arguments are quite strong, based on very well-understood physics.

A basic understanding of the interaction rates of neutrinos enables us to argue that neutrinos were once kept in equilibrium with the rest of the cosmic plasma. Since they are fermions, their distribution was Fermi–Dirac with zero chemical potential. At late times, they lost contact with the plasma because their interactions are *weak*. Nonetheless, their distribution remained Fermi–Dirac, with their temperature simply falling as a^{-1}. The main task therefore is to relate the neutrino temperature to the photon temperature today. The tricky part of this is the annihilation of electrons and positrons when the cosmic temperature was of order the electron mass. Neutrinos lost contact with the cosmic plasma slightly before this annihilation, so they did not inherit any of the associated energy. The photons, which did, are therefore hotter than the neutrinos.

We can account for the annhilation of electrons and positrons by using the fact that the total entropy density s (Eq. (2.66)) scales as a^{-3}. Massless bosons contribute $2\pi^2 T^3/45$ to the entropy density for each spin state, while massless fermions contribute 7/8 this, and massive particles contribute negligibly (Exercise 17). Before annihilation, the fermions are electrons (2 spin states), positrons (2), neutrinos (3 generations and one spin state) and anti-neutrinos (3). The bosons are photons (2 spin states). So at a_1 before annihilation,

$$s(a_1) = \frac{2\pi^2}{45}T_1^3 \left[2 + (7/8)(2 + 2 + 3 + 3)\right]$$

$$= \frac{43\pi^2}{90}T_1^3 \qquad (2.73)$$

where T_1 is the common temperature at a_1. After annihilation, the electrons and positrons have gone away and the photon and neutrino temperatures are no longer identical: we must distinguish between them. Therefore, the entropy density is

$$s(a_2) = \frac{2\pi^2}{45}\left[2T_\gamma^3 + \frac{7}{8}6T_\nu^3\right]. \qquad (2.74)$$

Equating $s(a_1)a_1^3$ with $s(a_2)a_2^3$ leads to

$$\frac{43}{2}\left(a_1 T_1\right)^3 = 4\left[\left(\frac{T_\gamma}{T_\nu}\right)^3 + \frac{21}{8}\right]\left(T_\nu(a_2)a_2\right)^3. \qquad (2.75)$$

But the neutrino temperature scales throughout as a^{-1}, so $a_1 T_1 = a_2 T_\nu(a_2)$. Therefore, the ratio of the two temperatures is

$$\frac{T_\nu}{T_\gamma} = \left(\frac{4}{11}\right)^{1/3}. \qquad (2.76)$$

We can now evaluate the energy density of neutrinos in the universe. Let's sum up what we know about the cosmic abundance of neutrinos

- One spin degree of freedom for neutrinos
- Neutrino has antiparticle
- Three generations of neutrinos
- Neutrinos are fermions → Fermi–Dirac distribution function
- Neutrino temperature is lower by a factor of $(4/11)^{1/3}$ since photons are heated by $e^+ e^-$ annihilation

The first three items on the list then imply that the degeneracy factor of neutrinos is equal to six. The fourth means we need to change the denominator in the integrand in Eq. (2.67) to $e^{p/T} + 1$. The Fermi-Dirac integral is then smaller by a factor of 7/8. Finally, since the energy density of a massless particle scales as T^4, the last item implies that the neutrino energy density is smaller than the photon density by $(4/11)^{4/3}$. Putting all these factors together leads to

$$\rho_\nu = 3 \, \frac{7}{8} \, \left(\frac{4}{11}\right)^{4/3} \rho_\gamma. \tag{2.77}$$

Equivalently, if there were three species of massless neutrinos today, then their contribution to the energy density would be

$$\Omega_\nu \equiv \left.\frac{\rho_\nu}{\rho_{\rm cr}}\right|_{\rm today} = \frac{1.68 \times 10^{-5}}{h^2} \qquad m_\nu = 0. \tag{2.78}$$

In reality, all the neutrinos do not appear to be massless. Observations of neutrinos from both the sun (Bahcall, 1989) and from our atmosphere (Fukuda *et al.*, 1998) strongly suggest that neutrinos of different flavors (generations) oscillate into each other. This can happen only if the neutrinos have mass. The atmospheric neutrino observations in particular imply that at least one neutrino has a mass larger than 0.05 eV.[10] The energy density of a massive neutrino is

$$\rho_\nu = 2 \int \frac{d^3p}{(2\pi)^3} \frac{\sqrt{p^2 + m_\nu^2}}{e^{p/T_\nu} + 1}. \tag{2.79}$$

At high temperatures, this reduces to Eq. (2.77) (without the 3), so when considering neutrinos in the early universe, it is often sufficient to use Eq. (2.77). Indeed, we will do this shortly when we come to esimtate the epoch at which the energy density of matter equals that of radiation. At late times, the massive neutrino energy density is $m_\nu n_\nu$, with the neutrino number density equal to $3n_\gamma/11$ (Exercise 18). As can be seen from Figure 2.6, the transition takes place when $T_\nu \sim m_\nu$. Therefore,

$$\Omega_\nu = \frac{m_\nu}{94h^2{\rm eV}} \qquad m_\nu \neq 0. \tag{2.80}$$

Those who trafficked in both astrophysics and particle physics (Gerstein and Zel'dovich, 1966; Marx and Szalay, 1972; Cowsik and McClelland, 1972) early on noted that the simple observation that the total density was not much greater than the critical density leads to constraints on neutrino mass, constraints much more stringent than those obtainable at accelerators. When the need for nonbaryonic dark matter first became evident, a number of cosmologists (e.g., Gunn *et al.*, 1978) proposed neutrinos as the natural candidate. Subsequent studies (Bond, Efstathiou, and Silk, 1980; White, Frenk, and Davis 1983) of the structure of the universe with neutrinos as the dominant dark matter component looked significantly different from the actual universe. Nonetheless, the possibility that neutrinos might make up a *fraction* of the total density reemerged in the 1990s. We can then hope to detect a trace amount — corresponding to masses smaller than an eV — by observing its effect on large-scale structure.

---

[10] The oscillation experiments are sensitive to mass *differences*, $m_2^2 - m_1^2$, so the actual constraint is that the mass squared difference is of order 10^{-3}. This could also be accomodated with two nearly degenerate masses with a small splitting.

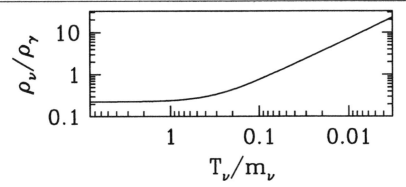

Figure 2.6. Energy density of one generation of massive neutrinos as compared with the density in the CMB. At high temperatures, the ratio is a fixed constant; at low temperatures, the neutrino behaves like nonrelativistic matter (scaling as a^{-3}) and so begins to dominate over the photon density (which scales as a^{-4}).

2.4.5 Dark Energy

There are two sets of evidence pointing toward the existence of something else, something beyond the radiation and matter itemized above. The first comes from a simple budgetary shortfall. The total energy density of the universe is very close to critical. We expect this theoretically (Chapter 6) and we observe it in the anisotropy pattern of the CMB (Chapter 8). Yet, the total matter density inferred from observations is only a third critical. The remaining two-thirds of the density in the universe must be in some smooth, unclustered form, dubbed[11] *dark energy*. The second set of evidence is more direct. Given the energy composition of the universe, one can compute a theoretical distance vs redshift diagram. This relation can then be tested observationally.

In 1998, two groups (Riess *et al.*, 1998, Perlmutter *et al.*, 1999) observing supernovae reported direct evidence for dark energy. The evidence is based on the difference between the luminosity distance in a universe dominated by dark matter and one dominated by dark energy. As Figure 2.3 indicates, the luminosity distance is larger for objects at high redshifts in a dark-energy-dominated universe. Therefore, objects of fixed intrinsic brightness will appear fainter if the universe is composed of dark energy.

---

[11]After the discovery, quite a bit of attention was focused on choosing an appropriate name. *Cosmological constant*, everyone's initial moniker, is too restrictive in that the energy density is constant at all times, and we do not yet know that this is true of the dark energy. *Variable cosmological constant* fixes that problems but introduces an inherent contradiction ("variable" and "constant"?). *Variable Lambda* using the Greek letter reserved for the cosmological constant is too obscure. *Quintessence* is a good choice: it expresses the fact that, after cosmological photons, baryons, neutrinos, and dark matter, there is a fifth essence in the universe. It seems to me that dark energy has become a bit more popular, with quintessence referring to the subset of models in which the energy density can be associated with a time-dependent scalar field.

More concretely, the luminosity distance of Eq. (2.50) can be used to find the apparent magnitude m of a source with absolute magnitude M. Magnitudes are related to fluxes via $m = -(5/2)\log(F)+$ constant. Since the flux scales as d_L^{-2}, the apparent magnitude $m = M + 5\log(d_L)+$ constant. The convention is that

$$m - M = 5\log\left(\frac{d_L}{10\text{pc}}\right) + K \tag{2.81}$$

where K is a correction for the shifting of the spectrum into or out of the wavelength range measured due to expansion.

The two groups measured the apparent magnitudes of dozens of Type Ia supernovae, which are known to be standard candles, i.e., they have nearly identical absolute magnitudes. Although they were able to place tight constraints on dark energy using the many supernovae that they detected, we can get a feel for the measurement by simply considering two of these. Consider then Supernova 1997ap, found at redshift $z = 0.83$ with apparent magnitude $m = 24.32$, and Supernova 1992P, found at low redshift $z = 0.026$ with apparent magnitude $m = 16.08$. Since the absolute magnitudes of these are the same, the difference in apparent magnitudes is due solely to the difference in luminosity distance:

$$24.32 - 16.08 = 5\log\left(d_L(z = 0.83)\right) - 5\log\left(d_L(z = 0.026)\right). \tag{2.82}$$

The nearby luminosity distance is independent of cosmology, simply equal to $z/H_0 = 0.026/H_0$. Therefore, the only unknown remaining in Eq. (2.82) is fixed by the observations to be

$$H_0 d_L(z = 0.83) = 1.16. \tag{2.83}$$

In a flat, matter-dominated universe ($\Omega_m = 1$), the luminosity distance out to $z = 0.83$ is equal to $0.95 H_0^{-1}$, whereas a universe with $\Omega_m = 0.3$ and a cosmological constant $\Omega_\Lambda = 0.7$ has a luminosity distance of $1.23 H_0^{-1}$. The apparent magnitude of this single distant supernova then suggests that dark energy pervades the universe.

Of course, the discussion of the previous paragraph does not account for uncertainties (typical uncertainties in the magnitudes are of order 0.2), nor does it do a careful fit to all known supernovae, allow for extinction by dust, or allow for the variation of the absolute magnitude correlated with the duration. The supernova groups did all of those things, and emerged with the constraints shown in Figure 2.7. The two free cosmological parameters are the matter density Ω_m and a cosmological constant Ω_Λ, something we now recognize as one possible form of dark energy, one in which the energy density is constant. Note that the "theorists' dream" universe — flat and matter-dominated ($\Omega_m = 1$) — is excluded with high confidence. Indeed, even a pure open universe with $\Omega_m = 0.3, \Omega_\Lambda = 0$ is strongly disfavored by the supernova data.

While highly popular, Figure 2.7 suffers from two drawbacks. It allows for too much freedom in one sector and too little in another. Most of the region in the figure is taken up by a universe with both dark energy and nonzero curvature (not flat).

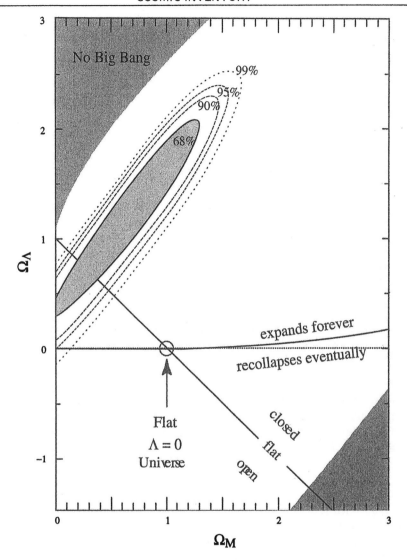

Figure 2.7. Constraints from Type Ia supernovae on the parameters $(\Omega_m, \Omega_\Lambda)$ (Perlmutter *et al.*, 1999). Flat, matter-dominated universe, the dot with $\Omega_m = 1, \Omega_\Lambda = 0$, is ruled out with high confidence. The line extending from upper left to lower right corresponds to a flat universe.

Although one or the other of these has been argued for, seldom have cosmologists suggested that the universe contains both. Thus, except for the "flat" line and the $\Omega_\Lambda = 0$ line, most of the region in Figure 2.7 is, at least aesthetically, unappealing: the figure allows for too much freedom. On the other hand, the only form of dark energy budgeted for is the cosmological constant. To open up other possibilities,

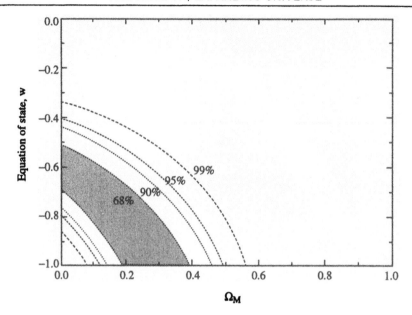

Figure 2.8. Constraints in a flat universe from Type Ia supernovae on the matter density (Ω_m) and equation of state of the dark energy (w) (Perlmutter *et al.*, 1999). Cosmological constant corresponds to $w = -1$, matter to $w = 0$.

consider Eq. (2.55) as applied to the cosmological constant. The only way for this equation to be satisfied with constant energy density is if the pressure is equal to $-\rho$. One might imagine energy with a slightly different pressure and therefore energy evolution. Define

$$w \equiv \frac{\mathcal{P}}{\rho}. \tag{2.84}$$

A cosmological constant corresponds to $w = -1$, matter to $w = 0$, and radiation to $w = 1/3$. With this new freedom, let's see what the supernova data imply for the equation of state of the dark energy if we fix the universe to be flat. Figure 2.8 shows that values of w greater than ~ -0.5 are disfavored; a cosmological constant is consistent with the data, but it is by no means the only possibility. Equation (2.55) can be integrated to find the evolution of the dark energy,

$$\rho_{\mathrm{de}} \propto \exp\left\{-3 \int^a \frac{da'}{a'}\left[1 + w(a')\right]\right\}. \tag{2.85}$$

Note that — if w is constant — this expression agrees with our knowledge of the cases explicated above $w = 1/3, 0, -1$.

2.4.6 Epoch of Matter–Radiation Equality

The epoch at which the energy density in matter equals that in radiation is called *matter–radiation equality*. It has a special significance for the generation of large-scale structure and for the development of CMB anisotropies, because perturbations grow at different rates in the two different eras. It is therefore a useful exercise to calculate the epoch of matter–radiation equality. To do this, we need to compute the energy density of both matter and radiation, and then find the value of the scale factor at which they were equal.

Using Eqs. (2.70) and (2.78), we see that the total energy density in radiation is

$$\frac{\rho_r}{\rho_{\mathrm{cr}}} = \frac{4.15 \times 10^{-5}}{h^2 a^4} \equiv \frac{\Omega_r}{a^4}. \tag{2.86}$$

To calculate the epoch of matter–radiation equality, we equate Equations (2.86) and (2.72) to find

$$a_{\mathrm{eq}} = \frac{4.15 \times 10^{-5}}{\Omega_m h^2}. \tag{2.87}$$

A different way to express this epoch is in terms of redshift z; the redshift of equality is

$$1 + z_{\mathrm{eq}} = 2.4 \times 10^4 \Omega_m h^2. \tag{2.88}$$

Note that — obviously — as the amount of matter in the universe today, $\Omega_m h^2$, goes up, the redshift of equality also goes up. For our purposes it will be very important that the redshift of equality is at least several times larger than the redshift when photons decouple from matter, $z_* \simeq 10^3$. Thus, we expect photons to decouple when the universe is already well into the matter-dominated era.

2.5 SUMMARY

The smooth universe can be described with the Friedmann–Robertson–Walker metric given in Eq. (2.4), which implies that physical distances are related to coordinate (comoving) distances with the time-dependent scale factor $a(t)$. The time dependence of the metric is determined by the Einstein equations. The time-time component of the Einstein equations reduces to Eq. (2.39) in a flat universe.

Measuring distances in the expanding universe is tricky, but all relevant distances can be obtained from the comoving distance between us and a source at redshift z:

$$\chi(z) = \int_0^z \frac{dz'}{H(z')}. \tag{2.89}$$

Another important distance is that light could have traveled since $t = 0$. This is usually expressed as a time, the conformal time,

$$\eta = \int_0^t \frac{dt'}{a(t')} = \int_z^\infty \frac{dz'}{H(z')}. \tag{2.90}$$

The conformal time will be the natural time variable when we come to consider the evolution of perturbations in the universe.

Photons in the universe have a Bose–Einstein distribution with zero chemical potential, so their energy density can be determined by measuring their temperature. Neutrinos have a Fermi-Dirac distribution, also probably with zero chemical potential, but there is some ambiguity in their energy density because of our ignorance of the neutrino masses. Early on, this ambiguity is irrelevant since the temperatures are so much larger than the masses and neutrinos behave relativistically. Thus, the uncertainty in neutrino mass does not affect Big Bang nucleosynthesis at temperatures of order 1 MeV and probably not even the epoch of matter radiation equality at temperatures of order 1 eV. The neutrino temperature is a factor of $(4/11)^{1/3}$ smaller than the photon temperature. This, and the difference in statistics, implies that a species of massless neutrinos has an energy density equal to 0.23 times that of photons. A single neutrino generation with mass m_ν contributes $\Omega_\nu = 0.01(m_\nu/0.94\,\text{eV}\,h^2)$. In addition to photons and neutrinos, the universe consists of baryons, best determined by nucleosynthesis to have $\Omega_b h^2 = 0.0205 \pm 0.0018$; dark matter ($\Omega_m \simeq 0.3$); and dark energy (with $\Omega_{\text{de}} \simeq 0.7$), a new form of energy with negative pressure.

There is significantly more energy today in nonrelativistic matter than in radiation. However, since the energy density of radiation scales as a^{-4} while that of matter as a^{-3}, the very early universe was radiation dominated. The epoch at which the matter density was equal to the radiation density delineates these two regimes: $a_{\text{eq}} = 4.15 \times 10^{-5}/\Omega_m h^2$.

SUGGESTED READING

My favorite book on general relativity at this level is *A First Course in General Relativity* (Schutz), which gives many simple examples to introduce the seemingly profound ideas of general relativity. Also very good are *Flat and Curved Spacetimes* (Ellis and Williams) and *Essential Relativity* (Rindler). Slightly more advanced is *Gravitation and Cosmology: Principles and Applications of the General Theory of Relativity* (Weinberg), which also has a nice discussion of the early universe in Chapter 15. Two more advanced books are *General Relativity* (Wald) and the classic *Gravitation* (Misner, Thorne, and Wheeler). Some of the thermodynamics and statistical mechanics introduced in this chapter is presented in *The Early Universe* (Kolb and Turner). The distance formulae of Section 2.2 are covered in all standard texts. Neutrinos and their relation to cosmology are covered in the standard texts as well, but there are also several other good books focused solely on neutrinos and astrophysics, *Neutrino Astrophysics* by the pioneer of the field, Bahcall, and *Massive Neutrinos in Physics and Astrophysics* by Mohapatra and Pal.

A number of papers treat the topics in this chapter at an accessible level. An especially coherent review of all the different distance measures is given by Hogg (1999). Fukugita, Hogan, and Peebles (1999) do the baryon inventory outlined in Section 2.4.2. Since the supernova discoveries in the late 1990s, many popular articles have appeared attempting to explain the dark energy. The two seminal articles though — Perlmutter *et al.* (1999) and Riess *et al.* (1998) — are extremely clear and well worth reading.

EXERCISES

Exercise 1. Convert the following quantities by inserting the appropriate factors of $c, \hbar$, and k_B:

- $T_0 = 2.725 \text{K} \rightarrow \text{eV}$
- $\rho_\gamma = \pi^2 T_0^4 / 15 \rightarrow \text{eV}^4$ and g cm^{-3}
- $1/H_0 \rightarrow \text{cm}$
- $m_{\text{Pl}} \equiv 1.2 \times 10^{19} \text{ GeV} \rightarrow \text{K, cm}^{-1}, \text{sec}^{-1}$

Exercise 2. Show that the geodesic equation gets the correct equations of motion for a particle traveling freely in two dimensions using polar coordinates. You can get the Christoffel symbols one of two ways (or both!) and then proceed to **(b)**.
(a) Get the Christoffel symbol either directly from the term in brackets in Eq. (2.17) or from the 2D metric

$$g_{ij} = \begin{pmatrix} 1 & 0 \\ 0 & r^2 \end{pmatrix} \tag{2.91}$$

using Eq. (2.19). Show that the only nonzero Christoffel symbols are

$$\Gamma^2_{12} = \Gamma^2_{21} = \frac{1}{r} \quad ; \quad \Gamma^1_{22} = -r \tag{2.92}$$

with $1, 2$ corresponding to r, θ.

(b) Write down the two components of the geodesic equation using these Christoffel symbols. Show that these give the proper equations of motion for a particle traveling in a plane.

Exercise 3. The metric for a particle traveling in the presence of gravitational field is $g_{\mu\nu} = \eta_{\mu\nu} + h_{\mu\nu}$ where $h_{00} = -2\phi$ where ϕ is the Newtonian gravitational potential; $h_{i0} = 0$; and $h_{ij} = -2\phi\delta_{ij}$. Find the equation of motion for a massive particle traveling in this field.
(a) Show that $\Gamma^0_{\ 00} = \partial\phi/\partial t$ and $\Gamma^i_{\ 00} = \delta^{ij}\partial\phi/\partial x^j$.
(b) Show that the time component of the geodesic equation implies that energy $p^0 + m\phi$ is conserved.
(c) Show that the space components of the geodesic equation lead to $d^2x^i/dt^2 = -m\delta^{ij}\partial\phi/\partial x^j$ in agreement with Newtonian theory. Use the fact that the particle is nonrelativistic so $p^0 \gg p^i$.

Exercise 4. Find how the energy of a *massive*, nonrelativistic particle changes as the universe expands. Recall that in the massless case we used the fact that $g_{\mu\nu}P^\mu P^\nu = 0$. In this case, it is equal not to zero, but to $-m^2$.

Exercise 5. Fill in some of the blanks left in our derivation of the Einstein equations.
(a) Compute the Christoffel symbol $\Gamma^i_{\ \alpha\beta}$ for a flat FRW metric.
(b) Compute the spatial components of the Ricci tensor in a flat FRW universe, R_{ij}. Show that the space-time component, R_{i0}, vanishes.

Exercise 6. Show that the space-space component of the Einstein equations in a flat universe is

$$\frac{d^2a/dt^2}{a} + \frac{1}{2}\left(\frac{da/dt}{a}\right)^2 = -4\pi G \mathcal{P} \tag{2.93}$$

where $\mathcal{P}$ is the pressure, the $T^i_{\ i}$ (no sum over i) component of the energy–momentum tensor.

Exercise 7. Find and apply the metric, Christoffel symbols, and Ricci scalar for a particle trapped on the surface of a sphere with radius r.
(a) Using coordinates t, θ, ϕ, the metric is

$$g_{\mu\nu} = \begin{pmatrix} -1 & 0 & 0 \\ 0 & r^2 & 0 \\ 0 & 0 & r^2\sin^2\theta \end{pmatrix}. \tag{2.94}$$

Show that the only nonvanishing Christoffel symbols are $\Gamma^\theta_{\ \phi\phi}, \Gamma^\phi_{\ \phi\theta}$, and $\Gamma^\phi_{\ \theta\phi}$. Express these in terms of θ.
(b) Use these and the geodesic equation to find the equations of motion for the particle.

(c) Find the Ricci tensor. Show that contraction of this tensor leads to

$$\mathcal{R} \equiv g^{\mu\nu} R_{\mu\nu} = \frac{2}{r^2}. \tag{2.95}$$

Exercise 8. Apply the Einstein equations to the case of an open universe. The interval in an open universe is

$$ds^2 = -dt^2 + a^2(t) \left\{ \frac{dr^2}{1 + \Omega_k H_0^2 r^2} + r^2 \left(d\theta^2 + \sin^2\theta d\phi^2 \right) \right\} \tag{2.96}$$

where r, θ, ϕ are the standard 3D spherical coordinates, and Ω_k is the curvature density.
(a) First compute the Christoffel symbols. Show that the only nonzero ones are equal to

$$\Gamma^i{}_{0j} = H\delta^i{}_j \qquad \Gamma^0{}_{ij} = g_{ij} H$$

$$\Gamma^i{}_{jk} = \frac{g^{il}}{2} \left[g_{lj,k} + g_{lk,j} - g_{jk,l} \right]. \tag{2.97}$$

(b) Show that the components of the Ricci tensor are

$$R_{00} = -3\frac{\ddot{a}}{a}$$

$$R_{ij} = g_{ij} \left[\frac{\ddot{a}}{a} + 2H^2 - \frac{2\Omega_k H_0^2}{a^2} \right]. \tag{2.98}$$

(c) From these, compute the Ricci scalar, and then derive the time-time component of Einstein equations.

Exercise 9. Show that the geodesic equation we derived in a flat universe implies that

$$\frac{d^2\vec{x}}{d\eta^2} = 0 \tag{2.99}$$

where η is the conformal time.

Exercise 10. Assume that there is only matter and radiation in the universe (no cosmological constant) and that the universe is flat ($\rho_0 = \rho_{\text{cr}}$). Integrate Eq. (1.2) to determine the times when the cosmic temperature was 0.1 MeV and 1/4 eV.

Exercise 11. Derive some simple expressions for the conformal time η as a function of a.
(a) Show that $\eta \propto a^{1/2}$ in a matter dominated universe and a in one dominated by radiation.

(b) Consider a universe with only matter and radiation, with equality at a_{eq}. Show that

$$\eta = \frac{2}{\sqrt{\Omega_m H_0^2}} \left[\sqrt{a + a_{eq}} - \sqrt{a_{eq}} \right]. \tag{2.100}$$

What is the conformal time today? At decoupling?

Exercise 12. Consider a galaxy of physical (visible) size 5 kpc. What angle would this galaxy subtend if situated at redshift 0.1? Redshift 1? Do the calculation in a flat universe, first matter-dominated and then with 30% matter and 70% cosmological constant.

Exercise 13. How is the energy density of a gas of photons with a blackbody spectrum related to the specific intensity of the radiation? That is, what is the relation between ρ_γ and I_ν defined in Eq. (1.8)?

Exercise 14. (a) Compute the pressure of a relativistic species in equilibrium with temperature T. Show that $\mathcal{P} = \rho/3$ for both Fermi–Dirac and Bose–Einstein statistics.
(b) Suppose the distribution function depends only on E/T as it does in equilibrium. Find $d\mathcal{P}/dT$. A simple way to do this is to rewrite df/dT in the integral as $-(E/T)df/dE$ and then integrate Eq. (2.62) by parts.

Exercise 15. The general relativistic expression for the energy–momentum tensor in terms of the distribution functions is given by

$$T^\mu{}_\nu(\vec{x}, t) \bigg|_{\text{species } i} = g_i \int \frac{dP_1 dP_2 dP_3}{(2\pi)^3} (-\det[g_{\alpha\beta}])^{-1/2} \frac{P^\mu P_\nu}{P^0} f_i(\vec{x}, \vec{p}, t) \tag{2.101}$$

where P_μ was defined in Eq. (2.24), g_i is the number of spin states for species i, and $\det[g_{\mu\nu}]$ is the determinant of the 4D matrix $g_{\mu\nu}$. Eliminate the comoving momenta P_μ in favor of the magnitude of the proper momentum defined via

$$p^2 \equiv g^{ij} P_i P_j \tag{2.102}$$

and the direction vector $\hat{p}$. Note that while the comoving momenta P_i remain constant as the universe expands, p falls off as a^{-1}. Show that the time-time component of Eq. (2.101) agrees with the expression for the energy density given in Eq. (2.59). Use the fact that $P^2 \equiv g_{\mu\nu} P^\mu P^\nu = -m^2$ for a particle of mass m.

Exercise 16. Plot $m - M$ as a function of redshift for a flat, matter-dominated universe (this can be done analytically) and for a flat universe with $\Omega_\Lambda = 0.7, \Omega_m = 0.3$ (for this you need to evaluate numerically a 1D integral). Neglect the K correction. Compare with Figure 1.7.

Exercise 17. Consider the entropy density, s, defined in Eq. (2.66). For a massless particle, you showed in Exercise 14 that $\mathcal{P} = \rho/3$, so $s = 4\rho/3T$. Express s as a function of T for both bosons and fermions (assumed massless) in equilibrium with zero chemical potential. Show that the entropy density for a massive particle in equilibrium $(T \ll m; \mu = 0)$ is exponentially small.

Exercise 18. Show that the number density of one generation of neutrinos and anti-neutrinos in the universe today is

$$ n_\nu = \frac{3}{11} n_\gamma = 112 \text{cm}^{-3}. $$

For this calculation, you will also have to compute the photon number density; both can be expressed in terms of Riemann zeta functions (Eq. (C.27)). Using this result, verify Eq. (2.80).

Exercise 19. We computed the epoch of equality in the event that all three neutrinos are massless. Suppose instead that two are massless, but the third has mass $m = 0.1$ eV. What is a_{eq} is this case?

3

BEYOND EQUILIBRIUM

The very early universe was hot and dense. As a result, interactions among particles occurred much more frequently than they do today. As an example, a photon today can travel across the observable universe without deflection or capture, so it has a mean free path greater than 10^{28} cm. When the age of the universe was equal to 1 sec, though, the mean free path of a photon was about the size of an atom. Thus in the time it took the universe to expand by a factor of 2, a given photon interacted many, many times. These multiple interactions kept the consitituents in the universe in equilibrium in most cases. Nonetheless, there were times when reactions could not proceed rapidly enough to maintain equilibrium conditions. These times are — perhaps not coincidentally — of the utmost interest to cosmologists today.

Indeed, we will see in this chapter that out-of-equilibrium phenomena played a role in (i) the formation of the light elements during Big Bang nucleosynthesis; (ii) recombination of electrons and protons into neutral hydrogen when the temperature was of order $1/4$ eV; and quite possibly in (iii) production of dark matter in the early universe. Each of these three periods, for obvious reasons, is the subject of intense study by many different groups. Often these studies are carried out in parallel, without the link among the three mentioned. I think it is important to understand that all three phenomena are the result of nonequilibrium physics and that all three can be studied with the same formalism: the Boltzmann equation. Section 3.1 introduces the Boltzmann equation and some approximations to it that are common to all three processes. The remaining three sections of the chapter are simply applications of this general formula.

Beyond the intrinsic importance of these nonequilibrium phenomena, this chapter also serves as a bridge between the smooth, homogeneous universe described in Chapter 2 and the inhomogeneous perturbations we will explore in the rest of the book. One way to think of this transition is in terms of phase-space distributions f of Eqs. (2.60) and (2.61). Until now, we have assumed that the chemical potentials are zero and temperatures uniform. In this chapter, we will have to abandon the idea of trivial chemical potentials in order to track abundances of particles losing contact with the plasma. In succeeding chapters, we will move beyond uniformity and explore temperatures which depend on both position and direction of propagation.

3.1 BOLTZMANN EQUATION FOR ANNIHILATION

The Boltzmann equation formalizes the statement that the rate of change in the abundance of a given particle is the difference between the rates for producing and eliminating that species. Suppose that we are interested in the number density n_1 of species 1. For simplicity, let's suppose that the only process affecting the abundance of this species is an annihilation with species 2 producing two particles, imaginatively called 3 and 4. Schematically, $1+2 \leftrightarrow 3+4$; i.e., particle 1 and particle 2 can annihilate producing particles 3 and 4, or the inverse process can produce 1 and 2. The Boltzmann equation for this system in an expanding universe is

$$a^{-3}\frac{d\left(n_1 a^3\right)}{dt} = \int \frac{d^3 p_1}{(2\pi)^3 2E_1} \int \frac{d^3 p_2}{(2\pi)^3 2E_2} \int \frac{d^3 p_3}{(2\pi)^3 2E_3} \int \frac{d^3 p_4}{(2\pi)^3 2E_4}$$

$$\times (2\pi)^4 \delta^3(p_1 + p_2 - p_3 - p_4)\delta(E_1 + E_2 - E_3 - E_4) |\mathcal{M}|^2$$

$$\times \{f_3 f_4 [1 \pm f_1][1 \pm f_2] - f_1 f_2 [1 \pm f_3][1 \pm f_4]\}. \tag{3.1}$$

In the absence of interactions, the left-hand side of Eq. (3.1) says that the density times the scale factor cubed is conserved. This reflects the nature of the expanding universe: as the comoving grid expands, the volume of a region containing a fixed number of particles grows as a^3. Therefore, the physical number density of these particles falls off as a^{-3}. Interactions are included in the right-hand side of the Boltzmann equation. Let's consider the interaction term starting from the last line and moving up. Putting aside the $1 \pm f$ terms on the last line, we see that the rate of producing species 1 is proportional to the occupation numbers of species 3 and 4, f_3 and f_4. Similarly the loss term is proportional to $f_1 f_2$. The $1 \pm f$ terms, with plus sign for bosons such as photons and minus sign for fermions such as electrons, represent the phenomena of Bose enhancement and Pauli blocking. If particles of type 1 already exist, a reaction producing more such particles is more likely to occur if 1 is a boson and less likely if a fermion. I have suppressed the momentum dependence of f, but of course all the occupation numbers depend on the corresponding momentum (e.g., $f_1 = f_1(p_1)$). Moving upward, the Dirac delta functions on the second line in Eq. (3.1) enforce energy and momentum conservation; the factors of 2π are the result of moving from discrete Kronecker δ's to the continuous Dirac version. The energies here are related to the momenta via $E = \sqrt{p^2 + m^2}$. The amplitude on the second line $\mathcal{M}$ is determined from the fundamental physics in question. For example, if we were interested in the Compton scattering of electrons off of photons, $\mathcal{M}$ would be proportional to α, the fine structure constant. In almost all cases of interest, this amplitude is *reversible*, identical for $1 + 2 \rightarrow 3 + 4$ and $3 + 4 \rightarrow 1 + 2$. Indeed, reversibility has been assumed in Eq. (3.1).

The last two lines of Eq. (3.1) depend on the momenta of the particles involved. To find the total number of interactions, we must sum over all momenta. The integrals on the first line do precisely that. As in Figure 2.4, the factors of $(2\pi)^3$ [really $(2\pi\hbar)^3$] represent the volume of one unit of phase space; we want to sum over all such units. Finally, the factors of $2E$ in the denominator arise because, relativistically, the phase space integrals should really be four-dimensional, over the three components of momentum and one of energy. However, these are constrained to lie on the 3-sphere fixed by $E^2 = p^2 + m^2$. In equations,

$$\int d^3p \int_0^\infty dE\, \delta(E^2 - p^2 - m^2) = \int d^3p \int_0^\infty dE\, \frac{\delta\left(E - \sqrt{p^2 + m^2}\right)}{2E}. \tag{3.2}$$

Performing the integral over E with the delta function yields the factor of $2E$.

Equation (3.1) is an integrodifferential equation for the phase space distributions. Further, in principle at least, it must be supplemented with similar equations for the other species. In practice, these formidable obstacles can be overcome for many practical cosmological applications. The first, most important realization is that scattering processes typically enforce *kinetic equilibrium*. That is, scattering takes place so rapidly that the distributions of the various species take on the generic Bose–Einstein/Fermi–Dirac forms (Eqs. (2.61) and (2.60)). This form condenses all of the uncertainty in the distribution into a single function of time μ. If annihilations were also in equilibrium, μ would be the chemical potential, and the sum of the chemical potentials in any reaction would have to balance. For example, the reaction $e^+ + e^- \leftrightarrow \gamma + \gamma$ would cause $\mu_{e^+} + \mu_{e^-} = 2\mu_\gamma$. In the out-of-equilibrium cases we will study, the system will not be in *chemical equilibrium* and we will have to solve a differential equation for μ. The great simplifying feature of kinetic equilibrium, though, is that this differential equation will be a single ordinary differential equation, as opposed to the very complicated form of Eq. (3.1).

We will typically be interested in systems at temperatures smaller than $E - \mu$. In this limit, the exponential in the Bose–Einstein or Fermi–Dirac distribution is large and dwarfs the ± 1 in the denominator. Thus, another simplification emerges: we can ignore the complications of quantum statistics. The distributions become

$$f(E) \to e^{\mu/T} e^{-E/T} \tag{3.3}$$

and the Pauli blocking/Bose enhancement factors in the Boltzmann equation can be neglected.

Under these approximations, the last line of Eq. (3.1) becomes

$$f_3 f_4 [1 \pm f_1][1 \pm f_2] - f_1 f_2 [1 \pm f_3][1 \pm f_4]$$

$$\to e^{-(E_1+E_2)/T} \left\{ e^{(\mu_3+\mu_4)/T} - e^{(\mu_1+\mu_2)/T} \right\}. \tag{3.4}$$

Here I have used energy conservation, $E_1 + E_2 = E_3 + E_4$. We will use the number densities themselves as the time-dependent functions to be solved for, instead of μ. The number density of species i is related to μ_i via

Table 3.1. Reactions in This Chapter: $1 + 2 \leftrightarrow 3 + 4$

	1	2	3	4
Neutron-Proton Ratio	n	ν_e or e^+	p	e^- or $\bar{\nu}_e$
Recombination	e	p	H	γ
Dark Matter Production	X	X	l	l

$$n_i = g_i e^{\mu_i/T} \int \frac{d^3p}{(2\pi)^3} e^{-E_i/T} \tag{3.5}$$

where g_i is the degeneracy of the species, e.g., equal to 2 for the two spin states of the photon. It will also be useful to define the species-dependent equilibrium number density as

$$n_i^{(0)} \equiv g_i \int \frac{d^3p}{(2\pi)^3} e^{-E_i/T} = \begin{cases} g_i \left(\frac{m_i T}{2\pi}\right)^{3/2} e^{-m_i/T} & m_i \gg T \\ g_i \frac{T^3}{\pi^2} & m_i \ll T \end{cases}. \tag{3.6}$$

With this defintion, $e^{\mu_i/T}$ can be rewritten as $n_i/n_i^{(0)}$, so the last line of Eq. (3.1) is equal to

$$e^{-(E_1+E_2)/T} \left\{ \frac{n_3 n_4}{n_3^{(0)} n_4^{(0)}} - \frac{n_1 n_2}{n_1^{(0)} n_2^{(0)}} \right\}. \tag{3.7}$$

With these approximations the Boltzmann equation now simplifies enormously. Define the thermally averaged cross section as

$$\langle \sigma v \rangle \equiv \frac{1}{n_1^{(0)} n_2^{(0)}} \int \frac{d^3p_1}{(2\pi)^3 2E_1} \int \frac{d^3p_2}{(2\pi)^3 2E_2} \int \frac{d^3p_3}{(2\pi)^3 2E_3} \int \frac{d^3p_4}{(2\pi)^3 2E_4} e^{-(E_1+E_2)/T}$$

$$\times (2\pi)^4 \delta^3(p_1 + p_2 - p_3 - p_4) \delta(E_1 + E_2 - E_3 - E_4) |\mathcal{M}|^2. \tag{3.8}$$

Then, the Boltzmann equation becomes

$$a^{-3} \frac{d(n_1 a^3)}{dt} = n_1^{(0)} n_2^{(0)} \langle \sigma v \rangle \left\{ \frac{n_3 n_4}{n_3^{(0)} n_4^{(0)}} - \frac{n_1 n_2}{n_1^{(0)} n_2^{(0)}} \right\}. \tag{3.9}$$

We thus have a simple ordinary differential equation for the number density. Although the details will vary from application to application (see Table 3.1), we will in the remainder of the chapter start from this equation when tracking abundances.

One qualitative note about Eq. (3.9). The left-hand side is of order n_1/t, or, since the typical cosmological time is H^{-1}, $n_1 H$. The right-hand side is of order $n_1 n_2 \langle \sigma v \rangle$. Therefore, if the reaction rate $n_2 \langle \sigma v \rangle$ is much larger than the expansion rate, then the terms on the right side will be much larger than the one on the left.

The only way to maintain equality then is for the individual terms on the right to cancel. Thus, when reaction rates are large,

$$\frac{n_3 n_4}{n_3^{(0)} n_4^{(0)}} = \frac{n_1 n_2}{n_1^{(0)} n_2^{(0)}}. \tag{3.10}$$

This equation, which follows virtually by inspection from the Boltzmann equation, goes under different names in different venues. The particle physics community, which first studied the production of heavy relics in the early universe, tends to call it *chemical equilibrium*. In the context of Big Bang nucleosynthesis, it is called *nuclear statistical equilibrium* (NSE), while students of recombination, the process of electrons and protons combining to form neutral hydrogen, use the terminology *Saha equation*.

3.2 BIG BANG NUCLEOSYNTHESIS

As the temperature of the universe cools to 1 MeV, the cosmic plasma consists of:

- **Relativistic particles in equilibrium: photons, electrons and positrons.** These are kept in close contact with each other by electromagnetic interactions such as $e^+ e^- \leftrightarrow \gamma\gamma$. Besides a small difference due to fermion/boson statistics, these all have the same abundances.
- **Decoupled relativistic particles: neutrinos.** At temperatures a little above 1 MeV, the rate for processes such as $\nu e \leftrightarrow \nu e$ which keep neutrinos coupled to the rest of the plasma drops beneath the expansion rate. Neutrinos therefore share the same temperature as the other relativistic particles, and hence are roughly as abundant, but they do not couple to them.
- **Nonrelativistic particles: baryons.** If there had been no asymmetry in the initial number of baryons and anti-baryons, then both would be completely depleted by 1 MeV. However, such an asymmetry did exist: $(n_b - n_{\bar{b}})/s \sim 10^{-10}$ initially,[1] and this ratio remains constant throughout the expansion. By the time the temperature is of order 1 MeV, all anti-baryons have annihilated away (Exercise 12) so

$$\eta_b \equiv \frac{n_b}{n_\gamma} = 5.5 \times 10^{-10} \left(\frac{\Omega_b h^2}{0.020} \right). \tag{3.11}$$

There are thus many fewer baryons than relativistic particles when $T \sim$ MeV.

Our task in this section will be to determine how the baryons end up. Were the system to remain in equilibrium throughout, the final state would be dictated solely by energetics, and all baryons would relax to the nuclear state with the lowest energy per baryon, iron (Figure 3.1). However, nuclear reactions, which scale as the second — or higher — power of the density, are too slow to keep the system in equilibrium as the temperature drops. So, in principle, we need to solve the

---

[1]s is the entropy density which scales as a^{-3}, as we saw in Chapter 2.

equivalent of Eq. (3.9) for all the nuclei, i.e., a set of coupled differential equations. In practice, at least for a qualitative understanding of the result, we can make use of two simplifications that obviate the need to solve the full set of differential equations.

Lightning Introduction to Nuclear Physics

A single proton is a hydrogen nucleus, referred to as ^{1}H or simply p; a proton and a neutron make up deuterium, ^{2}H or D; one proton and two neutrons make tritium, ^{3}H or T. Nuclei with two protons are helium; these can have one neutron (^{3}He) or two (^{4}He). Thus unique elements have a fixed number of protons, and isotopes of a given element have differing numbers of neutrons. The total number of neutrons and protons in the nucleus, the *atomic number*, is a superscript before the name of the element.

The total mass of a nucleus with Z protons and $A - Z$ neutrons differs slightly from the mass of the individual protons and neutrons alone. This difference is called the binding energy, defined as

$$B \equiv Zm_p + (A - Z)m_n - m \qquad (3.12)$$

where m is the mass of the nucleus. For example, the mass of deuterium is 1875.62 MeV while the sum of the neutron and proton masses is 1877.84 MeV, so the binding energy of deuterium is 2.22 MeV. Nuclear binding energies are typically in the MeV range, which explains why Big Bang nucleosynthesis occurs at temperatures a bit less than 1 MeV even though nuclear masses are in the GeV range.

Neutrons and protons can interconvert via weak interactions:

$$p + \bar{\nu} \leftrightarrow n + e^+ \quad ; \quad p + e^- \leftrightarrow n + \nu \quad ; \quad n \leftrightarrow p + e^- + \bar{\nu} \qquad (3.13)$$

where all the reactions can proceed in either direction. The light elements are built up via electromagnetic interactions. For example, deuterium forms from $p + n \rightarrow D + \gamma$. Then, $D + D \rightarrow n + {}^3$He, after which ^{3}He $+ D \rightarrow p + {}^4$He produces ^{4}He.

The first simplification is that essentially no elements heavier than helium are produced at appreciable levels.[2] So the only nuclei that need to be traced are hydrogen and helium, and their isotopes: deuterium, tritium, and ^{3}He. The second simplification is that, even in the context of this reduced set of elements, the physics splits up neatly into two parts since above $T \simeq 0.1$ MeV, no light nuclei form: only free protons and neutrons exist. Therefore, we first solve for the neutron/proton ratio and then use this abundance as input for the synthesis of helium and isotopes such as deuterium.

---

[2]An exception is lithium, produced at a part in 10^9–10^{10}, and this trace abundance may be observable today. See, e.g., Pinsonneault *et al.* (2001).

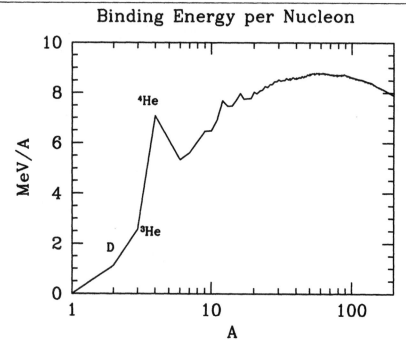

Figure 3.1. Binding energy of nuclei as a function of mass number. Iron has the highest binding energy, but among the light elements, ^{4}He is a crucial local maximum. Nucleosynthesis in the early universe essentially stops at ^{4}He because of the lack of tightly bound isotopes at $A = 5 - 8$. In the high-density environment of stars, three ^{4}He nuclei fuse to form ^{12}C, but the low baryon number precludes this process in the early universe.

Both of these simplifications — no heavy elements at all and only n/p above 0.01 MeV — rely on the physical fact that, at high temperatures, comparable to nuclear binding energies, any time a nucleus is produced in a reaction, it is destroyed by a high-energy photon. This fact is reflected in the fundamental equilibrium equation (3.10). To see how, let's consider this equation applied to deuterium production, $n + p \leftrightarrow D + \gamma$. Since photons have $n_\gamma = n_\gamma^{(0)}$, the equilibrium condition becomes

$$\frac{n_D}{n_n n_p} = \frac{n_D^{(0)}}{n_n^{(0)} n_p^{(0)}}. \tag{3.14}$$

The integrals on the right, as given in Eq. (3.6), lead to

$$\frac{n_D}{n_n n_p} = \frac{3}{4} \left(\frac{2\pi m_D}{m_n m_p T} \right)^{3/2} e^{[m_n + m_p - m_D]/T}, \tag{3.15}$$

the factor of 3/4 being due to the number of spin states (3 for D and 2 each for p and n). In the prefactor, m_D can be set to $2m_n = 2m_p$, but in the exponential the small difference between $m_n + m_p$ and m_D is important: indeed the argument of the exponential is by defintion equal to the binding energy of deuterium, $B_D = 2.22$ MeV. Therefore, as long as equilibrium holds,

$$\frac{n_D}{n_n n_p} = \frac{3}{4} \left(\frac{4\pi}{m_p T} \right)^{3/2} e^{B_D/T}. \tag{3.16}$$

Both the neutron and proton density are proportional to the baryon density, so roughly,

$$\frac{n_D}{n_b} \sim \eta_b \left(\frac{T}{m_p} \right)^{3/2} e^{B_D/T}. \tag{3.17}$$

As long as B_D/T is not too large, the prefactor dominates this expression. And the prefactor is very small because of the smallness of the baryon-to-photon ratio, Eq. (3.11).

The small baryon to photon ratio thus inhibits nuclei production until the temperature drops well beneath the nuclear binding energy. At temperatures above 0.1 MeV, then, virtually all baryons are in the form of neutrons and protons. Around this time, deuterium and helium are produced, but the reaction rates are by now too low to produce any heavier elements. We could have anticipated this by considering Figure 3.1. The lack of a stable isotope with mass number 5 implies that heavier elements cannot be produced via $^4\text{He} + p \rightarrow X$. In stars, the triple alpha process $^4\text{He} + ^4\text{He} + ^4\text{He} \rightarrow ^{12}\text{C}$ produces heavier elements, but in the early universe, densities are far too low to allow three nuclei to find one another on relevant time scales.

3.2.1 Neutron Abundance

We begin by solving for the neutron–proton ratio. Protons can be converted into neutrons via weak interactions, $p + e^- \rightarrow n + \nu_e$ for example. As we will see, reactions of this sort keep neutrons and protons in equilibrium until $T \sim$ MeV. Thereafter, one must solve the rate equation (3.9) to track the neutron abundance.

From Eq. (3.6), the proton/neutron equilibrium ratio in the nonrelativistic limit $(E = m + p^2/2m)$ is

$$\frac{n_p^{(0)}}{n_n^{(0)}} = \frac{e^{-m_p/T} \int dp \, p^2 \, e^{-p^2/2m_p T}}{e^{-m_n/T} \int dp \, p^2 \, e^{-p^2/2m_n T}}. \tag{3.18}$$

The integrals here are proportional to $m^{3/2}$, but the resulting ratio $(m_p/m_n)^{3/2}$ is sufficiently close to unity that we can neglect the mass difference. However, in the exponential the mass difference is very important, and we are left with

$$\frac{n_p^{(0)}}{n_n^{(0)}} = e^{\mathcal{Q}/T} \tag{3.19}$$

with $\mathcal{Q} \equiv m_n - m_p = 1.293$ MeV. Therefore, at high temperatures, there are as many neutrons as protons. As the temperature drops beneath 1 MeV, the neutron fraction goes down. If weak interactions operated efficiently enough to maintain equilibrium indefinitely, then it would drop to zero. The main task of this section is to find out what happens in the real world where weak interactions are not so efficient.

It is convenient to define

$$X_n \equiv \frac{n_n}{n_n + n_p},$$ (3.20)

that is, X_n is the ratio of neutrons to total nuclei. In equilibrium,

$$X_n \to X_{n,\mathrm{EQ}} \equiv \frac{1}{1 + (n_p^{(0)}/n_n^{(0)})}.$$ (3.21)

To track the evolution of X_n, let's start from Eq. (3.9), with 1 = neutron, 3 = proton, and 2, 4 = leptons in complete equilibrium ($n_l = n_l^{(0)}$). Then,

$$a^{-3} \frac{d\left(n_n a^3\right)}{dt} = n_l^{(0)} \langle \sigma v \rangle \left\{ \frac{n_p n_n^{(0)}}{n_p^{(0)}} - n_n \right\}.$$ (3.22)

We have already determined the ratio $n_n^{(0)}/n_p^{(0)} = e^{-\mathcal{Q}/T}$ and we can identify $n_l^{(0)} \langle \sigma v \rangle$ as λ_{np}, the rate for neutron $\to$ proton conversion since it multiplies n_n in the loss term. Also if we rewrite n_n on the left as $(n_n + n_p)X_n$, then the total density times a^3 can be taken outside the derivative, leaving

$$\frac{dX_n}{dt} = \lambda_{np} \left\{ (1 - X_n)e^{-\mathcal{Q}/T} - X_n \right\}.$$ (3.23)

Equation (3.23) is a differential equation for X_n as a function of time, but it contains the temperature T and the reaction rate λ_{np}, both of which have complicated time dependences. It is simplest therefore to recast the equation using as the evolution variable

$$x \equiv \frac{\mathcal{Q}}{T}.$$ (3.24)

The left-hand side of Eq. (3.23) then becomes $\dot{x}\, dX_n/dx$, so we need an expression for $dx/dt = -x\dot{T}/T$. Since $T \propto a^{-1}$,

$$\frac{1}{T} \frac{dT}{dt} = -H = -\sqrt{\frac{8\pi G \rho}{3}},$$ (3.25)

the second equality following from Eq. (2.39). Nucleosynthesis occurs in the radiation-dominated era, so the main contribution to the energy density ρ comes from relativistic particles. Recall from Chapter 2 that the contribution to the energy density from relativistic particles is

$$\rho = \frac{\pi^2}{30} T^4 \left[\sum_{i=\mathrm{bosons}} g_i + \frac{7}{8} \sum_{i=\mathrm{fermions}} g_i \right] \qquad (i \text{ relativistic})$$

$$\equiv g_* \frac{\pi^2}{30} T^4. \tag{3.26}$$

The effective numbers of relativistic degrees of freedom, g_*, is a function of temperature. At temperatures of order 1 MeV, the contributing species are: photons ($g_\gamma = 2$), neutrinos ($g_\nu = 6$), and electrons and positrons ($g_{e^+} = g_{e^-} = 2$). Adding up leads to $g_* \simeq 10.75$, roughly constant throughout the regime of interest. Then, Eq. (3.23) becomes

$$\frac{dX_n}{dx} = \frac{x\lambda_{np}}{H(x=1)} \left\{ e^{-x} - X_n\left(1 + e^{-x}\right) \right\} \tag{3.27}$$

with

$$H(x=1) = \sqrt{\frac{4\pi^3 G \mathcal{Q}^4}{45}} \times \sqrt{10.75} = 1.13 \text{ sec}^{-1}. \tag{3.28}$$

Finally, we need an expression for the neutron–proton conversion rate, λ_{np}. Under the approximations we are using, the rate is (Bernstein 1988 or Exercise 3)

$$\lambda_{np} = \frac{255}{\tau_n x^5} \left(12 + 6x + x^2\right), \tag{3.29}$$

with the neutron lifetime $\tau_n = 886.7$ sec. Thus, when $T = \mathcal{Q}$ (i.e., when $x = 1$), the conversion rate is 5.5 sec^{-1}, somewhat larger than the expansion rate. As the temperature drops beneath 1 MeV, though, the rate rapidly falls below the expansion rate, so conversions become inefficient.

We can now integrate Eq. (3.27) numerically to track the neutron abundance (Exercise 4). Figure 3.2 shows the results of this integration. Note that the result agrees extremely well at temperatures above ~ 0.1 MeV with the exact solution which includes proper statistics, nonzero electron mass, and changing g_*. The neutron fraction X_n does indeed fall out of equilibrium once the temperature drops below 1 MeV: it freezes out at 0.15 once the temperature drops below 0.5 MeV.

At temperatures below 0.1 MeV, two reactions we have not included yet become important: neutron decay ($n \to p + e^- + \bar{\nu}$) and deuterium production ($n + p \to D + \gamma$). Decays can be added trivially by adding in a factor of e^{-t/τ_n} to the results of Figure 3.2. By the time decays become important, electrons and positrons have annihilated, so g_* in Eq. (3.26) is 3.36 and the time–temperature relation is (Exercise 5):

$$t = 132 \sec \left(\frac{0.1 \text{MeV}}{T}\right)^2. \tag{3.30}$$

We will see shortly that production of deuterium, and other light elements, begins in earnest at $T \sim 0.07$ MeV. By then, decays have depleted the neutron fraction by a factor of $\exp[-(132/886.7)(0.1/0.07)^2] = 0.74$. So the neutron abundance at the onset of nucleosynthesis is 0.15×0.74, or

$$X_n(T_{\text{nuc}}) = 0.11. \tag{3.31}$$

We now turn to light element formation to understand the ramifications of this number.

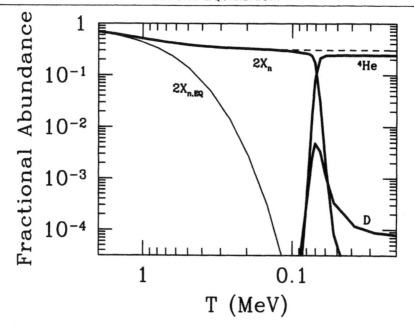

Figure 3.2. Evolution of light element abundances in the early universe. Heavy solid curves are results from Wagoner (1973) code; dashed curve is from integration of Eq. (3.27); light solid curve is twice the neutron equilibrium abundance. Note the good agreement of Eq. (3.27) and the exact result until the onset of neutron decay. Also note that the neutron abundance falls out of equilibrium at $T \sim$ MeV.

3.2.2 Light Element Abundances

A useful way to approximate light element production is that it occurs instantaneously at a temperature $T_{\rm nuc}$ when the energetics compensates for the small baryon to photon ratio. Let's consider deuterium production as an example, with Eq. (3.17) as our guide. The equilibrium deuterium abundance is of order the baryon abundance (i.e. if the universe stayed in equilibrium, all neutrons and protons would form deuterium) when Eq. (3.17) is of order unity, or

$$\ln(\eta_b) + \frac{3}{2}\ln(T_{\rm nuc}/m_p) \sim -\frac{B_D}{T_{\rm nuc}}. \qquad (3.32)$$

Equation (3.32) suggests that deuterium production takes place at $T_{\rm nuc} \sim 0.07$ MeV, with a weak logarithmic dependence on η_b.

    Since the binding energy of helium is larger than that of deuterium, the exponential factor $e^{B/T}$ favors helium over deuterium. Indeed, Figure 3.2 illustrates that helium is produced almost immediately after deuterium. Virtually all remaining neutrons at $T \sim T_{\rm nuc}$ then are processed into ^{4}He. Since two neutrons go into ^{4}He, the final ^{4}He abundance is equal to half the neutron abundance at $T_{\rm nuc}$. Often, results are quoted in terms of mass fraction; then,

$$X_4 \equiv \frac{4n_{4_{He}}}{n_b} = 2X_n(T_{\mathrm{nuc}}).$$ (3.33)

Figure 3.2 shows that this relation holds. Indeed, to find the final helium mass fraction, we need only take the neutron fraction at T_{nuc}, Eq. (3.31), and multiply by 2, so the final helium mass fraction is 0.22. This rough estimate, obtained by solving a single differential equation, is in remarkable agreement with the exact solution, which can be fit via (Olive, 2000; Kolb and Turner, 1990)

$$Y_p = 0.2262 + 0.0135\ln(\eta_b/10^{-10}).$$ (3.34)

One important feature of this result is that it depends only logarithmically on the baryon fraction. We saw in Eq. (3.32) that T_{nuc} has this dependence. You might think that the exponential sensitivity to T_{nuc} in the decay fraction would turn this into linear dependence. However, T_{nuc} is sufficiently early that only a small fraction of neutrons have decayed: the exponential in this regime is linear in the time. Therefore, the final helium abundance maintains only logarithmic dependence on the baryon density.

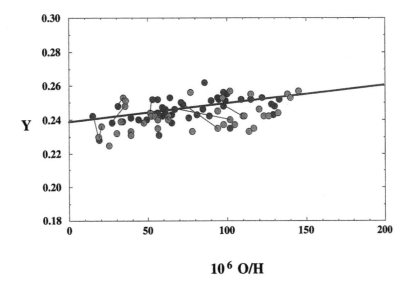

10^6 O/H

Figure 3.3. Helium abundance $(Y \equiv Y_p)$ as a function of oxygen/hydrogen ratio. Lower oxygen systems have undergone less processing, so the helium abundance in those systems is closer to primordial. Line, and extrapolation to $Y_p = 0.238$, from Olive (2000). Data from Pagel *et al.* (1992), Skillman and Kennicutt (1993), Skillman *et al.* (1994), and Izotov and Thuan (1998). Short lines connect the same region observed by different groups.

The prediction agrees well with the observations, as indicated in Figure 3.3. The best indication of the primordial helium abundance comes from the most unprocessed systems, typically identified by low metallicities. As Figure 3.3 indicates, the primordial abundance almost certainly lies between 0.22 and 0.25. Although there have been claims of discord in the past, the agreement remains one of the pillars of observational cosmology.

Figure 3.2 shows that not all of the deuterium gets processed into helium. A trace amount remains unburned, simply because the reaction which eliminates it $(D + p \rightarrow {}^3He + \gamma)$ is not completely efficient. Figure 3.2 shows that after T_{nuc}, deuterium is depleted via these reactions, eventually freezing out at a level of order 10^{-5}–10^{-4}. If the baryon density is low, then the reactions proceed more slowly, and the depletion is not as effective. Therefore, low baryon density inevitably results in more deuterium; the sensitivity is quite stark, as illustrated in Figure 1.8. As a result, deuterium is a powerful probe of the baryon density. Complementing this sensitivity is the possiblity of measuring deuterium in gas clouds as $z \sim 3$ by looking for absorbtion in the spectra of distant QSOs. For example, O'Meara *et al.* (2001) combine the measurements of primordial deuterium in four systems to obtain, D/H= $3.0 \pm 0.4 \times 10^{-5}$, corresponding to $\Omega_b h^2 = 0.0205 \pm 0.0018$.

3.3 RECOMBINATION

As the temperature drops to ~ 1 eV, photons remain tightly coupled to electrons via Compton scattering and electrons to protons via Coulomb scattering. It will come as no surprise that at these temperatures, there is very little neutral hydrogen. Energetics of course favors the production of neutral hydrogen with a binding energy of $\epsilon_0 = 13.6$ eV, but the high photon/baryon ratio ensures that any hydrogen atom produced will be instantaneously ionized. This phenomenon is identical to the delay in the production of light nuclei we saw above, replayed on the atomic scale.

As long as the reaction[3] $e^- + p \leftrightarrow H + \gamma$ remains in equilibrium, the condition in Eq. (3.10) (with $1 = e, 2 = p, 3 = H$) ensures that

$$\frac{n_e n_p}{n_H} = \frac{n_e^{(0)} n_p^{(0)}}{n_H^{(0)}}. \tag{3.35}$$

We can go further here by recognizing that the neutrality of the universe ensures that $n_e = n_p$. Let's define the free electron fraction

$$X_e \equiv \frac{n_e}{n_e + n_H} = \frac{n_p}{n_p + n_H}, \tag{3.36}$$

the denominator equal to the total number of hydrogen nuclei. Carrying out the integrals on the right of Eq. (3.35) leads then to

$$\frac{X_e^2}{1 - X_e} = \frac{1}{n_e + n_H} \left[\left(\frac{m_e T}{2\pi} \right)^{3/2} e^{-[m_e + m_p - m_H]/T} \right] \tag{3.37}$$

where we have made the familiar approximation of neglecting the small mass difference of H and p in the prefactor. The argument of the exponential is $-\epsilon_0/T$.

---

[3]Here p stands for free protons and H for neutral hydrogen, i.e., a proton with an electron attached.

Neglecting the relatively small number of helium atoms, the denominator $n_e + n_H$ (or $n_p + n_H$) is equal to the baryon density, $\eta_b n_\gamma \sim 10^{-9} T^3$. So when the temperature is of order ϵ_0, the right-hand side is of order $10^9 (m_e/T)^{3/2} \simeq 10^{15}$. In that case, Eq. (3.37) can be satisfied only if the denominator on the left is very small, that is if X_e is very close to 1: all hydrogen is ionized. Only when the temperature drops far below ϵ_0 does appreciable recombination take place. As X_e falls, the rate for recombination also falls, so that equilibrium becomes more difficult to maintain. Thus, in order to follow the free electron fraction accurately, we need to solve the Boltzmann equation, just as we did for the neutron–proton ratio.

In this case, Eq. (3.9) for the electron density becomes

$$a^{-3} \frac{d\left(n_e a^3\right)}{dt} = n_e^{(0)} n_p^{(0)} \langle \sigma v \rangle \left\{ \frac{n_H}{n_H^{(0)}} - \frac{n_e^2}{n_e^{(0)} n_p^{(0)}} \right\}$$

$$= n_b \langle \sigma v \rangle \left\{ (1 - X_e) \left(\frac{m_e T}{2\pi} \right)^{3/2} e^{-\epsilon_0/T} - X_e^2 n_b \right\} \qquad (3.38)$$

where the last line follows since the ratio $n_e^{(0)} n_p^{(0)} / n_H^{(0)}$ is equal to the term in square brackets in Eq. (3.37). Meanwhile, since $n_b a^3$ is constant it can be passed through the derivative on the left after expressing n_e as $n_b X_e$, so that

$$\frac{dX_e}{dt} = \left\{ (1 - X_e)\beta - X_e^2 n_b \alpha^{(2)} \right\} \qquad (3.39)$$

where the ionization rate is typically denoted

$$\beta \equiv \langle \sigma v \rangle \left(\frac{m_e T}{2\pi} \right)^{3/2} e^{-\epsilon_0/T} \qquad (3.40)$$

and the recombination rate

$$\alpha^{(2)} \equiv \langle \sigma v \rangle. \qquad (3.41)$$

The recombination rate has superscript $^{(2)}$ because recombination to the ground ($n = 1$) state is not relevant. Ground-state recombinations lead to production of an ionizing photon, and this photon immediately ionizes a neutral atom. The net effect of such a recombination is zero: no new neutral atoms are formed this way. The only way for recombination to proceed is via capture to one of the excited states of hydrogen; to a good approximation, this rate is

$$\alpha^{(2)} = 9.78 \frac{\alpha^2}{m_e^2} \left(\frac{\epsilon_0}{T} \right)^{1/2} \ln \left(\frac{\epsilon_0}{T} \right). \qquad (3.42)$$

The Saha approximation, Eq. (3.37), does a good job predicting the redshift of recombination, but fails as the electron fraction drops and the system goes out of equilibirium. Therefore, the detailed evolution of X_e must be obtained by a numerical integration of Eq. (3.39) (Exercise 8). Results from numerical integration of Eq. (3.39) are shown in Figure 3.4.

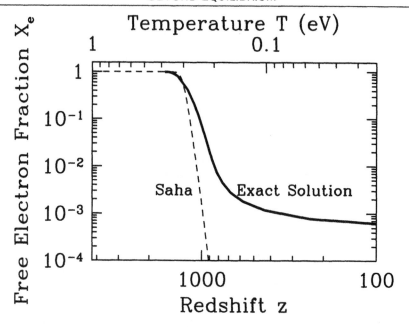

Figure 3.4. Free electron fraction as a function of redshift. Recombination takes place suddenly at $z \sim 1000$ corresponding to $T \sim 1/4$ eV. The Saha approximation, Eq. (3.37), holds in equilibrium and correctly identifies the redshift of recombination, but not the detailed evolution of X_e. Here $\Omega_b = 0.06, \Omega_m = 1, h = 0.5$.

The computation of the neutron/proton ratio affects the abundance of light elements today. Similarly, the evolution of the free electron abundance has major ramifications for observational cosmology. Recombination at $z_* \sim 1000$ is directly tied to the *decoupling* of photons from matter.[4] This decoupling, in turn, directly affects the pattern of anisotropies in the CMB that we observe today.

Decoupling occurs roughly when the rate for photons to Compton scatter off electrons becomes smaller than the expansion rate.[5] The scattering rate is

$$n_e \sigma_T = X_e n_b \sigma_T \tag{3.43}$$

where $\sigma_T = 0.665 \times 10^{-24}$ cm^2 is the Thomson cross section, and I continue to ignore helium, thereby assuming that the total number of hydrogen nuclei (free protons + hydrogen atoms) is equal to the total baryon number. Since the ratio of the

---

[4]Notice from Figure 1.4 that even though photons stop scattering off electrons at $z \sim 1000$, electrons do scatter many times off photons until much later. This is not a contradiction: there are many more photons than baryons. In any event, many cosmologists shy away from the word *decoupling* to describe what happens at $z \sim 1000$ for this reason.

[5]In Chapter 8 we will define a more precise measure of decoupling, making use of the *visibility function*, the probability that a photon last scattered at a given redshift. Using the visibility function, we will show that a CMB photon today most likely last scattered at a slightly higher redshift than inferred by the simple $n_e \sigma_T = H$ criterion.

baryon density to the critical density is $m_p n_{\rm b}/\rho_{\rm cr} = \Omega_b a^{-3}$, n_b can be eliminated in Eq. (3.43) in favor of Ω_b:

$$n_e \sigma_T = 7.477 \times 10^{-30} {\rm cm}^{-1} X_e \Omega_b h^2 a^{-3}. \qquad (3.44)$$

Dividing by the expansion rate leads to

$$\frac{n_e \sigma_T}{H} = 0.0692 a^{-3} X_e \Omega_b h \frac{H_0}{H}. \qquad (3.45)$$

The ratio on the right depends on the Hubble rate, which is given in Eq. (1.2). From that equation or from Figure 1.3, we see that at early times, the main contribution comes from either matter or radiation, so $H/H_0 = \Omega_m^{1/2} a^{-3/2}[1 + a_{\rm eq}/a]^{1/2}$. Then,

$$\frac{n_e \sigma_T}{H} = 113 X_e \left(\frac{\Omega_b h^2}{0.02}\right) \left(\frac{.15}{\Omega_m h^2}\right)^{1/2} \left(\frac{1+z}{1000}\right)^{3/2} \left[1 + \frac{1+z}{3600}\frac{0.15}{\Omega_m h^2}\right]^{-1/2}. \qquad (3.46)$$

Here I have normalized with "best fit" values for the baryon and matter densities. When the free electron fraction X_e drops below $\sim 10^{-2}$, photons decouple. From Figure 3.4, we see that X_e drops very quickly from unity to 10^{-3}. Therefore, decoupling takes place during recombination.

Let's forget all we just learned and ask what would happen if the universe remained ionzied throughout its history. In that hypothetical case, $X_e = 1$, and Eq. (3.46) can be trivially solved to find the redshift of decoupling. Setting the right hand side to 1 leads to

$$1 + z_{\rm decouple} = 43 \left(\frac{0.02}{\Omega_b h^2}\right)^{2/3} \left(\frac{\Omega_m h^2}{0.15}\right)^{1/3} \qquad \text{(no recombination).} \qquad (3.47)$$

Equation (3.47) tells us that even if the electrons remained ionized throughout the history of the universe, eventually the photons decoupled simply because expansion made it more difficult to find the increasingly dilute electrons. In theory, we do not expect the electrons to remain ionized throughout, so this calculation would appear academic. However, Eq. (3.47) is relevant for a more general reason. We do expect that at some late time, the electrons were reionized. We expect this because the universe we observe back to redshift $z \sim 6$ appears to be ionized. If the universe was reionized at very late times, much after the $z_{\rm decouple}$ of Eq. (3.47), there would not be a huge change in the CMB anisotropy pattern. However, if the universe was reionized earlier than this redshift, multiple scattering of the photons would dramatically alter the primordial anisotropy pattern set up at $z \sim 1000$. Observations of the most distant quasars (Becker et al., 2001; Fan et al., 2002) suggest that reionization took place at $z \simeq 6$, so the alteration is expected to be slight.

3.4 DARK MATTER

There is strong evidence for nonbaryonic dark matter in the universe, with $\Omega_{\rm dm} \simeq 0.3$. Perhaps the most plausible candidate for dark matter is a weakly interacting massive particle (WIMP), which was in close contact with the rest of the cosmic plasma at high temperatures, but then experienced *freeze-out* as the temperature dropped below its mass. Freeze-out is the inability of annihilations to keep the particle in equlibrium. Indeed, were it kept in equilibrium indefinitely, its abundance would be suppressed by $e^{-m/T}$: there would be no such particles in the observable universe. The purpose of this section, then, is to solve the Boltzmann equation for such a particle, determining the epoch of freeze-out and its relic abundance. The hope is that, by fixing its relic abundance so that $\Omega_{\rm dm} \simeq 0.3$, we will learn something about the fundamental properties of the particle, such as its mass and cross section. We then might use this knowledge to detect the particles in a laboratory.

In the generic WIMP scenario, two heavy particles X can annihilate producing two light (essentially massless) particles l. The light particles are assumed to be very tightly coupled to the cosmic plasma, so they are in complete equilibrium (chemical as well as kinetic), with $n_l = n_l^{(0)}$. There is then only one unknown, n_X, the abundance of the heavy particle. We can use Eq. (3.9) to solve for this abundance:

$$a^{-3} \frac{d\left(n_X a^3\right)}{dt} = \langle \sigma v \rangle \left\{ (n_X^{(0)})^2 - n_X^2 \right\}. \tag{3.48}$$

To go further, recall that the temperature typically scales as a^{-1}, so if we multiply and divide the factor of $n_X a^3$ inside the parentheses on the left by T^3, we can remove $(aT)^3$ outside the derivative, leaving $T^3 d(n_X/T^3)/dt$. Let's define then

$$Y \equiv \frac{n_X}{T^3}. \tag{3.49}$$

The differential equation for Y becomes

$$\frac{dY}{dt} = T^3 \langle \sigma v \rangle \left\{ Y_{\rm EQ}^2 - Y^2 \right\}, \tag{3.50}$$

with $Y_{\rm EQ} \equiv n_X^{(0)}/T^3$.

To go further, as in the neutron–proton case, it is convenient to introduce a new time variable,

$$x \equiv m/T \tag{3.51}$$

where m, the mass of the heavy particle, sets a rough scale for the temperature during the region of interest. Very high temperature corresponds to $x \ll 1$, in which case reactions proceed rapidly so $Y \simeq Y_{\rm EQ}$. Since the X particles are relativistic at these epochs, the $m \ll T$ limit of Eq. (3.6) implies that $Y \simeq 1$. For high x, the equilibrium abundance $Y_{\rm EQ}$ becomes exponentially suppressed (e^{-x}). Ultimately, X particles will become so rare because of this suppression that they will not be able to find each other fast enough to maintain the equilibrium abundance. This is the onset of freeze-out. To change from t to x, we need the Jacobian $dx/dt = Hx$. Dark matter production typically occurs deep in the radiation era where the energy density scales as T^4, so $H = H(m)/x^2$. Then the evolution equation becomes

$$\frac{dY}{dx} = -\frac{\lambda}{x^2} \left\{ Y^2 - Y_{\rm EQ}^2 \right\}, \tag{3.52}$$

where the ratio of the annihilation rate to the expansion rate is parameterized by

$$\lambda \equiv \frac{m^3 \langle \sigma v \rangle}{H(m)}. \tag{3.53}$$

In many theories λ is a constant, but in some, the thermally averaged cross section is temperature dependent; this leads to slight numerical changes in the following but unchanged qualitative solutions.

Equation (3.52) is a form of the Riccati equation, for which in general there are no analytic solutions. In this case, though, we can make use of our understanding of the freeze-out process to get an analytic expression for the final freeze-out abundance $Y_\infty \equiv Y(x = \infty)$. Let's review this understanding in the context of Eq. (3.52). The left-hand side is of order Y (for $x \sim 1$) while the right is of order $Y^2 \lambda$. We will see that λ is typically quite large, so as long as Y is not too small, the right-hand side must zero itself by setting $Y = Y_{\mathrm{EQ}}$. At late times, as Y_{EQ} drops precipitously, the terms on the right-hand side will no longer be much larger than the one on the left. In fact, well after freeze-out, Y will be much larger than Y_{EQ}: the X particles will not be able to annihilate fast enough to maintain equilibrium. Thus at late times,

$$\frac{dY}{dx} \simeq -\frac{\lambda Y^2}{x^2} \qquad (x \gg 1). \tag{3.54}$$

Integrate this analytically from the epoch of freeze-out x_f until very late times $x = \infty$ to get

$$\frac{1}{Y_\infty} - \frac{1}{Y_f} = \frac{\lambda}{x_f}. \tag{3.55}$$

Typically Y at freeze-out Y_f is significantly larger than Y_∞, so a simple analytic approximation is

$$Y_\infty \simeq \frac{x_f}{\lambda}. \tag{3.56}$$

This approximation is incomplete, in that it depends on the freeze-out temperature, which we have not determined. Although more precise determinations are possible (Exercise 10), a simple order-of-magnitude estimate for the dark matter problem is $x_f \sim 10$.

Figure 3.5 shows the numerical solution to Eq. (3.52) for several different values of λ. The abundances do track the equilibrium abundances until $m/T \sim 10$, after which they level off to a constant. The rough estimate $Y_\infty \sim 10/\lambda$ is seen to be a reasonable approximation for the relic abundance. Note that particles with larger cross sections (e.g. in the figure, $\lambda = 10^{10}$) freeze out later, and this later freeze-out carries along with it a lower relic abundance. Also note from the inset in Figure 3.5 that the distinction between Bose–Einstein, Fermi–Dirac, and Boltzmann statistics is important only at temperatures above the particle's mass. For temperatures relevant to the freeze-out process, our use of Boltzmann statistics is completely warranted.

There is one more piece of physics needed in order to determine the present-day abundance of these heavy particle relics. After freeze-out, the heavy particle density

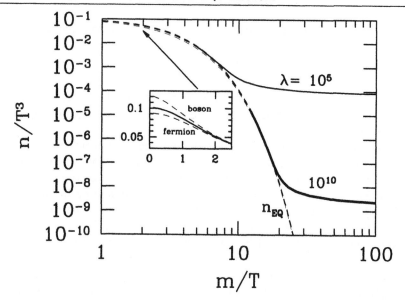

Figure 3.5. Abundance of heavy stable particle as the temperature drops beneath its mass. Dashed line is equilibrium abundance. Two different solid curves show heavy particle abundance for two different values of λ, the ratio of the annihilation rate to the Hubble rate. Inset shows that the difference between quantum statistics and Boltzmann statistics is important only at temperatures larger than the mass.

simply falls off as a^{-3}. So its energy density today is equal to $m(a_1/a_0)^3$ times its number density where a_1 corresponds to a time sufficiently late that Y has reached its aymptotic value, Y_∞. The number density at that time is $Y_\infty T_1^3$, so

$$\rho_X = mY_\infty T_0^3 \left(\frac{a_1 T_1}{a_0 T_0}\right)^3 \simeq \frac{mY_\infty T_0^3}{30}. \tag{3.57}$$

The second equality here is nontrivial. You might expect that aT remains constant through the evolution of the universe, so that the ratio $a_1 T_1/a_0 T_0$ would be unity. It is not, for the same reason that the CMB and neutrinos have different temperatures. We saw in Chapter 2 that photons are heated by $e^\pm$ annihilation, while neutrinos which have already decoupled are not. Similarly, as the universe expands, photons are heated by the annihilation of the zoo of particles with masses between 1 MeV and 100 GeV, so T does not fall simply as a^{-1}. You can show in Exercise 11 that as a result $(a_1 T_1/a_0 T_0)^3 \simeq 1/30$. Finally, to find the fraction of critical density today contributed by X, insert our expression for Y_∞ and divide by ρ_{cr}:

$$\Omega_X = \frac{x_f}{\lambda} \frac{mT_0^3}{30\rho_{cr}}$$

$$= \frac{H(m)x_f T_0^3}{30m^2\langle\sigma v\rangle\rho_{cr}}. \tag{3.58}$$

To find the present density of heavy particles, then, we need to compute the Hubble rate when the temperature was equal to the X mass, $H(m)$, for which we need the energy density when the temperature was equal to m. The energy density in the radiation era is given by Eq. (3.26) with g_* a function of temperature. Therefore,

$$\Omega_X = \left[\frac{4\pi^3 G g_*(m)}{45}\right]^{1/2} \frac{x_f T_0^3}{30\langle\sigma v\rangle\rho_{\rm cr}}. \qquad (3.59)$$

We see that Ω_X does not explicitly depend on the mass of the X particle.[6] So it is mainly the cross section which determines the relic abundance.

Let's now see what order of magnitude is needed to get dark matter today, i.e., to get $\Omega_X = \Omega_{\rm dm} \simeq 0.3$. At the temperatures of interest for dark matter production, $T \sim 100$ GeV, $g_*(m)$ includes contributions from all the particles in the standard model (three generations of quarks and leptons, photons, gluons, weak bosons, and perhaps even the Higgs boson) and so is of order 100. Normalizing $g_*(m)$ and x_f by their nominal values leads to

$$\Omega_X = 0.3 h^{-2} \left(\frac{x_f}{10}\right) \left(\frac{g_*(m)}{100}\right)^{1/2} \frac{10^{-39}\text{cm}^2}{\langle\sigma v\rangle}. \qquad (3.60)$$

The fact that this estimate is of order unity for cross sections of order 10^{-39} cm^2 is taken as a good sign: there are several theories which predict the existence of particles with cross sections this small.

Perhaps the most notable of these theories is supersymmetry, the theory which predicts that every particle has a partner with opposite statistics. For example, the supersymmetric partner of the spin zero Higgs boson is the spin-1/2 *Higgsino* (fermion). Initially it was hoped that the observed fermions could be the partners of the observed bosons, but this hope is not realized in nature. Instead, supersymmetry must be broken and all the supersymmetric partners of the known particles must be so massive that they have not yet been observed even in accelerators: they must have masses greater than 10 to 100 GeV. Which of the supersymmetric partners is the best candidate for the dark matter today? The particle must be neutral since the evidence points to dark matter that is truly dark, i.e., does not interact much with the known particles and especially does not emit photons. The particle must also be stable: if it could decay to lighter particles, then decays would have kept it in equilibrium throughout the early universe and there would be none left today. The first of these criteria restricts the dark matter to be the partner of one of the neutral particles, such as the Higgs or the photon.[7] The second requires the particle to be the lightest (LSP for lightest supersymmetric partner) of these, for any heavier particles could decay into the lightest one plus some ordinary particles.

---

[6]There is a slight implicit dependence on mass in the freeze-out temperature x_f and in g_*, which is to evaluated when $T = m$.

[7]The other neutral boson in the Standard Model is the Z vector boson. The lightest supersymmetric partner is likely a linear combination of the partners of all of these.

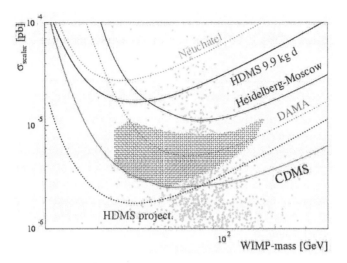

Figure 3.6. Constraints on supersymmetric dark matter (Baudis *et al.*, 2000). Region above the solid curves is excluded, while filled region is reported detection by DAMA (Bernabei *et al.*, 1999). Lowest curve (labled *HDMS project*) is only a projected limit based on one future experiment. The CDMS experiment (Abusaidi *et al.*, 2000) appears to rule out the DAMA detection. Points scattered throughout correspond to different parameter choices in a class of supersymmetric models. Note the limits on the cross section are in units of picobarns (1 picobarn $= 10^{-36}$ cm^2).

Not only would weakly interacting particles such as LSP's annihilate in the early universe, but if they were around today they would scatter off ordinary matter. Although it is difficult to detect these reactions because the rate is so low, a number of experiments have been performed searching for dark matter particles. Figure 3.6 shows the limits on the masses and cross sections of dark matter from these experiments (note that the scattering cross section, while related to the annihilation cross section, is not identically equal to it). Apart from a tantalizing detection from the DAMA experiment, so far we have only upper limits. However, as indicated in the figure, in the coming years the experiments are expected to pierce into the region predicted by supersymmetric theories.

3.5 SUMMARY

The light elements in the universe formed when the temperature of the cosmic plasma was of order 0.1 MeV. Roughly a quarter of the mass of the baryons is in the form of ^{4}He, the remaining in the form of free protons with only trace amounts of deuterium, ^{3}He, and lithium.

These elements remain ionized until the temperature of the universe drops well below the ionization energy of hydrogen. The epoch of recombination — at which time electrons and protons combine to form neutral hydrogen — is at redshift $z \sim 1000$ corresponding to a temperature $T \sim 0.25$ eV. Before recombination, photons and electrons and protons are tightly coupled with one another because of Compton and Coulomb scattering. After this time, photons travel freely through the universe without interacting, so the photons in the CMB we observe today offer an excellent snapshot of the universe at $z \sim 1000$. The importance of this snapshot cannot be overstated.

The details of both nucleosynthesis and recombination are heavily influenced by the fact that the reactions involved eventually become too slow to keep up with the expansion rate. This feature may also be responsible for the production of dark matter in the universe. We explored the popular scenario wherein a massive, neutral stable particle stops annihilating when the temperature drops significantly beneath its mass. The present-day abundance of such a particle can be deterimined in terms of its annihilation cross section, as in Eq. (3.60). Larger cross sections correspond to more efficient annhilation and therefore a lower abundance today. Roughly cross sections of order 10^{-40} are needed to get the dark matter abundance observed today. Such cross sections and the requisite stable, neutral particles emerge fairly natrually in extensions of the Standard Model of particle physics, such as supersymmetry, and may well be tested by accelerator experiments in the near future.

SUGGESTED READING

The Early Universe (Kolb and Turner) contains especially clear treatments of the heavy particle freeze-out problem and Big Bang nucleosynthesis, also based on the Boltzmann equation. *Kinetic Theory in an Expanding Universe* (Bernstein) offers some semianalytic solutions to these problems, as well as the requisite Boltzmann-ology.

Work on nucleosynthesis in the early universe dates back to Gamow and collaborators, summarized in Alpher, Follin, and Herman (1953). The first post-CMB papers were Peebles (1966) and Wagoner, Fowler, and Hoyle (1967); these got the basics right: 25% helium and roughly the correct amount of deuterium. Yang *et al.* (1984) helped many people of my generation understand that the baryonic density could be constrained with observations of the light elements. A nice review article on nucleosynthesis is Olive, Steigman, and Walker (2000).

As I tried to indicate in the text, the process of recombination is very rich; it involves some subtle physics. The original paper which worked through all the details was by Peebles (1968). Ma and Bertschinger (1996) however managed to describe the physics succinctly in just one page in their Section 5.8. Seager, Sasselov, and Scott (1999) have presented a more accurate treatment (although, as they emphasize, Peebles' more intuitive work holds up remarkably well), including many small effects previously neglected.

Jungman, Kamionkowski, and Griest (1996) is a comprehensive review of all aspects of supersymmetric dark matter. Many papers have explored limits on supersymmetric dark matter candidates from cosmology and accelerators. To mention just several: Roszkowski (1991) showed that the Higgsino is likely *not* the lightest supersymmetric partner; Nath and Arnowitt (1992) and Kane *et al.* (1994) showed that the *bino*, the partner of the initial gauge eigenstate B, is the most likely — both from the point of view of physics and cosmology — LSP; Ellis *et al.* (1997) combined accelerator constraints with those from cosmology to place a lower limit on the mass of the LSP; Edsjo and Gondolo (1997) included some subtle effects in the relic abundance calculation which affects the limits if the LSP is composed primarily of the partner of the Higgs boson; and Bottino *et al.* (2001) explored the consequences if the signal seen in the DAMA signal is due to dark matter particles.

EXERCISES

Exercise 1. Compute the equilibrium number density (i.e., zero chemical potential) of a species with mass m and degeneracy $g = 2$ in the limits of large and small m/T. Take these limits for all three types of statistics: Boltzmann, Bose–Einstein, and Fermi–Dirac. You will find Eqs. (C.26) and (C.27) helpful for the high-T Bose–Einstein and Fermi–Dirac limits.

Exercise 2. In the text, we treated $e^{\pm}$ as relevant to the energy density at temperatures above m_e, but irrelevant afterwards (Eq. (3.30) and the discussion leading up to it). Track the $e^\pm$ density through annihilation assuming $n_{e\pm} = n_{e\pm}^{(0)}$. This equality holds during the BBN epoch because electromagnetic interactions (e.g., $e^+ + e^- \leftrightarrow \gamma + \gamma$) are so strong. When does the density fall to 1% of the photon energy density? If $\eta_b \simeq 6 \times 10^{-10}$, at what temperature do you expect n_{e^-} to depart from $n_{e^-}^{(0)}$?

Exercise 3. Compute the rate for neutron-to-proton conversion, λ_{np}. Show that it is equal to Eq. (3.29). There are two processes which contribute to λ_{np}: $n + \nu_e \rightarrow p + e^-$ and $n + e^+ \rightarrow p + \bar{\nu}_e$. Assume that all particles can be described by Boltzmann statistics and neglect the mass of the electron. With these approximations the two rates are identical.
(a) Use Eq. (3.8) to write down the rate for $n + \nu_e \rightarrow p + e^-$. Perform the integrals over heavy particle momenta to get

$$\lambda_{np} = n_{\nu_e}^{(0)} \langle \sigma v \rangle = \frac{\pi}{4m^2} \int \frac{d^3 p_\nu}{(2\pi)^3 2p_\nu} e^{-p_\nu/T}$$

$$\times \int \frac{d^3 p_e}{(2\pi)^3 2p_e} \delta(\mathcal{Q} + p_\nu - p_e) |\mathcal{M}|^2 . \tag{3.61}$$

(b) The amplitude squared is equal to $|\mathcal{M}|^2 = 32 G_F^2 (1 + 3g_A^2) m_p^2 p_\nu p_e$, where g_A is the axial-vector coupling of the nucleon. The present best measurement of g_A is via the neutron lifetime, $\tau_n = \lambda_0 G_F^2 (1 + 3g_A^2) m_e^5 / (2\pi^3)$, where the phase space integral

$$\lambda_0 \equiv \int_1^{\mathcal{Q}/m_e} dx\, x (x - \mathcal{Q}/m_e)^2 (x^2 - 1)^{1/2} = 1.636 . \tag{3.62}$$

Carry out the integrals in Eq. (3.61) to get the rate, λ_{np} in terms of τ_n. Don't forget to multiply by 2 for the two different reactions.

Exercise 4. Solve the rate equation (3.27) numerically to determine the neutron fraction as a function of temperature. Ignore decays. There are (at least) two ways to perform this computation. The first is to treat it as a simple ordinary differential equation and solve numerically. The second is to proceed analytically and reduce the problem to an evaluation of a single numerical integral. This second method, which I'll lead you through here, is based on a numerical coincidence noted by Bernstein, Brown, and Feinberg (1988).
(a) Using standard differential equation techniques, show that a formal solution to Eq. (3.27) is

$$X_n(x) = \int_{x_i}^x dx' \frac{\lambda_{np}(x') e^{-x'}}{x' H(x')} e^{\mu(x') - \mu(x)} \tag{3.63}$$

where x_i is some initial, very high temperature, and

$$\mu(x) \equiv \int_{x_i}^x \frac{dx'}{x' H(x')} \lambda_{np}(x') \left[1 + e^{-x'} \right] . \tag{3.64}$$

(b) Use Eqs. (3.29) and (3.26) to compute the integrating factor μ analytically. Show that it is equal to

$$\mu = -\frac{255}{\tau_n Q} \left[\frac{4\pi^3 G Q^2 g_*}{45} \right]^{-1/2}$$

$$\times \left[\left(\frac{4}{x^3} + \frac{3}{x^2} + \frac{1}{x} \right) + \left(\frac{4}{x^3} + \frac{1}{x^2} \right) e^{-x} \right] \Bigg|_{x_i}^{x} . \qquad (3.65)$$

The simple form for μ is the result of numerical coincidence alone.
(c) With the results of part **(b)**, do the single numerical integral in **(a)** numerically. Compare the asymptotic result at $x = \infty$ with the result in the text, $X_n(x = \infty) = 0.15$.

Exercise 5. Integrate the Friedmann equation (1.2) to verify the time–temperature relation in Eq. (3.30) in the epoch after $e^{\pm}$ annihilation, but before matter domination.

Exercise 6. Determine η_b in terms of $\Omega_b h^2$. Show that it is given by Eq. (3.11).

Exercise 7. An important parameter for CMB anisotropies is the sound speed at decoupling. This is determined by the ratio of baryons to photons.
(a) Find

$$R \equiv \frac{3\rho_b}{4\rho_\gamma}$$

as a function of a. Evaluate it at decoupling. Your answer should depend on $\Omega_b h^2$.
(b) We will see in Chapter 8 that the sound speed of the combined photon/baryon fluid is

$$c_s = \sqrt{\frac{1}{3(1+R)}} . \qquad (3.66)$$

Use your answer from (a) to plot the sound speed at decoupling as a function of $\Omega_b h^2$.

Exercise 8. Solve for the evolution of the free electron fraction. Do not compare your results with Figure 3.4 until you finish part **(d)**. Throughout, take parameters $\Omega_m = 1, \Omega_b = 0.06, h = 0.5$.
(a) Use as an evolution variable $x \equiv \epsilon_0/T$ instead of time in Eq. (3.39). Rewrite the equation in terms of x and the Hubble rate at $T = \epsilon_0$.
(b) Using the methods of Section 3.4, find the final freeze-out abundance of the free electron fraction, $X_e(x = \infty)$.
(c) Numerically integrate the equation from **(a)** from $x = 1$ down to $x = 1000$. What is the final frozen-out X_e?
(d) Peebles (1968) argued that even captures to excited states would not be important except for the fraction of times that the $n = 2$ state decays into two photons or

expansion redshifts the Lyman alpha photon so that it cannot pump up a ground-state atom. Quantitatively, he multiplied the right-hand side of Eq. (3.39) by the correction factor,

$$C = \frac{\Lambda_\alpha + \Lambda_{2\gamma}}{\Lambda_\alpha + \Lambda_{2\gamma} + \beta^{(2)}} \tag{3.67}$$

where the two-photon decay rate is $\Lambda_{2\gamma} = 8.227 \text{sec}^{-1}$; Lyman alpha production is $\beta^{(2)} = \beta e^{3\epsilon_0/4T}$; and

$$\Lambda_\alpha = \frac{H(3\epsilon_0)^3}{(8\pi)^2}. \tag{3.68}$$

Do this and show how it changes your final answer. Now compare the freeze-out abundance with the result of **(c)** and the evolution with Figure 3.4.

Exercise 9. Find the redshift of decoupling as a function of Ω_b. If you do not have the evolution code of Exercise 8, use the Saha equation to determine X_e.

Exercise 10. Find an approximation to the freeze-out temperature of annihilating heavy particles by setting x_f such that $n^{(0)}(x_f)\langle\sigma v\rangle = H(x_f)$.

Exercise 11. Typically the temperature of the cosmic plasma cools as a^{-1} with the expansion. However, when particles annihilate, they deposit energy into the plasma, thereby slowing the cooling. (Scherrer and Turner, 1986, showed that the annihilations do not actually *heat* the universe: T never increases, it simply decreases more slowly than a^{-1}.) Use the fact that the entropy density (Eq. (2.66)) scales as a^{-3} to compute the ratio of $(aT)^3$ at $T = 10$ GeV (roughly the time when WIMPs decouple) to its present value today.

Exercise 12. Suppose that there were no baryon asymmetry so that the number density of baryons exactly equaled that of anti-baryons. Determine the final relic density of (baryons+anti-baryons). At what temperature is this asymptotic value reached?

Exercise 13. There is a fundamental limitation on the annihilation cross section of a particle with mass m. Because of unitarity, $\langle\sigma v\rangle$ must be less than or equal to $1/m^2$, give or take a factor of order unity. Determine Ω_X for a particle which saturates this bound, i.e., for a particle with $\langle\sigma v\rangle = 1/m^2$. For what value of m is Ω_X equal to 1? (Keep x_f and g_* equal to the nominal values given in Eq. (3.60).) Note that if m is greater than this critical value, $\Omega_X > 1$, which is ruled out. This is a strong argument against stable particles (and therefore dark matter candidates) with masses above this critical value.

4

THE BOLTZMANN EQUATIONS

We are interested in the anisotropies in the cosmic distribution of photons and inhomogeneities in the matter. Figure 4.1 shows why these are complicated to calculate. The photons are affected by gravity and by Compton scattering with free electrons. The electrons are tightly coupled to the protons. Both of these, of course, are also affected by gravity. The metric which determines the gravitational forces is influenced by all these components plus the neutrinos and the dark matter. Thus to solve for the photon and dark matter distributions, we need to simultaneously solve for all the other components.

There is a systematic way to account for all of these couplings. We write down a Boltzmann equation for each species in the universe. We have already encountered the Boltzmann equation in its integrated form in Chapter 3. There we were interested solely in the number density of the dark matter, the neutrons, and the free electrons. The number density is the integral over all momenta of the distribution function. Here we will be interested in more detailed information, not just the integrated number density, but the full distribution of photons, say, as a function of momentum. We then need a more primitive version of Eq. (3.1). Schematically, the unintegrated Boltzmann equation is

$$\frac{df}{dt} = C[f].\tag{4.1}$$

The right-hand side of the Boltzmann equation contains all possible collision terms. These terms in general are complicated functionals of the distribution functions of the various components. In the absence of collisions, the distribution function obeys $df/dt = 0$. This seemingly innocent equation says that the number of particles in a given element of phase space does not change with time. The catch is that the phase space elements themselves are moving in time in complicated ways due to the nontrivial metric. This catch makes the problem more difficult than it seems from Eq. (4.1). Nonetheless, we can still progress systematically by reexpressing the full derivative in terms of partial derivatives.

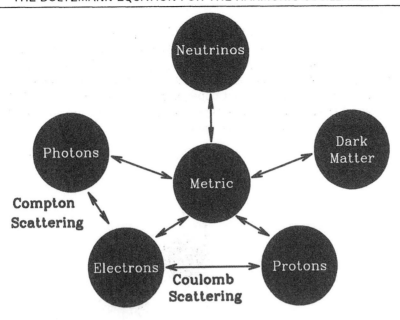

Figure 4.1. The ways in which the different components of the universe interact with each other. These connections are encoded in the coupled Boltzmann–Einstein equations.

In this chapter, we derive the Boltzmann equations for photons, eletrons, protons, dark matter, and massless neutrinos. This set of equations governs the evolution of perturbations in the universe.

4.1 THE BOLTZMANN EQUATION FOR THE HARMONIC OSCILLATOR

Before tackling the problem of interest — the Boltzmann equation for all species in an expanding universe — let us treat a much simpler example of the Boltzmann equation: the nonrelativistic harmonic oscillator. This simple example is very similar to the full general relativistic version we will encounter in the next section, but the algebra is much less cumbersome. So here the physics will be quite transparent. It will be useful to keep this example in mind when the algebra threatens to obscure the physics in the next section.

Consider a one-dimensional harmonic oscillator with energy

$$E = \frac{p^2}{2m} + \frac{1}{2}kx^2. \tag{4.2}$$

The distribution function of the harmonic oscillator depends on time t, position x, and momentum p. Thus, the full time derivative in Eq. (4.1) can be rewritten as

$$\frac{df(t,x,p)}{dt} = \frac{\partial f}{\partial t} + \frac{\partial f}{\partial x}\frac{dx}{dt} + \frac{\partial f}{\partial p}\frac{dp}{dt}. \tag{4.3}$$

Figure 4.2 illustrates the movement through phase space of a distribution of collisionless ($C = 0$) oscillators. The full time derivative df/dt vanishes since the number of particles in the bunch at t_1 equals that at t_2. What has changed is the location of the phase space elements $x(t)$ and $p(t)$ themselves. Alternatively, we can think of x and p as independent variables (not dependent on t) and take partial derivatives of f with respect to t, x, and p. All of these partial derivatives are nonzero, but the appropriate weighted sum of the three vanishes.

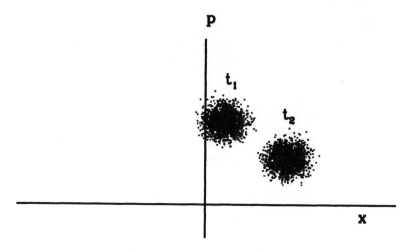

Figure 4.2. Distribution function for a set of collisionless harmonic oscillators. The initial distribution at t_1 moves in phase space by time t_2. The distribution function $f(t, x, p)$ remains constant as long as the evolution of $x(t)$ and $p(t)$ is accounted for.

To determine the coefficients dx/dt and dp/dt we must use the equations of motion. By the definition of momentum,

$$\frac{dx}{dt} \equiv \frac{p}{m}. \tag{4.4}$$

This equation will be generalized to a fully relativistic, three-dimensional version in the next section. Indeed we already got a preview of this when we defined $P^\mu \equiv dx^\mu/d\lambda$ in Chapter 2. Newton's equation governing the motion of the oscillator is

$$\frac{dp}{dt} = -kx. \tag{4.5}$$

The analogue of this familiar equation in the next section will be the geodesic equation of general relativity.

The collisionless Boltzmann equation for the harmonic oscillator is thus

$$\frac{\partial f}{\partial t} + \frac{p}{m}\frac{\partial f}{\partial x} - kx\frac{\partial f}{\partial p} = 0. \tag{4.6}$$

The second term here governs how rapidly the oscillator moves in real space; the coefficient in front is just the velocity, p/m. The last term governs how quickly particles lose momentum.

In order to solve the Boltzmann equation, we need to know the initial conditions on the distribution function. Even without these, though, the Boltzmann equation offers some useful physics. Consider the equilibrium distribution, wherein $\partial f/\partial t = 0$. A general solution for the equilibrium distribution is

$$f(p, x) = f_{EQ}(E); \tag{4.7}$$

that is, f is a function only of energy E. To see that this is indeed a solution, consider

$$\frac{p}{m}\frac{\partial f(E)}{\partial x} - kx\frac{\partial f(E)}{\partial p} = \frac{df}{dE}\left[\frac{p}{m}\frac{\partial E}{\partial x} - kx\frac{\partial E}{\partial p}\right]$$

$$= 0. \tag{4.8}$$

So any function of the energy alone is an equilibrium distribution. Of course, in general, there will be interactions, or collisions. The only way for the full Boltzmann equation to be satisfied is if the collision terms also vanish. This will in general drive f to one of the familiar equilibrium distributions, e.g., $e^{-E/T}$ for the classical Maxwell–Boltzmann distribution.

4.2 THE COLLISIONLESS BOLTZMANN EQUATION FOR PHOTONS

Let us begin then by considering the left hand side of Eq. (4.1) for massless photons. First we must specify the form of the metric, accounting for perturbations around the smooth universe described by Eq. (2.4). Whereas the smooth universe is characterized by a single function, $a(t)$, which depends only on time and not on space, the perturbed universe requires two more functions, Ψ and Φ, both of which depend on both space and time. In terms of them, the metric can be written as

$$g_{00}(\vec{x}, t) = -1 - 2\Psi(\vec{x}, t)$$

$$g_{0i}(\vec{x}, t) = 0$$

$$g_{ij}(\vec{x}, t) = a^2\delta_{ij}\left(1 + 2\Phi(\vec{x}, t)\right). \tag{4.9}$$

In the absence of Ψ and Φ, Eq. (4.9) is simply the FRW metric of the zero-order homogeneous, flat cosmology. Similarly, in the absence of expansion ($a = 1$) this metric describes a weak gravitational field (Exercise 2.3). The perturbations to the metric are Ψ, which corresponds to the Newtonian potential, and Φ, the perturbation to the spatial curvature. Since the perturbations in the universe are small at the times and scales of interest, we will treat these Ψ and Φ as small quantities, dropping all terms quadratic in them.

There are two technical points about the metric in Eq. (4.9) which you don't need to worry about for most of this book, but which nonetheless are important to be aware of, if only to better understand the literature. First, one can break up perturbations into those behaving as scalars, vectors, and tensors under a transformation from one 3D coordinate system to another. Equation (4.9) contains only scalar perturbations. In principle, it is possible that the metric of our universe also has vector or tensor perturbations. If so, $g_{\mu\nu}$ would require other functions besides Ψ and Φ to fully describe all perturbations. For example, the off-diagonal elements become nonzero if there are vector perturbations. Indeed, there are many cosmological theories wherein there are both tensor and vector perturbations. For example, inflation tends to predict that there will be tensor perturbations, while models based on topological defects tend to produce large vector perturbations. For now we focus solely on the scalar perturbations; these are the only ones that couple to matter perturbations and are the most important that couple to photon perturbations as well.

The other feature of Eq. (4.9) worth noting is that its form corresponds to a choice of *gauge*. The simplest way to understand this gauge freedom is to think back to electricity and magnetism. There, the vector potential A_μ and its derivatives contain all possible information about the electric and magnetic fields. Since the physical $\vec{E}$ and $\vec{B}$ fields remain unchanged if a constant is added to A_μ, there is some residual freedom in choosing the potential. (For example, one often chooses $A_0 = 0$ or $\partial_\mu A^\mu = 0$.) In our case of perturbations to the metric, a similar freedom exists. Even if only scalar perturbations are considered, there is still considerable freedom in the variables one chooses to describe the fluctuations. Although any physical results must be insensitive to the gauge choice, it is possible to use a gauge which looks quite different from Eq. (4.9) and still describes the same physics. For the record, the gauge in Eq. (4.9) is called the *conformal Newtonian* gauge.[1]

We want to reexpress the total derivative in Eq. (4.1) as a sum of partial derivatives. The distribution function depends on the space-time point $x^\mu = (t, \vec{x})$ and also on the momentum vector defined as

$$P^\mu \equiv \frac{dx^\mu}{d\lambda} \qquad (4.10)$$

where λ again parametrizes the particle's path, as in Eq. (2.18) (and again we will not need to specify λ explicitly). Thus, in principle, f is a function defined in an 8-dimensional space. However, not all the components of the momentum vector are independent since the masslessness of the photon implies that

$$P^2 \equiv g_{\mu\nu}P^\mu P^\nu = 0. \qquad (4.11)$$

---

[1]Historically, the initial ground-breaking work on the evolution of fluctuations was carried out in synchronous gauge (Peebles and Yu, 1970; Wilson and Silk, 1981; Peebles 1982; Bond and Szalay, 1983; Bond and Efstathiou, 1984). Recently, the physics of the anisotropies has been elucidated most clearly by using conformal Newtonian gauge (e.g., Hu and Sugiyama, 1995). Exercise 2 works out some of the relevant equations in synchronous gauge.

So there are only three independent components of the momentum vector. Before we choose which three we will use, let us enforce the constraint of Eq. (4.11), using the metric of Eq. (4.9).

$$P^2 = 0 = -(1 + 2\Psi)(P^0)^2 + p^2 = 0 \tag{4.12}$$

where I have defined

$$p^2 \equiv g_{ij}P^iP^j. \tag{4.13}$$

We can use the constraint equation then to eliminate the time component of P^μ:

$$P^0 = \frac{p}{\sqrt{1 + 2\Psi}} = p\,(1 - \Psi). \tag{4.14}$$

This last equality holds since we are doing first-order perturbation theory in the small quantity Ψ. With our sign convention, an overdense region has $\Psi < 0$. Therefore, in an overdense region, the term in parentheses on the right-hand side here is greater than one. Thus, Eq. (4.14) tells us photons lose energy —redshift— as they move out of a potential well.

Equation (4.14) is the generalization of the relativistic expression $E = pc$ to a perturbed Friedmann–Robertson–Walker metric. It allows us to eliminate P^0 whenever it occurs in favor of p, the generalized magnitude of the momentum. Recall that in the harmonic oscillator case, we did not include a term proportional to $\partial f/\partial E$ in Eq. (4.3). Here, too, we do not need to include a term proportional to $\partial f/\partial P^0$ when expanding the total time derivative. We need include only the dependence of f on the momentum: both the magnitude p and the angular direction. For the direction vector, we'll use the unit vector $\hat{p}^i = \hat{p}_i$, which by definition satisfies $\delta_{ij}\hat{p}^i\hat{p}^j = 1$.

We can now write Eq. (4.1) as

$$\frac{df}{dt} = \frac{\partial f}{\partial t} + \frac{\partial f}{\partial x^i} \cdot \frac{dx^i}{dt} + \frac{\partial f}{\partial p}\frac{dp}{dt} + \frac{\partial f}{\partial \hat{p}^i} \cdot \frac{d\hat{p}^i}{dt}. \tag{4.15}$$

The easiest term in Eq. (4.15) is the last one since it does not contribute at first order in perturbation theory. To see this, first recall that the zero-order distribution function is simply the Bose–Einstein function which depends only on p, not on the direction $\hat{p}^i$. Therefore, $\partial f/\partial \hat{p}^i$ is nonzero only if we consider the perturbation to the zero-order f; i.e., it is a first-order term. But so is the term which multiplies it, $d\hat{p}^i/dt$, for the direction of a photon changes only in the presence of potentials Φ and Ψ. In the absence of these potentials, a photon moves in a straight line. Thus the last term is the product of two first-order terms, rendering it a second-order term. We can neglect it.

Next let us reexpress the second term on the right-hand side of Eq. (4.15) by recalling that (Eq. (4.10)) $P^i \equiv dx^i/d\lambda$ and $P^0 \equiv dt/d\lambda$. Therefore,

$$\frac{dx^i}{dt} = \frac{dx^i}{d\lambda}\frac{d\lambda}{dt}$$

$$= \frac{P^i}{P^0} \cdot \cdot \tag{4.16}$$

We want to reexpress this ratio in terms of our favored variables p and $\hat{p}^i$. Equation (4.14) does this for P^0; let's do the same for the numerator P^i. The comoving momentum P^i is proportional to $\hat{p}^i$; call the proportionality constant C:

$$P^i \equiv C\hat{p}^i. \tag{4.17}$$

To determine the coefficient C, we can use Eq. (4.13):

$$p^2 = g_{ij}\hat{p}^i\hat{p}^j C^2$$

$$= a^2(1+2\Phi)\delta_{ij}\hat{p}^i\hat{p}^j C^2$$

$$= a^2(1+2\Phi)C^2 \tag{4.18}$$

where the last equality holds because the direction vector is a unit vector. Equation (4.18) tells us that $C = p(1-\Phi)/a$ so whenever we encounter P^i, we can always eliminate it in terms of $p, \hat{p}^i$ via

$$P^i = p\hat{p}^i\frac{1-\Phi}{a}. \tag{4.19}$$

From Eqs. (4.16) and (4.19), we see that

$$\frac{dx^i}{dt} = \frac{\hat{p}^i}{a}\left(1 + \Psi - \Phi\right). \tag{4.20}$$

An overdense region has $\Psi < 0$ and $\Phi > 0$, rendering the term in parentheses less than one. So, Eq. (4.20) says that a photon slows down (dx/dt becomes smaller) when traveling through an overdense region. This makes perfect sense: we expect the gravitational force of an overdense region to slow down even photons. Having said that, I now claim that we can neglect the potentials in Eq. (4.20). For, in the Boltzmann equation they multiply $\partial f/\partial x^i$ which is a first-order term. (Again, the zero-order distribution function does not depend on position.) So collecting terms up to this point, we have

$$\frac{df}{dt} = \frac{\partial f}{\partial t} + \frac{\hat{p}^i}{a}\frac{\partial f}{\partial x^i} + \frac{\partial f}{\partial p}\frac{dp}{dt}. \tag{4.21}$$

The remaining term to be calculated is dp/dt. Alas, unlike the harmonic oscillator, here $dp/dt \neq -kx$. Rather we will need the geodesic equation from general relativity and more fortitude to compute dp/dt for photons in a perturbed FRW metric.

To begin, let us recall that the time component of the geodesic equation (2.18) can be written as

$$\frac{dP^0}{d\lambda} = -\Gamma^0{}_{\alpha\beta}P^\alpha P^\beta. \tag{4.22}$$

We can rewrite the derivative with respect to λ as a derivative with respect to time multiplied by $dt/d\lambda = P^0$. Also, we can use Eq. (4.14) to eliminate P^0 in terms of our favored variable p. Then the geodesic equation reduces to

$$\frac{d}{dt}\left[p(1 - \Psi)\right] = -\Gamma^0{}_{\alpha\beta}\frac{P^\alpha P^\beta}{p}(1 + \Psi). \tag{4.23}$$

Expand out the time derivative to get

$$\frac{dp}{dt}(1 - \Psi) = p\frac{d\Psi}{dt} - \Gamma^0{}_{\alpha\beta}\frac{P^\alpha P^\beta}{p}(1 + \Psi). \tag{4.24}$$

Now we multiply both sides by $(1+\Psi)$; drop all terms quadratic in Ψ; and reexpress the total time derivative of Ψ in terms of partial derivatives so that

$$\frac{dp}{dt} = p\left\{\frac{\partial\Psi}{\partial t} + \frac{\hat{p}^i}{a}\frac{\partial\Psi}{\partial x^i}\right\} - \Gamma^0{}_{\alpha\beta}\frac{P^\alpha P^\beta}{p}(1 + 2\Psi). \tag{4.25}$$

In order to evaluate dp/dt then we need to evaluate the product $\Gamma^0{}_{\alpha\beta}P^\alpha P^\beta/p$. Recall that the Christoffel symbol is best written as a sum of derivatives of the metric (Eq. (2.19)). Here we are interested only in the $\Gamma^0{}_{\alpha\beta}$ component. It multiplies $P^\alpha P^\beta$, which is symmetric in α, β. Thus, the first two metric derivatives contribute equally, and we have

$$\Gamma^0{}_{\alpha\beta}\frac{P^\alpha P^\beta}{p} = \frac{g^{0\nu}}{2}\left[2\frac{\partial g_{\nu\alpha}}{\partial x^\beta} - \frac{\partial g_{\alpha\beta}}{\partial x^\nu}\right]\frac{P^\alpha P^\beta}{p}. \tag{4.26}$$

Now $g^{0\nu}$ is nonzero only when $\nu = 0$, in which case it is simply the inverse of g_{00}, so

$$\Gamma^0{}_{\alpha\beta}\frac{P^\alpha P^\beta}{p} = \frac{-1 + 2\Psi}{2}\left[2\frac{\partial g_{0\alpha}}{\partial x^\beta} - \frac{\partial g_{\alpha\beta}}{\partial t}\right]\frac{P^\alpha P^\beta}{p}. \tag{4.27}$$

Once again, $g_{0\alpha}$ in the first term in brackets is nonzero only when $\alpha = 0$, in which case its derivative is $-2\partial\Psi/\partial x^\beta$. The second term in brackets multiplied by the product of momenta is

$$-\frac{\partial g_{\alpha\beta}}{\partial t}\frac{P^\alpha P^\beta}{p} = -\frac{\partial g_{00}}{\partial t}\frac{P^0 P^0}{p} - \frac{\partial g_{ij}}{\partial t}\frac{P^i P^j}{p}$$

$$= 2\frac{\partial\Psi}{\partial t}p - a^2\delta_{ij}\left[2\frac{\partial\Phi}{\partial t} + 2H(1 + 2\Phi)\right]\frac{P^i P^j}{p}. \tag{4.28}$$

But, via Eq. (4.19), $\delta_{ij}P^i P^j = p^2(1 - 2\Phi)/a^2$, so we have

$$\Gamma^0{}_{\alpha\beta}\frac{P^\alpha P^\beta}{p} = \frac{-1 + 2\Psi}{2}\left[-4\frac{\partial\Psi}{\partial x^\beta}P^\beta + 2p\frac{\partial\Psi}{\partial t}\right.$$

$$-p\left\{2\frac{\partial\Phi}{\partial t}+2H(1+2\Phi)\right\}(1-2\Phi)\right]. \qquad (4.29)$$

The last line here simplifies since $(1+2\Phi)(1-2\Phi)\to 1$ at first order, and $1-2\Phi$ can be set to 1 when multiplying $\partial\Phi/\partial t$. Summing over the index β in the first term then leads to

$$\Gamma^0{}_{\alpha\beta}\frac{P^\alpha P^\beta}{p}=\frac{-1+2\Psi}{2}\left[-4\left(\frac{\partial\Psi}{\partial t}p+\frac{\partial\Psi}{\partial x^i}\frac{p\hat{p}^i}{a}\right)+2p\frac{\partial\Psi}{\partial t}-p\left\{2\frac{\partial\Phi}{\partial t}+2H\right\}\right]$$

$$=\{-1+2\Psi\}\left[-\frac{\partial\Psi}{\partial t}p-2\frac{\partial\Psi}{\partial x^i}\frac{p\hat{p}^i}{a}-p\left\{\frac{\partial\Phi}{\partial t}+H\right\}\right]. \qquad (4.30)$$

We can insert this into Eq. (4.25) to get

$$\frac{dp}{dt}=p\left\{\frac{\partial\Psi}{\partial t}+\frac{\hat{p}^i}{a}\frac{\partial\Psi}{\partial x^i}\right\}-\frac{\partial\Psi}{\partial t}p-2\frac{\partial\Psi}{\partial x^i}\frac{p\hat{p}^i}{a}-p\left\{\frac{\partial\Phi}{\partial t}+H\right\}. \qquad (4.31)$$

Collecting terms, we finally have

$$\frac{1}{p}\frac{dp}{dt}=-H-\frac{\partial\Phi}{\partial t}-\frac{\hat{p}^i}{a}\frac{\partial\Psi}{\partial x^i}. \qquad (4.32)$$

Equation (4.32) is what we were after. It describes the change in the photon momentum as it moves through a perturbed FRW universe. The first term accounts for the loss of momentum due to the Hubble expansion. To understand the significance of the next two terms in Eq. (4.32), we first need to remember that an overdense region has $\Phi>0$ and $\Psi<0$ with our sign conventions. Therefore, the second term says that a photon in a deepening gravitational well ($\partial\Phi/\partial t>0$) loses energy. This is understandable: the deepening well makes it more difficult for the photon to emerge, thereby increasing the magnitude of the redshift. Finally, a photon traveling into a well ($\hat{p}^i\partial\Psi/\partial x^i<0$) gains energy because it is being pulled toward the center. Conversely, as it leaves the well, it gets redshifted.

We are now in a position to write down the Boltzmann equation for photons. Using Eq. (4.32) in Eq. (4.21) leads to

$$\frac{df}{dt}=\frac{\partial f}{\partial t}+\frac{\hat{p}^i}{a}\frac{\partial f}{\partial x^i}-p\frac{\partial f}{\partial p}\left[H+\frac{\partial\Phi}{\partial t}+\frac{\hat{p}^i}{a}\frac{\partial\Psi}{\partial x^i}\right]. \qquad (4.33)$$

This equation incorporates much of the physics with which we are already familiar, such as the fact that photons redshift in an expanding universe. It also leads directly to the equations governing anisotropies. Working through the terms on the right, the first two are familiar from standard hydrodynamics; when integrated, they lead to the continuity and Euler equations (Exercise 1). The third term dictates that photons lose energy in an expanding universe. We saw some of this in Chapter 2 when considering geodesics. Shortly, we will see how the Boltzmann formalism enforces this result. Finally, the last two encode the effects of under-/overdense regions on the photon distribution function.

To go further we must now expand the photon distribution function f about its zero-order Bose–Einstein value. I will do this in a way that may seem odd at first. Let us write

$$f(\vec{x}, p, \hat{p}, t) = \left[\exp\left\{\frac{p}{T(t)[1 + \Theta(\vec{x}, \hat{p}, t)]}\right\} - 1\right]^{-1}. \qquad (4.34)$$

Here the zero-order temperature T is a function of time only (i.e., scales as a^{-1}), not space. The perturbation to the distribution function is characterized by Θ, which could also be called $\delta T/T$. In the smooth zero-order universe, photons are distributed homogeneously, so T is independent of $\vec{x}$, and isotropically, so T is independent of the direction of propagation $\hat{p}$. Now that we want to describe perturbations about this smooth universe, we need to allow for inhomogeneities in the photon distribution (so Θ depends on $\vec{x}$) and anisotropies (so Θ depends on $\hat{p}$). There is one assumption built into Eq. (4.34). I have explicitly written down that Θ depends on $\vec{x}, \hat{p}$, and t. This assumes that it does *not* depend on the magnitude of the momentum p. We will soon see that this is a valid assumption, following directly from that fact that the magnitude of the photon momentum is virtually unchanged during a Compton scatter. The perturbation Θ is small, so we can expand (again keeping only terms up to first order)

$$f \simeq \frac{1}{e^{p/T} - 1} + \left(\frac{\partial}{\partial T}\left[\exp\left\{\frac{p}{T}\right\} - 1\right]^{-1}\right)T\Theta$$

$$= f^{(0)} - p\frac{\partial f^{(0)}}{\partial p}\Theta. \qquad (4.35)$$

In the last line I have identified the zero-order distribution function as the Bose–Einstein distribution with zero chemical potential,

$$f^{(0)} \equiv \left[\exp\left\{\frac{p}{T}\right\} - 1\right]^{-1}, \qquad (4.36)$$

and made use of the fact that for this function $T\partial f^{(0)}/\partial T = -p\partial f^{(0)}/\partial p$.

4.2.1 Zero-Order Equation

We can now set about systematically collecting the terms of similar order in Eq. (4.33). Let us start with the zero-order terms, those with no Φ, Ψ, or Θ. These lead immediately to

$$\left.\frac{df}{dt}\right|_{\text{zero order}} = \frac{\partial f^{(0)}}{\partial t} - Hp\frac{\partial f^{(0)}}{\partial p} = 0. \qquad (4.37)$$

I have set df/dt here equal to zero, i.e., set the collision term on the right of Eq. (4.1) to zero. I could justify this by claiming that we are now looking only at the collisionless Boltzmann equation. But there is a much deeper justification. In fact, even when we come around to including collisions, we will see that there is no zero-order collision term. That is, the collision terms will be proportional to Θ and other perturbatively small quantities. There is a profound reason for this: the zero-order distribution function is set precisely by the requirement that the collision term vanishes. Another, perhaps more familiar, way of saying this is to point out that any collision term includes the rate for the given reaction and for its inverse. If the distribution functions are set to their equilibrium values, the rate for the reaction precisely cancels the rate for its inverse. If a given component is out of equilibrium, collisions will drive it toward its equilibrium distribution. This is the reason we expected a Bose–Einstein distribution in the first place. Its observation is convincing evidence that photons were at one point in the early universe tightly coupled to the electrons.

Returning to Eq. (4.37), we can rewrite the time derivative as

$$\frac{\partial f^{(0)}}{\partial t} = \frac{\partial f^{(0)}}{\partial T}\frac{dT}{dt} = -\frac{dT/dt}{T}p\frac{\partial f^{(0)}}{\partial p}$$

so that the zero-order equation becomes

$$\left[-\frac{dT/dt}{T} - \frac{da/dt}{a}\right]\frac{\partial f^{(0)}}{\partial p} = 0. \tag{4.38}$$

Thus $dT/T = -da/a$ or

$$T \propto \frac{1}{a}. \tag{4.39}$$

This is precisely what we expected from the heuristic argument about the photon's wavelength getting stretched as the universe expands (Section 1.1) and the more concrete argument of Section 2.1. It is reassuring to see this result emerge from the Boltzmann treatment.

4.2.2 First-Order Equation

We now return to Eq. (4.33) and extract the equation for the deviation of the photon temperature from its zero-order value, i.e., an equation for Θ. To do this, everywhere we encounter f in Eq. (4.33), we insert the expansion of Eq. (4.35):

$$\left.\frac{df}{dt}\right|_{\text{first order}} = -p\frac{\partial}{\partial t}\left[\frac{\partial f^{(0)}}{\partial p}\Theta\right] - p\frac{\hat{p}^i}{a}\frac{\partial\Theta}{\partial x^i}\frac{\partial f^{(0)}}{\partial p} + Hp\Theta\frac{\partial}{\partial p}\left[p\frac{\partial f^{(0)}}{\partial p}\right]$$

$$-p\frac{\partial f^{(0)}}{\partial p}\left[\frac{\partial\Phi}{\partial t} + \frac{\hat{p}^i}{a}\frac{\partial\Psi}{\partial x^i}\right]. \tag{4.40}$$

Consider the first term on the right-hand side here. The time derivative can be rewritten as a temperature derivative so

$$-p\frac{\partial}{\partial t}\left[\frac{\partial f^{(0)}}{\partial p}\Theta\right] = -p\frac{\partial f^{(0)}}{\partial p}\frac{\partial\Theta}{\partial t} - p\Theta\frac{dT}{dt}\frac{\partial^2 f^{(0)}}{\partial T\partial p}$$

$$= -p\frac{\partial f^{(0)}}{\partial p}\frac{\partial \Theta}{\partial t} + p\Theta\frac{dT/dt}{T}\frac{\partial}{\partial p}\left[p\frac{\partial f^{(0)}}{\partial p}\right]. \qquad (4.41)$$

The second line follows here since $\partial f^{(0)}/\partial T = -(p/T)\partial f^{(0)}/\partial p$. The second term on this second line cancels the third term on the right in Eq. (4.40), so we can finally write down the equation governing the perturbation Θ:

$$\left.\frac{df}{dt}\right|_{\text{first order}} = -p\frac{\partial f^{(0)}}{\partial p}\left[\frac{\partial \Theta}{\partial t} + \frac{\hat{p}^i}{a}\frac{\partial \Theta}{\partial x^i} + \frac{\partial \Phi}{\partial t} + \frac{\hat{p}^i}{a}\frac{\partial \Psi}{\partial x^i}\right]. \qquad (4.42)$$

The first two terms here account for "free streaming," which translates into anisotropies on increasingly small scales as the universe evolves. The last two account for the effect of gravity. Note that every time x appears it is multiplied by a, the scale factor. This must happen, for physical distances are ax.

4.3 COLLISION TERMS: COMPTON SCATTERING

Our task in this section is to determine the influence Compton scattering has on the photon distribution function. The scattering process of interest is

$$e^-(\vec{q}) + \gamma(\vec{p}) \leftrightarrow e^-(\vec{q}\,') + \gamma(\vec{p}\,'), \qquad (4.43)$$

where I have explicitly indicated the momentum of each particle.

We are interested in the change of distribution of photons with momentum $\vec{p}$ (with magnitude p and direction $\hat{p}$). Therefore we must sum over all other momenta $(\vec{q}, \vec{q}\,', \vec{p}\,')$ which affect $f(\vec{p})$. Schematically, then, the collision term is

$$C[f(\vec{p})] = \sum_{\vec{q},\vec{q}\,',\vec{p}\,'} |\text{Amplitude}|^2 \{f_e(\vec{q}\,')f(\vec{p}\,') - f_e(\vec{q})f(\vec{p})\}. \qquad (4.44)$$

The amplitude is reversible so it multiplies both the reaction and its inverse. The products of the electron distribution function f_e and the photon distribution function simply count the number of particles with the given momenta. I have neglected stimulated emission and Pauli blocking, which would lead to factors of $1 + f$ and $1 - f_e$ with the appropriate momenta. At first order this turns out to be a valid assumption. If one were to go to second order, though, stimulated emission would have to be included. Pauli blocking is never important after electron–positron annihilation because the occupation numbers f_e are very small (Exercise 4).

Unfortunately, the collision term becomes a bit messier than the schematic version when we put in all the factors of 2π to properly account for the sums over phase space. Explicitly, the collision term is[2]

$$C[f(\vec{p})] = \frac{1}{p}\int\frac{d^3q}{(2\pi)^3 2E_e(q)}\int\frac{d^3q'}{(2\pi)^3 2E_e(q')}\int\frac{d^3p'}{(2\pi)^3 2E(p')}|\mathcal{M}|^2(2\pi)^4$$

---

[2]Most of the phase space factors here follow from our discussion in Section 3.1, the exception being the factor of $1/p$ in front. You may have wondered about one other feature of the Boltzmann equation presented in this chapter: I started at the outset taking df/dt; does not general relativity require us to take the derivative with respect to the affine parameter λ? Exercise 5 illustrates how these problems solve each other.

$$\times \delta^3 [\vec{p} + \vec{q} - \vec{p}\,' - \vec{q}\,'] \, \delta \left[E(p) + E_e(q) - E(p') - E_e(q') \right]$$

$$\times \left\{ f_e(\vec{q}\,') f(\vec{p}\,') - f_e(\vec{q}) f(\vec{p}) \right\}. \tag{4.45}$$

Here the delta functions enforce energy momentum conservation. The energies at this order are the relativistic limit for photons and non-relativistic limit for electrons: $E(p) = p$ and $E_e(q) = m_e + q^2/(2m_e)$. Note the similarity between this collision term and the general one (Eq. (3.1)) we considered in Chapter 3. The only difference is that I have not integrated this collision term over all photon momenta $\vec{p}$, so there are only three momentum integrals. Again, this reflects our need to understand how photons traveling in different directions interact: we will see that the collision term depends on $\hat{p}$.

Since the kinetic energy of the electrons is very small at the epochs of interest compared with their rest energy, the factors of E_e in the denominator of Eq. (4.45) may be replaced with m_e. Then using the three-dimensional momentum delta function to do the $\vec{q}\,'$ integral, we have

$$C[f(\vec{p})] = \frac{\pi}{4m_e^2 p} \int \frac{d^3q}{(2\pi)^3} \int \frac{d^3p'}{(2\pi)^3 p'} \, \delta \left[p + \frac{q^2}{2m_e} - p' - \frac{(\vec{q} + \vec{p} - \vec{p}\,')^2}{2m_e} \right]$$

$$\times |\mathcal{M}|^2 \left\{ f_e(\vec{q} + \vec{p} - \vec{p}\,') f(\vec{p}\,') - f_e(\vec{q}) f(\vec{p}) \right\}. \tag{4.46}$$

To go further, we need to understand the kinematics of nonrelativistic Compton scattering. The most important feature of this process for our purposes is that very little energy is transferred. In particular,

$$E_e(q) - E_e(\vec{q} + \vec{p} - \vec{p}\,') = \frac{q^2}{2m_e} - \frac{(\vec{q} + \vec{p} - \vec{p}\,')^2}{2m_e}$$

$$\simeq \frac{(\vec{p}\,' - \vec{p}) \cdot \vec{q}}{m_e}, \tag{4.47}$$

where the last approximate equality holds since q is much larger than $\vec{p}$ and $\vec{p}\,'$. In nonrelativistic Compton scattering, $p' \simeq p$, scattering is nearly elastic. Therefore, $\vec{p}\,' - \vec{p}$ is of order p, of order the ambient temperature T. So the right-hand side of Eq. (4.47) is of order $Tq/m_e \sim Tv_b$ where the baryonic velocity v_b is very small. The change in the electron energy due to Compton scattering is therefore of order Tv_b. Since the typical kinetic energy of the electrons is also of order T, the fractional energy change in a single Compton collision is very small, of order v_b. It makes sense, therefore, to expand the final electron kinetic energy $(\vec{q} + \vec{p} - \vec{p}\,')^2/(2m_e)$ around its zero-order value of $q^2/(2m_e)$. The delta function can be expanded as

$$\delta \left[p + \frac{q^2}{2m_e} - p' - \frac{(\vec{q} + \vec{p} - \vec{p}\,')^2}{2m_e} \right] \simeq \delta(p - p')$$

$$+ (E_e(q') - E_e(q)) \left. \frac{\partial \delta \left(p + E_e(q) - p' - E_e(q') \right)}{\partial E_e(q')} \right|_{E_e(q) = E_e(q')}$$

$$= \delta(p - p') + \frac{(\vec{p} - \vec{p}') \cdot \vec{q}}{m_e} \frac{\partial \delta(p - p')}{\partial p'} \quad (4.48)$$

where the second equality makes use of the fact that for a general function f of the sum of two variables, $\partial f(x - y)/\partial x = -\partial f(x - y)/\partial y$. This formal expansion appears ill-defined at present, but when integrating over momenta, the derivatives of delta functions will be handled by integrating by parts. With this expansion, and using the fact that $f_e(\vec{q} + \vec{p} - \vec{p}') \simeq f_e(\vec{q})$, the collision term becomes

$$C[f(\vec{p})] = \frac{\pi}{4m_e^2 p} \int d^3q \frac{f_e(\vec{q})}{(2\pi)^3} \int \frac{d^3p'}{(2\pi)^3 p'} |\mathcal{M}|^2$$

$$\times \left\{ \delta(p - p') + \frac{(\vec{p} - \vec{p}') \cdot \vec{q}}{m_e} \frac{\partial \delta(p - p')}{\partial p'} \right\} \{ f(\vec{p}') - f(\vec{p}) \}. \quad (4.49)$$

To proceed further, we need the amplitude for Compton scattering. This can be computed using Feynman rules as explicated for example in Bjorken and Drell (1965). We will take it to be constant:

$$|\mathcal{M}|^2 = 8\pi \sigma_T m_e^2 \quad (4.50)$$

where σ_T is the Thomson cross-section. This is wrong, and it is wrong for two reasons. First of all, the amplitude squared has an angular dependence $\propto (1 + \cos^2[\hat{p} \cdot \hat{p}\,'])$. Ignoring this angular dependence, as I now propose to do, makes a small difference in the final collision term. It needs to be included in calculations which aspire to 1% accuracy. But it would simply distract us here, so let us ignore it for the present. The second reason a constant amplitude is wrong is a little more subtle and, when properly accounted for, opens up a whole new branch of CMB study. In particular, the amplitude squared has a polarization dependence ($\propto |\hat{\epsilon} \cdot \hat{\epsilon}'|^2$, where $\hat{\epsilon}$ and $\hat{\epsilon}'$ are the polarizations of the incoming and outgoing photons) which I have implicitly summed over here. The dependence on polarization means that at a small level the CMB will be polarized due to Compton scattering (Bond and Efstathiou, 1984; Polnarev, 1985). It turns out that the information carried by the polarization spectrum is as valuable as that carried by the temperature spectrum (Seljak, 1997; Seljak and Zaldarriaga, 1997; Kamionkowski, Kosowsky, and Stebbins, 1997a,b). We will devote considerable time in Chapter 10 to understanding polarization. Even if we were not concerned with polarization, the temperature anisotropies are coupled to the polarization field, so an accurate determination of the former requires a treatment of the latter. Again, though, I will neglect this small effect here in the derivation of the collision term. It is straightforward to include both the effects of polarization and the angular dependence of Compton scattering using the same formalism we are now in the midst of. The algebra is simply a bit more tedious.

Once we have assumed that $|\mathcal{M}|^2$ is constant, we can multiply out the terms in brackets in Eq. (4.49) keeping only terms first order in energy transfer. Also, the $\vec{q}$ integral simply gives a factor of n_e (or $n_e \vec{v}_b$ for the term which has a factor of $\vec{q}/m_e$). So,

$$C[f(\vec{p})] = \frac{2\pi^2 n_e \sigma_T}{p} \int \frac{d^3 p'}{(2\pi)^3 p'} \left\{ \delta(p - p') + (\vec{p} - \vec{p}') \cdot \vec{v}_{\rm b} \frac{\partial \delta(p - p')}{\partial p'} \right\}$$

$$\times \left\{ f(\vec{p}') - f(\vec{p}) - p' \frac{\partial f^{(0)}}{\partial p'} \Theta(\hat{p}') + p \frac{\partial f^{(0)}}{\partial p} \Theta(\hat{p}) \right\}$$

$$= \frac{n_e \sigma_T}{4\pi p} \int_0^\infty dp' p' \int d\Omega' \left[\delta(p - p') \left(-p' \frac{\partial f^{(0)}}{\partial p'} \Theta(\hat{p}') + p \frac{\partial f^{(0)}}{\partial p} \Theta(\hat{p}) \right) \right.$$

$$\left. + (\vec{p} - \vec{p}') \cdot \vec{v}_{\rm b} \frac{\partial \delta(p - p')}{\partial p'} (f^{(0)}(p') - f^{(0)}(p)) \right] \tag{4.51}$$

where Ω' is the solid angle spanned by the unit vector $\hat{p}'$. On the first line, I have broken up the difference $f(\vec{p}') - f(\vec{p})$ into a zero-order piece,[3] which doesn't contribute when multiplying $\delta(p - p')$, and a first-order part which can be neglected when multiplying the velocity term.

There are only two terms in Eq. (4.51) which depend on $\hat{p}'$ and therefore which must be accounted for when integrating over solid angle Ω'. First, there is the perturbation to the distribution function, $\Theta(\hat{p}')$. It is convenient at this stage to introduce the notation

$$\Theta_0(\vec{x}, t) \equiv \frac{1}{4\pi} \int d\Omega' \, \Theta(\hat{p}', \vec{x}, t). \tag{4.52}$$

So Θ_0 does not depend on the direction vector; it is an integral of the perturbation over all directions. In other words, it is the *monopole* part of the perturbation. Note that we *cannot* absorb this monopole into the definition of the zero order temperature since the latter is constant over all space. The perturbation Θ_0 therefore represents the deviation of the monopole at a given point in space from its average in all space. Later on we will generalize Eq. (4.52) to all other multipoles.

The second term in Eq. (4.51) which depends on $\hat{p}'$ is the explicit factor $\hat{p}' \cdot \vec{v}_{\rm b}$. This term integrates to zero since $\vec{v}_{\rm b}$ is a fixed vector. Thus, the integration over solid angle leaves

$$C[f(\vec{p})] = \frac{n_e \sigma_T}{p} \int_0^\infty dp' p' \left[\delta(p - p') \left(-p' \frac{\partial f^{(0)}}{\partial p'} \Theta_0 + p \frac{\partial f^{(0)}}{\partial p} \Theta(\hat{p}) \right) \right.$$

$$\left. + \vec{p} \cdot \vec{v}_{\rm b} \frac{\partial \delta(p - p')}{\partial p'} (f^{(0)}(p') - f^{(0)}(p)) \right]. \tag{4.53}$$

Now the p' integral can be done: in the first line by trivially integrating over the delta function and in the second by integrating by parts. We are left with

---

[3] Note that we are expanding in two small quantities simultaneously, the small perturbations and the small energy transfer. Here, we are breaking up $f(\vec{p}') - f(\vec{p})$ into terms zero and first order in the small perturbations.

$$C[f(\vec{p})] = -p\frac{\partial f^{(0)}}{\partial p} n_e \sigma_T \left[\Theta_0 - \Theta(\hat{p}) + \hat{p} \cdot \vec{v}_b\right]. \qquad (4.54)$$

Already, we can anticipate the effect of Compton scattering on the photon distribution. In the absence of a bulk velocity for the electrons ($v_b = 0$), the collision terms serve to drive Θ to Θ_0. That is, when Compton scattering is very efficient, only the monopole perturbation survives; all other moments are washed out (Figure 4.3). Intuitively, strong scattering means that the mean free path of a photon is very small. Therefore, photons arriving at a given point in space last scattered off very nearby electrons if Compton scattering is efficient. These nearby electrons most likely had a temperature very similar to the point of observation. Therefore, photons from all directions have the same temperature. This is the characteristic signature of a monopole distribution: the temperature on the sky is uniform.

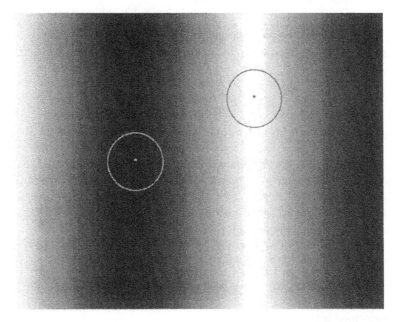

Figure 4.3. A plane wave perturbation in the matter and its effect on tightly coupled photons. Dark (white) regions represent hot (cold) spots in the electron temperature. If Compton scattering is very efficient then photons last scattered very near the point of observation. Circles denote last scattering surfaces for observation points indicated by stars. The temperature on these surfaces is very close to uniform, so the distribution is almost all monopole. Note though that different circles (corresponding to different observers) have different temperatures due to the perturbation. So the monopole varies in space.

The situation changes slightly if the electrons carry a bulk velocity. In that case, the photons will also have a dipole moment, fixed by the amplitude and direction of the electron velocity. Even in this case, though, all higher moments will vanish. Thus Compton scattering produces a photon distribution which is extremely simple to categorize: it has only a nonvanishing monopole and dipole. This is equivalent to saying that the photons behave like a fluid. Indeed, strong scattering, or tight coupling, produces a situation wherein the photons and electrons behave as a single fluid.

4.4 THE BOLTZMANN EQUATION FOR PHOTONS

We can now collect the left- and right-hand sides of the Boltzmann equations from the previous two sections. A few more definitions will complete the first goal of this chapter, a linear equation for the perturbation to the photon distribution. Equating Eqs. (4.42) and (4.54) leads to

$$\frac{\partial \Theta}{\partial t} + \frac{\hat{p}^i}{a}\frac{\partial \Theta}{\partial x^i} + \frac{\partial \Phi}{\partial t} + \frac{\hat{p}^i}{a}\frac{\partial \Psi}{\partial x^i} = n_e \sigma_T \left[\Theta_0 - \Theta + \hat{p}\cdot\vec{v}_{\rm b}\right]. \tag{4.55}$$

At this point, it is convenient to reintroduce the conformal time η, defined in Eq. (2.41), as our time variable. In terms of the conformal time, the Boltzmann equation becomes

$$\dot{\Theta} + \hat{p}^i \frac{\partial \Theta}{\partial x^i} + \dot{\Phi} + \hat{p}^i \frac{\partial \Psi}{\partial x^i} = n_e \sigma_T a \left[\Theta_0 - \Theta + \hat{p}\cdot\vec{v}_{\rm b}\right]. \tag{4.56}$$

Here, and from now on, overdots represent derivatives with respect to conformal time.

Equation (4.56) is a partial differential linear equation coupling Θ to other variables Φ, Ψ, and $\vec{v}_{\rm b}$ which also evolve linearly. If we Fourier transform all these variables, so that $\partial/\partial x^i \rightarrow k_i (\equiv k^i)$, the resulting Fourier amplitudes obey ordinary differential equations, which are much simpler to solve. In the case of small perturbations around a smooth universe, there is an added benefit of Fourier transforming. Since the background is smooth, the only $\vec{x}$ dependence in Eq. (4.56) is hidden in the perturbation variables themselves. In general, an equation of the form

$$aA(\vec{x}) = bB(\vec{x}) \tag{4.57}$$

gets transformed into

$$a\tilde{A}(\vec{k}) = b\tilde{B}(\vec{k}). \tag{4.58}$$

That is, every Fourier mode evolves independently: $A(\vec{k}_1)$ can be evolved even if we know nothing of $A(\vec{k}_2)$. So the Fourier transform of Eq. (4.56) produces a set of ordinary differential equations for the Fourier modes, and this set of equations is uncoupled. Instead of solving an infinite number of coupled equations, we can solve for one k-mode at a time.

Note that this simplification arises because the perturbations are small (equivalently the equations are linear). In this case, the different Fourier modes all evolve independently. Perturbations to the CMB remain small at all cosmological epochs, so Fourier transforms are very useful. In contrast, perturbations to matter are more complicated. Initially they are small, and they remain small until relatively recently.

The largest scales today are still in the linear regime, so Fourier transforming is certainly useful for the matter perturbations as well. However, to completely characterize the matter field today requires accounting for nonlinearities, and for this purpose, Fourier transforms lose much of their appeal. Different Fourier modes couple when nonlinear behavior becomes important, so the codes which follow matter perturbations all the way until today work in real space. Even these codes, however, start at $z \sim 20$ with the initial conditions set by linear evolution.

Our Fourier convention will be

$$\Theta(\vec{x}) = \int \frac{d^3k}{(2\pi)^3} e^{i\vec{k}\cdot\vec{x}} \tilde{\Theta}(\vec{k}). \tag{4.59}$$

We will often characterize a mode by the magnitude of its wavevector[4] : $k = \sqrt{k^i k^i}$.

Before rewriting Eq. (4.56) in terms of Fourier modes, let us make two final definitions. First, define the cosine of the angle between the wavenumber $\vec{k}$ and the photon direction $\hat{p}$ to be

$$\mu \equiv \frac{\vec{k} \cdot \hat{p}}{k}. \tag{4.60}$$

From now on, μ will be the variable describing the direction of photon propagation. A good way to think of μ is to go back to Figure 4.3. The wavevector $\vec{k}$ is pointing in the direction in which the temperature is changing, so it is perpendicular to the gradient ($\vec{k}$ is horizontal in the figure). When $\mu = 1$ then the photon direction is aligned with $\vec{k}$, so the photon is traveling in the direction along which the temperature is changing. A photon traveling in a direction in which the temperature remains the same (vertically in the figure) has $\mu = 0$. We will typically assume that the velocity points in the same direction as $\vec{k}$ (this is equivalent to saying that the velocity is irrotational), so $\vec{v}_b \cdot \hat{p} = \tilde{v}_b \mu$. Next, we define the optical depth

$$\tau(\eta) \equiv \int_\eta^{\eta_0} d\eta' \, n_e \sigma_T a. \tag{4.61}$$

At late times, the free electron density is small, so $\tau \ll 1$, while at early times, it is very large. Note that I have defined the limits of integration in such a way that

$$\dot{\tau} \equiv \frac{d\tau}{d\eta} = -n_e \sigma_T a. \tag{4.62}$$

With these definitions, we are finally left with

$$\dot{\tilde{\Theta}} + ik\mu\tilde{\Theta} + \dot{\tilde{\Phi}} + ik\mu\tilde{\Psi} = -\dot{\tau}\left[\tilde{\Theta}_0 - \tilde{\Theta} + \mu\tilde{v}_b\right]. \tag{4.63}$$

---

[4]Note that k^i is a 3D vector in Eudclidean space so that $k_i = k^i$; you do not need a factor of g_{ij} to go back and forth. The same goes for the velocity v_b^i.

4.5 THE BOLTZMANN EQUATION FOR COLD DARK MATTER

We can apply the formalism developed in the previous sections to derive the Boltzmann equation for any other constituent in the universe. Of particular importance is the evolution of the dark matter. In almost all currently popular models of structure formation, dark matter plays an important role in structure formation and in determining the gravitational field in the universe.

It is perhaps simplest to derive the evolution equations for dark matter by imposing conservation of the energy–momentum tensor, as we did in Chapter 2 in the homogeneous case. Unlike the photons, the dark matter always behaves like a fluid so can always be described completely by $T_{\mu\nu}$. Nonetheless, here we will sacrifice simplicity and use the Boltzmann formalism to derive the dark matter equations. This will (i) reinforce the calculations of the previous sections and also (ii) pave the way for the electron/proton equations of the next section.

There are several ways in which the dark matter distribution differs from that of the photons. First, by definition, "dark" matter does not interact with any of the other constituents in the universe. Thus we need not deal with any collision terms. Second, cold dark matter, in contrast to the photons, is nonrelativistic. So we need to redo some of the kinematics which led to the left side of the Boltzmann equation. In particular, the constraint Eq. (4.11) now becomes

$$g_{\mu\nu}P^\mu P^\nu = -m^2 \tag{4.64}$$

where m is the mass of the dark matter particle. It is also useful to define the energy as

$$E \equiv \sqrt{p^2 + m^2}, \tag{4.65}$$

where p is defined exactly as in Eq. (4.13): $p^2 = g_{ij}P^i P^j$. In the massless case, of course, Eq. (4.65) says that $E = p$, so E is superfluous. Here it will be convenient to let E replace p as one of the variables on which the distribution function depends (in addition to position $\vec{x}$, time t, and the direction vector $\hat{p}$). We can now derive the equivalent of equations (4.14) and (4.19) for the four-momentum of a massive particle:

$$P^\mu = \left[E(1 - \Psi), p\hat{p}^i \frac{1 - \Phi}{a} \right]. \tag{4.66}$$

Only the time component is different from that of a massless particle, with E replacing p.

Using E as one of the dependent variables means that the total time derivative of the dark matter distribution function $f_{\rm dm}$ is

$$\frac{df_{\rm dm}}{dt} = \frac{\partial f_{\rm dm}}{\partial t} + \frac{\partial f_{\rm dm}}{\partial x^i}\frac{dx^i}{dt} + \frac{\partial f_{\rm dm}}{\partial E}\frac{dE}{dt} + \frac{\partial f_{\rm dm}}{\partial \hat{p}^i}\frac{d\hat{p}^i}{dt}. \tag{4.67}$$

Once again, the last term here vanishes since it is the product of two first-order terms. Because of the change in the constraint equation, the coefficients of the derivatives of the distribution function with respect to x^i and E are slightly different than they were in the massless case. Working through the algebra, which is otherwise identical to the calculation presented in Section 4.2, leads to the collisionless Boltzmann equation for nonrelativistic matter:

$$\frac{\partial f_{\rm dm}}{\partial t} + \frac{\hat{p}^i}{a}\frac{p}{E}\frac{\partial f_{\rm dm}}{\partial x^i} - \frac{\partial f_{\rm dm}}{\partial E}\left[\frac{da/dt}{a}\frac{p^2}{E} + \frac{p^2}{E}\frac{\partial \Phi}{\partial t} + \frac{\hat{p}^i p}{a}\frac{\partial \Psi}{\partial x^i}\right] = 0. \qquad (4.68)$$

Equation (4.68) reduces to Eq. (4.33) in the massless limit as it must. The main difference between the two is the presence of factors of p/E, or velocity. For dark matter particles, these velocity factors suppress any free streaming, as we will shortly see.

In the massless case, to proceed further we used our knowledge of the distribution function. Namely, we knew that the zero-order distribution function was Bose–Einstein, and we perturbed around this zero-order solution. For cold dark matter particles, we do not need such detailed information about the zero-order distribution function. All we need to know is that these particles are very nonrelativistic. So we can neglect the thermal motion of the dark matter (Exercise 9). We cannot however neglect p/m completely, because the density perturbations themselves induce velocity flows in the dark matter via the continuity equation. These coherent flows give rise to $p/m \sim v$ terms, which must be retained. What we can do, however, in our linear treatment, is to neglect terms second-order in p/E.

Instead of assuming a form for $f_{\rm dm}$, we will take moments of Eq. (4.68). First, multiply both sides by the phase space volume $d^3p/(2\pi)^3$ and integrate. This leads to

$$\frac{\partial}{\partial t}\int\frac{d^3p}{(2\pi)^3}f_{\rm dm} + \frac{1}{a}\frac{\partial}{\partial x^i}\int\frac{d^3p}{(2\pi)^3}f_{\rm dm}\frac{p\hat{p}^i}{E} - \left[\frac{da/dt}{a} + \frac{\partial \Phi}{\partial t}\right]\int\frac{d^3p}{(2\pi)^3}\frac{\partial f_{\rm dm}}{\partial E}\frac{p^2}{E}$$

$$-\frac{1}{a}\frac{\partial \Psi}{\partial x^i}\int\frac{d^3p}{(2\pi)^3}\frac{\partial f_{\rm dm}}{\partial E}\hat{p}^i p = 0. \qquad (4.69)$$

Note that, since they are independent variables, the integral over p passes through the partial derivatives with respect to x^i and t. The last term here can be neglected since the integral over the direction vector is nonzero only for the perturbed part of $f_{\rm dm}$. Thus the integral is first order and it multiplies the first-order term $\partial \Psi/\partial x^i$. The rest of the terms are all relevant, though. To simplify, let us recall that the dark matter density is[5]

$$n_{\rm dm} = \int\frac{d^3p}{(2\pi)^3}f_{\rm dm} \qquad (4.70)$$

while the velocity is defined as

$$v^i \equiv \frac{1}{n_{\rm dm}}\int\frac{d^3p}{(2\pi)^3}f_{\rm dm}\frac{p\hat{p}^i}{E}. \qquad (4.71)$$

---

[5]Here I have incorporated the spin degeneracy $g_{\rm dm}$ into the phase space distribution $f_{\rm dm}$. We implicitly did the same thing in the last section for the electrons.

The first two terms in Eq. (4.69), then, can be simply expressed in terms of the velocity and the density. The third term is a bit more subtle; to relate it to the density, we need to integrate by parts. Since $dE/dp = p/E$, the integrand can be reexpressed as $p\,\partial f_{\rm dm}/\partial p$. Thus, the integral becomes

$$\int \frac{d^3p}{(2\pi)^3} p \frac{\partial f_{\rm dm}}{\partial p} = \frac{4\pi}{(2\pi)^3} \int_0^\infty dp\, p^3 \frac{\partial f_{\rm dm}}{\partial p}$$

$$= -3\frac{4\pi}{(2\pi)^3} \int_0^\infty dp\, p^2 f_{\rm dm}$$

$$= -3n_{\rm dm}. \qquad (4.72)$$

So the zeroth moment of the Boltzmann equation leads to the cosmological generalization of the continuity equation:

$$\frac{\partial n_{\rm dm}}{\partial t} + \frac{1}{a}\frac{\partial(n_{\rm dm}v^i)}{\partial x^i} + 3\left[\frac{da/dt}{a} + \frac{\partial\Phi}{\partial t}\right] n_{\rm dm} = 0. \qquad (4.73)$$

The first two terms here are the standard continuity equation from fluid mechanics. The last term arises due to the FRW metric and its perturbations.

To go further, we can collect zero-order and first-order terms in Eq. (4.73). The velocity is first order as is Φ, so the only zero-order terms are

$$\frac{\partial n_{\rm dm}^{(0)}}{\partial t} + 3\frac{da/dt}{a} n_{\rm dm}^{(0)} = 0 \qquad (4.74)$$

where $n_{\rm dm}^{(0)}$ is the zero-order, homogeneous part of the density. Equivalently, we have

$$\frac{d(n_{\rm dm}^{(0)}a^3)}{dt} = 0 \implies n_{\rm dm}^{(0)} \propto a^{-3}, \qquad (4.75)$$

a relation we anticipated back in Chapter 1 as an obvious ramification of the expansion. We also proved this scaling in Chapter 2 by using the conservation of the energy momentum tensor.

Now let us extract the first-order part of Eq. (4.73). All factors of $n_{\rm dm}$ multiplying the first-order quantities v and Φ may be set to $n_{\rm dm}^{(0)}$. Everywhere else, we need to expand $n_{\rm dm}$ out to include a first-order perturbation. In particular, we will set

$$n_{\rm dm} = n_{\rm dm}^{(0)}\left[1 + \delta(\vec{x}, t)\right], \qquad (4.76)$$

which defines the first-order piece as $n_{\rm dm}^{(0)}\delta$. Since the energy density of matter is equal to mass times n, δ is also the fractional overdensity, $\delta\rho/\rho$, of the dark matter. After dividing by $n_{\rm dm}^{(0)}$, the first-order equation is therefore

$$\frac{\partial\delta}{\partial t} + \frac{1}{a}\frac{\partial v^i}{\partial x^i} + 3\frac{\partial\Phi}{\partial t} = 0. \qquad (4.77)$$

As it stands, we have introduced two new perturbation variables for the dark matter, the density perturbation δ and the velocity $\vec{v}$. Equation (4.77) is only one equation, though, for these two variables. We need another. To get it, we return to the unintegrated Boltzmann equation (4.68). We have just taken its zeroth moment; to extract a second equation, let us take its first moment. In particular, multiply Eq. (4.68) by $d^3p(p/E)\hat{p}^j/(2\pi)^3$ and then integrate. The first moment equation is then

$$0 = \frac{\partial}{\partial t} \int \frac{d^3p}{(2\pi)^3} f_{\rm dm} \frac{p\hat{p}^j}{E} + \frac{1}{a}\frac{\partial}{\partial x^i} \int \frac{d^3p}{(2\pi)^3} f_{\rm dm} \frac{p^2\hat{p}^i\hat{p}^j}{E^2}$$

$$- \left[\frac{da/dt}{a} + \frac{\partial\Phi}{\partial t}\right] \int \frac{d^3p}{(2\pi)^3} \frac{\partial f_{\rm dm}}{\partial E} \frac{p^3\hat{p}^j}{E^2} - \frac{1}{a}\frac{\partial\Psi}{\partial x^i} \int \frac{d^3p}{(2\pi)^3} \frac{\partial f_{\rm dm}}{\partial E} \frac{\hat{p}^i\hat{p}^j p^2}{E}. (4.78)$$

The first two terms are straightforward: the first is the time derivative of $n_{\rm dm}v^i$ while the second can be safely neglected since it is of order $\langle (p/E)^2 \rangle$. The last sets of terms must be handled more carefully, though, because of the partial derivatives. Since $(p/E)\partial/\partial E = \partial/\partial p$ the third term is actually of order p/E while the last is independent of velocity. Let us do the integration by parts explicitly in the third term. The integral is:

$$\int \frac{d^3p}{(2\pi)^3} \frac{\partial f_{\rm dm}}{\partial p} \frac{p^2\hat{p}^j}{E} = \int \frac{d\Omega\,\hat{p}^j}{(2\pi)^3} \int_0^\infty dp \frac{p^4}{E} \frac{\partial f_{\rm dm}}{\partial p}$$

$$= - \int \frac{d\Omega\,\hat{p}^j}{(2\pi)^3} \int_0^\infty dp f_{\rm dm} \left(\frac{4p^3}{E} - \frac{p^5}{E^3}\right). \qquad (4.79)$$

The p^5/E^3 term is completely negligible, so the only relevant contribution to the integral comes from the $-4p^3/E$ term: its integral is $-4n_{\rm dm}v^j$. The same steps carry through for the last term in Eq. (4.78); the one additional fact we need is that

$$\int d\Omega\,\hat{p}^i\hat{p}^j = \delta^{ij}\frac{4\pi}{3}. \qquad (4.80)$$

So the first moment of the Boltzmann equation is

$$\frac{\partial(n_{\rm dm}v^j)}{\partial t} + 4\frac{da/dt}{a}n_{\rm dm}v^j + \frac{n_{\rm dm}}{a}\frac{\partial\Psi}{\partial x^j} = 0. \qquad (4.81)$$

This equation has no zero-order parts, since the velocity is a first-order quantity. Therefore, we need extract only the first-order terms, which allows us to set $n_{\rm dm} \to n_{\rm dm}^{(0)}$ everywhere. Using the time dependence we found in Eq. (4.75) we arrive at

$$\frac{\partial v^j}{\partial t} + \frac{da/dt}{a}v^j + \frac{1}{a}\frac{\partial\Psi}{\partial x^j} = 0. \qquad (4.82)$$

Equations (4.77) and (4.82) are the two equations governing the evolution of the density and the velocity of the cold, dark matter. The momentum conservation equation (4.82) does not have the standard $(\vec{v}\cdot\nabla)\vec{v}$ term, since any term with two factors of v is manifestly second order. An interesting feature of the two equations is generic to this process of integrating the Boltzmann equations to get the fluid equations. Note that the equation for the density depends on the next highest moment, the velocity. This is general: the integrated Boltzmann equation for the lth moment depends on the $l + 1$ moment. In principle, then, this process of integrating leads to an infinite heirarchy of equations for the moments of the distribution function. Indeed, we will see that this is one way of solving the Boltzmann equation for the photons, Eq. (4.63), which we have not yet integrated over. One might expect, then, that the velocity equation would depend on the next highest moment, the quadrupole, of the dark matter distribution. Why doesn't it? The answer lies in our assumption that the dark matter is *cold*. We have explicitly dropped all terms of order $(p/E)^2$ and higher. These terms correspond to the higher moments of the distribution, but since we are dealing with cold, dark matter they are irrelevant. Thus, the set of two equations, (4.77) and (4.82), are a closed set of equations for the cold, dark matter distribution.[6] If we were interested in dark matter particles with much smaller masses, such as massive neutrinos, we would need to keep these higher moments.

Let us finally rewrite Eqs. (4.77) and (4.82) in terms of conformal time η and the Fourier transforms. The density equation becomes

$$\dot{\tilde{\delta}} + ik\tilde{v} + 3\dot{\tilde{\Phi}} = 0 \tag{4.83}$$

where I have assumed that the velocity is irrotational so $\tilde{v}^i = (k^i/k)\tilde{v}$. The velocity equation is

$$\dot{\tilde{v}} + \frac{\dot{a}}{a}\tilde{v} + ik\tilde{\Psi} = 0. \tag{4.84}$$

4.6 THE BOLTZMANN EQUATION FOR BARYONS

The final components of the universe which require a set of Boltzmann equations are the electrons and protons. These components are often grouped together and called *baryons*, nomenclature which is obviously ridiculous (electrons are leptons, not baryons) but nonetheless common.

Electrons and protons are coupled by Coulomb scattering ($e + p \rightarrow e + p$). The Coulomb scattering rate is much larger than the expansion rate at all epochs of interest (Exercise 12). This tight coupling forces the electron and proton overdensities to a common value:

$$\frac{\rho_e - \rho_e^{(0)}}{\rho_e^{(0)}} = \frac{\rho_p - \rho_p^{(0)}}{\rho_p^{(0)}} \equiv \delta_b \tag{4.85}$$

where we bow to common usage with the subscript b. Similarly the velocities of the two species are forced to a common value,

---

[6]Of course, we still need equations for the gravitational potentials Φ and Ψ. These come from Einstein's equations, as does the zero-order equation for a.

$$\vec{v}_e = \vec{v}_p \equiv \vec{v}_{\rm b}. \tag{4.86}$$

We need to derive equations then for δ_b and $v_{\rm b}$. The starting point will be the unintegrated equations for electrons and protons:

$$\frac{df_e(\vec{x}, \vec{q}, t)}{dt} = \langle c_{ep} \rangle_{QQ'q'} + \langle c_{e\gamma} \rangle_{pp'q'} \tag{4.87}$$

$$\frac{df_p(\vec{x}, \vec{Q}, t)}{dt} = \langle c_{ep} \rangle_{qq'Q'}. \tag{4.88}$$

The notation here is more compact, and therefore more deceiving, than that in previous sections. We will need this compactness in what follows, so let's walk through it slowly. First, notice that initial and final momenta for the photon are $\vec{p}$ and $\vec{p}'$; for electron $\vec{q}$ and $\vec{q}'$; and the proton has been assigned $\vec{Q}$ and $\vec{Q}'$. Consider the Compton collision term in the equation for the electron distribution function. I have defined the unintegrated part of the collision term as

$$c_{e\gamma} \equiv (2\pi)^4 \delta^4(p + q - p' - q') \frac{|\mathcal{M}|^2}{8E(p)E(p')E_e(q)E_e(q')} \{ f_e(q') f_\gamma(p') - f_e(q) f_\gamma(p) \} \tag{4.89}$$

and the angular brackets denote integration over all momenta in the subscripts:

$$\langle (\ldots) \rangle_{pp'q'} \equiv \int \frac{d^3p}{(2\pi)^3} \int \frac{d^3p'}{(2\pi)^3} \int \frac{d^3q'}{(2\pi)^3} (\ldots). \tag{4.90}$$

The Coulomb collision term is similar, the main difference being the amplitude for the two processes.

In principle, Eq. (4.88) should contain a term accounting for scattering of protons off photons. In practice, though, the cross section for this process is much smaller than for Compton scattering off electrons (in each case the cross section is inversely proportional to the mass squared). So the interactions of the combined electron–proton fluid with the photons is driven by Compton scattering of electrons, and the proton–photon process can be ignored. Also, in principle, we should include ionization and recombination terms in Eqs. (4.87) and (4.88). These however would merely distract us here, so we treat all electrons as ionized.

With this notation defined, we can now proceed and derive equations for δ_b and $v_{\rm b}$. First, multiply both sides of Eq. (4.87) by the phase space volume $d^3q/(2\pi)^3$ and integrate. The left-hand side then becomes identical to the left-hand side we derived for dark matter in Eq. (4.73). So we can immediately write

$$\frac{\partial n_e}{\partial t} + \frac{1}{a}\frac{\partial (n_e v_{\rm b}^i)}{\partial x^i} + 3\left[\frac{da/dt}{a} + \frac{\partial \Phi}{\partial t} \right] n_e = \langle c_{ep} \rangle_{QQ'q'q} + \langle c_{e\gamma} \rangle_{pp'q'q}. \tag{4.91}$$

Both terms on the right vanish. The mathematical way to see this is to realize that the integration measure in the first term on the right, e.g., is completely symmetric under the interchange of $Q \leftrightarrow Q'$ and $q \leftrightarrow q'$. Because of the factors of the distribution function, the integrand — c_{ep} — is antisymmetric under this interchange. So the full integral vanishes. More intuitively, the processes we are considering conserve electron number so they certainly cannot contribute to dn/dt. That is, the integral over $f_e(q')f_p(Q')$ counts the total number of electrons that are produced in Coulomb scattering. But this is obviously equal to the integral over $f_e(q)f_p(Q)$, which counts the number of electrons lost in Coulomb scattering. More generally, any time we multiply an unintegrated collision term by a conserved quantity and then integrate we will get zero.

The perturbed version of Eq. (4.91) equation is therefore identical to Eq. (4.77). Switching to Fourier space and using conformal time leads to

$$\dot{\tilde{\delta}}_b + ik\tilde{v}_b + 3\dot{\tilde{\Phi}} = 0. \tag{4.92}$$

The second equation for the baryons is obtained by taking the first moments of both Eqs. (4.87) and (4.88) and adding them together. We did something similar for the dark matter; there we first multiplied by $\vec{p}/E$ and then integrated over all momenta. Here we will take the moments by first multiplying the unintegrated equations by $\vec{q}$ (and $\vec{Q}$ for the protons) instead of by $\vec{q}/E$. Therefore, our results from the dark matter case carry over as long as we multiply them by a factor of m. The left-hand side of the integrated electron equation, for example, will look exactly like the left-hand side of Eq. (4.81) except it will be multiplied by m_e. The proton equation will be multiplied by m_p. Since the proton mass is so much larger than the electron mass, the sum of the two left-hand sides will be dominated by the protons. So, following Eq. (4.81), we have

$$m_p \frac{\partial(n_b v_b^j)}{\partial t} + 4\frac{da/dt}{a} m_p n_b v_b^j + \frac{m_p n_b}{a}\frac{\partial\Psi}{\partial x^i}$$
$$= \langle c_{ep}(q^j + Q^j)\rangle_{QQ'q'q} + \langle c_{e\gamma}q^j\rangle_{pp'q'q}. \tag{4.93}$$

The right-hand side here is the sum of that from both the electron and proton equations. Both equations have the Coulomb term, so it is weighted by $\vec{q}$ (by which we multiplied the electron equation) $+\vec{Q}$ (from the proton equation). Only the electron equation has the Compton term, so there is only the factor of $\vec{q}$ there. Once again we can use a conservation law, this time conservation of momentum, to argue that the integral of $c_{ep}(\vec{q} + \vec{Q})$ over all momenta vanishes. So dividing both sides by[7] $\rho_b = m_p n_b^{(0)}$, we are left with

$$\frac{\partial v_b^j}{\partial t} + \frac{da/dt}{a}v_b^j + \frac{1}{a}\frac{\partial\Psi}{\partial x^j} = \frac{1}{\rho_b}\langle c_{e\gamma}q^j\rangle_{pp'q'q}. \tag{4.94}$$

Here I have used the by now familiar $n_b^{(0)} \propto a^{-3}$ scaling to eliminate the $n_b^{(0)}$ time derivative and three of the four factors of the da/dt term on the left.

---

[7]Note that convention here, which I will stick with, that ρ_b is the zero order baryon energy density. The total baryon density is therefore $\rho_b(1 + \delta_b)$.

The final step is to evaluate the average momentum $\vec{q}$ in Compton scattering. As before, we can use the conservation of total momentum $\vec{q} + \vec{p}$ to argue that

$$\langle c_{e\gamma} \vec{q} \rangle_{pp'q'q} = -\langle c_{e\gamma} \vec{p} \rangle_{pp'q'q}. \tag{4.95}$$

Now switch to Fourier space and multiply both sides of Eq. (4.94) by $\hat{k}^j$. Since $\hat{k} \cdot \vec{p} = p\mu$, the right-hand side of Eq. (4.94) becomes $-\langle c_{e\gamma} p\mu \rangle_{pp'q'q} / \rho_b$. We have already computed $\langle c_{e\gamma} \rangle_{p'q'q}$ in Eq. (4.54). We need simply multiply this by $p\mu$ and integrate over all $\vec{p}$ to find the right-hand side of Eq. (4.94):

$$-\frac{\langle c_{e\gamma} p\mu \rangle_{pp'q'q}}{\rho_b} = \frac{n_e \sigma_T}{\rho_b} \int \frac{d^3p}{(2\pi)^3} p^2 \frac{\partial f^{(0)}}{\partial p} \mu \left[\tilde{\Theta}_0 - \tilde{\Theta}(\mu) + \tilde{v}_b \mu \right]$$

$$= \frac{n_e \sigma_T}{\rho_b} \int_0^\infty \frac{dp}{2\pi^2} p^4 \frac{\partial f^{(0)}}{\partial p} \int_{-1}^1 \frac{d\mu}{2} \mu \left[\tilde{\Theta}_0 - \tilde{\Theta}(\mu) + \tilde{v}_b \mu \right]. \tag{4.96}$$

The integral over p can be done by integrating by parts: it is $-4\rho_\gamma$. The μ-integration over the first and third terms is straightforward (first term vanishes and second gives $v_b/3$). The second term is the first moment of the perturbation Θ. Recall that the zeroth moment was defined as Θ_0. It makes sense therefore to define the first moment as

$$\Theta_1 \equiv i \int_{-1}^1 \frac{d\mu}{2} \mu \Theta(\mu) \tag{4.97}$$

where the factor of i is a convention and the definition holds in either real or Fourier space.

We now have an expression for the collision term which can be inserted into Eq. (4.94), and after switching to conformal time, we have:

$$\dot{\tilde{v}}_b + \frac{\dot{a}}{a} \tilde{v}_b + ik\tilde{\Psi} = \dot{\tau} \frac{4\rho_\gamma}{3\rho_b} \left[3i\tilde{\Theta}_1 + \tilde{v}_b \right]. \tag{4.98}$$

Why is there a factor of ρ_b in the denominator? That is, since photons scatter primarily off electrons, why does the total baryon density (which is dominated by protons) appear in this velocity equation? Physically, it arises from the fact that moving electrons is difficult because they are tightly coupled to protons via Coulomb scattering. If the proton was infinitely heavy, so $\rho_b \to \infty$, Compton scattering would not change the electron velocity at all; it would not have any impact on the combined proton–electron fluid. We derived Eq. (4.98) by setting $n_e = n_p = n_b$, but it turns out to be valid even if there is an appreciable amount of neutral hydrogen, so that $n_e \neq n_b$. Indeed after recombination, most protons are bound in neutral hydrogen atoms. And even before recombination, a small fraction are in helium atoms or ions. You might be tempted therefore to replace ρ_b in the denominator of Eq. (4.98) by the density of free protons. In fact, though, even neutral hydrogen and helium are tightly coupled to electrons and protons (see Exercise 12), so all baryons should be included. Equation (4.98) quite generally governs the evolution of the baryon velocity.

4.7 SUMMARY

The constituents of the universe are not distributed completely uniformly in space. For the nonrelativistic components such as the dark matter and the baryons, this means that some regions are more dense than others and that there are small coherent velocities. For the dark matter we denote the fractional overdensity as $\delta(\vec{x}, t)$ and the velocity as $\vec{v}(\vec{x}, t)$. The equivalent perturbations for the baryons are $\delta_{\mathrm{b}}(\vec{x}, t)$ and $\vec{v}_{\mathrm{b}}(\vec{x}, t)$. In solving the linear evolution equations, it is simplest to work with Fourier transforms of all of these. It turns out that the evolution of a mode associated with wavevector $\vec{k}$ depends only on the magnitude of $\vec{k}$, so we have equations for $\tilde{\delta}(k, t)$. We have found it convenient to use conformal time η as the evolution variable. Also, it is conventional[8] in the literature to drop the ˜s over Fourier transformed variables, so our equations will be for $\delta(k, \eta), \delta_{\mathrm{b}}(k, \eta), v(k, \eta)$, and $v_{\mathrm{b}}(k, \eta)$. The scalar velocities here are the components parallel to $\vec{k}$; these are the only ones that are cosmologically relevant.

Relativistic particles such as photons and neutrinos require more information to characterize. They have not only a monopole perturbation (the equivalent of δ) and a dipole (the equivalent of a velocity), but also a quadrupole, octopole, and higher moments as well. In other words, the photon distribution depends not only on $\vec{x}$ and time but also on the direction of propagation of the photon, $\hat{p}$. In Fourier space, therefore, the photon perturbations depend not only on k and η but also on $\hat{p} \cdot \hat{k}$, which we defined as μ. Thus, the photon perturbation variable is $\Theta(k, \mu, \eta)$, the Fourier transform of $\delta T/T$, the fractional temperature difference. Neutrino perturbations require a separate variable with the same dependence; let's call it $\mathcal{N}(k, \mu, \eta)$.

We found it useful to define the monopole (Eq. (4.52)) and dipole (Eq. (4.97)) of the photon distribution. These moments, $\Theta_0(k, \eta)$ and $\Theta_1(k, \eta)$, do not completely characterize the photon distribution. More generally, it is useful to define the lth multipole moment of the temperature field as

$$\Theta_l \equiv \frac{1}{(-i)^l} \int_{-1}^{1} \frac{d\mu}{2} \mathcal{P}_l(\mu) \Theta(\mu), \qquad (4.99)$$

where $\mathcal{P}_l$ is the Legendre polynomial of order l. The quadrupole corresponds to $l = 2$, octopole to $l = 3$, etc. The higher Legendre polynomials have structure on smaller scales (see Figure 4.4), so the higher moments capture information about the small scale structure of the temperature field. So the photon perturbations can be described either by $\Theta(k, \mu, \eta)$ or by a whole hierarchy of moments, $\Theta_l(k, \eta)$. And of course similar freedom applies to the neutrino distribution.

I have postponed a discussion of polarization until Chapter 10, but I mentioned in Section 4.3 that a completely accurate treatment of anisotropies in the temperature requires us to incorporate polarization effects. Again, waiting until Chapter 10 for more formal definitions, let's call the strength of the polarization Θ_P. It

---

[8]Conventional, but "abominable" according to one early reviewer.

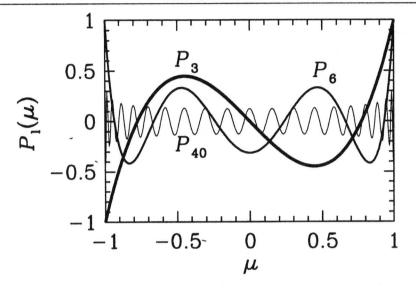

Figure 4.4. Some Legendre polynomials. Note that the higher order ones vary on smaller scales than do the low-order ones. In general $\mathcal{P}_l$ crosses zero l times between -1 and 1.

describes the change in the polarization field in space. Upon Fourier transforming, it too depends on $k, \mu,$ and η.

We now collect the equations we have derived for the photons, dark matter, and baryons and supplement them with a trivial extension to massless neutrinos:

$$\dot{\Theta} + ik\mu\Theta = -\dot{\Phi} - ik\mu\Psi - \dot{\tau}\left[\Theta_0 - \Theta + \mu v_{\mathrm{b}} - \frac{1}{2}\mathcal{P}_2(\mu)\Pi\right] \qquad (4.100)$$

$$\Pi = \Theta_2 + \Theta_{P2} + \Theta_{P0} \qquad (4.101)$$

$$\dot{\Theta}_P + ik\mu\Theta_P = -\dot{\tau}\left[-\Theta_P + \frac{1}{2}(1 - \mathcal{P}_2(\mu))\Pi\right] \qquad (4.102)$$

$$\dot{\delta} + ikv = -3\dot{\Phi} \qquad (4.103)$$

$$\dot{v} + \frac{\dot{a}}{a}v = -ik\Psi \qquad (4.104)$$

$$\dot{\delta}_{\mathrm{b}} + ikv_{\mathrm{b}} = -3\dot{\Phi} \qquad (4.105)$$

$$\dot{v}_{\mathrm{b}} + \frac{\dot{a}}{a}v_{\mathrm{b}} = -ik\Psi + \frac{\dot{\tau}}{R}\left[v_{\mathrm{b}} + 3i\Theta_1\right] \qquad (4.106)$$

$$\dot{\mathcal{N}} + ik\mu\mathcal{N} = -\dot{\Phi} - ik\mu\Psi. \qquad (4.107)$$

Equation (4.100) is the Boltzmann equation for photons we have derived. The one change from our derivation is the last term $\mathcal{P}_2\Pi/2$, which requires some explanation. First, note that it is proportional to the second Legendre polynomial, $\mathcal{P}_2(\mu) = (3\mu^2 - 1)/2$. From Eq. (4.101), one of the new terms then is $\mathcal{P}_2\Theta_2/2$; this term accounts for the angular dependence of Compton scattering, which we ignored in Section 4.3. The other parts of Π represent the fact that the temperature field is also coupled to the strength of the polarization field Θ_P which obeys Eq. (4.102). Note that Θ_P is sourced by the quadrupole, Θ_2, and none of the other temperature moments.

In the equation for the baryon velocity (4.106), the ratio of photon to baryon density has been defined as

$$\frac{1}{R} \equiv \frac{4\rho_\gamma^{(0)}}{3\rho_b^{(0)}}. \tag{4.108}$$

Equation (4.107) governs perturbations to the neutrino distribution, $\mathcal{N}$. It is identical to the photon equation except that there is no scattering term since neutrinos interact only very weakly. Here I have assumed that the neutrinos are massless. If any of the neutrinos had appreciable mass, then Eq. (4.107) would have to be amended to account for this. Exercise 11 discusses the question of how large a mass is interesting.

SUGGESTED READING

In the 1960s a national magazine ran a cartoon showing dozens of businessmen and -women walking the streets of Manhattan looking very important and serious. Thought bubbles over each head revealed their true focus: each was imagining a raucus sex scene. In at least some ways, the Boltzmann equation plays a similar role for physicists and astronomers: no one ever talks about it, but everyone is always thinking about it.

Two excellent astronomy textbooks which do make abundant use of the Boltzmann equation — either explicitly or implicitly — are *Radiative Processes in Astrophysics* (Rybicki and Lightman) and *Galactic Dynamics* (Binney and Tremaine). In the context of cosmology, in addition to the books mentioned in Chapter 1, *The Large Scale Structure of the Universe* (Peebles), written by the field's pioneer, uses the Boltzmann equation extensively, working in synchronous gauge. If you struggled through Section 4.3, you will be amused (angered?) to see §92 of Peebles' book, where he takes much less space to derive terms due to Compton scattering.

A number of papers deriving the Boltzmann equation for cosmological perturbations are well worth reading. There is the path-breaking work by Lifshitz (1946), Peebles and Yu (1970), and Bond and Szalay (1983). A nice review was written by Efstathiou (1990). The treatment of Compton scattering presented here is based on Dodelson and Jubas (1995). If you were to read just one paper in this area, I would recommend Ma and Bertschinger (1995), which skips many of the steps presented here but has all the relevant formulae and the added virtue of equations in both conformal Newtonian and synchronous gauges. For derivation of the polarization terms in the Boltzmann equations, see Kosowsky (1996). The first paper to present the Boltzmann equation for tensors was Crittenden *et al.* (1993).

We will not spend too much time in this book on different gauges or on the decomposition of perturbations into scalar, vector, and tensor parts. Two excellent review articles which discuss both of these topics in detail are Mukhanov, Feldman, and Brandenberger (1992) and Kodama and Sasaki (1984). Both of these are also very good on the subjects of the next two chapters, the perturbed Einstein equations and inflation.

EXERCISES

Exercise 1. Derive the fluid equations for the collisionless, one-dimensional harmonic oscillator by taking the moments of Eq. (4.6). The relevant quantities are the number density and the velocity defined as integrals over the distribution function:

$$n \equiv \int_{-\infty}^{\infty} \frac{dp}{2\pi} f \qquad ; \qquad v \equiv \frac{1}{n} \int_{-\infty}^{\infty} \frac{dp}{2\pi} \frac{p}{m} f. \qquad (4.109)$$

Exercise 2. The metric in a synchronous gauge is

$$g_{00}(\vec{x}, t) = -1$$

$$g_{0i}(\vec{x}, t) = 0$$

$$g_{ij}(\vec{x}, t) = a^2 \left[\delta_{ij} + h_{ij} \right],$$ (4.110)

with perturbations

$$h_{ij} = \begin{pmatrix} -2\tilde{\eta} & 0 & 0 \\ 0 & -2\tilde{\eta} & 0 \\ 0 & 0 & h + 4\tilde{\eta} \end{pmatrix}$$ (4.111)

where $\tilde{\eta}$ has nothing to do with conformal time. Here I have chosen the wavevector $\vec{k}$ to lie in the $\hat{z}$ direction. Derive the equivalent of Eq. (4.63) in synchronous gauge:

$$\dot{\Theta} + ik\mu\Theta - \frac{\mu^2 \dot{h}}{2} - \mathcal{P}_2(\mu)\dot{\tilde{\eta}} = -\dot{\tau} \left[\Theta_0 - \Theta + \mu v \right].$$ (4.112)

Exercise 3. Start from the zero-order unintegrated Boltzmann equation (4.37). Integrate this equation over all momenta to show that the number density falls off as a^3. In the course of this, you will have justified the left-hand side of Eq. (3.1).

Exercise 4. Show that the Pauli blocking factor $1 - f_e$ can be set to 1 for all epochs of interest. First find f_e as a function of temperature and number density using the results/approximations of Section 3.1 (i.e. assume that $T_e \ll m_e$). Then, show that as long as the temperature is much less than m_e, f_e is much less than 1.

Exercise 5. Suppose we started this chapter by writing

$$\frac{df}{d\lambda} = C'.$$ (4.113)

Change from this form to the one in Eq. (4.1) (with df/dt) on the left. How is the collision term here, C' related to C in Eq. (4.1)? Argue that the first-order perturbations in the factor relating the two collision terms can be dropped since the collision terms themselves are first-order.

Exercise 6. Derive Eq. (4.68), the unintegrated Boltzmann equation for a massive particle.

Exercise 7. Account for the angular dependence of Compton scattering. Start from Eq. (4.49) but instead of assuming the amplitude is constant, take

$$|\mathcal{M}|^2 = 6\pi\sigma_T m_e^2 (1 + \cos^2[\hat{p} \cdot \hat{p}\,']).$$

Show that correctly accounting for the angular dependence introduces the factor of $(1/2)\mathcal{P}_2(\mu)\Theta_2$ presented in Eq. (4.100).

Exercise 8. Show that the temperature of nonrelativistic matter scales as a^{-2} in the absence of interactions. Start from the zero-order part of Eq. (4.68) and assume a form $f_{\rm dm} \propto e^{-E/T} = e^{-p^2/2mT}$. Note that this argument does *not* apply to electrons and protons: as long as they are coupled to the photons, their temperature scales as a^{-1}.

Exercise 9. In Exercise 8, you showed that a thermal distribution of nonrelativistic particles which do not interact has a temperature which scales as a^{-2}, as opposed to that of relativistic particles which we have seen scales as a^{-1}. So $T_{\rm dm} \propto T^2$. Fix the normalization by requiring $T_{\rm dm} = T$ when each is equal to the dark matter mass. Estimate the typical thermal velocity of a dark matter particle with mass equal to 100 GeV when the photon temperature is 1 eV.

Exercise 10. The purpose of this problem is to derive the results of Section 2.3 using the Boltzmann equation. Multiply the zero-order part of Eq. (4.68) by $d^3pE(p)/(2\pi)^3$ and integrate. Show that the resulting equation is identical to Eq. (2.55).

Exercise 11. Consider the effect of a massive neutrino on the evolution equations.
(a) Start from the Boltzmann equation for a massive particle (4.68). Turn it into an equation for $\mathcal{N}$, the perturbation to the massive neutrino distribution function. Use the fact that to first order the neutrino distribution function is

$$f_\nu = f_\nu^{(0)} + \frac{\partial f_\nu^{(0)}}{\partial T_\nu} T_\nu \mathcal{N} \qquad (4.114)$$

where $f_\nu^{(0)} = [e^{p/T_\nu} + 1]^{-1}$. Express the final equation in Fourier space using conformal time as the evolution variable.
(b) Recent experiments measuring the atmospheric neutrino flux suggest that the mass of the tau neutrino is 0.07 eV, far larger than either the electron or muon neutrino. Find the contribution of a 0.07-eV neutrino to the energy density today. You may assume it is nonrelativistic.
(c) Consider the following two scenarios. Each has energy density equal to the critical density divided up between only two components: a cold, dark matter particle and a neutrino. The neutrino in each case has the standard abundance and temperature. The only difference between the two scenarios is in one the neutrino is massless while in the other it has a mass of 0.07 eV. Plot the energy density as a function of scale factor in each of the these scenarios. Note that they should agree very early on (in each case there is only a relativistic neutrino early on) and very late. The only difference comes in the middle.

Exercise 12. Show that ordinary matter is tightly coupled during the relevant epochs in the early universe.
(a) Compute the ratio of the Coulomb scattering rate to the Hubble rate. You may assume that all electrons and protons are ionized.
(b) Show that the rate for neutral hydrogen to scatter off ionized protons is always much larger than the expansion rate even when the ionization fraction is on the

order of 10^{-4}.

Exercise 13. Consider tensor perturbations to the metric. These do not perturb $g_{00}(=-1)$ or $g_{0i}(=0)$. However, the spatial part of the metric is now

$$g_{ij} = a^2 \begin{pmatrix} 1 + h_+ & h_\times & 0 \\ h_\times & 1 - h_+ & 0 \\ 0 & 0 & 1 \end{pmatrix}.$$

Derive the equation for the photon distribution function in the presence of tensor perturbations. Unlike scalar perturbations, tensor perturbations induce an azimuthal dependence in Θ_l, so decompose the anisotropy due to tensors into

$$\Theta^T(k, \mu, \phi) = \Theta_+^T(k, \mu)(1 - \mu^2)\cos(2\phi) + \Theta_\times^T(k, \mu)(1 - \mu^2)\sin(2\phi). \qquad (4.115)$$

Show that both the + and the × component satisfy

$$\frac{d\Theta_i^T}{d\eta} + ik\mu\Theta_i^T + \frac{1}{2}\frac{dh_i}{d\eta} = \dot{\tau}\left[\Theta_i^T - \frac{1}{10}\Theta_{i,0}^T - \frac{1}{7}\Theta_{i,2}^T - \frac{3}{70}\Theta_{i,4}^T\right] \qquad (4.116)$$

where i stands for either × or +, and the moments are defined as were the scalar moments, in Eq. (4.99).

5

EINSTEIN EQUATIONS

The previous chapter set up the formalism to describe how perturbations in the gravitational field affect particle distributions. This formalism led to the set of equations (4.100) -(4.107). We need to supplement these equations with an account of how the perturbations to the particle distributions affect the gravitational field. For this, we need the Einstein equations of general relativity. The calculation detailed in this chapter expands the Einstein equations perturbatively around the zero-order homogeneous solution. Far from being subtle or complex as one might expect from general relativity's reputation, this calculation is completely straightforward, although a bit long. Still, working through it is a "must-do-once" exercise, so the steps are presented in some detail.

5.1 THE PERTURBED RICCI TENSOR AND SCALAR

The fundamental equation of general relativity (2.30) is a 4D tensor equation, so in principle it represents 16 separate equations. However, since both sides of the equation are symmetric tensors, only 10 of these are distinct. We are interested though in only two of the 10, since the metric we are focusing on has only two independent functions, Φ and Ψ.

Evaluating the left-hand side of the Einstein equation requires three pre-steps:

- Compute the Christoffel symbols, $\Gamma^{\mu}{}_{\alpha\beta}$, for the perturbed metric of Eq. (4.9).
- From these, form the Ricci tensor, $R_{\mu\nu}$, using Eq. (2.31).
- Contract the Ricci tensor to form the Ricci scalar, $\mathcal{R} \equiv g^{\mu\nu}R_{\mu\nu}$.

Note that, unfortunately, even if we are interested in only several components of the Einstein equations, we need to compute all the elements of the Ricci tensor. For all the components of the Einstein tensor $G_{\mu\nu} \equiv R_{\mu\nu} - g_{\mu\nu}\mathcal{R}/2$ depend on the Ricci scalar, which depends on all elements of $R_{\mu\nu}$.

5.1.1 Christoffel Symbols

We have already computed the zero-order Christoffel symbols in Eqs. (2.22) and (2.23). Now we need to look at the first-order terms, those that are linear in Φ and/or Ψ. First let us consider $\Gamma^0{}_{\mu\nu}$, which by definition is

$$\Gamma^0{}_{\mu\nu} = \frac{1}{2} g^{0\alpha} \left[g_{\alpha\mu,\nu} + g_{\alpha\nu,\mu} - g_{\mu\nu,\alpha} \right] \tag{5.1}$$

where again $,\alpha$ means the derivative with respect to x^α. The only nonzero component of $g^{0\alpha}$ is the time component,[1] which is the inverse of $g_{00} = -1 - 2\Psi$. So, to first-order in the perturbations, $g^{00} = -1 + 2\Psi$, and

$$\Gamma^0{}_{\mu\nu} = \frac{-1 + 2\Psi}{2} \left[g_{0\mu,\nu} + g_{0\nu,\mu} - g_{\mu\nu,0} \right]. \tag{5.2}$$

Take each component in turn: first the one with $\mu = \nu = 0$. Each of the terms in square brackets is identical, so the brackets give $g_{00,0} = -2\Psi_{,0}$. Since we are interested only in first-order terms the factor of 2Ψ out in front can be dropped and we are left with

$$\Gamma^0{}_{00} = \Psi_{,0}. \tag{5.3}$$

The next possibility is that one of the indices μ or ν is spatial and the other time. It doesn't matter which, since the Christoffel symbol is symmetric in its lower indices. In this case, only one of terms in brackets in Eq. (5.2) is nonzero, $g_{00,i} = -2\Psi_{,i}$. Once again since this is first-order, we can drop the factor of 2Ψ in front, leading to

$$\Gamma^0{}_{0i} = \Gamma^0{}_{i0} = \Psi_{,i} = i k_i \Psi. \tag{5.4}$$

The final equality here moves to Fourier space, where we will stay from now on. Recall our convention of not using ˜s for Fourier transformed variables: Ψ on the far right is really $\tilde\Psi$.

Finally, if both lower indices in Eq. (5.2) are spatial, the first two terms in brackets vanish since $g_{0i} = 0$ and only the last term survives, leaving

$$\Gamma^0{}_{ij} = \frac{1 - 2\Psi}{2} \frac{\partial}{\partial t} \left[\delta_{ij} a^2 (1 + 2\Phi) \right]. \tag{5.5}$$

There is a zero-order term here, the one we computed in Eq. (2.22), and three first-order terms:

$$\Gamma^0{}_{ij} = \delta_{ij} a^2 \left[H + 2H(\Phi - \Psi) + \Phi_{,0} \right] \tag{5.6}$$

with $H = (da/dt)/a$.

Computing the Christoffel symbols, $\Gamma^i{}_{\mu\nu}$, will be left as an exercise. They are

$$\Gamma^i{}_{00} = \frac{i k^i}{a^2} \Psi$$

---

[1] We will do the calculation with $x^0 = t$, not conformal time. Therefore, $\Psi_{,0}$ for example means derivative with respect to time. Since our convention is now $\dot\Psi \equiv \partial\Psi/\partial\eta$, $\Psi_{,0} = \dot\Psi/a$.

$$\Gamma^i{}_{j0} = \Gamma^i{}_{0j} = \delta_{ij}\left(H + \Phi_{,0}\right)$$

$$\Gamma^i{}_{jk} = i\Phi\left[\delta_{ij}k_k + \delta_{ik}k_j - \delta_{jk}k_i\right]. \tag{5.7}$$

Note that the only nonvanishing zero-order component is $\Gamma^i{}_{j0}$, in agreement with Eq. (2.23). Also remember that both δ_{ij} and the 3-vector k_i live in Euclidean space, so we can freely interchange their upper and lower indices.

5.1.2 Ricci Tensor

The Ricci tensor is most easily expressed in terms of the Christoffel symbols, as in Eq. (2.31). First, consider the time-time component:

$$R_{00} = \Gamma^\alpha{}_{00,\alpha} - \Gamma^\alpha{}_{0\alpha,0} + \Gamma^\alpha{}_{\beta\alpha}\Gamma^\beta{}_{00} - \Gamma^\alpha{}_{\beta0}\Gamma^\beta{}_{0\alpha}. \tag{5.8}$$

All of these terms contribute at first-order. One simplification comes from considering the $\alpha = 0$ part of all these terms. The first and second terms are equal and opposite to each other as are the last two. So the sum over the index α contributes only when α is spatial. Let's consider each of the terms one by one.

- The first is

$$\Gamma^i{}_{00,i} = \frac{-k^2}{a^2}\Psi, \tag{5.9}$$

  using the first of equations (5.7).
- The second term in Eq. (5.8) is

$$-\Gamma^i{}_{0i,0} = -3\left(\frac{d^2a/dt^2}{a} - H^2 + \Phi_{,00}\right) \tag{5.10}$$

  using the second of equations (5.7). The factor of 3 in front comes from the implicit sum in δ_{ii}.
- The next term is $\Gamma^i{}_{i\beta}\Gamma^\beta{}_{00}$. Note that $\Gamma^\beta{}_{00}$ is first order no matter what β is, so we need keep only the zero-order part of $\Gamma^i{}_{i\beta}$. However, the last of equations (5.7) shows that $\Gamma^i{}_{i\beta}$ is first-order unless $\beta = 0$. So to first-order,

$$\Gamma^i{}_{i\beta}\Gamma^\beta{}_{00} = \Gamma^i{}_{i0}\Gamma^0{}_{00}$$

$$= 3H\Psi_{,0}. \tag{5.11}$$

- Finally the last term is $-\Gamma^i{}_{\beta0}\Gamma^\beta{}_{0i}$. In this case, if $\beta = 0$ both Γ's are first-order, so their product is second-order and can be neglected. Therefore, only spatial β need be considered, leading to

$$-\Gamma^i{}_{\beta0}\Gamma^\beta{}_{0i} = -\Gamma^i{}_{j0}\Gamma^j{}_{0i}$$

$$= -3\left(H^2 + 2H\Phi_{,0}\right). \tag{5.12}$$

Collecting these four sets of terms gives

$$R_{00} = -3\frac{d^2a/dt^2}{a} - \frac{k^2}{a^2}\Psi - 3\Phi_{,00} + 3H(\Psi_{,0} - 2\Phi_{,0}).\tag{5.13}$$

Note that the zero-order term agrees with Eq. (2.34).

The space-space part of the Ricci tensor is left as an exercise. It is

$$R_{ij} = \delta_{ij}\left[\left(2a^2H^2 + a\frac{d^2a}{dt^2}\right)(1 + 2\Phi - 2\Psi)\right.$$

$$\left. + a^2H\left(6\Phi_{,0} - \Psi_{,0}\right) + a^2\Phi_{,00} + k^2\Phi\right] + k_ik_j(\Phi + \Psi).\tag{5.14}$$

We can now contract the indices on the Ricci tensor and find the Ricci scalar:

$$\mathcal{R} \equiv g^{\mu\nu}R_{\mu\nu} = g^{00}R_{00} + g^{ij}R_{ij}$$

$$= [-1 + 2\Psi]\left[-3\frac{d^2a/dt^2}{a} - \frac{k^2}{a^2}\Psi - 3\Phi_{,00} + 3H(\Psi_{,0} - 2\Phi_{,0})\right]$$

$$+ \left[\frac{1 - 2\Phi}{a^2}\right]\left[3\left\{\left(2a^2H^2 + a\frac{d^2a}{dt^2}\right)\left(1 + 2\Phi - 2\Psi\right)\right.\right.$$

$$\left.\left. + a^2H\left(6\Phi_{,0} - \Psi_{,0}\right) + a^2\Phi_{,00} + k^2\Phi\right\} + k^2(\Phi + \Psi)\right].\tag{5.15}$$

First let us check the zero-order part of $\mathcal{R}$. Combining terms, we find that it is $6(H^2 + \frac{d^2a/dt^2}{a})$, in agreement with Eq. (2.37). To get the first-order part, $\delta\mathcal{R}$, we go through the by-now-familiar routine of multiplying terms, keeping only those first-order in Φ and Ψ. This gives

$$\delta\mathcal{R} = -6\Psi\frac{d^2a/dt^2}{a} + \frac{k^2}{a^2}\Psi + 3\Phi_{,00} - 3H(\Psi_{,0} - 2\Phi_{,0})$$

$$- 6\Psi\left(2H^2 + \frac{d^2a/dt^2}{a}\right) + 3H\left(6\Phi_{,0} - \Psi_{,0}\right)$$

$$+ 3\Phi_{,00} + 4\frac{k^2\Phi}{a^2} + \frac{k^2\Psi}{a^2},\tag{5.16}$$

where the first line contains the terms from R_{00} (the second line in Eq. (5.15)) and the last two from R_{ij} (the last two lines in Eq. (5.15)). Combining these leads to

$$\delta\mathcal{R} = -12\Psi\left(H^2 + \frac{d^2a/dt^2}{a}\right) + \frac{2k^2}{a^2}\Psi + 6\Phi_{,00}$$

$$- 6H(\Psi_{,0} - 4\Phi_{,0}) + 4\frac{k^2\Phi}{a^2}.\tag{5.17}$$

5.2 TWO COMPONENTS OF THE EINSTEIN EQUATIONS

We can now derive the evolution equations for Φ and Ψ, the perturbations to the Friedmann–Robertson–Walker metric. There is some freedom here because the Einstein equations

$$G^\mu{}_\nu = 8\pi G T^\mu{}_\nu \tag{5.18}$$

have 10 components and we need only two. All of the other eight components will either be zero at first-order or be redundant.[2]

The first component we will use is the time-time component. Thus we need to evaluate

$$G^0{}_0 = g^{00}\left[R_{00} - \frac{1}{2}g_{00}\mathcal{R}\right]$$

$$= (-1 + 2\Psi)R_{00} - \frac{\mathcal{R}}{2}. \tag{5.19}$$

Here one of the indices has been raised by multiplying G_{00} by g^{00} (recall that g^{0i} vanish). This turns out to simplify the energy–momentum tensor (see Exercise 3) which sources the Einstein tensor. Also note that the second line follows from the first since $g^{00}g_{00} = 1$. We have computed the time-time component of the Ricci tensor (Eq. (5.13)) and the perturbed Ricci scalar (Eq. (5.17)), so the first-order part of the time-time component of the Einstein tensor is

$$\delta G^0{}_0 = -6\Psi\frac{d^2a/dt^2}{a} + \frac{k^2}{a^2}\Psi + 3\Phi_{,00} - 3H(\Psi_{,0} - 2\Phi_{,0})$$

$$+ 6\Psi\left(H^2 + \frac{d^2a/dt^2}{a}\right) - \frac{k^2}{a^2}\Psi - 3\Phi_{,00}$$

$$+ 3H(\Psi_{,0} - 4\Phi_{,0}) - 2\frac{k^2\Phi}{a^2}. \tag{5.20}$$

Combining terms leads to

$$\delta G^0{}_0 = -6H\Phi_{,0} + 6\Psi H^2 - 2\frac{k^2\Phi}{a^2}. \tag{5.21}$$

Einstein's equation equates $G^0{}_0$ with $8\pi G T^0{}_0$ where $T_{\mu\nu}$ is the energy–momentum tensor. To complete our derivation of the first evolution equation for Φ and Ψ, therefore, we need to compute the first-order part of the source term, $T^0{}_0$. Recall from Section 2.3 that $-T^0{}_0$ is the energy density of all the particles

---

[2]This is true for scalar perturbations. When we come to consider tensor perturbations, some of the other components will be useful.

in the universe, and that the contribution from each species is an integral over the distribution function (Eq. (2.59)),

$$T^0{}_0(\vec{x}, t) = -\sum_{\text{all species } i} g_i \int \frac{d^3p}{(2\pi)^3}\, E_i(p) f_i(\vec{p}, \vec{x}, t). \qquad (5.22)$$

Recall also that g_i is the spin degeneracy of the species (has nothing to do with the metric); $E_i = \sqrt{p^2 + m_i^2}$ is the energy of a particle with momentum p and mass m_i; and f_i is the distribution function. In Section 2.3, we considered the zero-order distributions of the smooth universe. To get the first-order part of the energy–momentum tensor, we must naturally consider the first-order part of the distribution functions, i.e. the perturbation variables we defined in Chapter 4 for the photons, neutrinos, dark matter, and baryons. This is easiest for the dark matter and baryons. For we defined the right-hand side as $-\rho_i(1+\delta_i)$ where i labels either dark matter or baryons. For photons, a little more care is required. Using the definition of Θ in Eq. (4.35), we have

$$T^0{}_0 = -2 \int \frac{d^3p}{(2\pi)^3}\, p \left[f^{(0)} - p \frac{\partial f^{(0)}}{\partial p} \Theta \right] \qquad \text{(photons)}. \qquad (5.23)$$

The first term here is just the zero-order photon energy density, ρ_γ. To reduce the second term, we first do the angular integral, which picks out the monopole Θ_0 from Θ. Then, we do the radial integral by parts. This changes the sign and introduces a factor of 4 since $\partial p^4/\partial p = 4p^3$, leading to

$$T^0{}_0 = -\rho_\gamma \left[1 + 4\Theta_0 \right] \qquad \text{(photons)}. \qquad (5.24)$$

The factor of 4 here is obvious in retrospect. The perturbation variable Θ is the fractional temperature change, while the energy momentum tensor is interested in the perturbed energy density, $\delta\rho$. We should have expected that since $\rho \propto T^4$, $\delta\rho/\rho = 4\delta T/T$. In any event, it falls out of the algebra. I harp on it only to warn those who turn to the literature that authors are virtually split between those who define Θ as $\delta\rho/\rho$ and those who opt for the convention we use here. Finally, note that the first-order contribution from massless neutrinos is identical in form,

$$T^0{}_0 = -\rho_\nu \left[1 + 4\mathcal{N}_0 \right] \qquad \text{(neutrinos)}. \qquad (5.25)$$

In principle, we should also include a term for the perturbation to the dark energy. In practice, though, most models of the dark energy predict that (i) it should be smooth and (ii) it should be important only very recently. Both of these features are inherent in the cosmological constant model for example. There are some models which deviate from one or both of these conditions, but for the most part we are justified in neglecting the dark energy as a source of perturbations to the metric.

Returning to Einstein's equation, we equate Eq. (5.21) with $8\pi G$ times the first-order part of the time-time component of the energy–momentum tensor. Dividing both sides by 2 leads to

$$-3H\Phi_{,0} + 3\Psi H^2 - \frac{k^2\Phi}{a^2} = -4\pi G \left[\rho_{\mathrm{dm}}\delta + \rho_b\delta_{\mathrm{b}} + 4\rho_\gamma\Theta_0 + 4\rho_\nu\mathcal{N}_0 \right]. \qquad (5.26)$$

It is again useful to write the equation in terms of conformal time. This introduces an extra factor of $1/a$ every time a time derivative appears, so

$$k^2\Phi + 3\frac{\dot{a}}{a}\left(\dot{\Phi} - \Psi\frac{\dot{a}}{a} \right) = 4\pi Ga^2 \left[\rho_{\mathrm{dm}}\delta + \rho_b\delta_{\mathrm{b}} + 4\rho_\gamma\Theta_0 + 4\rho_\nu\mathcal{N}_0 \right]. \qquad (5.27)$$

This is our first evolution equation for Φ and Ψ. In the limit of no expansion ($a = \mathrm{constant}$), Eq. (5.27) reduces to the ordinary Poisson equation for gravity (in Fourier space). The left-hand side is $-\nabla^2\Phi$ while the right-hand side is $4\pi G\delta\rho$. The terms proportional to $\dot{a}$ account for expansion and are typically important for modes with wavelengths ($\sim a/k$) comparable to, or larger than, the Hubble radius, H^{-1}. We need this general relativistic expression when we consider the evolution of parturbations, because almost all modes of interest today used to have wavelengths larger than the Hubble radius. More on this in Chapter 6.

We now obtain a second evolution equation for Φ and Ψ. Since we have already dealt with the time-time component of the Einstein tensor, let's focus on the spatial part of $G^\mu{}_\nu$,

$$G^i{}_j = g^{ik}\left[R_{kj} - \frac{g_{kj}}{2}\mathcal{R} \right] = \frac{\delta^{ik}(1 - 2\Phi)}{a^2}R_{kj} - \frac{\delta_{ij}}{2}\mathcal{R}. \qquad (5.28)$$

From Eq. (5.14), we see that most of the terms in R_{kj} are proportional to δ_{kj}. When contracted with δ^{ik} this will lead to a host of terms proportional to δ_{ij}, in addition to the last term here, the one proportional to $\mathcal{R}$. Therefore,

$$G^i{}_j = A\delta_{ij} + \frac{k_i k_j (\Phi + \Psi)}{a^2} \qquad (5.29)$$

where A has close to a dozen terms which we would rather not write down. Since all of these terms are proportional to δ_{ij} they all contribute to the trace of $G^i{}_j$. To avoid dealing with these terms, consider the *longitudinal, traceless* part of $G^i{}_j$, which can be extracted by contracting $G^i{}_j$ with $\hat{k}_i\hat{k}^j - (1/3)\delta_i^j$, a *projection* operator. That is, it picks out the piece which is longitudinal, traceless and only that part (Exercise 4). This projection operator kills all terms proportional to δ_{ij}, leaving only

$$\left(\hat{k}_i\hat{k}^j - (1/3)\delta_i^j \right) G^i{}_j = \left(\hat{k}_i\hat{k}^j - (1/3)\delta_i^j \right)\left(\frac{k^i k_j(\Phi + \Psi)}{a^2} \right) = \frac{2}{3a^2}k^2(\Phi + \Psi). \qquad (5.30)$$

This is to be equayed with the longitudinal, traceless part of the energy–momentum tensor, extracted in the same fashion:

$$\left(\hat{k}_i\hat{k}^j - (1/3)\delta_i^j\right)T^i{}_j = \sum_{\text{all species i}} g_i \int \frac{d^3p}{(2\pi)^3} \frac{p^2\mu^2 - (1/3)p^2}{E_i(p)} f_i(\vec{p}). \tag{5.31}$$

We can immediately recognize the combination $\mu^2 - 1/3$ as proportional to the second Legendre polynomial, more precisely equal to $(2/3)\mathcal{P}_2(\mu)$. Therefore, the integral picks out the quadrupole part of the distribution. Of course the zero-order part of the distribution function has no quadrupole, so the source term is first order, proportional to Θ_2, which is nonzero only for neutrinos and photons. The integral in Eq. (5.31) for photons is

$$-2\int \frac{dp\,p^2}{2\pi^2} p^2 \frac{\partial f^{(0)}}{\partial p} \int_{-1}^{1} \frac{d\mu}{2} \frac{2\mathcal{P}_2(\mu)}{3}\Theta(\mu) = 2\frac{2\Theta_2}{3}\int \frac{dp\,p^2}{2\pi^2} p^2 \frac{\partial f^{(0)}}{\partial p}$$

$$= -\frac{8\rho^{(0)}\Theta_2}{3} \tag{5.32}$$

where the first equality follows from the definition of the quadrupole and the second from an integration by parts. This component of the energy–momentum tensor is called the *anisotropic stress*. Nonrelativistic particles, such as baryons and dark matter, do not contribute anisotropic stress.

For the second Einstein equation, we therefore equate Eq. (5.30) with $8\pi G$ times the photon and neutrino anisotropic stresses:

$$k^2(\Phi + \Psi) = -32\pi Ga^2 \left[\rho_\gamma\Theta_2 + \rho_\nu\mathcal{N}_2\right]. \tag{5.33}$$

That is, the two gravitational potentials are equal and opposite unless the photons or neutrinos have appreciable quadrupole moments. In practice, the photons' quadrupole contributes little to this sum, because it is very small during the time when it has appreciable energy density. [Recall the argument after Eq. (4.54).] Only the collisionless neutrino has an appreciable quadrupole moment early on when radiation dominates the universe.

5.3 TENSOR PERTURBATIONS

Until now, we have focused almost exclusively on *scalar* perturbations to the homogeneous FRW universe. Formally, this means that the perturbations $\Phi(\vec{x}, t)$ and $\Psi(\vec{x}, t)$ transform as scalars as $\vec{x} \to \vec{x}'$; i.e., they remain unchanged under a spatial coordinate transformation. This focus is reasonable: as we have seen, scalar perturbations to the metric are sourced by density fluctuations and vice versa. For the most part, the density fluctuations that led to the structure of the universe are our primary interest.

Nonetheless, many theories of structure formation produce, in addition to scalar fluctuations, *tensor* perturbations to the metric. These are potentially detectable because they produce observable distortions in the CMB, especially on large scales. Sprinkled throughout the book, therefore, are exercises (with hints) relating to tensor perturbations. The tools needed to study these are precisely those we crafted when studying scalar perturbations. For the most part, therefore, I regard the evolution of tensor perturbations as one rather large homework problem, one which introduces no new physics.

One question which naturally arises when working out these exercises, though, is why consider scalar and tensor perturbations separately? To answer this question (and to alleviate the homework load) this section derives Einstein's equations for tensor perturbations. We will see that scalar and tensor perturbations *decouple*; that is, they evolve completely independently. So the presence of tensor perturbations does not affect the scalars and vise versa. Contrast this with Φ and Ψ. We have just shown that they are quite tightly coupled to each other. It is impossible to learn about Φ without also solving for Ψ. The decoupling of scalars and tensors is a manifestation of the *decomposition theorem*. Needless to say, it is much more instructive to work out an example of this theorem than to prove it abstractly. Incidentally, as you would expect, the same theorem can be applied to *vector* perturbations. These too are produced by some early-universe models (but not as ubiquitously as tensors) and can be treated completely independently.

Tensor perturbations can be characterized by a metric with $g_{00} = -1$, zero space-time components $g_{0i} = 0$, and spatial elements

$$g_{ij} = a^2 \begin{pmatrix} 1 + h_+ & h_\times & 0 \\ h_\times & 1 - h_+ & 0 \\ 0 & 0 & 1 \end{pmatrix}. \tag{5.34}$$

That is, the perturbations to the metric are described by two functions, h_+ and $h_\times$, assumed small. For definiteness, I have chosen the perturbations to be in the x-y plane. This corresponds to an implicit choice of axes; in particular, it corresponds to choosing the z-axis to be in the direction of the wavevector, $\vec{k}$. More generally, h_+ and $h_\times$ are two components of a *divergenceless, traceless, symmetric tensor*. If this perturbation tensor is written as $\mathcal{H}_{ij}$, *divergenceless* means that $k^i \mathcal{H}_{ij} = k^j \mathcal{H}_{ij} = 0$. This is clearly satisfied by Eq. (5.34) since there are no components in the $\hat{k} = \hat{z}$ direction. Tracelessness is also satisfied since the sum of the perturbations along the diagonal vanishes.

Once the metric in Eq. (5.34) has been written down, we can blast away and derive the Einstein equations. Once again the derivation proceeds in three steps: (i) Christoffel symbols, (ii) Ricci tensor, and (iii) Ricci scalar.

5.3.1 Christoffel Symbols for Tensor Perturbations

First consider $\Gamma^0_{\alpha\beta}$. The metric we are considering in Eq. (5.34) has constant g_{00} and vanishing g_{0i}. Recall that the Christoffel symbol is a sum of derivatives of the metric. The only terms that will be nonzero are those which involve derivatives of the spatial part of the metric, $g_{ij,\alpha}$. Therefore, we can immediately argue that

$$\Gamma^0_{00} = \Gamma^0_{i0} = 0. \tag{5.35}$$

The term with two lower spatial indices is

$$\Gamma^0{}_{ij} = -\frac{g^{00}}{2} g_{ij,0}$$

$$= \frac{1}{2} g_{ij,0}. \tag{5.36}$$

Let's use the notation mentioned earlier: the 3D matrix $\mathcal{H}_{ij}$ contains the perturbations, which in this basis (with $\hat{k}$ in the $\hat{z}$ direction) is equal to

$$\mathcal{H}_{ij} = \begin{pmatrix} h_+ & h_\times & 0 \\ h_\times & -h_+ & 0 \\ 0 & 0 & 0 \end{pmatrix} \tag{5.37}$$

so that $g_{ij} = a^2 (\delta_{ij} + \mathcal{H}_{ij})$. Therefore,

$$g_{ij,0} = 2H g_{ij} + a^2 \mathcal{H}_{ij,0} \tag{5.38}$$

where the Hubble rate H is not to be confused with tensor perturbations $\mathcal{H}$. The first nonzero Christoffel symbol is therefore

$$\Gamma^0{}_{ij} = H g_{ij} + \frac{a^2 \mathcal{H}_{ij,0}}{2}. \tag{5.39}$$

When both lower indices on Γ are 0, the Christoffel symbol vanishes. The two remaining components are $\Gamma^i{}_{0j}$ and $\Gamma^i{}_{jk}$. The former is

$$\Gamma^i{}_{0j} = \frac{g^{ik}}{2} g_{jk,0}. \tag{5.40}$$

The time derivative of g_{jk} acts on both the scale factor and on the perturbations $h_{+,\times}$, as in Eq. (5.38), so

$$\Gamma^i{}_{0j} = \frac{g^{ik}}{2} \left[2H g_{jk} + a^2 \mathcal{H}_{jk,0} \right]. \tag{5.41}$$

But $g^{ik} g_{jk} = \delta_{ij}$, so the first term here is simply $\delta_{ij} H$. To get the second, we can set $g^{jk} = \delta_{jk}/a^2$ (i.e., neglect first-order terms) since it multiplies the first-order $\mathcal{H}$. So,

$$\Gamma^i{}_{0j} = H \delta_{ij} + \frac{1}{2} \mathcal{H}_{ij,0}, \tag{5.42}$$

where I have used the fact that $\mathcal{H}_{ij}$ is symmetric.

The last Christoffel symbol we need is $\Gamma^i{}_{jk}$. In Exercise 7 you will show that

$$\Gamma^i{}_{jk} = \frac{i}{2} \left[k_k \mathcal{H}_{ij} + k_j \mathcal{H}_{ik} - k_i \mathcal{H}_{jk} \right]. \tag{5.43}$$

5.3.2 Ricci Tensor for Tensor Perturbations

Following the same steps as in the scalar perturbation case, we now combine these Christoffel symbols to form the Ricci tensor. First we compute the time-time component of the Ricci tensor:

$$R_{00} = \Gamma^{\alpha}{}_{00,\alpha} - \Gamma^{\alpha}{}_{0\alpha,0} + \Gamma^{\alpha}{}_{\beta\alpha}\Gamma^{\beta}{}_{00} - \Gamma^{\alpha}{}_{\beta 0}\Gamma^{\beta}{}_{0\alpha}. \tag{5.44}$$

We have shown that the Christoffel symbol vanishes for tensor perturbations when the two lower indices are time-time. Therefore, the first and third terms here are zero. Using the same argument, the indices α and β in the second and fourth terms must be spatial, so

$$R_{00} = -\Gamma^{i}{}_{0i,0} - \Gamma^{i}{}_{j0}\Gamma^{j}{}_{0i}. \tag{5.45}$$

Using Eq. (5.42) for $\Gamma^{i}{}_{j0}$ which is the only element appearing, we find that

$$R_{00} = -3\frac{dH}{dt} - \frac{1}{2}\mathcal{H}_{ii,00}$$
$$- \left(H\delta_{ij} + \frac{1}{2}\mathcal{H}_{ij,0}\right)\left(H\delta_{ij} + \frac{1}{2}\mathcal{H}_{ij,0}\right). \tag{5.46}$$

On the first line, the trace $\mathcal{H}_{ii}$ vanishes since h_{+} appears in the metric with opposite signs along the diagonal. Expanding the second line out to first-order leads to a similar cancellation: $\mathcal{H}_{ij}$ is multiplied by δ_{ij}, so there are no first-order terms. The zero-order terms combine to form

$$R_{00} = -3\frac{d^2a/dt^2}{a}, \tag{5.47}$$

an equation in which we are by now quite confident since this is the third time we have derived it (see equations (2.34) and (5.13)). Of course the big news here is not that we have correctly derived the zero-order term, but rather that tensor perturbations do not appear at first-order in R_{00}. Looking ahead, we will soon see that the Ricci scalar also has no tensor contribution (even though R_{ij} does). Therefore, we can anticipate that the time-time component of Einstein's equations contains no tensor perturbations. This is important for it tells us that density perturbations— which form the right hand side of the time-time component as shown in Eq. (5.26)—do *not* induce any tensor perturbations. We are beginning therefore to get a glimmer of the decomposition theorem. Density perturbations and scalar perturbations to the metric are coupled; indeed their names are often used as synonyms. Tensor perturbations, however, are decoupled from these and evolve on their own.

The spatial components of the Ricci tensor do depend on the tensor perturbation variables. We now turn to

$$R_{ij} = \Gamma^{\alpha}{}_{ij,\alpha} - \Gamma^{\alpha}{}_{i\alpha,j} + \Gamma^{\alpha}{}_{\alpha\beta}\Gamma^{\beta}{}_{ij} - \Gamma^{\alpha}{}_{\beta j}\Gamma^{\beta}{}_{i\alpha}. \tag{5.48}$$

Let's consider the first two terms together. Expanding out leads to

$$\Gamma^\alpha{}_{ij,\alpha} - \Gamma^\alpha{}_{i\alpha,j} = \Gamma^0{}_{ij,0} + \Gamma^k{}_{ij,k} - \Gamma^k{}_{ik,j} \tag{5.49}$$

since $\alpha = 0$ does not contribute in $\Gamma^\alpha{}_{i\alpha,j}$ because of Eq. (5.35). The hardest (i.e., longest) term here is the first, which involves multiple time derivatives. Let's postpone its calculation by recalling that $\Gamma^0{}_{ij} = g_{ij,0}/2$ so that the first term can be written in shorthand as $g_{ij,00}/2$. The last term in Eq. (5.49) vanishes since $\Gamma^k{}_{ik} = 0$ for tensor perturbations. Combining the other terms then leads to

$$\Gamma^\alpha{}_{ij,\alpha} - \Gamma^\alpha{}_{i\alpha,j} = \frac{g_{ij,00}}{2} + \frac{1}{2}\left[-k_i k_k \mathcal{H}_{jk} - k_j k_k \mathcal{H}_{ik} + k^2 \mathcal{H}_{ji}\right]. \tag{5.50}$$

Recall that we chose $\vec{k}$ to be along the z-axis. Therefore, the indices i and k in the first term in brackets must be equal to 3. But these multiply $\mathcal{H}_{jk} = \mathcal{H}_{j3} = 0$ so this term and its cousin $k_j k_k \mathcal{H}_{ik}$ must both vanish. Therefore,

$$\Gamma^\alpha{}_{ij,\alpha} - \Gamma^\alpha{}_{i\alpha,j} = \frac{g_{ij,00}}{2} + \frac{k^2}{2}\mathcal{H}_{ji}. \tag{5.51}$$

The third term in Eq. (5.48), $\Gamma^\alpha{}_{\alpha\beta}\Gamma^\beta{}_{ij}$, is nonzero only when the index α is spatial, so

$$\Gamma^\alpha{}_{\alpha\beta}\Gamma^\beta{}_{ij} = \Gamma^k{}_{k0}\Gamma^0{}_{ij} + \Gamma^k{}_{kl}\Gamma^l{}_{ij}. \tag{5.52}$$

But each of the Christoffel symbols in the second term here are first-order, so their product vanishes. In the first term, the sum over k makes the first-order terms go away, so $\Gamma^k{}_{k0}$ is purely zero-order, $3H$. Therefore,

$$\Gamma^\alpha{}_{\alpha\beta}\Gamma^\beta{}_{ij} = \frac{3}{2}H g_{ij,0}. \tag{5.53}$$

The final term in Eq. (5.48) will be left as an exercise; it is

$$\Gamma^\alpha{}_{\beta j}\Gamma^\beta{}_{i\alpha} = 2H^2 g_{ij} + 2a^2 H \mathcal{H}_{ij,0}. \tag{5.54}$$

We can now combine all four terms in Eq. (5.48) to get

$$R_{ij} = \frac{g_{ij,00}}{2} + \frac{k^2}{2}\mathcal{H}_{ji} + \frac{3}{2}H g_{ij,0}$$

$$- 2H^2 g_{ij} - 2a^2 H \mathcal{H}_{ij,0}. \tag{5.55}$$

We now need to expand out the time derivatives of the metric. Using Eq. (5.38), one finds

$$g_{ij,00} = 2g_{ij}\left(\frac{d^2a/dt^2}{a} + H^2\right) + 4a^2 H \mathcal{H}_{ij,0} + a^2 \mathcal{H}_{ij,00}. \tag{5.56}$$

Therefore the Ricci tensor is

$$R_{ij} = g_{ij}\left(\frac{d^2a/dt^2}{a} + 2H^2\right) + \frac{3}{2}a^2 H \mathcal{H}_{ij,0}$$

$$+ a^2 \frac{\mathcal{H}_{ij,00}}{2} + \frac{k^2}{2}\mathcal{H}_{ij}. \tag{5.57}$$

Again we see that we have successfully recaptured the zero-order part of the Ricci tensor. Remarkably, we will see that the first-order parts — when used in Einstein's equations — do not couple to the scalar perturbations.

First, though, we must compute the Ricci scalar:

$$\mathcal{R} = g^{00}R_{00} + g^{ij}R_{ij}. \tag{5.58}$$

The time-time product is all zero-order, so we can neglect it when considering the first-order piece $\delta\mathcal{R}$. The space-space contraction has two types of terms. First, there are the terms in Eq. (5.57) proportional to the metric g_{ij}. But $g^{ij}g_{ij} = 3$, so there are no first-order terms here. All the other terms in Eq. (5.57) are proportional to $\mathcal{H}_{ij}$, so when contracting them we can set g^{ij} to its zero-order value, δ_{ij}/a^2. This corresponds to taking the trace of the first-order terms in Eq. (5.57). Since all first-order terms are proportional to $\mathcal{H}_{ij}$, the trace vanishes. Therefore, tensor perturbations do not affect (at first order) the Ricci scalar.

5.3.3 Einstein Equations for Tensor Perturbations

Now let's read off the perturbations to the Einstein tensor induced by tensor modes. Since the Ricci scalar is unperturbed by tensor perturbations, the first-order Einstein tensor is simply

$$\delta G^i{}_j = \delta R^i{}_j. \tag{5.59}$$

To get $R^i{}_j$, we contract $g^{ik}R_{kj}$, using the Ricci tensor we computed in Eq. (5.57). The first term, proportional to the contraction of $g^{ik}g_{kj} = \delta^i_j$, has no first-order piece; the remaining terms are explicitly first-order in $\mathcal{H}$, so we can set $g^{ik} = \delta^{ik}/a^2$, leading to

$$\delta G^i{}_j = \delta^{ik}\left[\frac{3}{2}H\mathcal{H}_{kj,0} + \frac{\mathcal{H}_{kj,00}}{2} + \frac{k^2}{2a^2}\mathcal{H}_{kj}\right]. \tag{5.60}$$

We can now derive a set of equations governing the evolution of the tensor variables, h_+ and $h_\times$.

To derive an equation for h_+, let us consider the difference between the $^1{}_1$ and $^2{}_2$ components of the Einstein tensor. The Einstein tensor in Eq. (5.60) is proportional to $\mathcal{H}_{ij}$ and its derivatives. Since $\mathcal{H}_{11} = -\mathcal{H}_{22} = h_+$, $\delta G^1{}_1$ is equal and opposite to $\delta G^2{}_2$. Therefore,

$$\delta G^1{}_1 - \delta G^2{}_2 = 3Hh_{+,0} + h_{+,00} + \frac{k^2 h_+}{a^2}. \tag{5.61}$$

Change to conformal time so that $h_{+,0} = \dot{h}_+/a$ and $h_{+,00} = \ddot{h}_+/a^2 - (\dot{a}/a^3)\dot{h}_+$. Then,

$$a^2\left[\delta G^1{}_1 - \delta G^2{}_2\right] = \ddot{h}_+ + 2\frac{\dot{a}}{a}\dot{h}_+ + k^2 h_+. \tag{5.62}$$

The right-hand side of this component of Einstein's equations is zero (Exercise 8), and $h_\times$ obeys the same equation (Exercise 9), so the tensor modes are governed by

$$\ddot{h}_\alpha + 2\frac{\dot{a}}{a}\dot{h}_\alpha + k^2 h_\alpha = 0 \tag{5.63}$$

where $\alpha = +, \times$. Equation (5.63) is a wave equation, and the corresponding solutions are called *gravity waves*. For example, if we neglect the expansion of the universe so that the damping term in Eq. (5.63) vanishes, we immediately see that the two solutions are $h_\alpha \propto e^{\pm ik\eta}$. In real space, then the perturbation to the metric is of the form

$$h_\alpha(\vec{x}, \eta) = \int d^3 k e^{i\vec{k}\cdot\vec{x}} \left[A e^{ik\eta} + B e^{-ik\eta}\right] \qquad \text{(no expansion)}. \tag{5.64}$$

The two modes here corresponds to waves traveling in the $\pm\hat{z}$ direction at the speed of light.

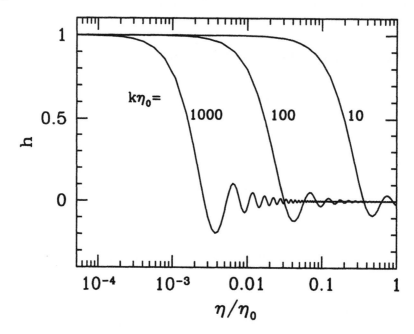

Figure 5.1. Evolution of gravity waves as a function of conformal time. Three different modes are shown, labeled by their wave numbers. Smaller scale modes decay earlier.

Equation (5.63) is a generalization of the wave equation to an expanding universe. Exercise 12 illustrates that if the universe is purely radiation or matter dominated, exact analytic solutions can be obtained. These are oscillatory, like the simple ones in Eq. (5.64), but also damp out. Figure 5.1 shows the evolution of h_α for three different wavelength modes. The large-scale mode (with $k\eta_0 = 10$) remains constant at early times when its wavelength is larger than the horizon $k\eta < 1$. Once its wavelength becomes comparable to the horizon, the amplitude begins to die off, oscillating several times until the present. The small-scale mode $k\eta_0 = 1000$ shown in Figure 5.1 also begins to decay when its wavelength becomes comparable to the horizon. Its entry into the horizon occurs much earlier, though, so the decay is much more efficient. By today, the amplitude is extremely small.

An important point about the effect of gravity waves on the CMB anisotropy spectrum can be gleaned from Figure 5.1. Because small-scale modes decay earlier than large-scale modes, at decoupling (at $\eta/\eta_0 \simeq 0.02$) only modes with $k\eta_0$ less than about 100 persist. All smaller scale modes can be neglected. Therefore, anisotropies on small angular scales will not be affected by gravity waves. Only the large-scale anisotropies are impacted by gravity waves.

5.4 THE DECOMPOSITION THEOREM

The decomposition theorem states that perturbations to the metric can be divided up into three types: *scalar, vector,* and *tensor.* Each of these types of perturbations evolves independently. That is, if some physical process in the early universe sets up tensor perturbations, these do not induce scalar perturbations. Conversely, to determine the evolution of scalar perturbations, we will not have to worry about possible vector or tensor perturbations.

Now that we have computed the contributions to the Einstein tensor $G_{\mu\nu}$ from scalars and tensors, we can demonstrate the decomposition of these two types of perturbations. To do this, remember that we obtained the scalar equations by considering the two components of Einstein's tensor:

$$G^0{}_0 \quad ; \quad \left(\hat{k}_i\hat{k}_j - (1/3)\delta_{ij}\right)G^i{}_j. \tag{5.65}$$

Inserting these components into Einstein's equations led to equations (5.27) and (5.33). If we can show that tensor perturbations do not contribute to these two components, then we will have convinced ourselves of at least part of the decomposition theorem, namely that the equations governing scalar equations are not affected by tensors.

It is easy to see that tensor perturbations do not contribute to $G^0{}_0$. For $G^0{}_0$ depends on R_{00} and $\mathcal{R}$. But we have seen that both of these do not depend on h_+ or $h_\times$.

Now let's show that $(\hat{k}_i\hat{k}_j - \delta_{ij}/3)G^i{}_{\bar{j}}$ also does not pick up a contribution from tensor perturbations. Multiply Eq. (5.60) by the projection operator:

$$\left(\hat{k}_i\hat{k}_j - (1/3)\delta_{ij}\right)\delta G^i{}_j = (\delta_{i3}\delta_{j3} - (1/3)\delta_{ij})$$

$$\times \left[\frac{3}{2}H\mathcal{H}_{ij,0} + \frac{\mathcal{H}_{ij,00}}{2} + \frac{k^2}{2a^2}\mathcal{H}_{ij}\right] \tag{5.66}$$

where the equality holds since we have chosen $\hat{k}$ to lie in the $\hat{z}$ direction. The terms in which indices i and j are set to 3 vanish since $\mathcal{H}_{33} = 0$. The only remaining terms are those proportional to δ_{ij}. But the Kronecker delta instructs us to take the trace of $\mathcal{H}$. This too vanishes. The scalar equations we derived in the previous section are therefore unchanged by the presence of tensor modes. This is a manifestation of the decomposition theorem.

5.5 FROM GAUGE TO GAUGE

Let's go back to scalar perturbations. Until now, we have characterized these with Ψ and Φ in the form of Eq. (4.9). This corresponds to a choice of *gauge* or a choice of a coordinate system with which to study the space-time. If we changed the coordinate system we use, we would get a metric of a different form, i.e., a different gauge. Although we will work almost exclusively in the gauge corresponding to Eq. (4.9), the conformal Newtonian gauge, historically many other gauges have been used, and for different parts of the "cosmological perturbation" problem, different gauges have their advantages. Indeed, we will see in Section 6.5.3 that people who work on the theory of inflation sometimes prefer a gauge with *spatially flat slicing* (g_{ij} unperturbed), since the equations for the perturbations generated by inflation simplify considerably. Also, the code currently used by most people to compute anisotropies and inhomogeneities in the universe uses *synchronous gauge*, partly because the equations are better behaved numerically in that gauge. So the ability to move back and forth between different gauges is useful, and I want to spend a few pages describing how to do this.

Most generally, scalar perturbations to the metric can be written down as

$$g_{00} = -(1 + 2A)$$

$$g_{0i} = -aB_{,i}$$

$$g_{ij} = a^2 \left(\delta_{ij} \left[1 + 2\psi \right] - 2E_{,ij} \right). \tag{5.67}$$

There are four functions which characterize scalar perturbations to the metric: A, B, ψ, and E. They all depend on space and time, and they are all scalars. For example, the g_{0i} components are the derivatives of a scalar function, not an independent vector function with its own orientation. In conformal Newtonian gauge, $A = \Psi$ and $\psi = \Phi$, while $B = E = 0$.

How do we transform from one gauge to another? The invariant distance of Eq. (2.2) does not change if different coordinates $\tilde{x}$ are used instead of x. Therefore,

$$\tilde{g}_{\alpha\beta}(\tilde{x})d\tilde{x}^\alpha d\tilde{x}^\beta = g_{\mu\nu}(x)dx^\mu dx^\nu, \tag{5.68}$$

where I have used a different set of dummy indices on both sides to make the upcoming few lines clearer. One of the differentials on the left-hand side can be written as $d\tilde{x}^\alpha = (\partial \tilde{x}^\alpha / \partial x^\mu) dx^\mu$ and similarly with the other differential, so equating coeffients of $dx^\mu dx^\nu$ leads to

$$\tilde{g}_{\alpha\beta}(\tilde{x}) \frac{\partial \tilde{x}^\alpha}{\partial x^\mu} \frac{\partial \tilde{x}^\beta}{\partial x^\nu} = g_{\mu\nu}(x). \tag{5.69}$$

This equation is what we are after: a prescription for how the metric changes under a coordinate transformation.

The most general coordinate transformation is generated by

$$t \to \tilde{t} = t + \xi^0(t, \vec{x})$$

$$x^i \to \tilde{x}^i = x^i + \delta^{ij} \xi_{,j}(t, \vec{x}), \tag{5.70}$$

where we take ξ^0 and ξ to be small perturbations of the same order as the variables characterizing the perturbations. Let's examine how the metric changes under such a transformation. I'll work out one component explicitly and leave the rest as an exercise. Consider the $_{00}$ component of Eq. (5.69):

$$\tilde{g}_{\alpha\beta}(\tilde{x}) \frac{\partial \tilde{x}^\alpha}{\partial t} \frac{\partial \tilde{x}^\beta}{\partial t} = -[1 + 2A]. \tag{5.71}$$

I claim that the only term that contributes to the left-hand side is the one with $\alpha = \beta = 0$. Consider for example $\alpha = 0$ and $\beta = i$. The off-diagonal component of the metric $\tilde{g}_{0i}$ is proportional to $\tilde{B}_{,i}$ a first-order perturbation. But $\partial \tilde{x}^i / \partial t$ is proportional to the first-order variable ξ, so the product is second-order and can be neglected. A similar argument holds for the $\alpha = i; \beta = j$ terms. Therefore, the left-hand side is simply

$$-\left[1 + 2\tilde{A}\right] \left(\frac{\partial \tilde{t}}{\partial t}\right)^2 = -\left[1 + 2\tilde{A}\right] \left(1 + \frac{\partial \xi^0}{\partial t}\right)^2$$

$$\simeq -1 - 2\tilde{A} - 2 \frac{\partial \xi^0}{\partial t}. \tag{5.72}$$

Equating this with g_{00} leads to

$$-2\tilde{A} - 2 \frac{\partial \xi^0}{\partial t} = -2A, \tag{5.73}$$

so under the coordinate transformation specified by Eq. (5.70)

$$A \to \tilde{A} = A - \frac{1}{a} \dot{\xi}^0. \tag{5.74}$$

In a similar vein, the other components of the metric transform into

$$\tilde{\psi} = \psi - H\xi^0$$

$$\tilde{B} = B - \frac{\xi^0}{a} + \dot{\xi}$$

$$\tilde{E} = E + \xi. \tag{5.75}$$

One technical point: Eqs. (5.74) and (5.75) describe how the components of the metric tensor transform under general coordinate changes. These equations, which have become standard, are a bit misleading, though, because each of the individual functions, A for example, transforms as a scalar, i.e., does not change under a spatial coordinate transformation. Here we have transformed the metric and accomodated the resulting changes in new definitions of A, B, ψ, and E. This is not the same thing as seeing how A by itself changes under a transformation.

To sum up, then, there are four functions which characterize scalar perturbations, but these can be manipulated with two other functions which characterize coordinate transformations. For example, starting with a metric in which $E \neq 0$, it is trivial to make a transformation to eliminate E: simply choose $\xi = -E$, and the resulting $\tilde{E} = 0$. Thus, there are really only $4 - 2 = 2$ functions which matter. Indeed, this is the reason that we had only Φ and Ψ in conformal Newtonian gauge. More generally, one might hope to construct two *gauge invariant* variables, those which remain unchanged under a general coordinate transformation. Bardeen (1980) first identified two such variables:

$$\Phi_A \equiv A + \frac{1}{a} \frac{\partial}{\partial \eta} \left[a(\dot{E} - B) \right]$$

$$\Phi_H \equiv -\psi + aH(B - \dot{E}). \tag{5.76}$$

In conformal Newtonian gauge, in which $E = B = 0$, $\Phi_A = \Psi$ and $\Phi_H = -\Phi$. These invariants are very useful: if equations simplify in a particular gauge, then one can do calculations in that gauge, form the gauge-invariant variables, and then turn these into the perturbations in any other gauge. We will do precisely this in Section 6.5.3. In other words, Φ_A and Φ_H are useful shortcuts or recipes for transforming from one gauge to another.

Under a general coordinate transformation, the components of the energy–momentum tensor $T_{\mu\nu}$ also change. In exact analogy with the metric tensor,

$$\tilde{T}_{\mu\nu}(\tilde{x}) = \frac{\partial x^\alpha}{\partial \tilde{x}^\mu} \frac{\partial x^\beta}{\partial \tilde{x}^\nu} T_{\alpha\beta}(x). \tag{5.77}$$

Again, though, Bardeen found combinations of the components of $T_{\mu\nu}$ which remain invariant and therefore facilitate mapping from one gauge to another. In particular, in Fourier space

$$v \equiv ikB + \frac{\hat{k}^i T^0{}_i}{(\rho + \mathcal{P})a} \tag{5.78}$$

remains invariant under a coordinate transformation. In conformal Newtonian gauge, for matter, v is indeed equal to the v we defined in Chapter 4. For radiation, $v = -3i\Theta_{r,1}$, i.e., proportional to the dipole, again in conformal Newtonian gauge. A second invariant is the generalized perturbation to the energy density,

$$\epsilon_m \equiv -1 - \frac{T^0{}_0}{\rho} + \frac{3H}{k^2 \rho} k^i T^0{}_i. \tag{5.79}$$

For matter, in conformal Newtonian gauge, $\epsilon_m = \delta + (3aHv/k)$; i.e., it reduces to the ordinary overdensity δ on scales smaller than the horizon. For radiation, $\epsilon_m = 4\Theta_{r,0} - 12i\Theta_{r,1}aH/k$, again reducing to the standard overdensity on small scales.

5.6 SUMMARY

The Einstein equations relate perturbations in the metric to perturbations in the matter and radiation. Taking two components of the Einstein equations $G_{\mu\nu} = 8\pi G T_{\mu\nu}$, we found equations governing the evolution of the two functions which describe scalar metric perturbations, Φ and Ψ of Eq. (4.9). It is easiest to write these equations in Fourier space. Again recalling our convention of dropping the ~s on transformed variables, we can write

$$k^2 \Phi + 3 \frac{\dot{a}}{a} \left(\dot{\Phi} - \Psi \frac{\dot{a}}{a} \right) = 4\pi G a^2 \left[\rho_m \delta_m + 4\rho_r \Theta_{r,0} \right] \tag{5.27}$$

$$k^2 (\Phi + \Psi) = -32\pi G a^2 \rho_r \Theta_{r,2}. \tag{5.33}$$

Here subscript m includes all matter such as baryons and dark matter and subscript r all radiation such as neutrinos and photons. More precisely

$$\rho_m \delta_m \equiv \rho_{\rm dm} \delta + \rho_b \delta_{\rm b} \qquad ; \qquad \rho_r \Theta_{r,0} \equiv \rho_\gamma \Theta_0 + \rho_\nu \mathcal{N}_0$$

$$\rho_m v_m \equiv \rho_{\rm dm} v + \rho_b v_{\rm b} \qquad ; \qquad \rho_r \Theta_{r,1} \equiv \rho_\gamma \Theta_1 + \rho_\nu \mathcal{N}_1. \tag{5.80}$$

Some of the other components of Einstein's equation are redundant; they add no new information about the evolution of Φ and Ψ. An example is the time-space component, which you can derive in Exercise 5. At times, though, one form of the evolution equation will be more useful than another. For example, one combination (Exercise 6) of these equations leads to an algebraic equation for the potential,

$$k^2 \Phi = 4\pi G a^2 \left[\rho_m \delta_m + 4\rho_r \Theta_{r,0} + \frac{3aH}{k} \left(i\rho_m v_m + 4\rho_r \Theta_{r,1} \right) \right]. \tag{5.81}$$

Other components of Einstein's equation contain information not about the scalar perturbations Φ and Ψ, but about vector and tensor perturbations. Scalar, vector, and tensor perturbations are decoupled: each evolves independently of the others. We will see in Chapter 6 that inflation can produce tensor perturbations, so it is important to know what the Einstein equation says about their evolution. We showed that there are two functions which can characterize tensor perturbations, h_+ and $h_\times$; each of these evolves independently and satisfies

$$\ddot{h}_\alpha + 2 \frac{\dot{a}}{a} \dot{h}_\alpha + k^2 h_\alpha = 0 \tag{5.63}$$

where α denotes $+, \times$. In an expanding universe, the amplitude of a gravity wave described by Eq. (5.63) falls off once the mode enters the horizon.

SUGGESTED READING

Most cosmology books offer some treatment of the perturbed Einstein equations in cosmology. Again *The Large Scale Structure of the Universe* (Peebles) is a useful reference, especially for synchronous gauge. *Cosmological Inflation and Large Scale Structure* (Liddle and Lyth) has a very nice treatment which, among other virtues, explains the physics of gauge choices. Probably the two most comprehensive works are the review articles by Mukhanov, Feldman, and Brandenberger (1992) and Kodama and Sasaki (1984), with the former slightly more accessible and the latter more general. These are both based on the seminal Bardeen (1980) article which is remarkable for its clarity and conciseness in its treatment of gauge invariant variables.

The general relativity books mentioned in Chapter 2 all have good discussions of gravity waves. Before turning to any of the technical literature, though, you must read *Black Holes and Time Warps* (Thorne), a wonderful mixture of the history, science, and personalities associated with 20th-century general relativity. It is the best popular science book I have ever read.

EXERCISES

Exercise 1. Derive the Christoffel symbols, $\Gamma^i{}_{\mu\nu}$, given in Eq. (5.7). When doing this, you will need g^{ij}. Show that it is equal to $\delta_{ij}(1 - 2\Phi)/a^2$.

Exercise 2. Show that R_{ij} is given by Eq. (5.14).

Exercise 3. Use the full general relativistic expression for the energy momentum tensor given in Eq. (2.101), which holds even in the presence of metric perturbations. Show that, with scalar perturbations to the metric, the phase space integral for the time-time component reduces to that in Eq. (5.22). Show that the contribution from species α to $T^0{}_i$ is

$$T^0{}_i = g_\alpha a \int \frac{d^3p}{(2\pi)^3} p_i f_\alpha(\vec{p}, \vec{x}, t).$$ (5.82)

Note the extra factor of a.

Exercise 4. Consider a 3D matrix with components $G_{ij} = (\hat{k}_i \hat{k}_j - \delta_{ij}/3)G^L$. Show that this form is traceless and satisfies $\epsilon_{ijk} G_{kl,jl} = 0$ so it is the proper form for the longitudinal component.

Exercise 5. Compute the time-space component of the Einstein tensor. Show that, in Fourier space,

$$G^0{}_i = 2ik_i \left(\frac{\dot{\Phi}}{a} - H\Psi \right).$$ (5.83)

Combine with the energy–momentum tensor derived in Exercise 3 to show that

$$\dot{\Phi} - aH\Psi = \frac{4\pi G a^2}{ik}\left[\rho_{\mathrm{dm}}v + \rho_b v_b - 4i\rho_\gamma\Theta_1 - 4i\rho_\nu\mathcal{N}_1\right]. \qquad (5.84)$$

The time-space component of Einstein's equations adds no new information once we already have the two equations derived in the text. Deciding which two to use is a matter of convenience.

Exercise 6. Take the Newtonian limit of Einstein's equations. Combine the time-time equation (5.27) with the time-space equation of Exercise 5 to obtain the algebraic (i.e., no time derivatives) equation for the potential given in Eq. (5.81). Show that this reduces to Poisson's equation (with the appropriate factors of a) when the wavelength is much smaller than the horizon ($k\eta \gg 1$).

Exercise 7. Fill in the blanks in the derivation of the tensor equation.
(a) Show that $\Gamma^i{}_{jk}$ is given by Eq. (5.43) in the presence of tensor perturbations.
(b) Show that the last term in Eq. (5.48) is given by Eq. (5.54).

Exercise 8. We defined the perturbations to the photon distribution function via Eq. (4.34). Show that, if Θ depends only on μ, the cosine of the angle between $\hat{k}(\equiv \hat{z}$ here) and $\hat{p}$, then $T^1{}_1 - T^2{}_2$ vanishes. This is indeed the dependence we have been dealing with so far. This is yet another aspect of the decomposition theorem: the terms Θ that source the scalar perturbations (and are sourced by them) do not affect tensor perturbations. Anisotropies induced by tensor perturbations will have Θ of the form

$$\Theta(\mu,\phi) = (1 - \mu^2)\cos(2\phi)\Theta_+(\mu) \qquad (5.85)$$

for those perturbations generated by h_+ and a similar expression for $h_\times$ with the cos replaced by a sin. These, however, have a negligible impact on the evolution of the gravity waves, so we are justified in setting the right-hand side of Eq. (5.63) to zero.

Exercise 9. Use the $^1{}_2$ component of the Einstein equations to show that $h_\times$ obeys the same equation as does h_+.

Exercise 10. Show that scalar perturbations (Φ and Ψ) do not contribute to either $G^1{}_1 - G^2{}_2$ or to $G^1{}_2$. This completes the demonstration of the decomposition theorem for scalars and tensors.

Exercise 11. Consider vector perturbations to the metric. These can be described by two function h_{xz} and h_{yz} where again only the spatial part of the metric is perturbed. The perturbative part of g_{ij} is

$$h^V_{ij} = \begin{pmatrix} 0 & 0 & h_{xz} \\ 0 & 0 & h_{yz} \\ h_{xz} & h_{yz} & 0 \end{pmatrix}. \qquad (5.86)$$

Show that h_{xz} and h_{yz} do not affect any of the equations we have derived so far for scalar or tensor evolution—yet another aspect of the decomposition theorem.

Exercise 12. Solve the wave equation (5.63) if the universe is purely matter dominated. Do the same for the radiation-dominated case.

Exercise 13. Define the *transfer function* for gravity wave evolution as

$$T(k) \equiv \frac{h_\alpha(k, \eta)}{h_\alpha(k, \eta = 0)} \left(\frac{k\eta}{3j_1(k\eta)} \right). \tag{5.87}$$

You should recognize the term in parentheses as the inverse of the matter-dominated solution you derived in Exercise 12. Solve Eq. (5.63) numerically and compute the transfer function. Compare your solution with the fit of Turner, White, and Lidsey (1993),

$$T(y) = \left[1 + 1.34y + 2.5y^2 \right]^{1/2} \tag{5.88}$$

where $y \equiv (k\eta_0/370h)$ (with h parametrizing the Hubble constant). Assume the universe today is flat and matter dominated, but account for transition from matter to radiation.

Exercise 14. Derive the transformations in the metric components given by Eq. (5.75). Show that Φ_A and Φ_H do not change under a general coordinate transformation.

6

INITIAL CONDITIONS

In order to understand structure in the universe, we have derived the equations governing perturbations around a smooth background. Before we start solving these equations, we need to know the initial conditions. This quest for initial conditions will lead to an entirely new realm of physics, the theory of inflation. Inflation was introduced (Guth, 1981; Linde 1982; Albrecht and Steinhardt, 1982) partly to explain how regions which could not have been in causal contact with each other have the same temperature. It was soon realized (Starobinsky, 1982; Guth and Pi, 1982; Hawking, 1982; Bardeen, Steinhardt, and Turner, 1983; Brandenberger, Kahn, and Press, 1983; Guth and Pi, 1985) that the very mechanism that explains the uniformity of the temperature in the universe can also account for the origin of perturbations in the universe. Therefore, in order to produce a set of initial conditions, we will need to detour into the world of inflation. One warning: we are not sure that inflation is the mechanism that generated the initial perturbations. It is very difficult to test a theory based on energy scales well beyond the reach of accelerators. Nonetheless, it is by far the most plausible explanation. Indeed, one of the current problems in cosmology is that there is really no viable alternative to inflation. Also, the next generation of CMB and large-scale structure observations will put inflation to some stringent tests.

6.1 THE EINSTEIN–BOLTZMANN EQUATIONS AT EARLY TIMES

Chapters 4 and 5 contain nine first-order differential equations for the nine perturbation variables we need to track. In principle, we need initial conditions for all of these variables. In practice, though, a combination of arguments will relate many of these variables to each other, and we need only determine the initial conditions for one of these. This section determines the way all variables depend on the gravitational potential Φ at early times; the remaining sections work out the initial conditions for Φ.

Let us consider first the Boltzmann equations (4.100)–(4.107) at very early times. In particular, we want to consider times so early that for any k-mode of

139

interest, $k\eta \ll 1$. This inequality immediately leads to several important simplifications. Consider the terms $\dot{\Theta}$ and $ik\mu\Theta$ in Eq. (4.100). The first term is of order Θ/η, while the second is of order $k\Theta$. Therefore, the first is larger than the second by a factor of order $1/(k\eta)$, which, by assumption, is much greater than 1. In a similar way, we can argue that all terms in the Boltzmann equations multiplied by k can be neglected at early times. Physically, this means that, at early times, all perturbations of interest have wavelengths ($\sim k^{-1}$) much larger than the distance over which causal physics operates. A hypothetical observer then who sees only photons from within his causal horizon will see a uniform sky. Thus higher multipoles ($\Theta_1, \Theta_2, \ldots$) are much smaller than the monopole, Θ_0. Therefore, the perturbations to the photon and neutrino temperatures evolve according to

$$\dot{\Theta}_0 + \dot{\Phi} = 0$$

$$\dot{\mathcal{N}}_0 + \dot{\Phi} = 0. \tag{6.1}$$

The same principles can be applied to the matter distributions. The overdensity equations reduce to

$$\dot{\delta} = -3\dot{\Phi}$$

$$\dot{\delta}_b = -3\dot{\Phi}. \tag{6.2}$$

The velocities are comparable to the first moments of the radiation distributions, so they are smaller than the overdensities by a factor of order $k\eta$ and may be set to zero initially. In fact, the baryon velocity is not only comparable to the photon first moment, Θ_1: it is equal to it by virtue of the strength of Compton scattering. That is, the largeness of $\dot{\tau}$ in Eq. (4.106) ensures that $v_b = -3i\Theta_1$. We will use this later when reexamining the Boltzmann equations closer to decoupling. For now, we are interested in times so early that the only relevant fact is that higher moments are all negligibly small.

Now let us turn to the Einstein equations at early times. First consider Eq. (5.27). The first term there contains a factor of k^2 so may be neglected. Also the two matter terms on the right are negligible at early times since radiation dominates. Therefore, we have

$$3\frac{\dot{a}}{a}\left(\dot{\Phi} - \frac{\dot{a}}{a}\Psi\right) = 16\pi Ga^2\left(\rho_\gamma\Theta_0 + \rho_\nu\mathcal{N}_0\right). \tag{6.3}$$

But since radiation dominates, $a \propto \eta$ (recall Eq. (2.100) and the discussion immediately afterward) so $\dot{a}/a = 1/\eta$. Therefore,

$$\frac{\dot{\Phi}}{\eta} - \frac{\Psi}{\eta^2} = \frac{16\pi G\rho a^2}{3}\left(\frac{\rho_\gamma}{\rho}\Theta_0 + \frac{\rho_\nu}{\rho}\mathcal{N}_0\right)$$

$$= \frac{2}{\eta^2}\left(\frac{\rho_\gamma}{\rho}\Theta_0 + \frac{\rho_\nu}{\rho}\mathcal{N}_0\right) \tag{6.4}$$

where the last equality follows by virtue of the zero-order Einstein equation.

To simplify further, we can define the ratio of neutrino energy density to the total radiation density as

$$f_\nu \equiv \frac{\rho_\nu}{\rho_\gamma + \rho_\nu}. \tag{6.5}$$

Then multiplying Eq. (6.4) by η^2 leads to

$$\dot{\Phi}\eta - \Psi = 2\left([1 - f_\nu]\Theta_0 + f_\nu \mathcal{N}_0\right). \tag{6.6}$$

Recall that Eq. (6.1) relates the derivative of the monopoles to the derivative of the potential. We can therefore eliminate both monopoles from Eq. (6.6) by differentiating both right- and left-hand sides. Then,

$$\ddot{\Phi}\eta + \dot{\Phi} - \dot{\Psi} = -2\dot{\Phi} \tag{6.7}$$

where the right-hand side follows since both $\dot{\Theta}_0$ and $\dot{\mathcal{N}}_0$ are equal to $-\dot{\Phi}$ for these large scale modes.

So far we have used only one Einstein equation. The second, Eq. (5.33), describes how the higher moments of the photon and neutrino distributions cause $\Psi + \Phi$ to be nonzero. Let us here neglect these higher order moments, which cause the sum of the gravitational potentials to be slightly nonzero.[1] Under this approximation, we can eliminate Ψ everywhere by simply setting it to $-\Phi$. Then,

$$\ddot{\Phi}\eta + 4\dot{\Phi} = 0. \tag{6.8}$$

Setting $\Phi = \eta^p$ leads to the algebraic equation

$$p(p - 1) + 4p = 0 \tag{6.9}$$

which allows two solutions: $p = 0, -3$. The $p = -3$ mode is the decaying mode. If it is excited very early on, it will quickly die out and have no impact on the universe. The $p = 0$ mode, on the other hand, does not decay if excited. It is the mode we are interested in. If some mechanism can be found which excites this mode, this mechanism may well be responsible for the perturbations in the universe.

Focusing therefore on only the $p = 0$ mode, we see that Eq. (6.6) relates the gravitational potential to the neutrino and photon overdensities:

$$\Phi = 2\left([1 - f_\nu]\Theta_0 + f_\nu \mathcal{N}_0\right). \tag{6.10}$$

Both Θ_0 and $\mathcal{N}_0$ are also constant in time. In most models of structure formation, they are equal since whatever causes the perturbations tends not to distinguish between photons and neutrinos. Therefore, we will set

$$\Theta_0(k, \eta_i) = \mathcal{N}_0(k, \eta_i) \tag{6.11}$$

which leads to

$$\Phi(k, \eta_i) = 2\Theta_0(k, \eta_i) \tag{6.12}$$

where I have explicitly written the k-dependence of all these variables and the fact that we are setting up the initial conditions at some early time η_i.

---

[1] See Exercise 2 for a careful accounting of the effect of the neutrino quadrupole; the photon quadrupole is kept minuscule by Compton scattering, so it really does not contribute to Eq. (5.33).

The initial conditions for matter, both δ and δ_b, depend upon the nature of the primordial perturbations. Combining the first of equations (6.1) and (6.2) leads to

$$\delta = 3\Theta_0 + \text{constant} \qquad (6.13)$$

for the dark matter overdensity, with an identical equation for the baryon overdensity. Primordial perturbations are often divided into those for which the constant in Eq. (6.13) is zero (*adiabatic* perturbations) and those for which the constant is nonzero (*isocurvature* perturbations). Adiabatic perturbations have a constant matter-to-radiation ratio everywhere since

$$\frac{n_{\text{dm}}}{n_\gamma} = \frac{n_{\text{dm}}^{(0)}}{n_\gamma^{(0)}} \left[\frac{1+\delta}{1+3\Theta_0} \right]. \qquad (6.14)$$

The prefactor, the ratio of zero-order number densities, is a constant in both space and time. For the ratio of matter to radiation number density to be uniform, therefore, the combination inside the brackets which linearizes to $1 + \delta - 3\Theta_0$ must be independent of space. So the perturbations must sum to zero,

$$\delta = 3\Theta_0, \qquad (6.15)$$

for adiabatic perturbations. By similar arguments for the baryons, $\delta_b = 3\Theta_0$. There are models based on isocurvature perturbations, but these have not been very successful to date; we will focus on adiabatic initial conditions.

For the most part, velocities and dipole moments are negligibly small in the very early universe. However, we will encounter situations where we need to know the initial conditions for these as well. You will show in Exercise 3 that the appropriate initial conditions are

$$\Theta_1 = \mathcal{N}_1 = \frac{iv_b}{3} = \frac{iv}{3}$$

$$= -\frac{k\Phi}{6aH}. \qquad (6.16)$$

6.2 THE HORIZON

If this book were a novel or a biography, a better title for this section might be *Midlife Crisis*. The main character would have attended a good high school, studied hard, and gone on to a solid university. There he fell in love with an exciting, but sensible, woman; upon graduating, he set up some interviews, and got a good job downtown. He married his college girlfriend, and after several years in the city, they moved to the suburbs and had three kids. Our hero contributed to the community and was recognized all over town as a solid citizen. He was moving up fast in his company and there was talk about a political position. Just when he was about to declare his candidacy, he began to have doubts. "What have I been doing with my life? What is really important? Were all those years of study and work simply a 'track'? Did I take this path just because everyone else was moving in the same direction? Where is the innovation and the signature that my life is mine?" And worse, he has a secret, an underlying feeling that everything he has built is based on a fallacy.

OK, maybe it wouldn't be a bestseller, but it does serve as a useful metaphor for our study of perturbations in the universe. Until now, we have done everything in a systematic, proper way. We reviewed the standard Big Bang cosmology. We expanded about this zero-order smooth universe, getting evolution equations for the perturbations to the particle distributions and to the gravitational fields. We realized that these coupled differential equations needed initial conditions so in the last section we set those up. However, now we must ask, What caused those initial perturbations? It is one thing to say that $\Phi = 2\Theta_0$ initially. It is quite another to explain what caused Φ to be nonzero in the first place.

And it is worse than that. To understand why let us recall the physical meaning of the conformal time η: it is the maximum comoving distance traveled by light since the beginning of the universe. Equivalently, objects separated by comoving distances larger than η today were not ever in causal contact: there is simply no way information could have propagated over distances larger than η. For this reason, η is called the comoving horizon.

With this in mind, we can now revisit the condition used in the previous section that $k\eta \ll 1$. The wavenumber k is roughly equal to the inverse of the wavelength of the mode in question (give or take a factor of 2π). Therefore $k\eta$ is the ratio of the comoving horizon to the comoving wavelength of the perturbation. If this ratio is much smaller than 1, then the mode in question has a wavelength so large that no causal physics could possibly have affected it. A picture worth remembering is shown in Figure 6.1. The horizon grows as the scale factor increases. On the other hand, comoving wavelengths remain constant. All modes of cosmological interest therefore had wavelengths much larger than the horizon early on. Eventually these cosmological modes *enter* the horizon; after that, causal physics begins to operate on them.

The truly disturbing feature of this realization is most apparent when considering the microwave background today. On all scales observed the CMB is very close to isotropic. How can this be? The largest scales observed have entered the horizon just recently, long after decoupling. (An example is the scale corresponding to the quadrupole moment of the CMB, shown in Figure 6.1.) Before decoupling, the wavelengths of these modes are so large that no causal physics could force deviations from smooothness to go away. After decoupling, the photons do not interact at all; they simply freestream. So even though it is technically possible that photons reaching us today from opposite directions had a chance to communicate with each other and equilibrate to the same temperature, practically this could not have happened. Why then is the CMB temperature so uniform? This is a profound problem that we have glossed over by simply assuming that the temperature is uniform and that perturbations about the zero-order temperature are small.

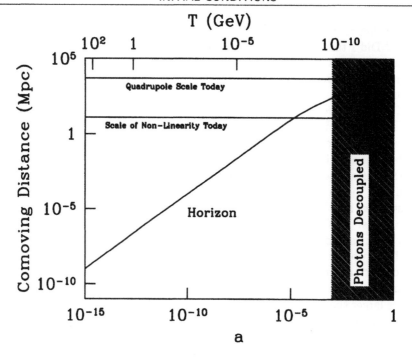

Figure 6.1. The comoving horizon as a function of the scale factor. Also shown are two comoving wavelengths, which remain constant with time. Early in the history of the universe, both of these modes — as well as all other modes of cosmological interest — had wavelengths much larger than the horizon. The CMB comes from the last scattering surface at $a \simeq 10^{-3}$. At that time, the largest scales (e.g., the one labeled "quadrupole") were still outside the horizon. The horizon problem asks how regions separated by distances larger than the horizon at the last scattering surface can have the same temperature.

A more intuitive picture of the horizon problem is shown in Figure 6.2. At any given time, the region within the cone is causally connected to us (at the center). Photons that we observe today from the last scattering surface were well outside our horizon when they were first emitted. The most disturbing aspect of this is the observation of large-angle isotropy, an indication that photons apparently separated by many horizons at the last scattering surface nonetheless shared the same temperature (to a part in 10^5).

6.3 INFLATION

This section describes a beautiful solution to the horizon problem outlined in the previous section. First, we explore a logical way out of the previous argument by realizing that an early epoch of rapid expansion solves the horizon problem. Then we consider the Einstein equations to tell us what type of energy is needed in order to produce this rapid expansion, showing that negative pressure is required.

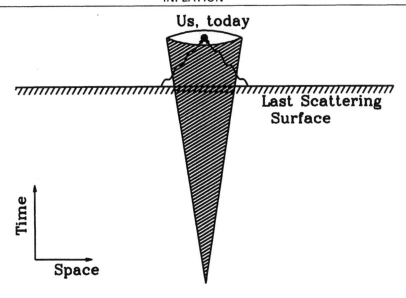

Figure 6.2. The horizon problem. The region inside the cone at any time is causally connected to us (at the center). Photons emitted from the last scattering surface (at redshift ∼ 1000) started outside of this region. Therefore, at the last scattering surface, they were not in causal contact with us and certainly not with each other. Yet their temperatures are almost identical.

Finally, we consider a scalar field theory and show that negative pressure is easy to accommodate in such a theory.

Two comments about the field theory implementation. First, field theory has a reputation as a difficult subject. It is, but the part we will need for inflation is decidedly simple. Indeed, almost all we will need to know about field theory we've already used in the previous chapter on general relativity. The second point is that there is no known scalar field which can drive inflation. (A skeptic might point out that there is no known fundamental scalar field at all!) Therefore, it may well be true that the idea of inflation is correct but it is driven by something other than a scalar field. Having said that, there are a number of reasons to work with scalar fields, as we will do whenever we need to specify the source of inflation. Almost all fundamental particle physics theories contain scalar fields. In fact, historically it was particle physicists studying high-energy extensions of the Standard Model (in particular Grand Unified Theories) who proposed the idea of inflation driven by a scalar field as a natural byproduct of some of these extensions. Indeed, almost all current work on inflation is based on a scalar field (or sometimes two). The alternative from a particle physics point of view is to use a vector field (such as the electromagnetic potential) or a set of fermions (similar to the way condensates induce superconductivity) to drive inflation. Neither of these choices works very well, but they both complicate things severely, so we will stick to a scalar field.

6.3.1 A Solution to the Horizon Problem

To motivate a solution to the horizon problem, let me rewrite the comoving horizon as

$$\eta = \int_0^a \frac{da'}{a'} \frac{1}{a' H(a')}. \tag{6.17}$$

The comoving horizon then is the logarithmic integral of the *comoving Hubble radius*, $1/aH$. The Hubble radius is the distance over which particles can travel in the course of one expansion time, i.e., roughly the time in which the scale factor doubles. So the Hubble radius is another way of measuring whether particles are causally connected with each other: if they are separated by distances larger than the Hubble radius, then they cannot currently communicate. There is a subtle distinction between the comoving horizon η and the comoving Hubble radius $(aH)^{-1}$. If particles are separated by distances greater than η, they *never* could have communicated with one another; if they are separated by distances greater than $(aH)^{-1}$, they cannot talk to each other *now*. It is therefore possible that η could be much larger than $(aH)^{-1}$ now, so that particles cannot communicate today but were in causal contact early on. This might happen if the comoving Hubble radius early on was much larger than it is now so that η got most of its contribution from early times. This could happen, but it does not happen during matter- or radiation-dominated epochs. In those cases, the comoving Hubble radius increases with time, so typically we expect the largest contribution to η to come from the most recent times. Indeed, this is precisely what Figure 6.1 indicates.

Look again at Figure 6.1. On top of the figure I have drawn an axis which depicts the temperature of the cosmic plasma for the given value of the scale factor. We know quite a bit about physics going up to the limits on the plot, several hundred GeV. Beneath these energies, the standard model of particle physics works very well. Beyond those energies, although we have ideas, there is no experimental reason to prefer one theory over another. Since the energy content of the universe determines $a(t)$, when you mentally extrapolate the horizon in Figure 6.1 back to $a = 0$, or equivalently to infinitely high temperatures, you are really making an assumption. You are assuming that nothing strange happened early on, in particular that the universe was always radiation dominated at early times. If this were true, then it does indeed follow that the comoving horizon received a negligible contribution from the very early universe, that photons can travel only very small distances in the first fraction of a second after the Big Bang.

This suggests a solution to the horizon problem: perhaps early on, the universe was not dominated by either matter or radiation. Perhaps, for at least a brief time, the comoving Hubble radius decreased. Then, we would have the situation depicted in Figure 6.3. The comoving Hubble radius would decrease dramatically during this epoch. In that case, the comoving horizon would get most of its contribution not from recent times, but rather from primordial epochs before the rapid expansion of the grid. Particles separated by many Hubble radii today, for example those outside the circle in the bottom panel of the figure, were in causal contact — were inside the Hubble circle in the top panel — before this epoch of rapid expansion.

How must the scale factor evolve in order to solve the horizon problem? We can first answer this question qualitatively. If the comoving Hubble radius is to decrease, then aH must increase. That is,

$$\frac{d}{dt}\left[a\frac{da/dt}{a}\right] = \frac{d^2a}{dt^2} > 0. \qquad (6.18)$$

So to solve the horizon problem, the universe must go through a period in which it is accelerating, expanding ever more rapidly. This is the origin of the term *inflation*. To understand the epoch of inflation more quantitatively, let me give away the punchline that most inflationary models typically operate at energy scales of order 10^{15} GeV or larger. How big was the comoving Hubble radius when the temperature was 10^{15} GeV? We can get an order of magnitude estimate by ignoring the relatively brief epoch of recent matter domination and assuming that the universe has been radiation dominated since the end of inflation (you can correct this assumption in Exercise 6). Then H scales as a^{-2} so $a_0 H_0/a_e H_e = a_e$ where a_e is the scale factor at the end of inflation. If a_e corresponds to a time at which the temperature was 10^{15} GeV, then $a_e \simeq T_0/10^{15}$ GeV $\simeq 10^{-28}$. So the comoving Hubble radius at the end of inflation was 28 orders of magnitude smaller than it is today. For inflation to work, the comoving Hubble radius at the onset of inflation had to be larger than the largest scales observable today, i.e., larger than the current comoving Hubble radius. So during inflation, the comoving Hubble radius had to decrease by some 28 orders of magnitude.

The most common way to arrange this is to construct a model wherein H is constant during inflation. In that case, since $da/a = Hdt$, the scale factor evolves as

$$a(t) = a_e e^{H(t-t_e)} \qquad t < t_e \qquad (6.19)$$

where t_e is the time at the end of inflation. The decrease in the comoving Hubble radius $(aH)^{-1}$ is now due solely to the exponential increase in the scale factor. For the scale factor to increase by a factor of 10^{28}, the argument of the exponential must be of order $\ln(10^{28}) \sim 64$ (but remember the corrections in Exercise 6), so inflation can solve the horizon problem if the universe expands exponentially for more than 60 e-folds.

Thus, consider Figure 6.4, which shows the comoving Hubble radius as a function of the scale factor. The right side of this plot is virtually identical to Figure 6.1, which tells us that the comoving scales of interest to us were much larger than the Hubble radius in the standard cosmology. The left-hand side of the plot though shows that an inflationary epoch reduces the comoving Hubble radius dramatically. This makes sense: since the scale factor is inflating very rapidly, it becomes increasingly difficult for photons to move along the comoving grid (which is itself expanding with a). Before inflation started, the comoving Hubble radius was very large, larger than any scale of cosmological interest today, so all such scales were well within the horizon.

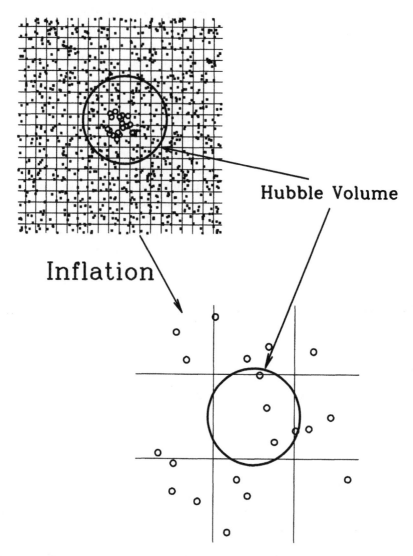

Figure 6.3. Particles on the comoving grid before (top) and after (bottom) inflation. Open circles are the same particles on top and bottom. Before inflation, the comoving Hubble radius was quite large, encompassing dozens of cells on the grid. After inflation, the comoving Hubble radius has shrunk to just one cell. (In this caricature, the scale factor has grown by a factor of order 7; during inflation the scale factor increases by greater than e^{60}.) The shrinkage of the comoving Hubble radius means that particles which were initially in causal contact with one another (within the large circle at top) can now no longer communicate. Note that the physical Hubble radius, depicted by large circles on the top and bottom grids, remains roughly constant during inflation.

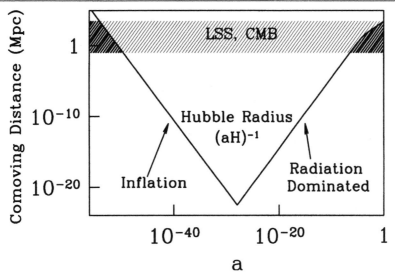

Figure 6.4. The comoving Hubble radius as a function of scale factor. Scales of cosmological interest (shaded band) were larger than the Hubble radius until $a \sim 10^{-5}$. Dark shaded regions show when these scales were smaller than the Hubble radius, and therefore susceptible to microphysical processes. Very early on, before inflation operated, all scales of interest were smaller than the Hubble radius and therefore susceptible to microphysical processing. Similarly, at very late times, scales of cosmological interest came back within the Hubble radius.

Note the symmetry in Figure 6.4. Scales just entering the horizon today—roughly 60 e-folds after the end of inflation—left the horizon 60 e-folds before the end of inflation. The amplitude of the perturbations on these scales remained constant as long as they were super-horizon. So, when we measure them today, we are actually seeing them as they were when they first left the horizon during the inflationary era (modulo whatever processing has taken place since they reentered the horizon, processing we will study in great detail in Chapters 7 and 8). To explain the structure in the universe today, then, it is clearly important to understand the generation of perturbations during inflation.

We have until now discussed inflation in comoving coordinates. But it is also profitable to think of the exponential expansion in physical coordinates. The idea that the horizon blows up early on is depicted in Figure 6.5. The physical size (a times the comoving size) of a causally connected region blows up exponentially quickly during inflation. So regions that we observe to be astronomical today were actually microscopically small before inflation, and they were in causal contact with each other.

The total comoving horizon ceases to be an effective time parameter after inflation because it becomes large very early on, and then changes very little as the universe expands during the matter- and radiation-dominated eras. A simple way to rectify this is to subtract off its primordial part η_{prim}, and redefine η as

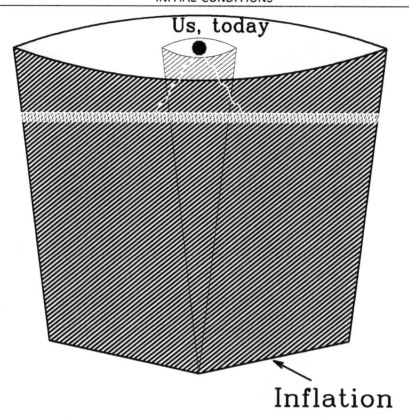

Figure 6.5. Inflationary solution to the horizon problem. Larger cone shows the true horizon in an inflationary model; smaller inner cone shows the horizon without inflation. During inflation, the physical horizon blows up very rapidly. All scales in the shaded region were once in causal contact so it is not surprising that the temperature is uniform.

$$\eta = \int_{t_e}^{t} \frac{dt'}{a(t')}, \tag{6.20}$$

so that the total comoving horizon is $\eta_{\mathrm{prim}} + \eta$. This is the convention we will follow; note that this means that during inflation, η is negative, but always monotonically increasing. A scale leaves the horizon in the sense of Figure 6.4 when $k|\eta|$ becomes less than 1, and returns at late times when $k\eta$ becomes larger than 1.

To sum up, inflation — an epoch in which the universe accelerates — solves the horizon problem. During the accelerated expansion the physical Hubble radius remains fixed, so particles initially in causal contact with one another can no longer communicate. Regions which are separated by vast distances today were actually in causal contact before and during inflation. At that time, these regions were given the necessary initial conditions, the smoothness we observe today, but also, as we will soon see, the small perturbations about smoothness that eventually grew into galaxies and other structure in the universe.

6.3.2 Negative Pressure

We have shown that an accelerating universe can solve the horizon problem. Since general relativity ties the expansion of the universe to the energy in it, we now need to ask what type of energy can produce acceleration. We can get an immediate answer if we appeal to the time-time and space-space components of the zero-order Einstein equations. They are (Eqs. (2.39) and (2.93))

$$\left(\frac{da/dt}{a} \right)^2 = \frac{8\pi G}{3} \rho$$

$$\frac{d^2 a/dt^2}{a} + \frac{1}{2} \left(\frac{da/dt}{a} \right)^2 = -4\pi G \mathcal{P}. \tag{6.21}$$

Multiplying the first of these by one-half and then subtracting one from the other eliminates the first derivative of a, leaving

$$\frac{d^2 a/dt^2}{a} = -\frac{4\pi G}{3} \left(\rho + 3\mathcal{P} \right). \tag{6.22}$$

Acceleration is defined to mean that $d^2 a/dt^2$ is positive. For this to happen, the terms in parentheses on the right must be negative. So inflation requires

$$\mathcal{P} < -\frac{\rho}{3}. \tag{6.23}$$

Since the energy density is always positive, the pressure must be negative. This result is perhaps not surprising: we saw back in Chapter 2 that the accelerated expansion which cause supernovae to appear very faint can be caused only by dark energy with negative pressure. Inflation was apparently driven by the a similar form of energy, one with $\mathcal{P} < 0$. To reiterate what we emphasized in Chapter 2, negative pressure is not something with which we have any familiarity. Nonrelativistic matter has small positive pressure proportional to temperature divided by mass, while a relativistic gas has $\mathcal{P} = +\rho/3$, again positive. So whatever it is that drives inflation is not ordinary matter or radiation.

6.3.3 Implementation with a Scalar Field

We have become familiar with the fields $\Psi(\vec{x}, t)$ and $\Phi(\vec{x}, t)$, deriving equations for them which govern their evolution (the Einstein equations) and the evolution of particles which are affected by them. These two fields are parts of the multicomponent field, the metric $g_{\mu\nu}$. The metric is one of the fundamental fields in physics, but there are others. Every elementary particle—the electron, neutrino, quarks, photon, etc.—is associated with its own field. It would be wonderful if one of these fields, the electromagnetic potential associated with photons say, could serve as the source for an inflationary model. Unfortunately we do not yet have such a concrete model. Instead, I will discuss inflation in terms of a generic scalar field (not a fermion like the quarks and leptons or a vector like the electromagnetic field). The simplest version of the standard model does indeed have within it one such field, the Higgs field. But, again unfortunately, we know too much about the Higgs of the standard model. Its interactions and properties are constrained enough for us to know that it cannot serve as the source for inflation. So we will drop any pretensions of connecting the generic scalar field which drives inflation to known physics. Making this connection is left as a homework problem for a future Nobel laureate.

We want to know if a scalar field — which I will call $\phi(\vec{x}, t)$, not to be confused with the metric perturbation $\Phi(\vec{x}, t)$ — can have negative $\rho + 3\mathcal{P}$. So our first task is to write down the energy–momentum tensor for ϕ. This is

$$T^{\alpha}{}_{\beta} = g^{\alpha\nu} \frac{\partial \phi}{\partial x^{\nu}} \frac{\partial \phi}{\partial x^{\beta}} - g^{\alpha}{}_{\beta} \left[\frac{1}{2} g^{\mu\nu} \frac{\partial \phi}{\partial x^{\mu}} \frac{\partial \phi}{\partial x^{\nu}} + V(\phi) \right]. \tag{6.24}$$

Here $V(\phi)$ is the potential for the field. For example a free field with mass m has a potential $V(\phi) = m^2 \phi^2 / 2$. A warning about signs: if you delve into the literature you will invariably find different signs than those in Eq. (6.24). These are dictated by the choice of metric. Although our metric signature $(-, +, +, +)$ is probably most common in the context of cosmology, it is probably not as common in particle physics. Beware. We will assume that ϕ is mostly homogeneous, consisting of a zero-order part, $\phi^{(0)}(t)$, and a first-order perturbation, $\delta\phi(\vec{x}, t)$. In this section we will derive information about the zero-order homogeneous part, its energy density and pressure, and its time evolution. Later we will consider its perturbations, $\delta\phi$, and how they are generated.

For the homogeneous part of the field, only time derivatives of ϕ are relevant so the indices α and β in the first term in Eq. (6.24) and μ, ν in the second must be equal to zero. The energy–momentum tensor then reduces to

$$T^{(0)\alpha}{}_{\beta} = -g^{\alpha}{}_{0} g^{0}{}_{\beta} \left(\frac{d\phi^{(0)}}{dt} \right)^2 + g^{\alpha}{}_{\beta} \left[\frac{1}{2} \left(\frac{d\phi^{(0)}}{dt} \right)^2 - V(\phi^{(0)}) \right]. \tag{6.25}$$

The time-time component of $T^0{}_0$ is equal to $-\rho$, so the energy density is

$$\rho = \frac{1}{2} \left(\frac{d\phi^{(0)}}{dt} \right)^2 + V(\phi^{(0)}). \tag{6.26}$$

The first term here is the kinetic energy density of the field, the second its potential energy density. A homogeneous scalar field therefore has much the same dynamics as a single particle moving in a potential [think of $\phi^{(0)}(t)$ as the position of the particle $x(t)$]. In fact this analogy dominates even the language used to describe inflation. The pressure for the homogeneous field is $\mathcal{P} = T^{(0)i}{}_i$ (no sum over spatial index i), so

$$\mathcal{P} = \frac{1}{2} \left(\frac{d\phi^{(0)}}{dt} \right)^2 - V(\phi^{(0)}). \tag{6.27}$$

A field configuration with negative pressure is therefore one with more potential energy than kinetic. An example is shown in Figure 6.6, in which a field is trapped in a *false vacuum*, i.e., a local, but not the global, minimum of the potential.

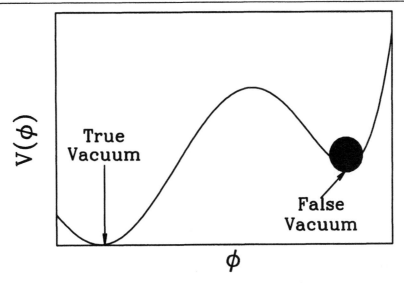

Figure 6.6. A scalar field trapped in a false vacuum. Since it is trapped, it has little kinetic energy. The potential energy is nonzero, however, so the pressure is negative. The global minimum of the potential is called the true vacuum, since a homogeneous field sitting at the global minimum of the potential is in the ground state of the system.

There is something important to notice about a field trapped in a false vacuum. Since $\phi^{(0)}$ is constant, its energy density, which is all potential, remains constant with time. Constant energy density is much different than anything with which we are familiar. The densities of both matter and radiation, for example, fall off very rapidly as the universe expands. Therefore, even if the universe initially contains a mixture of matter, radiation, and false vacuum energy, it will quickly become dominated by the vacuum energy. For a trapped field, it is trivial to determine the evolution of the scale factor. Since the energy density is constant, Einstein's equation for the evolution of a is

$$\frac{da/dt}{a} = \sqrt{\frac{8\pi G\rho}{3}} = \text{constant.} \tag{6.28}$$

We immediately see that a field trapped in a false vacuum produces exponential expansion as in Eq. (6.19), with $H \propto \rho^{1/2}$ constant. The primordial comoving horizon, that generated before the end of inflation, is then obtained by integrating the inverse of Eq. (6.19) over time,

$$\eta_{\text{prim}} = \frac{1}{H_e a_e}\left(e^{H(t_e - t_b)} - 1\right), \tag{6.29}$$

where t_b is the beginning of inflation. So if the field is trapped for at least 60 e-foldings ($H(t_e - t_b) > 60$), the horizon problem is solved.

Guth's (1981) initial formulation of inflation used a scalar field trapped in a false minimum of the potential, but it was quickly realized (Guth and Weinberg, 1983; Hawking, Moss, and Stewart, 1982) that such a scenario is not viable. The only way for the field to evolve to its true minimum is similar to the way an alpha particle migrates out of the potential barrier in a nucleus: it tunnels quantum mechanically. Thus, initially small localized regions tunnel from the false to the true vacuum. These bubbles of the true vacuum state must coalesce in order for the universe as a whole to move to the true vacuum state. Careful calculations showed that these bubbles would never coalesce, that the regions of false vacuum would expand rapidly and remain, so that the true vacuum state of the universe would never be attained.

To avoid the problem of the universe never reaching its true vacuum state, subsequent models of inflation (Linde, 1982; Albrecht and Steinhardt, 1982) made use of a scalar field slowly rolling toward its true ground state. The energy density of such a field is also very close to constant (if the potential is not too steep) so it quickly comes to dominate. To determine the evolution of $\phi^{(0)}$ in general when the field is not trapped, we return to the Einstein equations as given in Eq. (6.21). Consider the first of these. If the dominant component in the universe is ϕ, then the energy density on the right-hand side becomes $(d\phi^{(0)}/dt)^2/2 + V$. Differentiating this first equation therefore leads to

$$2\frac{da/dt}{a}\left[\frac{d^2a/dt^2}{a} - \left(\frac{da/dt}{a}\right)^2\right] = \frac{8\pi G}{3}\left[\left(\frac{d\phi^{(0)}}{dt}\right)\left(\frac{d^2\phi^{(0)}}{dt^2}\right) + V'\frac{d\phi^{(0)}}{dt}\right] \quad (6.30)$$

where V' is defined as the derivative of V with respect to the field $\phi^{(0)}$. We can replace the first term in brackets on the left by $-4\pi G(\rho/3+\mathcal{P})$ as in Eq. (6.22). Similarly the second term on the left is $8\pi G\rho/3$. The left-hand side therefore becomes

$$\frac{da/dt}{a}8\pi G\left[-(\rho/3) - \mathcal{P} - 2\rho/3\right] = -8\pi GH\left(\frac{d\phi^{(0)}}{dt}\right)^2. \quad (6.31)$$

Equating this to the right side of Eq. (6.30) leads to the evolution equation for a homogeneous scalar field in an expanding universe,

$$\frac{d^2\phi^{(0)}}{dt^2} + 3H\frac{d\phi^{(0)}}{dt} + V' = 0. \quad (6.32)$$

A more useful form for us will be with the conformal time η as the time variable; then it is straightforward (Exercise 8) to show that

$$\ddot{\phi}^{(0)} + 2aH\dot{\phi}^{(0)} + a^2V' = 0 \quad (6.33)$$

where overdots still denote derivatives with respect to conformal time η.

Most models of inflation are *slow roll* models, in which the zero-order field, and hence the Hubble rate, vary slowly. Therefore, a simple relation between the conformal time η and the expansion rate holds. In particular, during inflation

$$\eta \equiv \int_{a_e}^{a} \frac{da}{Ha^2}$$

$$\simeq \frac{1}{H} \int_{a_e}^{a} \frac{da}{a^2}$$

$$\simeq \frac{-1}{aH} \tag{6.34}$$

where the rough equality on the second line holds because H is nearly constant, and the one on the third because the scale factor at the end of inflation is much larger than in the middle ($a_e \gg a$). To quantify slow roll, cosmologists typically define two variables which vanish in the limit that ϕ remains constant. First, define

$$\epsilon \equiv \frac{d}{dt}\left(\frac{1}{H}\right) = \frac{-\dot{H}}{aH^2}. \tag{6.35}$$

Since H is always decreasing, ϵ is always positive. During inflation, it is typically small, whereas it is equal to 2 during a radiation era. In fact, one definition of an inflationary epoch is one in which $\epsilon < 1$. A complementary variable which also quantifies how slowly the field is rolling is:

$$\delta \equiv \frac{1}{H}\frac{d^2\phi^{(0)}/dt^2}{d\phi^{(0)}/dt} = \frac{-1}{aH\dot{\phi}^{(0)}}\left[aH\dot{\phi}^{(0)} - \ddot{\phi}^{(0)}\right]$$

$$= \frac{-1}{aH\dot{\phi}^{(0)}}\left[3aH\dot{\phi}^{(0)} + a^2 V'\right]. \tag{6.36}$$

Here the paucity of Greek letters becomes a hindrance. The second slow-roll parameter is more conventionally defined as η, but we obviously cannot follow that convention as η is our conformal time. (Early universe cosmologists use τ for conformal time, freeing up η, but we do not have that luxury since we need τ for optical depth.) My choice of δ is also fairly common, but we need to bear in mind that this has nothing to do with the overdensities introduced in Chapter 4. The second line here follows from Eq. (6.33). Again, in most models δ is small. We will see in Section 6.6 that some unique features of inflation, deviations from the simplest possible spectrum and the production of gravity waves, are proportional to ϵ and δ. If these features are one day measured, they will not only be unique signatures of inflation but also allow us to learn something about the physics driving inflation.

6.4 GRAVITY WAVE PRODUCTION

Inflation does more than solve the horizon problem. The power of inflation is its ability to correlate scales that would otherwise be disconnected. The zero-order scheme outlined in the previous section ensures that the universe will be uniform on all scales of interest today. There are perturbations about this zero-order scheme, though, and these perturbations — produced early on when the scales are causally connected — persist long after inflation has terminated.

We are most interested in *scalar* perturbations to the metric since these couple to the density of matter and radiation and ultimately are responsible for most of the inhomogeneities and anisotropies in the universe. In Section 6.5 we will study these in detail. In addition to scalar perturbations, though, inflation also generates *tensor* fluctuations in the gravitational metric, so-called gravity waves. As we saw in Chapter 5, these are not coupled to the density and so are not responsible for the large-scale structure of the universe, but they do induce fluctuations in the CMB. In fact, these fluctuations turn out to be a unique signature of inflation and offer the best window on the physics driving inflation, so they are clearly worthy of our study. I choose to study the production of tensor perturbations first before scalar perturbations for a subtle technical reason. Tensor perturbations to the metric are not coupled to any of the other perturbation variables,[2] so when we consider them, we will be looking at the fluctuations in a single field. Scalar perturbations to the metric couple to energy density fluctutations. The coupled fields fluctuate together and this coupling requires a bit of work to understand. This work, while important, is *not* the main point: the most important idea is that quantum mechanical fluctuations during inflation are responsible for the variations around the smooth background that so fascinate us. This idea is best introduced in the much simpler context of a single field, so we start with tensor perturbations.

During inflation, the universe consists primarily of a uniform scalar field and a uniform background metric. Against this background, the fields fluctuate quantum mechanically. At any given time, the average fluctuation is zero, because there are regions in which the field is slightly larger than its average value and regions in which it is smaller. The average of the square of the fluctuations (the variance), however, is not zero. Our goal is to compute this variance and see how it evolves as inflation progresses. Looking ahead, once we know this variance, we can draw from a distribution with this variance to set the initial[3] conditions.

6.4.1 Quantizing the Harmonic Oscillator

In order to compute the quantum fluctuations in the metric, we need to quantize the field. The way to do this, in the case of both tensor and scalar perturbations, is to rewrite the problem so that it looks like a simple harmonic oscillator. Once that is done, we will appeal to our knowledge of this simple system. Therefore, let's first record some basic facts about the quantization of the harmonic oscillator.

- A simple harmonic oscillator with unit mass and frequency ω is governed by the equation

$$\frac{d^2 x}{dt^2} + \omega^2 x = 0. \tag{6.37}$$

- Upon quantization, x becomes a quantum operator

$$\hat{x} = v(\omega, t)\hat{a} + v^*(\omega, t)\hat{a}^\dagger \tag{6.38}$$

---

[2]This is not quite true. The quadrupole moments act as sources for tensor perturbations, but these vanish if a scalar field drives inflation. See Exercise 10

[3]*Initial* here means those when the modes of interest are still far outside the horizon. This is well before any processing can take place, but well after inflation has generated them.

where $\hat{a}$ is a quantum operator which acts on the state of the system, and v is a solution to Eq. (6.37), $v \propto e^{-i\omega t}$.

- In particular, $\hat{a}$ annihilates the *vacuum* state $|0\rangle$, in which there are no particles. It also satisfies the commutation relation

$$[\hat{a}, \hat{a}^\dagger] \equiv \hat{a}\hat{a}^\dagger - \hat{a}^\dagger\hat{a} = 1. \tag{6.39}$$

Other commutators vanish: $[\hat{a}, \hat{a}] = [\hat{a}^\dagger, \hat{a}^\dagger] = 0$. It is straightforward to show (Exercise 9) that these commutation relations are equivalent to the (perhaps more familiar) relations between $\hat{x}$ and its momentum $\hat{p}$:

$$[\hat{x}, \hat{p}] = i, \tag{6.40}$$

as long as v is normalized via

$$v(\omega, t) = \frac{e^{-i\omega t}}{\sqrt{2\omega}}. \tag{6.41}$$

These facts enable us to compute the quantum fluctuations of the operator $\hat{x}$ in the ground state $|0\rangle$:

$$\langle |\hat{x}|^2 \rangle \equiv \langle 0|\hat{x}^\dagger \hat{x}|0\rangle$$
$$= \langle 0| \left(v^* \hat{a}^\dagger + v\hat{a}\right)\left(v\hat{a} + v^*\hat{a}^\dagger\right)|0\rangle. \tag{6.42}$$

Since $\hat{a}|0\rangle = 0$, the first term in the second set of parentheses vanishes. Similarly, $\langle 0|\hat{a}^\dagger = (a|0\rangle)^\dagger = 0$, so we are left with

$$\langle |\hat{x}|^2 \rangle = |v(\omega, t)|^2 \langle 0|\hat{a}\hat{a}^\dagger|0\rangle$$
$$= |v(\omega, t)|^2 \langle 0|[\hat{a}, \hat{a}^\dagger] + \hat{a}^\dagger \hat{a}|0\rangle. \tag{6.43}$$

The second term again vanishes since $\hat{a}$ annihilates the vacuum, while the first is unity, so the variance in $\hat{x}$ is

$$\langle |\hat{x}|^2 \rangle = |v(\omega, t)|^2, \tag{6.44}$$

in this case $1/2\omega$. This is (almost) all we need to know about quantum fluctuations in order to compute the fluctuations in the early universe generated by inflation.

6.4.2 Tensor Perturbations

Recall that tensor perturbations to the metric are described by two functions h_+ and $h_\times$, each of which obeys Eq. (5.63),

$$\ddot{h} + 2\frac{\dot{a}}{a}\dot{h} + k^2 h = 0. \tag{6.45}$$

We would like to massage this equation into the form of a harmonic oscillator, so that h can be easily quantized. To do this, define[4]

$$\tilde{h} \equiv \frac{ah}{\sqrt{16\pi G}}. \tag{6.46}$$

Derivatives of h with respect to conformal time can be rewritten as

$$\frac{\dot{h}}{\sqrt{16\pi G}} = \frac{\dot{\tilde{h}}}{a} - \frac{\dot{a}}{a^2}\tilde{h} \tag{6.47}$$

and

$$\frac{\ddot{h}}{\sqrt{16\pi G}} = \frac{\ddot{\tilde{h}}}{a} - 2\frac{\dot{a}}{a^2}\dot{\tilde{h}} - \frac{\ddot{a}}{a^2}\tilde{h} + 2\frac{(\dot{a})^2}{a^3}\tilde{h}. \tag{6.48}$$

Inserting these into Eq. (6.45), and multiplying by $\sqrt{16\pi G}$, gives

$$\frac{\ddot{\tilde{h}}}{a} - 2\frac{\dot{a}}{a^2}\dot{\tilde{h}} - \frac{\ddot{a}}{a^2}\tilde{h} + 2\frac{(\dot{a})^2}{a^3}\tilde{h} + 2\frac{\dot{a}}{a}\left(\frac{\dot{\tilde{h}}}{a} - \frac{\dot{a}}{a^2}\tilde{h}\right) + k^2\frac{\tilde{h}}{a}$$

$$= \frac{1}{a}\left[\ddot{\tilde{h}} + \left(k^2 - \frac{\ddot{a}}{a}\right)\tilde{h}\right] = 0. \tag{6.49}$$

This is precisely the form we know how to use. It has no damping term ($\propto \dot{\tilde{h}}$) so we can immediately write down an expression for the quantum operator

$$\hat{\tilde{h}}(\vec{k}, \eta) = v(k, \eta)\hat{a}_{\vec{k}} + v^*(k, \eta)a^\dagger_{\vec{k}}, \tag{6.50}$$

where the coefficients of the creation and annihilation operators satisfy the equation

$$\ddot{v} + \left(k^2 - \frac{\ddot{a}}{a}\right)v = 0. \tag{6.51}$$

We will shortly solve Eq. (6.51), but first let's see how the eventual solution determines the power spectrum of the fluctuations of the tensor perturbations.

---

[4]Regarding the factor of $\sqrt{16\pi G}$ here, the only way I know of deriving this is to write down the action for the fields $h_{+,\times}$. The kinetic term is then multiplied by a factor of $1/32\pi G$. A canonical scalar field has prefactor equal to a half. So the additional $16\pi G$ must be absorbed into a redefinition of the field. The hard part of this is writing down the action to second order in perturbation variables. We have seen that even first-order perturbations are cumbersome to track. On the other hand, by dimensional analysis—the fact that $h(\vec{x})$ is dimensionless while a canonical scalar field has dimensions equal to mass—we could have guessed that the factor of $m_{\rm Pl} = G^{-1/2}$ is required. Note that this prefactor does not affect the equation for $\tilde{h}$; it simply provides the normalization that becomes important when trying to determine the amplitude of the gravity-wave spectrum.

Using our harmonic oscillator analogy, we can write the variance of perturbations in the $\tilde{h}$ field as

$$\langle \hat{\tilde{h}}^{\dagger}(\vec{k},\eta)\hat{\tilde{h}}(\vec{k}',\eta)\rangle = |v(\vec{k},\eta)|^2 (2\pi)^3 \delta^3(\vec{k}-\vec{k}').$$ (6.52)

There is one difference between this expression and the analogous expression for the one-dimensional harmonic oscillator in Eq. (6.44). A quantum field is defined in all space, so it can be considered as a collection, an infinite collection, of oscillators, each at a different spatial position (or, in Fourier space, at different values of $\vec{k}$). The quantum fluctuations in each of these oscillators are independent (as long as the equations are linear) so $\hat{\tilde{h}}(\vec{k})$ is completely uncorrelated with $\hat{\tilde{h}}(\vec{k}')$ if $\vec{k} \neq \vec{k}'$. The Dirac delta function in Eq. (6.52) enforces this independence; the $(2\pi)^3$ allows for the fact that we have moved to the continuum limit. Recalling that $\tilde{h} = ah/\sqrt{16\pi G}$, we see that

$$\langle \hat{h}^{\dagger}(\vec{k},\eta)\hat{h}(\vec{k}',\eta)\rangle = \frac{16\pi G}{a^2}\, |v(k,\eta)|^2\, (2\pi)^3 \delta^3(\vec{k}-\vec{k}')$$

$$\equiv (2\pi)^3 P_h(k)\delta^3(\vec{k}-\vec{k}')$$ (6.53)

where the second line defines the *power spectrum* of the primordial perturbations to the metric. Conventions for the power spectrum abound in the literature; the one I've chosen in Eq. (6.53) is not the most popular in the early universe community. Often a factor of k^{-3} is added so that the power spectrum is dimensionless. I prefer to omit this factor to be consistent with the large scale structure community which likes its power spectra to have dimensions of k^{-3}. In any event, with this definition,

$$P_h(k) = 16\pi G\, \frac{|v(k,\eta)|^2}{a^2}.$$ (6.54)

We have now reduced the problem of determining the spectrum of tensor perturbations produced during inflation to one of solving a second-order differential equation for $v(k,\eta)$, Eq. (6.51). To solve this equation, we first need to evaluate $\ddot{a}/a$ during inflation. Recall that overdots denote derivative with respect to conformal time, so $\dot{a} = a^2 H \simeq -a/\eta$ by virtue of Eq. (6.34). Therefore, the second derivative of a in Eq. (6.51) is

$$\frac{\ddot{a}}{a} \simeq -\frac{1}{a}\frac{d}{d\eta}\left(\frac{a}{\eta}\right)$$

$$\simeq \frac{2}{\eta^2}.$$ (6.55)

So the equation for v is

$$\ddot{v} + \left(k^2 - \frac{2}{\eta^2}\right)v = 0.$$ (6.56)

The initial conditions necessary to solve this equation come from considering v at very early times before inflation has done most of its work. At that time, $-\eta$ is large, of order η_{prim}, so the k^2 term dominates, and the equation reduces precisely to that of the simple harmonic oscillator. In that case, we know (Eq. (6.41)) that the properly normalized solution is $e^{-ik\eta}/\sqrt{2k}$. This knowledge enables us to choose (Exercise 11) the proper solution to Eq. (6.56),

$$v = \frac{e^{-ik\eta}}{\sqrt{2k}} \left[1 - \frac{i}{k\eta} \right]. \tag{6.57}$$

This obviously goes into the correct solution when the mode is well within the horizon ($k|\eta| \gg 1$). Even if you don't work through Exercise 11 (which arrives at the relatively simple solution of Eq. (6.57) in a rather tortured way), you should at least check that the v given here is indeed a solution to Eq. (6.56).

After inflation has worked for many e-folds $k|\eta|$ becomes very small. Now that v has been normalized, we can determine the amplitude of v, and hence the variance of the super-horizon gravitational wave amplitude, by taking the small argument limit of Eq. (6.57):

$$\lim_{-k\eta \to 0} v(k, \eta) = \frac{e^{-ik\eta}}{\sqrt{2k}} \frac{-i}{k\eta}. \tag{6.58}$$

Figure 6.7 shows the evolution of $h \propto v/a$ during inflation. At early times h falls as $1/a$ as inflation reduces the amplitude of the modes. Once $-k\eta$ becomes smaller than unity, the mode leaves the horizon, after which h remains constant.

The primordial power spectrum for tensor modes, which scales as $|v|^2/a^2$, is therefore constant in time after inflation has stretched the mode to be larger than the horizon. This constant determines the initial conditions for the gravity waves, those with which to start off $h_{+,\times}$ at early times (where in this context "early" means well after inflation has ended but before decoupling). Equations (6.54) and (6.58) show that this constant is

$$P_h(k) = \frac{16\pi G}{a^2} \frac{1}{2k^3\eta^2}$$

$$= \frac{8\pi G H^2}{k^3}. \tag{6.59}$$

The second line here follows from Eq. (6.34). We have assumed that H is constant in deriving this result; more generally, H is to be evaluated at the time when the mode of interest leaves the horizon. This is our final expression for the primordial power spectrum of gravity waves. Detection of these waves would, quite remarkably, measure the Hubble rate during inflation. Since potential energy usually dominates kinetic energy in inflationary models, a measure of H would be tantamount to measuring the potential V, again quite remarkable in view of the likelihood that inflation was generated by physics at energy scales above 10^{15} GeV, 12 orders of magnitude beyond the capacity of present-day accelerators. There is no guarantee that gravity waves produced during inflation will be detectable. Indeed, since $H^2 \propto \rho/m_{\mathrm{Pl}}^2$, the power spectrum is proportional to ρ/m_{Pl}^4, the energy density at the time

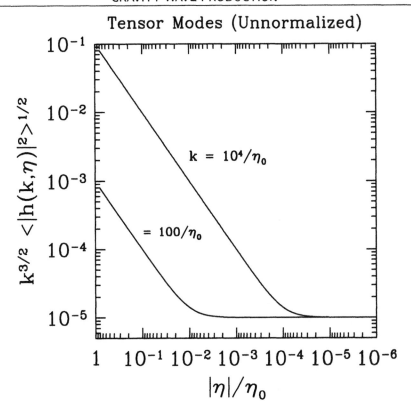

Figure 6.7. The root mean square fluctuations in the tensor field during inflation for two different k-modes. Time evolves from left to right: η is negative but gets closer to zero during inflation. Once a mode "leaves the horizon" ($\eta \sim -1/k$), its RMS amplitude remains constant. Note that after a mode has left the horizon, its RMS amplitude times $k^{3/2}$ is the same for all modes. This is called a scale-free spectrum, strictly true only if the Hubble rate is constant when the scale of interest leaves the horizon (the choice here).

of inflation in units of the Planck mass. If inflation takes place at scales sufficiently smaller than the Planck scale, then primordial gravity waves will not be detected. Later in the book, we will develop the machinery necessary to answer the question, How small can the gravity wave component be and still be detected?

Two final technical points are in order regarding Eq. (6.59). Although I have not emphasized this feature of the spectrum of these primordial perturbations, the fluctuations in h are Gaussian, just as are the quantum-mechanical fluctuations of the simple harmonic oscillator. Gaussianity is a fairly robust prediction of inflation; as such, many studies have been undertaken searching for signs of primordial non-Gaussianity in CMB and large-scale structure data, signs that would jeopardize the inflationary picture. Although there have been some hints, none have held up under greater scrutiny, so this prediction of inflation too appears to be verified. Second,

Eq. (6.59) is the power spectrum for h_+ and $h_\times$ separately; these are uncorrelated, so the power spectrum for all modes must be multiplied by a factor of 2.

6.5 SCALAR PERTURBATIONS

The goal of this chapter is to find the perturbation spectrum of Ψ (or Φ; we assume throughout that they are equal in magnitude) emerging from inflation. With that spectrum, we can use the relations in Section 6.1 to determine the spectrum of the other perturbation variables. Finding the spectrum for Ψ ,however, turns out to be complicated, more so than was the tensor case considered earlier. The primary complication is the presence of perturbations in the scalar field driving inflation, perturbations which are coupled to Ψ.

To deal with this problem, we will first ignore it: in Section 6.5.1, we compute the spectrum of perturbations in the scalar field ϕ generated during inflation, neglecting Ψ. This turns out to be relatively simple to do, since it is virtually identical to the tensor calculation we went through above. Why are we justified in neglecting Ψ and how do the perturbations get transferred from ϕ to Ψ? The next two subsections take turns answering this question from two different points of view. First, Section 6.5.2 argues that — in a sense to be defined there — until a mode moves far outside the horizon, Ψ is indeed negligibly small. Once it is far outside the horizon, this no longer holds, but we will find that a linear combination of Ψ and $\delta\phi$ (the perturbations to the scalar field driving inflation) is conserved. This will allow us to convert the initial spectrum for $\delta\phi$ into a final spectrum for Ψ. The second way of justifying the neglect of perturbations to the metric is to switch gauges and work in a gauge in which the spatial part of the metric is unperturbed, a so-called *spatially flat slicing*. In such a gauge, the calculation of Section 6.5.1 is exact; the only question remaining is how to convert back to conformal Newtonian gauge to move on with the rest of the book. In Section 6.5.3, we identify a *gauge-invariant variable*, one which does not change upon a gauge transformation, which is proportional to $\delta\phi$ in a spatially flat slicing. It is then a simple matter to determine this variable in conformal Newtonian gauge, thereby linking Ψ in conformal Newtonian gauge to $\delta\phi$ in spatially flat slicing. Note that the two solutions to the coupling problem, as worked out in Sections 6.5.2 and 6.5.3, are simply alternative approaches to the same problem. If you are comfortable with gauge transformations, Section 6.5.3 is probably a more elegant approach; the more brute-force approach of Section 6.5.2 gives the same answer though and requires less formalism and background.

6.5.1 Scalar Field Perturbations around a Smooth Background

Let's decompose the scalar field into a zero-order homogeneous part and a perturbation,

$$\phi(\vec{x}, t) = \phi^{(0)}(t) + \delta\phi(\vec{x}, t), \qquad (6.60)$$

and find an equation governing $\delta\phi$ in the presence of a smoothly expanding universe, i.e., with metric $g_{00} = -1; g_{ij} = \delta_{ij}a^2$. Consider the conservation of the energy–momentum tensor,

$$T^\mu{}_{\nu;\mu} = \frac{\partial T^\mu{}_\nu}{\partial x^\mu} + \Gamma^\mu{}_{\alpha\mu}T^\alpha{}_\nu - \Gamma^\alpha{}_{\nu\mu}T^\mu{}_\alpha = 0. \tag{6.61}$$

The $\nu = 0$ component of this equation, expanded out to first order, gives the desired equation for $\delta\phi$. Since we are assuming a smooth metric, the only first-order pieces are perturbations in the energy–momentum tensor. All the Γ's are either zero-order ($\Gamma^0{}_{ij} = \delta_{ij}a^2H$ and $\Gamma^i{}_{0j} = \Gamma^i{}_{j0} = \delta_{ij}H$) or zero (the rest of the components), as we found in Eqs. (2.22) and (2.23). So, writing the perturbed part of the energy–momentum tensor as $\delta T^\mu{}_\nu$ and considering the $\nu = 0$ component of the perturbed conservation equation leads to

$$0 = \frac{\partial \delta T^0{}_0}{\partial t} + ik_i \delta T^i{}_0 + 3H \delta T^0{}_0 - H \delta T^i{}_i. \tag{6.62}$$

It remains to determine the perturbations to the energy–momentum tensor in terms of the perturbations to the scalar field.

First let's compute $\delta T^i{}_0$. Since the time-space components of the scalar metric are zero, the second set of terms in Eq. (6.24), those with prefactor $g^\alpha{}_\beta$, vanish. Therefore,

$$T^i{}_0 = g^{i\nu}\phi_{,\nu}\phi_{,0} \tag{6.63}$$

where I have returned to using $_{,\nu}$ to denote the derivative with respect to x^ν. Since $g^{i\nu} = a^{-2}\delta_{i\nu}$, the index ν must be equal to i. Recall that the zero-order field $\phi^{(0)}$ is homogeneous, so $\phi^{(0)}_{,i} = 0$. The space-time component of the energy–momentum tensor therefore has no zero-order piece. To extract the first-order piece, we can set $\phi_{,i}$ to $\delta\phi_{,i} = ik_i\delta\phi$. Then, setting all other factors to their zero-order values leads to

$$\delta T^i{}_0 = \frac{ik_i}{a^3}\dot{\phi}^{(0)}\delta\phi. \tag{6.64}$$

The additional factor of a enters the denominator here because $\phi^{(0)}_{,0} = \dot{\phi}^{(0)}/a$ (recall that $\dot{\ }$ is derivative with respect to conformal time).

The time-time component of the energy–momentum tensor is a little more difficult:

$$T^0{}_0 = g^{00}(\phi_{,0})^2 - \frac{1}{2}g^{\alpha\beta}\phi_{,\alpha}\phi_{,\beta} - V. \tag{6.65}$$

Setting $\phi = \phi^{(0)} + \delta\phi$ leads to

$$T^0{}_0 = \frac{-1}{2}\left(\phi^{(0)}_{,0} + \delta\phi_{,0}\right)^2 - \frac{1}{2a^2}\delta\phi_{,i}\delta\phi_{,i} - V(\phi^{(0)} + \delta\phi). \tag{6.66}$$

The spatial derivatives come in pairs, and pairs of first-order variables ($\delta\phi_{,i}$) lead to second-order terms. These may therefore be neglected. The potential may be expanded as a zero-order term, $V(\phi^{(0)})$ plus a first-order correction, $V'\delta\phi$, so the first-order correction to the energy–momentum tensor is

$$\delta T^0{}_0 = -\phi^{(0)}_{,0}\delta\phi_{,0} - V'\delta\phi$$

$$= \frac{-\dot\phi^{(0)}\dot{\delta\phi}}{a^2} - V'\delta\phi. \tag{6.67}$$

Similarly, you can show that the space-space component is

$$\delta T^i{}_j = \delta_{ij}\left(\frac{\dot\phi^{(0)}\dot{\delta\phi}}{a^2} - V'\delta\phi\right). \tag{6.68}$$

Therefore, the conservation equation (6.62) becomes

$$\left(\frac{1}{a}\frac{\partial}{\partial\eta} + 3H\right)\left(\frac{-\dot\phi^{(0)}\dot{\delta\phi}}{a^2} - V'\delta\phi\right) - \frac{k^2}{a^3}\dot\phi^{(0)}\delta\phi - 3H\left(\frac{\dot\phi^{(0)}\dot{\delta\phi}}{a^2} - V'\delta\phi\right) = 0. \tag{6.69}$$

Carrying out the time derivatives (the only subtle one is $\partial V'/\partial\eta = V''\dot\phi^{(0)}$), multiplying by a^3, and collecting terms leads to

$$-\dot\phi^{(0)}\ddot{\delta\phi} + \dot{\delta\phi}\left(-\ddot\phi^{(0)} - 4aH\dot\phi^{(0)} - a^2V'\right) + \delta\phi\left(-a^2V''\dot\phi^{(0)} - k^2\dot\phi^{(0)}\right) = 0. \tag{6.70}$$

The V'' term here is typically small, proportional to the slow-roll variables ϵ and δ (Exercise 14), so it can be neglected. The coefficient of $\dot{\delta\phi}$, the first set of parentheses, is equal to $-2aH\dot\phi^{(0)}$ using the zero-order equation (6.33), so after dividing by $-\dot\phi^{(0)}$, we are left with

$$\ddot{\delta\phi} + 2aH\dot{\delta\phi} + k^2\delta\phi = 0. \tag{6.71}$$

This equation for perturbations to $\delta\phi$ is identical to Eq. (6.45) for tensor perturbations to the metric. Thus we can trivially copy our result from Section 6.4.2 and write immediately that the power spectrum of fluctuations in $\delta\phi$ is equal to

$$P_{\delta\phi} = \frac{H^2}{2k^3}. \tag{6.72}$$

Compare this with Eq. (6.57). It is identical apart from a factor of $16\pi G$. Recall that we inserted this factor (with a bit of hand-waving; see the footnote on page 158) in the tensor case to turn the dimensionless h into a field with dimensions of mass. To get the result for $\delta\phi$ which is already a scalar field with the proper dimensions, we simply remove this factor.

6.5.2 Super-Horizon Perturbations

Until now, we have neglected the metric perturbations. When the wavelength of the perturbation is of order the horizon or smaller, this approximation is valid, as we will shortly see. In the process of seeing this, we will also find that, by the end of inflation, the metric perturbation has become important. So, although the inflation-induced perturbations start out all-"$\delta\phi$," they end up as a linear combination of Ψ and $\delta\phi$ or more generally as a linear combination of Ψ and perturbations to the energy–momentum tensor. The trick is to find the linear combination which is conserved outside the horizon. The value of this conserved linear combination is determined by $\delta\phi$ at horizon crossing; we can then evaluate it after inflation solely in terms of Ψ. The resulting equation will be of the form $\Psi \propto \delta\phi$ with the left-hand side the post-inflation metric perturbation and the right the scalar field perturbation produced during inflation (the power spectrum for which we have calculated above). We can then finally relate P_Ψ (and the spectra of all other perturbation variables using the results of Section 6.1) to the $P_{\delta\phi}$ of Eq. (6.72).

Let's begin by rewriting the equation for conservation of energy, this time in the presence of the metric perturbation. It is straightforward to show that Eq. (6.62) gets generalized to

$$\frac{\partial \delta T^0_{\ 0}}{\partial t} + ik_i \delta T^i_{\ 0} + 3H\delta T^0_{\ 0} - H\delta T^i_{\ i} = -3(\mathcal{P} + \rho)\frac{\partial \Psi}{\partial t} \qquad (6.73)$$

where $\mathcal{P}$ and ρ are the zero-order pressure and energy density. Were we correct to neglect Ψ in the last section? We were, as long as the right-hand side is significantly smaller than the individual terms on the left. Taking the first term on the left as an example, we require

$$\Psi \ll \frac{\delta T^0_{\ 0}}{\mathcal{P} + \rho}. \qquad (6.74)$$

A simple way to see that this inequality holds is to use the Einstein equations we derived in Chapter 5. The most convenient for these purposes is the time-time (Eq. (5.21)) component:

$$k^2\Psi + 3aH(\dot{\Psi} + aH\Psi) = 4\pi G a^2 \delta T^0_{\ 0}. \qquad (6.75)$$

Here I have simply copied the results from Chapter 5, replacing Φ with $-\Psi$. The left-hand side here is of order $k^2\Psi \sim a^2 H^2 \Psi$ for modes which are crossing the horizon. Therefore,

$$\Psi \sim G\frac{\delta T^0_{\ 0}}{H^2} \sim \frac{\delta T^0_{\ 0}}{\rho}$$

$$= \frac{\mathcal{P} + \rho}{\rho}\left(\frac{\delta T^0_{\ 0}}{\mathcal{P} + \rho}\right). \qquad (6.76)$$

The left-hand side must be much less than the term in parentheses; equivalently, the prefactor $(\mathcal{P} + \rho)/\rho$ must be small. In fact, during inflation, the pressure is almost equal to minus the energy density, so this prefactor is very small. In terms of the slow-roll parameters, it is equal to $2\epsilon/3$. So, at least in slow-roll models of inflation, we are justified in neglecting metric perturbations when computing the spectrum of $\delta\phi$.

The above argument holds only for modes that have not yet passed outside the horizon. Super-horizon modes, on the other hand, require more careful treatment. Indeed, it is inevitable that the inequality of Eq. (6.74) will break down sometime before the end of inflation. To see this, recall that after inflation, when the universe is dominated by radiation, $\delta T^0{}_0 = -4\rho_r \Theta_0$ and $\mathcal{P} + \rho = 4\rho_r/3$. Therefore, after inflation, the right-hand side of Eq. (6.74) is $-3\Theta_0$. According to Eq. (6.12), $\Psi = -2\Theta_0$ right after inflation, so it is certainly not true that the inequality of Eq. (6.74) is satisfied for all times. At some point before inflation ends, perturbations to Ψ must grow in importance relative to those in the energy–momentum tensor.

One way to deal with the coupling between the metric perturbations and those to the energy density is to define

$$\zeta \equiv -\frac{ik_i \delta T^0{}_i H}{k^2(\rho + \mathcal{P})} - \Psi. \tag{6.77}$$

For sub-horizon modes and those which have just left the horizon, Ψ is negligible; $P + \rho = (\dot\phi^{(0)}/a)^2$ from Eqs. (6.26) and (6.27); and Eq. (6.64) fixes the numerator of the first term in ζ. We are left with

$$\zeta = -aH\delta\phi/\dot\phi^{(0)} \tag{6.78}$$

around the time of horizon crossing. After inflation ends, $ik_i \delta T^0{}_i = 4ak\rho_r \Theta_1$, proportional to the dipole of the radiation. Since the pressure of radiation is equal to a third of the energy density,

$$\zeta = -\frac{3aH\Theta_1}{k} - \Psi$$

$$= -\frac{3}{2}\Psi \qquad \text{(post inflation)}. \tag{6.79}$$

The second equality follows from the initial conditions relating the dipole to the potential (Eq. (6.16)).

The variable ζ is so important because it is conserved when the perturbation moves outside the horizon (Figure 6.8). We will show that ζ is conserved shortly, but first let's appreciate the importance of this conservation. Since we know that, after inflation, $\zeta = -3\Psi/2$, we can immediately relate Ψ coming out of inflation to the $\delta\phi$ at horizon crossing,

$$\Psi\Big|_{\text{post inflation}} = \frac{2}{3}aH\frac{\delta\phi}{\dot\phi^{(0)}}\Big|_{\text{horizon crossing}}. \tag{6.80}$$

Equivalently, the post-inflation power spectrum of Ψ is simply related to the horizon-crossing spectrum of $\delta\phi$:

$$P_\Psi = \frac{4}{9}\left(\frac{aH}{\dot\phi^{(0)}}\right)^2 P_{\delta\phi}\Big|_{aH=k}$$

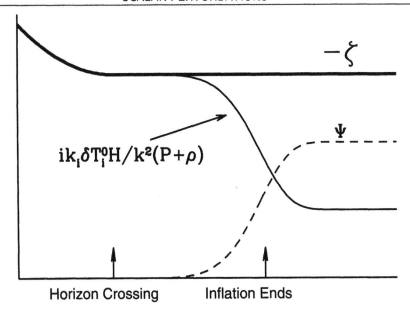

$$ik_i\delta T^0_iH/k^2(P+\rho)$$

Horizon Crossing   Inflation Ends

Time →

Figure 6.8. Cartoon view of the evolution of scalar, adiabatic perturbations during inflation in conformal Newtonian gauge. When a mode is sub-horizon, quantum-mechanical fluctuations are set up in the scalar field driving inflation $(ik_i\delta T^0{}_iH/k^2(\rho + \mathcal{P}) = aH\delta\phi/\dot{\phi}^{(0)})$. Scalar perturbations to the metric are negligible at this time. Once the mode leaves the horizon, the linear combination $\zeta = -ik_i\delta T^0{}_iH/k^2(\rho + \mathcal{P}) - \Psi$ is conserved. Well after inflation has ended, the metric perturbation has grown in importance, but the linear combination ζ remains unchanged.

$$= \frac{2}{9k^3}\left(\frac{aH^2}{\dot{\phi}^{(0)}}\right)^2\Bigg|_{aH=k}, \tag{6.81}$$

the second line following from Eq. (6.72). Another way to express the power spectrum of scalar perturbations is to eliminate $\dot{\phi}^{(0)}$ in favor of the slow-roll parameter ϵ. You will show (Exercise 12) that $(aH/\dot{\phi}^{(0)})^2 = 4\pi G/\epsilon$, so

$$P_\Psi = P_\Phi(k) = \frac{8\pi G}{9k^3}\frac{H^2}{\epsilon}\Bigg|_{aH=k}. \tag{6.82}$$

The first equality here follows from our ubiquitous assumption that anisotropic strsses are small, so that $\Psi = -\Phi$. Comparing to Eq. (6.59), we see that the ratio of scalar to tensor modes is of order $1/\epsilon$; that is, we expect scalar modes to dominate. Finally, another way of writing the scalar power spectrum is to eliminate ϵ in favor of the potential and its derivative, using the result of Exercise 14,

$$P_\Phi(k) = \frac{128\pi^2 G^2}{9k^3} \left(\frac{H^2 V^2}{V'^2} \right) \bigg|_{aH=k}. \tag{6.83}$$

It remains to prove that ζ is conserved on super-horizon scales. To see this, let's turn to the conservation equation, Eq. (6.73). On large scales, $k_i \delta T^i{}_0$ is proportional to k^2 and so can be ignored, leaving

$$\frac{\partial \delta T^0{}_0}{\partial t} + 3H\delta T^0{}_0 - H\delta T^i{}_i = -3(\mathcal{P} + \rho)\frac{\partial \Psi}{\partial t}. \tag{6.84}$$

On large scales, you will show (Exercise 13) that the energy–momentum tensor satisfies

$$\frac{ik_i \delta T^0{}_i H}{k^2} = \frac{\delta T^0{}_0}{3}. \tag{6.85}$$

Therefore, on large scales

$$\zeta = -\Psi - \frac{1}{3}\frac{\delta T^0{}_0}{\rho + \mathcal{P}}. \tag{6.86}$$

Eliminating Ψ in favor of ζ in the conservation equation leads to

$$\frac{\partial \delta T^0{}_0}{\partial t} + 3H\delta T^0{}_0 - H\delta T^i{}_i = 3(\mathcal{P} + \rho)\frac{\partial \zeta}{\partial t} + (\rho + \mathcal{P})\frac{\partial}{\partial t}\left[\frac{\delta T^0{}_0}{\rho + \mathcal{P}} \right]. \tag{6.87}$$

The partial derivative on the right acting on $\delta T^0{}_0$ cancels the first term on the left, leaving

$$\delta T^0{}_0 \left[3H + \frac{1}{\rho + \mathcal{P}}\left(\frac{d\rho}{dt} + \frac{d\mathcal{P}}{dt} \right) \right] - H\delta T^i{}_i = 3(\mathcal{P} + \rho)\frac{\partial \zeta}{\partial t}. \tag{6.88}$$

Recall from Eq. (2.55) that $d\rho/dt = -3H(\rho + \mathcal{P})$, so the first term in brackets cancels the second, and

$$\frac{\partial \zeta}{\partial t} = \frac{-1}{3(\rho + \mathcal{P})^2}\left[H(\rho + \mathcal{P})\delta T^i{}_i - \delta T^0{}_0 \frac{d\mathcal{P}}{dt} \right]. \tag{6.89}$$

I claim that the two terms in brackets on the right cancel for the class of perturbations we are considering. To see this, first rewrite $H(\rho + \mathcal{P})$ as $-(1/3)d\rho/dt$. Thus, the terms in brackets are proportional to

$$\frac{\delta T^i{}_i}{3} + \frac{d\mathcal{P}}{d\rho}\delta T^0{}_0 = \delta\mathcal{P} - \frac{d\mathcal{P}}{d\rho}\delta\rho \tag{6.90}$$

since $-\delta T^0{}_0$ is the perturbation to the energy density, while $\delta T^i{}_i/3$ is the perturbation to the pressure. If we know the background pressure–energy density relation $d\mathcal{P}/d\rho$, then given an overdensity $\delta\rho$, we expect the pressure perturbation to be proportional to the overdensity with coefficient $d\mathcal{P}/d\rho$. Indeed, this is the characteristic feature of adiabatic perturbations, precisely those set up during inflation. Thus, ζ is indeed conserved on large scales.

6.5.3 Spatially Flat Slicing

The treatment of the previous subsection is complete, but it is not the most elegant way to understand scalar perturbations in inflation. A much simpler way is to move back and forth between different gauges, making use along the way of the concept of a gauge-invariant variable, one which does not change under these transformations. Here I outline this method, leaving some of the more detailed calculations as problems.

We saw earlier that one of the major complications in conformal Newtonian gauge was that perturbations to the scalar field $\delta\phi$ are coupled to the potential Ψ. It would obviously be nice to transform to a gauge in which these perturbations decoupled. Consider a gauge with *spatially flat slicing*, with the spatial part of the metric $g_{ij} = \delta_{ij}a^2$. In this gauge the line element is

$$ds^2 = -(1 + 2A)dt^2 - 2aB_{,i}dx^i dt + \delta_{ij}a^2 dx^i dx^j, \tag{6.91}$$

i.e., there are two functions A and B characterizing the perturbations. In this case, the equation for $\delta\phi$ is given exactly (Exercise 16) by Eq. (6.71): the perturbations in the scalar field do not couple to those in the gravitational metric. Therefore, without having to neglect any couplings, we can identify the power spectrum for $\delta\phi$ as given by Eq. (6.72).

The next step is to identify a gauge-invariant variable, one which remains the same when transforming from one gauge to the next. Bardeen (1980) identified several such variables, two characterizing scalar perturbations to the metric and two characterizing perturbations to the matter. Of course any linear combination of these is still gauge invariant. We would like to identify the combination that is proportional to $\delta\phi$ in the gauge with spatially flat slicing. In this gauge, Bardeen's velocity (Eq. (5.78)) is

$$v = ikB - \frac{ik\dot{\phi}^{(0)}\delta\phi}{(\rho + \mathcal{P})a^2} \qquad \text{(spatially flat slicing)} \tag{6.92}$$

where I have evaluated $\delta T^0{}_i$ with Eq. (6.64). Thus, we can create a gauge-invariant variable proportional to $\delta\phi$ in a spatially flat slicing if we subtract off the kB term. Bardeen's Φ_H (Eq. (5.76)) is simply equal to aHB, so the combination

$$\zeta \equiv -\Phi_H - \frac{iaH}{k}v \tag{6.93}$$

is gauge invariant and in spatially flat slicing is equal to

$$\zeta = -\frac{aH}{\dot{\phi}^{(0)}}\delta\phi \qquad \text{(spatially flat slicing)}. \tag{6.94}$$

We can immediately relate the power in ζ to the power in $\delta\phi$,

$$P_\zeta = \left(\frac{aH}{\dot{\phi}^{(0)}}\right)^2 P_{\delta\phi}. \tag{6.95}$$

We know $P_{\delta\phi}$ from Eq. (6.72) and the prefactor is $4\pi G/\epsilon$, so

$$P_\zeta = \left.\frac{2\pi G H^2}{\epsilon k^3}\right|_{aH=k}. \tag{6.96}$$

Equation (6.96) is very useful, for it expresses the power spectrum of a gauge invariant quantity. Although we computed it in a gauge of the form in Eq. (6.91), once we have this answer, we can compute ζ in any gauge and then relate the power in the perturbation variables of that gauge to P_ζ.

Throughout this book, we have been working in conformal Newtonian gauge. In this gauge, $\Phi_H = -\Phi$, so ζ as defined in Eq. (6.93) is indeed given by Eq. (6.77). We argued in Section 6.5.2 that in conformal Newtonian gauge, after inflation, $\zeta = 3\Phi/2$, so $P_\Phi = 4P_\zeta/9$, or using Eq. (6.96),

$$P_\Phi = \left.\frac{8\pi G H^2}{9\epsilon k^3}\right|_{aH=k} \tag{6.97}$$

in exact agreement with our earlier calculation finalized in Eq. (6.82).

This is the end of the calculation, but not quite the end of the story. Bardeen and others have argued that Φ_H has a nice geometrical interpretation, one shared by ζ in certain gauges. In particular, the curvature of the three-dimensional space at fixed time is equal to $4k^2\Phi_H/a^2$. Therefore, perturbations in Φ_H represent *curvature perturbations*: even though the zero-order space is flat, perturbations induce a curvature which varies from place to place. In conformal Newtonian gauge or in a spatially flat slicing this interpretation would seem irrelevant to perturbations in ζ, since ζ is a combination of both Φ_H and the velocity. However, if one moves to a *comoving gauge*, one in which the velocities vanish, then ζ is equal to Φ_H. In comoving gauges, then, it is clear that a perturbation to ζ is a curvature perturbation, and indeed the scalar perturbations generated during inflation are often called curvature perturbations.

6.6 SUMMARY AND SPECTRAL INDICES

In order to understand how scales which should be uncorrelated today are observed to have almost identical temperatures, we are virtually forced into the theory of inflation. In addition to explaining away the nagging fine-tuning problems of the standard cosmology, inflation is also a mechanism for generating primordial perturbations over the smooth universe.

Inflation predicts that quantum-mechanical perturbations in the very early universe are first produced when the relevant scales are causally connected. Then these scales are whisked outside the horizon by inflation, only to reenter much later to serve as initial conditions for the growth of structure and anisotropy in the universe. The perturbations are best described in terms of the Fourier modes. The mean of a given Fourier mode, for example for the gravitational potential, is zero:

$$\langle \Phi(\vec{k}) \rangle = 0. \tag{6.98}$$

Further, the perturbations to one Fourier mode are uncorrelated with those to another. However, a given mode has nonzero variance, so

$$\langle \Phi(\vec{k}) \Phi^*(\vec{k}') \rangle = (2\pi)^3 P_\Phi(k) \delta^3(\vec{k} - \vec{k}'), \tag{6.99}$$

the Dirac delta function enforcing the independence of the different modes. In the case of scalar perturbations, the ones of most importance for us, the power spectrum is given by Eq. (6.82). Perturbations to the tensor part of the metric are also produced and are also Gaussian with mean zero; the power spectrum of tensor modes is given by Eq. (6.59). The scalar spectrum depends on the slow roll parameter ϵ, defined in Eq. (6.35), which is proportional to the derivative of the Hubble rate. Since the Hubble rate is close to constant during inflation — because of the dominance of potential energy — ϵ is typically small.

A spectrum in which $k^3 P_\Phi(k)$ is constant (i.e., does not depend on k) is called a *scale-invariant* or *scale-free* spectrum. Apart from small deviations encoded in the slow-roll parameters, both the scalar and the tensor perturbations are scale free. This is both a blessing and a curse. It is good because it is a fairly definite prediction, easy to test. It is unfortunate because a scale-free spectrum is what one might have expected even without the complex machinery of inflation. Indeed, a scale-free spectrum is also referred to as a Harrison–Zel'dovich–Peebles spectrum, crediting the smart people who first proposed it as the appropriate distribution for the initial conditions, a proposal that predates inflation by many years. This really is too bad, because if we observe a scale-free spectrum, and most present observations are consistent with this, then inflation cannot fairly claim all the credit. However, if we observe a small mixture of tensor modes and/or a small deviation from a scale-free spectrum, then this will go a long way toward convincing skeptics that inflation is responsible for the primordial perturbations.

To quantify the deviations from scale invariance, it is conventional to write the primordial power spectra as

$$P_\Phi(k) = \frac{8\pi}{9k^3} \frac{H^2}{\epsilon m_{\mathrm{Pl}}^2} \bigg|_{aH=k} \equiv \frac{50\pi^2}{9k^3} \left(\frac{k}{H_0} \right)^{n-1} \delta_H^2 \left(\frac{\Omega_m}{D_1(a=1)} \right)^2$$

$$P_h(k) = \frac{8\pi}{k^3} \frac{H^2}{m_{\mathrm{Pl}}^2} \bigg|_{aH=k} \equiv A_T k^{n_T - 3}. \tag{6.100}$$

These equations serve to define the scalar and tensor amplitudes, δ_H (subscript H for amplitude at horizon crossing) and A_T, and the spectral indices, n and n_T. Note that this convention — which has become common — says that a scale-free scalar spectrum corresponds to $n = 1$, while $n_T = 0$ for a scale-free tensor spectrum. The factor of $\Omega_m/D_1(a=1)$, where Ω_m is the fraction of the critical density in matter today and D_1 is the *growth function* which will be defined in Chapter 7 (Eqs. (7.4) and (7.77)), is part of this convention. It is inconvenient at this stage because we have not even encountered the growth function yet, but it has become standard to include in the definition of δ_H (Liddle and Lyth, 1993; Bunn and White, 1997). The resulting expression for the matter power spectrum today looks much simpler when these factors are included here. We pay the price of complexity now for the benefit of simplicity later.

We can relate the primordial spectral indices n and n_T to the slow-roll parameters ϵ and δ. Consider first the tensor spectrum. By virtue of the definition in Eq. (6.100),

$$\frac{d\ln[P_h]}{d\ln k} = n_T - 3. \tag{6.101}$$

The logarithmic derivative has two terms, first the trivial one $d\ln(k^{-3})/d\ln(k)$ which cancels the -3 here, leaving $n_T = 2d\ln H/d\ln(k)$. The logarithmic derivative of the Hubble rate at horizon crossing is a bit subtle:

$$\left.\frac{d\ln H}{d\ln k}\right|_{aH=k} = \frac{k}{H}\frac{dH}{d\eta} \times \left.\frac{d\eta}{dk}\right|_{aH=k}. \tag{6.102}$$

By definition (Eq. (6.35)), $\dot{H} = -aH^2\epsilon$, and $d\eta|_{aH=k}/dk = -d(aH)^{-1}|_{aH=k}/dk = 1/k^2$, so

$$\left.\frac{d\ln H}{d\ln k}\right|_{aH=k} = -\left.\frac{k}{H}\frac{aH^2\epsilon}{k^2}\right|_{aH=k} = -\epsilon. \tag{6.103}$$

Therefore, the primordial spectral index of tensor perturbations produced by inflation is

$$n_T = -2\epsilon. \tag{6.104}$$

The scalar spectral index follows from a similar argument. Taking the logarithmic derivative of P_Φ leads to

$$n - 1 = \frac{d}{d\ln(k)}\left[\ln(H^2) - \ln(\epsilon)\right]. \tag{6.105}$$

The derivative of H again gives -2ϵ while the logarithmic derivative of ϵ is $-2(\epsilon+\delta)$ (Exercise 12). So,

$$n = 1 - 4\epsilon - 2\delta. \tag{6.106}$$

The fact that the tensor index n_T is proportional to ϵ leads to one of the robust predictions of inflation. Many inflationary models have been proposed which offer different predictions for ϵ and δ. Almost all of these, however, maintain the feature that the ratio of tensor to scalar modes (which we saw earlier was proportional to ϵ) is directly related to the tensor spectral index (here also seen to be directly proportional to ϵ). As you progress through this book, moving from the evolution of anisotropies to their analyses, try to bear in mind the crucial question of whether this prediction can be put to the observational test.

The slow-roll parameters are a convenient way to summarize the predictions of an inflationary model. However, ultimately we are interested in the physics, so we are interested in how these parameters relate back to the fundamental entity, the potential V of the scalar field responsible for inflation. You will show in Exercise 14 that these parameters can be expressed in terms of the potential and its derivatives. Therefore, extracting the values of ϵ and δ from the data is tantamount to probing the potential of the field driving inflation. Given that the expected scale of this potential is on the order of 10^{15} GeV (Exercise 18), this is quite an impressive probe!

SUGGESTED READING

The 30 or so pages on inflation in this chapter, which were heavily slanted toward production of perturbations, offer but a glimpse into the many facets of this remarkable theory. Recently, Guth wrote a popular account of his discovery of inflation, *The Inflationary Universe*. One of the other originators of the theory, Linde, has a more technical book, *Inflation and Quantum Cosmology*, which emphasizes model building much more than I have here. As I mentioned earlier, *The Early Universe* (Kolb and Turner) has an excellent chapter on inflation. The recent *Cosmological Inflation and Large Scale Structure* (Liddle and Lyth) is most similar in spirit to this book, with a heavy emphasis on perturbations. The discussion there of the perturbation spectrum is laden with less algebra than the one in Section 6.5 so is worth reading. (Beware that their Planck mass is our $m_{\rm Pl}/\sqrt{8\pi}$.)

An extremely clear and deep look into inflation is given in *300 Years of Gravitation* (ed. Hawking and Israel) in the article by Blau and Guth. Many other articles in that thick compilation volume are also fascinating. The initial article by Guth (1981) is completely accessible and as clear a statement possible of the problems that led to inflation and the initial attempt (old inflation) to solve them. Indeed, I would recommend reading Guth's initial article because this chapter motivates inflation with the horizon problem, while Guth had several different problems in mind, including the monopole problem and the flatness problem (Exercise 4).

There have been many papers reviewing the production of perturbations during inflation. Two clear reviews are Lidsey *et al.* (1997) and Lyth and Riotto (1999). The former focuses on methods for going beyond the predictions elucidated here, which are accurate only to first order in the slow-roll parameters ϵ and δ, and on extracting the potential V from observations. The latter summarizes efforts to tie inflation to realistic particle physics models. The eight-page paper of Stewart and Lyth (1993) is a remarkably concise treatment of the techniques used to go beyond the first-order slow-roll approximation. Hollands and Wald (2002) have written a thoughtful critique of inflation, which is a refreshing antidote to some of the euphoria emanating from the discoveries of the late 1990s. Besides the importance of this critique in its own right, the paper has one of the clearest qualitative descriptions of perturbation generation during inflation that I have ever read.

The initial conditions relating the various perturbations described in Section 6.1 are perhaps most clearly discussed in the review article by Efstathiou (1990). Isocurvature perturbations, for the most part ignored here, are treated in detail there.

I have ignored the possibility of perturbations produced by topological defects. These theories, while fascinating, have not succeeded in making robust predictions; to the extent that predictions can be extracted from them, they are wrong. Nonetheless the numerics involved in their study is sufficiently complicated that I would not be shocked to see them make a comeback some day. There exist many books with comprehensive discussions of topological defects. Among them are *Cosmic Strings and Other Topological Defects* (Vilenkin and Shellard) and *The Formation and Evolution of Cosmic Strings* (ed. Gibbons, Hawking, and Vachaspati).

EXERCISES

Exercise 1. Find the ratio of neutrino to radiation energy density, f_ν. Assume that there are three species of massless neutrinos.

Exercise 2. Account for the neutrino quadrupole moment when setting up initial conditions.

(a) Start with Eq. (4.107). This is an equation for $\mathcal{N}(\mu)$. Turn this into a hierarchy of equations for the neutrino moments:

$$\dot{\mathcal{N}}_0 + k\mathcal{N}_1 = -\dot{\Phi}$$

$$\dot{\mathcal{N}}_1 - \frac{k}{3}\left(\mathcal{N}_0 - 2\mathcal{N}_2\right) = \frac{k}{3}\Psi$$

$$\dot{\mathcal{N}}_2 - \frac{2}{5}k\mathcal{N}_1 = 0. \tag{6.107}$$

To do this, you need to recall the definition of these moments, which is equivalent to that for photons, Eq. (4.99). A good way to reduce Eq. (4.107) into this hierarchy is to multiply it first by $\mathcal{P}_0$ and then integrate over $\int_{-1}^{1} d\mu$. This leads to the first equation above. Then multiply Eq. (4.107) by $\mathcal{P}_1$ to get the second and $\mathcal{P}_2$ to get the third. More details are given in Section 8.3, where we go through the same exercise for the photon moments. In the third equation you may neglect $\mathcal{N}_3$ because it is smaller than $\mathcal{N}_2$ by a factor of order $k\eta$ (prove this!).

(b) Eliminate $\mathcal{N}_1$ from these equations and show that

$$\ddot{\mathcal{N}}_2 = \frac{2k^2}{15}\left(\Psi + \mathcal{N}_0 - 2\mathcal{N}_2\right). \tag{6.108}$$

Drop $\mathcal{N}_2$ on the right-hand side because it is much smaller than $\Psi + \mathcal{N}_0$.

(c) Rewrite Einstein's equation (5.33) as

$$\mathcal{N}_2 = -(k\eta)^2 \frac{\Phi + \Psi}{12 f_\nu}. \tag{6.109}$$

This neglects the photon quadrupole. Argue that Compton scattering sets $\Theta_2 \ll \mathcal{N}_2$ so this is a reasonable assumption.

(d) Now differentiate this form of Einstein's equation twice to get an expression for $\ddot{\mathcal{N}}_2$. Equate this to the expression for $\ddot{\mathcal{N}}_2$ derived in part (b). (You may drop all derivatives of Φ and Ψ when doing this since the mode of interest is the $p = 0$ constant mode.) Use this equation to express $\mathcal{N}_0$ in terms of Φ and Ψ.

(e) Finally assume that $\Theta_0 = \mathcal{N}_0$ and use your expression for $\mathcal{N}_0$ to rewrite Eq. (6.12) as a relation between the two gravitational potentials. Show that this relation is

$$\Phi = -\Psi\left(1 + \frac{2 f_\nu}{5}\right). \tag{6.110}$$

Exercise 3. Show that the initial conditions for the velocities and dipoles of matter and radiation are as given in Eq. (6.16).

Exercise 4. Inflation also solves the *flatness* problem. This is the question of why the energy density today is so close to critical.
(a) Suppose that

$$\Omega(t) \equiv \frac{8\pi G \rho(t)}{3H^2(t)} \tag{6.111}$$

is equal to 0.3 today, where ρ counts the energy density in matter and radiation (assume zero cosmological constant). From Eq. (1.2), plot $\Omega(t) - 1$ as a function of the scale factor. How close to one would $\Omega(t)$ have been back at the Planck epoch (assuming no inflation took place so that the scale factor at the Planck epoch was of order 10^{-32})? This fine-tuning of the initial conditions is the flatness problem. If not for the fine tuning, an open universe would be *obviously* open (i.e., Ω would be almost exactly zero) today.
(b) Now show that inflation solve the flatness problem. Extrapolate $\Omega(t) - 1$ back to the end of inflation, and then through 60 e-folds of inflation. What is $\Omega(t) - 1$ right before these 60 e-folds of inflation?

Exercise 5. Another way of looking at the problems that inflation solves is to consider the entropy within our Hubble volume. This is proportional to the total number of particles in the volume, with a proportionality constant of order unity. How many photons are there within our Hubble volume today? Explain how inflation produces entropy this large.

Exercise 6. We showed that, if the universe was always dominated by ordinary matter or radiation early on, then the comoving horizon when the scale factor was a_e (very small) was $a_0 H_0 / a_e H_e$ times the comoving Hubble radius today. Compute this ratio assuming that the temperature was equal to 10^{15} GeV at a_e. Account for the radiation-to-matter transition at $a \sim 10^{-4}$.

Exercise 7. Consider a free, homogeneous scalar field with mass m. The potential for this field is $V = m^2 \phi^2 / 2$. Show that, if $m \gg H$, the scalar field oscillates with frequency equal to its mass. Also show that its energy density falls off as a^{-3}, so it behaves exactly like ordinary nonrelativistic matter.

Exercise 8. Show that Eq. (6.33) follows from Eq. (6.32) by changing variables from t to η.

Exercise 9. Compute some well-known properties of the quantized harmonic oscillator.
(a) The momentum of the harmonic oscillator with unit mass is $p = dx/dt$. Compute

$$[\hat{x}, \hat{p}]$$

and show that it is equal to i. You can obtain the operator $\hat{p}$ by differentiating $\hat{x}$ (Eq. (6.38)) with respect to time.

(b) Compute the zero-point energy of the harmonic oscillator with unit mass. Do this by quantizing the energy

$$E = \frac{p^2}{2} + \frac{\omega^2 x^2}{2}$$

and then computing its expectation value in the ground state: $\langle 0|\hat{E}|0\rangle$.

Exercise 10. Show that gravity waves are not sourced by the scalar field during inflation. To do this, recall that the right-hand side of Eq. (6.45) is

$$\delta T^1{}_1 - \delta T^2{}_2$$

where δT is the perturbation to the energy–momentum tensor (assumed to be dominated by ϕ) and, as in the derivation of Eq. (5.63), I have chosen $\vec{k}$ to be in the $\hat{z}$ direction. Show that this right-hand side is indeed zero for the scalar field.

Exercise 11. Show that Eq. (6.57) is the appropriate solution to Eq. (6.56).
(a) Define $\tilde{v} = v/\eta$ and rewrite Eq. (6.56) in terms of $\tilde{v}$.
(b) The resulting equation is the spherical Bessel equation. Write down the general solution to this as a linear combination of two functions of $k\eta$.
(c) Use the boundary conditions of Eq. (6.58) to determine the coefficients of part (b). Show that Eq. (6.57) is the correct solution for these boundary conditions.

Exercise 12. Derive some useful identities involving the slow-roll parameters during inflation.
(a) Show that

$$\frac{d}{d\eta}\left(\frac{1}{aH}\right) = \epsilon - 1.$$

(b) Show that

$$4\pi G(\dot{\phi}^{(0)})^2 = \epsilon a^2 H^2. \tag{6.112}$$

(c) Using the definitions of ϵ and δ, show that

$$\frac{d\epsilon}{d\eta} = -2aH\epsilon(\epsilon + \delta). \tag{6.113}$$

Use this to show that $d\ln\epsilon|_{aH=k}/d\ln(k) = -2(\epsilon + \delta)$.

Exercise 13. Show that on large scales Eq. (6.85) holds. One way to do this is to combine Einstein's equations, the time-time (5.27) and time-space (Exercise 5 of Chapter 5) components, and take the large-scale limit.

Exercise 14. Express the slow-roll parameters ϵ and η in terms of the potential V and its derivatives with respect to ϕ. Show that, to lowest order,

$$\epsilon = \frac{1}{16\pi G}\left(\frac{V'}{V}\right)^2$$

and

$$\delta = \epsilon - \frac{1}{8\pi G}\frac{V''}{V}$$

where primes denote derivatives with respect to $\phi^{(0)}$.

Exercise 15. There are a number of ways of describing pressure in the universe and of relating the pressure to the energy density. One was introduced back in Chapter 2, the *equation of state*,

$$w \equiv \frac{\mathcal{P}}{\rho}. \tag{6.114}$$

The second is the *sound speed*,

$$c_s^2 \equiv \frac{d\mathcal{P}}{d\rho}. \tag{6.115}$$

The way to compute c_s^2 is to differentiate both $\mathcal{P}$ and ρ with respect to time and take the ratio. Finally, there is the ratio of perturbations in the energy density to those in the pressure,

$$\frac{\delta\mathcal{P}}{\delta\rho} = \frac{-3\delta T^0{}_0}{\delta T^i{}_i}, \tag{6.116}$$

where the minus sign accounts for the fact the the time-time component of the energy–momentum tensor is minus the energy density with our convention, and the factor of 3 negates the sum over the three spatial indices. For adiabatic perturbations, $\delta\mathcal{P}/\delta\rho = c_s^2$. Show that this holds for three separate cases: matter, radiation, and a single scalar field during inflation at the time of horizon crossing. For the last case, it is enough to show that the difference $\delta\mathcal{P}/\delta\rho - c_s^2$ is of order the slow-roll parameters ϵ and δ.

Exercise 16. Show that in a gauge given by Eq. (6.91), the equation governing the perturbations to a scalar field $\delta\phi$ is Eq. (6.71).
(a) Bardeen's equation for the gauge-invariant density in the absence of anisotropic stress is

$$\frac{d}{d\eta}\left(a^3\rho\epsilon_m\right) = -(\rho + \mathcal{P})a^3 kv \tag{6.117}$$

with gauge-invariant density defined as

$$\rho\epsilon_m = -\rho - \delta T^0{}_0 + \frac{3iH}{k^2}k_i\delta T^0{}_i \tag{6.118}$$

and velocity v via Eq. (6.92). Compute $\rho\epsilon_m$ for a scalar field in a gauge with spatially flat slicing. Show that, to lowest order in slow-roll parameters ϵ (not ϵ_m) and δ, the equation reduces to

$$\frac{d}{d\eta}\left(a^3\dot{\phi}^{(0)}\delta\phi\right) = -k^2 a\dot{\phi}^{(0)}\delta\phi. \tag{6.119}$$

(b) Again using the slow-roll approximation reduce Eq. (6.119) to the form of Eq. (6.45).

Exercise 17. Show that the curvature in conformal Newtonian gauge is equal to $4k^2\Phi/a^2$. To do this, compute the three-dimensional Ricci scalar arising from the spatial part of the metric $g_{ij} = \delta_{ij}a^2(1+2\Phi)$.

Exercise 18. Determine the predictions of an inflationary model with a quartic potential,

$$V(\phi) = \lambda\phi^4.$$

(a) Compute the slow roll parameters ϵ and δ in terms of ϕ.
(b) Determine ϕ_e, the value of the field at which inflation ends, by setting $\epsilon = 1$ at the end of inflation.
(c) To determine the spectrum, you will need to evaluate ϵ and δ at $-k\eta = 1$. Choose the wavenumber k to be equal to $a_0 H_0$, roughly the horizon today. Show that the requirement $-k\eta = 1$ then corresponds to

$$e^{60} = \int_0^N dN' \frac{e^{N'}}{(H(N')/H_e)}$$

where H_e is the Hubble rate at the end of inflation, and N is defined to be the number of e-folds before the end of inflation:

$$N \equiv \ln\left(\frac{a_e}{a}\right).$$

(d) Take the Hubble rate to be a constant in the above with H/H_e equal to 1. This implies that $N \simeq 60$. Turn this into an expression for ϕ. The simplest way to do this is to note that $N = \int_t^{T_e} dt' H(t')$ and assume that H is dominated by potential energy. Show that this mode leaves the horizon when $\phi^2 = 60m_{\rm Pl}^2/\pi$.
(e) Determine the predicted values of n and n_T.
(f) Estimate the scalar amplitude in terms of λ. As a rough estimate, assume that $k^3 P_\Phi(k)$ for this mode is equal to 10^{-8} (we will find a more precise value when we normalize to large-angle anisotropies in Chapter 8). What value does this imply for λ?

This model illustrates many of the features of contemporary models. In it, (i) the field is of order — even greater than — the Planck scale, but (ii) the energy scale V is much smaller because of (iii) the very small coupling constant.

7

INHOMOGENEITIES

Having set up the system of equations to be solved and the initial conditions for the perturbations, we can now calculate the inhomogeneities and anisotropies in the universe. In this first solutions chapter, we start with the perturbations to the dark matter. In principle these are coupled to all other perturbations. In practice, though, perturbations to the dark matter depend very little on the details of the radiation perturbations. Dark matter, by definition, is affected by radiation only indirectly, through the gravitational potentials. At late times, when the universe is dominated by matter, these potentials are independent of the radiation. At early times, while it is true that the potentials are determined by the radiation, it is also true that the radiation perturbations are relatively simple, so that all moments beyond the monopole and dipole can be neglected. The converse is not true, as we will see in the next chapter: To treat the anisotropies properly we will need to know how the matter perturbations behave.

The ultimate goal of this exercise is to compare theory with observations. We will solve for the evolution of each Fourier mode, $\delta(k, \eta)$. Given this solution, and the initial power spectrum generated by inflation, we can construct the power spectrum of matter today. At least on large scales, this is the most important observable. On small scales, comparison with observation today is more difficult: one must worry about nonlinearities and gas dynamics when comparing with the galaxy distribution. Nonetheless, even on small scales, the linear power spectrum, which we compute in this chapter, is often the starting point for any quantitative statement about the distribution of matter.

7.1 PRELUDE

Gravitational instability is a powerful idea, easy to understand, and most likely responsible for the structure in our universe. As time evolves, matter accumulates in initially overdense regions. It doesn't matter how small the initial overdensity was (e.g., in typical cosmological scenarios, the overdensity was of order 1 part in 10^5); eventually enough matter will be attracted to the region to form structure.

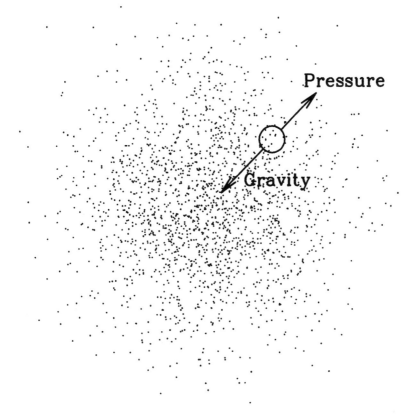

Figure 7.1. Gravitational instability. Mass near an overdense region is attracted to the center by gravity but repelled by pressure. If the region is dense enough, gravity wins and the overdensity grows with time.

The $F = ma$ of gravitational instability is the equation governing overdensities δ. Schematically, it reads

$$\ddot{\delta} + [\text{Pressure} \ - \ \text{Gravity}] \, \delta = 0. \tag{7.1}$$

These basic forces, depicted in Figure 7.1, act in opposite directions. Gravity acts to increase overdensities, grabbing more matter into the region. Since there are more particles in an overdense region, random thermal motion causes a net loss of mass in an overdense region. Therefore, if pressure is strong, inhomogeneities do not grow. As indicated by the cartoon equation (7.1), if pressure is low, δ grows exponentially; if it is large, δ oscillates with time.

We will see many manifestations of the simple form of gravitational instability depicted in Eq. (7.1). Different ambient cosmological conditions alter the growth rate. For example, in a matter-dominated universe, δ grows only as a power of time, not exponentially, whereas in a radiation-dominated universe, the growth is but logarithmic. We will treat super-horizon versions of this equation as well as the more familiar sub-horizon version. When going though the math, though, it is useful to bear in mind the dueling concepts of gravity and pressure.

7.1.1 Three Stages of Evolution

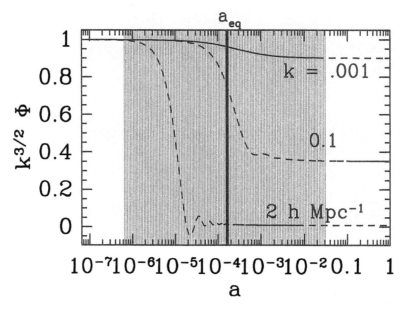

Figure 7.2. The linear evolution of the gravitional potential Φ. Dashed line denotes that the mode has entered the horizon. Evolution through the shaded region is described by the transfer function. The potential is unnormalized, but the relative normalization of the three modes is as it would be for scale-invariant perturbations. Here baryons have been neglected, $\Omega_m = 1$, and $h = 0.5$.

The evolution of cosmological perturbations breaks up naturally into three stages. To see this, let's cheat and look at the solutions for several different modes. Figure 7.2 shows the gravitational potential as a function of scale factor for long-, medium-, and short-wavelength modes. Early on, all of the modes are outside the horizon ($k\eta \ll 1$) and the potential is constant. At intermediate times (shaded in the figure), two things happen: the wavelengths fall within the horizon and the universe evolves from radiation domination ($a \ll a_{eq}$) to matter domination ($a \gg a_{eq}$). Without getting into the details, we see that the order of these epochs (a_{eq} and the epoch of horizon crossing) greatly affects the potential. The large-scale mode, which enters the horizon well after a_{eq}, evolves much differently than the small-scale mode, which enters the horizon before equality. Finally, at late times, all the modes evolve identically again, in this case (where $\Omega_m = 1$) remaining constant.

We are able to observe the distribution of matter predominantly at late epochs, in the third stage of evolution, when all modes are evolving identically. If we wish to relate the potential during these times to the primordial potential set up during inflation, and we do, we can write schematically

$$\Phi(\vec{k}, a) = \Phi_p(\vec{k}) \times \Big\{\text{Transfer Function}(k)\Big\} \times \Big\{\text{Growth Function}(a)\Big\}. \qquad (7.2)$$

where Φ_p is the primordial value of the potential, set during inflation. The transfer function describes the evolution of perturbations through the epochs of horizon cossing and radiation/matter transition (the shaded region in Figure 7.2), while the growth factor describes the wavelength-independent growth at late times. This schematic equation is indeed roughly how the growth factor and the transfer function are defined, with two caveats, both due to convention. Notice from Figure 7.2 that even the largest wavelength perturbations decline slightly as the universe passes through the epoch of equality. This decline is conventionally removed so that the transfer function on large scales is equal to 1. Therefore, the transfer function is defined as

$$T(k) \equiv \frac{\Phi(k, a_{\text{late}})}{\Phi_{\text{Large-Scale}}(k, a_{\text{late}})} \qquad (7.3)$$

where a_{late} denotes an epoch well after the transfer function regime and the *Large-Scale* solution is the primordial Φ decreased by a small amount. We will derive in Section 7.2 that — neglecting anisotropic stresses — this factor is equal to $(9/10)$. The second caveat concerns the growth function. The ratio of the potential to its value right after the transfer function regime is defined to be

$$\frac{\Phi(a)}{\Phi(a_{\text{late}})} \equiv \frac{D_1(a)}{a} \qquad (a > a_{\text{late}}), \qquad (7.4)$$

where D_1 is called the growth function. In the flat, matter-dominated case depicted in Figure 7.2, then, the potential is constant so $D_1(a) = a$. With these conventions, we have

$$\Phi(\vec{k}, a) = \frac{9}{10}\Phi_p(\vec{k})T(k)\frac{D_1(a)}{a} \qquad (a > a_{\text{late}}). \qquad (7.5)$$

The easiest way to probe the potential is to measure the matter distribution. Figure 7.3 shows the evolution of the matter overdensity for three different modes. Notice that at late times — when the potential is constant and all the modes are within the horizon — the overdensity grows with the scale factor ($\delta \propto a$). This explains the seemingly odd nomenclature above (Why is it called a *growth* function if the potential remains constant?): D_1 describes the growth of the matter perturbations at late times. This growth is completely consistent with our intuition that as time evolves, overdense regions attract more and more matter, thereby becoming more overdense.

We can now express the power spectrum of the matter distribution in terms of the primordial power spectrum generated during inflation, the transfer function, and the growth function. The simplest way to relate the matter overdensity to the

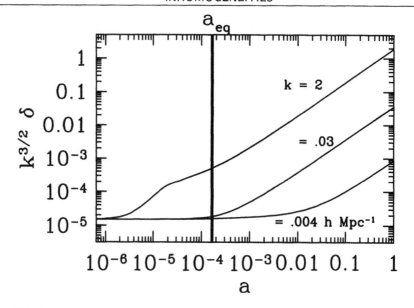

Figure 7.3. The evolution of perturbations to the dark matter in the same model as plotted in Figure 7.2. Amplitude starts to grow upon horizon entry (different times for the three different modes shown here). Well after a_{eq}, all sub-horizon modes evolve identically, scaling as the growth factor. In the case plotted, a flat, matter dominated universe, the growth factor is simply equal to a.

potential at late times is to use Poisson's equation (the large-k, no-radiation limit of Eq. (5.81))

$$\Phi = \frac{4\pi G \rho_m a^2 \delta}{k^2} \qquad (a > a_{\text{late}}). \qquad (7.6)$$

The background density of matter is $\rho_{\rm m} = \Omega_m \rho_{\rm cr}/a^3$, and $4\pi G \rho_{\rm cr} = (3/2)H_0^2$, so

$$\delta(\vec{k}, a) = \frac{k^2 \Phi(\vec{k}, a) a}{(3/2)\Omega_m H_0^2} \qquad (a > a_{\text{late}}). \qquad (7.7)$$

This, together with Eq. (7.5), allows us to relate the overdensity today to the primordial potential

$$\delta(\vec{k}, a) = \frac{3}{5} \frac{k^2}{\Omega_m H_0^2} \Phi_{\rm p}(\vec{k}) T(k) D_1(a) \qquad (a > a_{\text{late}}). \qquad (7.8)$$

Equation (7.8) holds regardless of how the initial perturbation $\Phi_{\rm p}$ was generated. In the context of inflation, $\Phi_{\rm p}(\vec{k})$ is drawn from a Gaussian distribution with mean zero and variance (Eq. (6.100)) $P_\Phi = (50\pi^2/9k^3)(k/H_0)^{n-1}\delta_H^2(\Omega_m/D_1(a = 1))^2$. So the power spectrum of matter at late times is

$$P(k, a) = 2\pi^2 \delta_H^2 \frac{k^n}{H_0^{n+3}} T^2(k) \left(\frac{D_1(a)}{D_1(a = 1)}\right)^2 \qquad (a > a_{\text{late}}). \qquad (7.9)$$

The power spectrum has dimensions of (length)^3. If we want to express the power as a dimensionless function, then, we must multiply by k^3. More precisely, one often associates $d^3kP(k)/(2\pi)^3$ with the excess power in a bin of width dk centered at k. After integrating over all orientations of $\vec{k}$, this becomes $(dk/k)\Delta^2(k)$, with

$$\Delta^2(k) \equiv \frac{k^3 P(k)}{2\pi^2}. \qquad (7.10)$$

Small Δ then corresponds to small inhomogeneities, while large Δ indicates nonlinear perturbations. Note that, with our conventions, a Harrison–Zel'dovich–Peebles spectrum today has $\Delta^2 = \delta_H^2$ on a horizon-sized scale $(k = H_0)$.

Figure 7.4 shows the power spectrum today for two different models. Note that in both of the models $P \propto k$ on large scales, where the transfer function is unity. This behavior is apparent from Eq. (7.9) and corresponds to the simplest inflationary model, wherein $n = 1$. On small scales the power spectrum turns over. To understand this, look back at Figure 7.2. The small-scale mode there $(k = 2\,h$ Mpc$^{-1})$ enters the horizon well before matter/radiation equality. During the radiation epoch the potential decays, so the transfer function is much smaller than unity. The effect of this on matter perturbations can be seen in Figure 7.3, where the growth of δ is retarded starting at $a \simeq 10^{-5}$ after the mode has entered the horizon and ending at $a \simeq 10^{-4}$ when the universe becomes matter dominated. Modes that enter the horizon even earlier undergo more suppression. Thus, the power spectrum is a decreasing function of k on small scales.

This leads to the realization that there will be a turnover in the power spectrum at a scale corresponding to the one which enters the horizon at matter/radiation equality. The power of this realization is apparent in Figure 7.4, which shows two different models: one corresponding to a flat, matter-dominated universe today (often called standard Cold Dark Matter or sCDM) and the other a universe with a cosmological constant today (Lambda Cold Dark Matter or ΛCDM). The major difference between the two models is that sCDM has more matter $(\Omega_m = 1)$ and hence an earlier $a_{\rm eq}$. An earlier $a_{\rm eq}$ means only the very small scales enter the horizon during the radiation-dominated epoch, and therefore the turnover occurs on smaller scales. Finally, another important scale to keep in mind is the scale above which nonlinearities cannot be ignored. This is roughly set by $\Delta(k_{\rm nl}) \simeq 1$, which corresponds to $k_{\rm nl} \simeq 0.2\,h$ Mpc^{-1} in most models. The power spectra shown in Figure 7.4 are the linear power spectra today. On scales smaller than $k_{\rm nl}$, one cannot blindly compare the spectra from Figure 7.4 with the matter distribution today.

7.1.2 Method

What are the evolution equations for the dark matter overdensity? In principle, these are the full set of Boltzmann equations derived in Chapter 4 and the pair of Einstein equations from Chapter 5. In practice, though, the full set of equations is not needed. To understand why, recall that early on (before recombination at

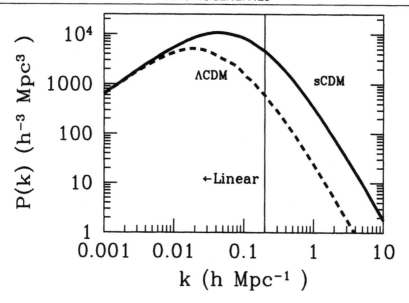

Figure 7.4. The power spectrum in two Cold Dark Matter models, with (ΛCDM) and without (sCDM) a cosmological constant. The spectra have been normalized to agree on large scales. The spectrum in the cosmological constant model turns over on larger scales because of a later $a_{\rm eq}$. Scales to the left of the vertical line are still evolving linearly.

$a = a_*$), the photon distribution can be characterized by only two moments, the monopole Θ_0 and the dipole Θ_1. All other moments are suppressed because the photons are tightly coupled to the electron/proton gas. After decoupling this ceases to be true, and to completely characterize the photon distribution we will need to follow high moments. However, for the purposes of the matter distribution, what the photons are doing after a_* is irrelevant. For, by that time, which is typically well into the matter era, the potential is dominated by the dark matter itself. To sum up then, we can neglect all photon moments except for the monopole and dipole when we are considering the evolution of the matter distribution.

Neglecting the higher radiation moments, the four relevant Boltzmann equations (Section 4.7) become

$$\dot{\Theta}_{r,0} + k\Theta_{r,1} = -\dot{\Phi} \tag{7.11}$$

$$\dot{\Theta}_{r,1} - \frac{k}{3}\Theta_{r,0} = \frac{-k}{3}\Phi \tag{7.12}$$

$$\dot{\delta} + ikv = -3\dot{\Phi} \tag{7.13}$$

$$\dot{v} + \frac{\dot{a}}{a}v = ik\Phi. \tag{7.14}$$

Even with the assumption that only the monopole and dipole are retained, getting from Eq. (4.100) to Eqs. (7.11) and (7.12) requires some explanation and work. First, the explanation: The subscript $_r$ here refers to radiation, both neutrinos and photons. Both species contribute to the gravitational potential (which is our interest in this chapter) and both start out with the same initial conditions. It is not quite as obvious that both follow the same evolution equations (the $\dot{\tau}$ terms can be neglected in Eq. (4.100)) or that these evolution equations are the ones given in (7.11) and (7.12). But it is true, at least in the limit of small baryon density, and again only for the purposes of following the matter evolution. You can work out the details in Exercise 1, and we will explore the full photon evolution equation in the next chapter.

To close the set of equations for the dark matter density, we need an equation for the gravitational potential Φ. You may have noticed that in Eqs. (7.11)-(7.14), I set $\Psi \to -\Phi$, an approximation valid in the limit that there are no quadrupole moments (Eq. (5.33)). Since some of the Einstein equations are redundant, we have several choices for one last equation relating Φ to the radiation and matter overdensities. We can use the time-time component, Eq. (5.27),

$$k^2\Phi + 3\frac{\dot{a}}{a}\left(\dot{\Phi} + \frac{\dot{a}}{a}\Phi\right) = 4\pi G a^2\left[\rho_{\mathrm{dm}}\delta + 4\rho_r\Theta_{r,0}\right]. \tag{7.15}$$

Here, again I have set Ψ to $-\Phi$, neglected the baryons,[1] and merged the neutrino and photon contributions to the potential. The alternative is to use the algebraic (no time derivatives) equation (5.81):

$$k^2\Phi = 4\pi G a^2\left[\rho_{\mathrm{dm}}\delta + 4\rho_r\Theta_{r,0} + \frac{3aH}{k}\left(i\rho_{\mathrm{dm}}v + 4\rho_r\Theta_{r,1}\right)\right]. \tag{7.16}$$

Both of these equations will be useful to us at various times, although only one is necessary to close the set of equations for the five variables $\delta, v, \Theta_{r,0}, \Theta_{r,1}$, and Φ.

At this stage, the simplest thing to do is solve the set of five coupled equations numerically (Exercise 2). If Eq. (7.15) is used, there are no numerical difficulties, and with very little work, you can have a code which computes the transfer function (in the absence of baryons) in less than a second.

Analytic solutions for the dark matter density are harder to come by. I know of no analytic solution valid on all scales at all times. To make progress, we will have to take some limits which reduce the full set of five equations to a more managable two or three. The cost is that these limits will be valid only for certain scales at certain times. Patching these analytic solutions together to obtain a reasonable transfer function is as much art as science.

As a guide to this analytic work which will occupy us much of the rest of this chapter, consider Figure 7.5. The solid curve is the comoving horizon (conformal time), which increases with time, equal to about $30\,h^{-1}$ Mpc at the epoch of

---

[1]This is a fairly good approximation since in most models, the baryon density is much smaller than the dark matter density. We will explore the effects of baryons in Section 7.6.

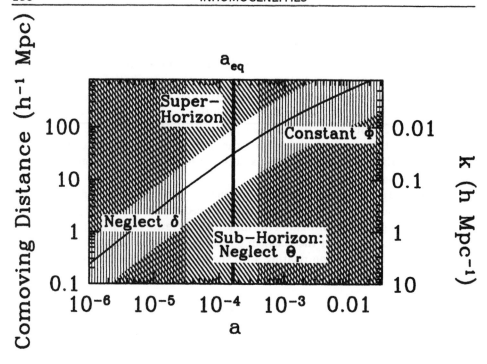

Figure 7.5. Physics of the transfer function. Hatched regions show where analytic expressions exist. The gaps in the center show that no analytic solutions exist to capture the full evolution of intermediate scale modes. The curve monotonically increasing from bottom left to top right is the comoving horizon.

equality.[2] A given comoving scale remains constant with time. Take for example, a comoving distance of $10\,h^{-1}$ Mpc, corresponding to wavenumber $k = 0.1\,h$ Mpc^{-1}. At early times ($a < 10^{-5}$) this distance is larger than the horizon, so $k\eta \ll 1$. We can then drop all terms proportional to k in the evolution equations. In Section 7.2.1, we will derive an exact solution for the potential in this super-horizon limit. Unfortunately, Figure 7.5 indicates that, for the mode in question, this super-horizon solution is valid only until $a \simeq 10^{-5}$. At much later times ($a > 10^{-3}$) the mode is well within the horizon and the radiation perturbations have become irrelevant (since the universe is matter dominated). We will see in Section 7.3.2 that, under these conditions, another analytic solution can be found. The difficulty is matching the super-horizon solution to the sub-horizon solution.

The problem of matching the super-horizon solution to the sub-horizon solution can be solved for very large scale ($k < 0.01\,h$ Mpc^{-1}) and very small scale ($k > 0.5\,h$ Mpc^{-1}) modes. In the large-scale case, we will see in Section 7.2.2 that once the universe becomes matter dominated, $\Phi = $ constant is a solution to the evolution

---

[2]This is model dependent; the plot shows sCDM, with $h = 0.5$.

equations even as the mode crosses the horizon. This fact serves as a bridge between the super- and sub-horizon solutions, both of which have constant Φ in the matter-dominated regime. In the small-scale case, we can neglect matter perturbations as the mode crosses the horizon, since these modes cross the horizon when the universe is deep in the radiation era. Then, once the mode is sufficiently within the horizon, radiation perturbations decay away, and we can match on to the sub-horizon, no-radiation perturbation solution of Section 7.3.2.

With analytic expressions on both large and small scales, we can obtain a good fit to the transfer function by splining the two solutions together. We will see in Section 7.4 that this works, primarily because the transfer function is so smooth, monotonically decreasing from unity on large scales.

7.2 LARGE SCALES

On very large scales, we can get analytic solutions for the potential first through the matter–radiation transition and then through horizon crossing. We start with the super-horizon solution valid through the matter–radiation transition. The results of Section 7.2.1 will be that the potential drops by a factor of 9/10 as the universe goes from radiation to matter domination.

7.2.1 Super-horizon Solution

For modes that are far outside the horizon, $k\eta \ll 1$ and we can drop all terms in the evolution equations dependent on k. From Eqs. (7.11) and (7.13), we see that, in this limit, the velocities (v and $\Theta_{r,1}$) decouple from the evolution equations. This immediately reduces the number of equations to solve from five to three. For the third equation, we notice that Eq. (7.16) has terms inversely propor-

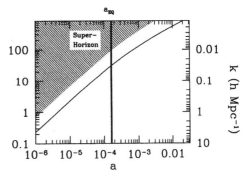

tional to k. These will be difficult to deal with, so let us choose Eq. (7.15) instead. We are left with

$$\dot{\Theta}_{r,0} = -\dot{\Phi} \tag{7.17}$$

$$\dot{\delta} = -3\dot{\Phi} \tag{7.18}$$

$$3\frac{\dot{a}}{a}\left(\dot{\Phi} + \frac{\dot{a}}{a}\Phi\right) = 4\pi G a^2 \left[\rho_{dm}\delta + 4\rho_r\Theta_{r,0}\right]. \tag{7.19}$$

We can go a step further by realizing that the first two equations require $\delta - 3\Theta_{r,0}$ to be constant. Further, we know that this constant is zero (these are the initial conditions). So let us use the dark matter equation (7.18) and the Einstein equation with $\Theta_{r,0}$ set to $\delta/3$. The Einstein equation is then

$$3\frac{\dot{a}}{a}\left(\dot{\Phi} + \frac{\dot{a}}{a}\Phi\right) = 4\pi G a^2 \rho_{\rm dm}\delta\left[1 + \frac{4}{3y}\right]. \tag{7.20}$$

Here I have introduced

$$y \equiv \frac{a}{a_{\rm eq}} = \frac{\rho_{\rm dm}}{\rho_r} \tag{7.21}$$

which we will use as an evolution variable instead of η or a. Again I emphasize that we are ignoring baryons, so $a_{\rm eq}$ is determined solely by $\rho_{\rm dm}$; in the real world, the numerator in the last term in Eq. (7.21) would be $\rho_{\rm m}$, accounting for all matter including baryons.

Equations (7.18) and (7.20) are two first-order equations for the two variables δ and Φ. The stategy will be to turn these two first-order equations into one second-order equation and then solve. First, though, let us rewrite the equations in terms of the new variable y. The derivative with respect to y is related to that with respect to η via the Jacobian,

$$\frac{d}{d\eta} = \frac{dy}{d\eta}\frac{d}{dy}$$

$$= aHy\frac{d}{dy}, \tag{7.22}$$

where the second line follows from the definition of y and the fact that $\dot{a} = a^2 H$. In terms of y then, the Einstein equation becomes

$$y\Phi' + \Phi = \frac{y}{2(y+1)}\delta\left[1 + \frac{4}{3y}\right]$$

$$= \frac{3y+4}{6(y+1)}\delta \tag{7.23}$$

where prime denotes derivatives with respect to y and the right side of the first line follows since $8\pi G\rho_{\rm dm}/3 = (8\pi G\rho/3)y/(y+1) = H^2 y/(y+1)$.

In general, to turn two first-order equations into one second-order equation, the trick is to differentiate one of them. Here, to simplify the algebra, we first rewrite Eq. (7.23) as an expression for δ; then differentiate with respect to y; and finally set δ' to $-3\Phi'$ thanks to the dark matter equation (7.18). This leads to

$$-3\Phi' = \frac{d}{dy}\left\{\frac{6(y+1)}{3y+4}\left[y\Phi' + \Phi\right]\right\}. \tag{7.24}$$

Carrying out the derivative is tedious but straightforward. We are left with

$$\Phi'' + \frac{21y^2 + 54y + 32}{2y(y+1)(3y+4)}\Phi' + \frac{\Phi}{y(y+1)(3y+4)} = 0. \tag{7.25}$$

Remarkably, Kodama and Sasaki (1984) found an analytic solution to Eq. (7.25). They introduced a new variable

$$u \equiv \frac{y^3}{\sqrt{1+y}} \Phi. \tag{7.26}$$

In terms of this variable, you will show (Exercise 4) that Eq. (7.25) becomes

$$u'' + u' \left[\frac{-2}{y} + \frac{3/2}{1+y} - \frac{3}{3y+4} \right] = 0. \tag{7.27}$$

That is, there is no term proportional to u. Instead of a second-order equation for Φ, then, we have a first-order equation for u'. Fortunately, this first-order equation is integrable. Starting from

$$\frac{du'}{u'} = dy \left[\frac{2}{y} - \frac{3/2}{1+y} + \frac{3}{3y+4} \right], \tag{7.28}$$

we can integrate to get

$$\ln(u') = \text{constant} + 2\ln(y) - (3/2)\ln(1+y) + \ln(3y+4). \tag{7.29}$$

Then exponentiating gives

$$u' = A \frac{y^2(3y+4)}{(1+y)^{3/2}} \tag{7.30}$$

where A is a constant to be determined.

We are one integral away from an analytic expression for the gravitational potential. Remembering the definition of u (Eq. (7.26)), we can integrate Eq. (7.30) to obtain

$$\frac{y^3}{\sqrt{1+y}} \Phi = A \int_0^y dy' \frac{y'^2(3y'+4)}{(1+y')^{3/2}}. \tag{7.31}$$

Note that there should be another constant, $u(0)$, here. However, since $y^3 \Phi \to 0$ early on, this constant is vanishes. By similar logic, we can determine the constant A even before performing the integral. For small y, the integrand becomes $4y'^2$, so for small y, Eq. (7.31) becomes $\Phi = 4A/3$. Therefore, $A = 3\Phi(0)/4$. The integral can be done analytically (Exercise 4 again) leaving

$$\Phi = \frac{\Phi(0)}{10} \frac{1}{y^3} \left[16\sqrt{1+y} + 9y^3 + 2y^2 - 8y - 16 \right]. \tag{7.32}$$

Equation (7.32) is our final expression for the potential on super-horizon scales. Although it is not obvious, at small y this expression sets $\Phi = \Phi(0)$, a constant. This must be so, since we chose the two constants of integration with precisely this condition. At large y, once the universe has become matter-dominated, the y^3 term in the brackets dominates, so $\Phi \to (9/10)\Phi(0)$. This is precisely the result we were after: the potential on even the largest scales drops by 10% as the universe passes through the epoch of equality.

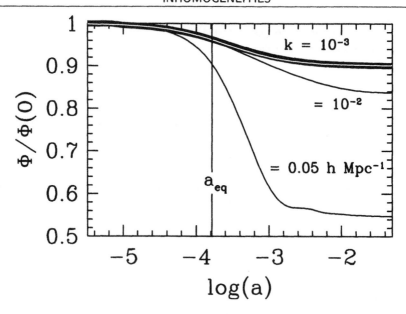

Figure 7.6. Super-horizon evolution of the potential in a CDM model with no baryons, $h = 0.5$ and $\Omega_m = 1$. Thick solid line shows the analytic result of Eq. (7.32), valid only on large scales. White curve within is for the mode $k = 0.001\,h$ Mpc^{-1}. Two other smaller scale modes are shown.

Let us compare this analytic result, valid only when modes are super-horizon, with the numerical results. Figure 7.6 shows that the solution works perfectly on the largest scales and even tolerably well (better than 10%) for scales as small as $k = 0.01\,h$ Mpc^{-1}. This is slightly better than we had anticipated from a crude estimate of where the super-horizon solution is valid (Figure 7.5) and will be important for us later on when spline together the large and small scales solutions. A feature of the analytic solution which may be surprising to you is that, although it is true that the (large scale) potentials are constant in both the matter and radiation epochs, the transition between the pure matter and pure radiation eras is quite long. For example, and this is an important example for the purposes of the CMB as we will see in the next chapter, the potentials, even for the largest scale modes, are still decaying as late as $a \simeq 10^{-3}$, significantly after $a_{\rm eq}$. In models with less matter, $a_{\rm eq}$ is pushed even closer to 10^{-3} so the decay of the potentials becomes even more apparent at the time of recombination.

7.2.2 Through Horizon Crossing

One interesting feature of Figure 7.6 which you should take note of is that large-scale potential (the numerical solution) becomes constant at very late times ($a \gtrsim 10^{-2}$). For $k = 10^{-3}\,h$ Mpc^{-1}, the mode enters the horizon at $\eta \sim k^{-1} = 1000\,h^{-1}$ Mpc which corresponds to $a \sim 0.03$ in the flat, matter-dominated universe depicted in Figure 7.6. The potential remains constant as the mode crosses the horizon.

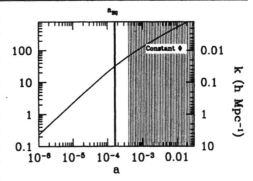

This result is valid as long as the universe is matter dominated. We now set out to prove it.

We are interested then in our set of five equations in the limit that radiation is not important. The potential depends only on the matter inhomogeneities, so we can neglect the two radiation equations, (7.11) and (7.12). In addition to the two matter equations, we now keep the second of Einstein's equations (7.16). This is an algebraic equation, meaning that we could in principle eliminate Φ in the two matter equations and be left with a system of two first-order differential equations. These two first-order equations in general have two solutions. Instead of solving them directly, though, we can cheat using our knowledge of the initial conditions. Here is the idea: we just learned that, deep in the matter epoch, super-horizon potentials are constant. Therefore, the initial conditions for our problem are that the potential is constant ($\dot{\Phi} = 0$). If we can show that constant Φ is one of the two general solutions to the set of matter-dominated equations, then we don't care what the other solution is. For, the initial conditions ensure that the constant Φ solution will be *the* solution.

We want to see, then, if the set of equations

$$\dot{\delta} + ikv = 0 \tag{7.33}$$

$$\dot{v} + aHv = ik\Phi \tag{7.34}$$

$$k^2\Phi = \frac{3}{2}a^2H^2\left[\delta + \frac{3aHiv}{k}\right] \tag{7.35}$$

admits a solution with Φ a constant in time. We can use the algebraic equation (7.35) to eliminate δ from the other two equations. In the matter dominated era, $H \propto a^{-3/2}$, so $d(aH)/d\eta = -a^2H^2/2$. Replacing δ in Eq. (7.33) with Φ and v therefore leads to

$$\frac{2k^2\dot{\Phi}}{3a^2H^2} + \frac{2k^2\Phi}{3aH} - \frac{3aHi\dot{v}}{k} + \frac{3a^2H^2iv}{2k} + ikv = 0. \tag{7.36}$$

We now have two first-order equations for Φ and v. The strategy is to turn these two equations into one second-order equation for Φ. First eliminate $\dot{v}$ from Eq. (7.36) by using the velocity equation. This leaves

$$\frac{2k^2\dot{\Phi}}{3a^2H^2} + \left[\frac{iv}{k} + \frac{2\Phi}{3aH}\right]\left(\frac{9a^2H^2}{2} + k^2\right) = 0. \tag{7.37}$$

If the second-order equation is of the form $\alpha\ddot{\Phi} + \beta\dot{\Phi} = 0$, that is, if it has no terms proportional to Φ, then $\Phi = $ constant is a solution to the equations. So we differentiate Eq. (7.37) with respect to η but consider only the terms proportional to Φ, dropping all terms proportional to derivatives of Φ. Using the fact that $(d/d\eta)(aH)^{-1} = 1/2$, we see that the remaining terms are

$$\left[\frac{iv}{k} + \frac{\Phi}{3}\right]\left(\frac{9a^2H^2}{2} + k^2\right) + \left[\frac{iv}{k} + \frac{2\Phi}{3aH}\right]\frac{d}{d\eta}\frac{9a^2H^2}{2}$$

$$= -\left[\frac{iaHv}{k} + \frac{2\Phi}{3}\right]\left(9a^2H^2 + k^2\right) \tag{7.38}$$

where I have eliminated $\dot{v}$ by using the velocity equation again. But Eq. (7.37) tells us that the term in square brackets on the right here is proportional to $\dot{\Phi}$. So there are no terms in the second-order equation proportional to Φ. Constant potentials are therefore a solution in the matter-dominated era. Since the initial conditions pick out this mode, constant potential is *the* solution in the matter-dominated era.

Potentials remain constant as long as the universe is matter dominated. At much later times ($a > 1/10$), it is conceivable that the universe becomes dominated by some other form of energy — dark energy for example — or, less likely, by curvature. If so, then the potentials will decay. This decay is described by the growth function, though (Section 7.5), and does not affect the transfer function. The main result of this section is that the transfer function as defined in Eq. (7.3) is very close to unity on all scales that enter the horizon after the universe becomes matter dominated. That is, it is unity for all $k \ll a_{eq}H(a_{eq})$, the inverse comoving Hubble radius at equality. You will show in Exercise 5 that the relevant scale is

$$k_{eq} = 0.073\text{Mpc}^{-1}\Omega_m h^2. \tag{7.39}$$

In the limit in which we are working, where baryons and anisotropic stresses are neglected, the transfer function depends only on k/k_{eq}. To get a feel for when the large-scale approximations of this section are valid, look back at Figure 7.6, plotted for the standard CDM model with $\Omega_m = 1$ and $h = 0.5$. The transfer function for the curve labeled 10^{-2} is 7% lower (0.84/0.9) than unity. For that mode, $k/k_{eq} = 0.01/(0.073h) = 0.27$. So if we are interested in 10% accuracy in the transfer function, then we can use the large-scale approximation for $k \lesssim k_{eq}/3$.

7.3 SMALL SCALES

We were able to solve for the evolution of large-scale perturbations in the previous section because the modes crossed the horizon well *after* the epoch of equality. Therefore, the problem neatly divided into (i) super-horizon modes passing through the epoch of equality and then (ii) modes in the matter-dominated era which cross the horizon. The converse is true for the small-scale modes considered in this section. They cross the horizon when the universe is deep in the radiation era. So the problem divides neatly into (i) modes in the radiation era crossing the horizon and then (ii) sub-horizon modes passing through the epoch of equality. Step (i) we treat in Section 7.3.1, step (ii) in Section 7.3.2. Notice that we are unable to treat analytically modes which enter the horizon around the epoch of equality.

7.3.1 Horizon Crossing

When the universe is radiation dominated, the potential is determined by perturbations to the radiation. The dark matter perturbations—the ones we are interested in in this chapter—are influenced by the potential, but do not themselves influence the potential. So the situation is as depicted in Figure 7.7. To solve for matter perturbations in this epoch, therefore, is a two-step problem. First, we must solve the coupled equations for $\Theta_{r,0}, \Theta_{r,1}$, and Φ. Then we solve the equation for matter evolution using the potential as an external driving force.

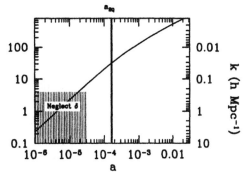

Figure 7.7. Coupling of perturbations in the radiation era. Radiation perturbations and the gravitational potential affect each other. Matter perturbations do not affect the potential but are driven by it.

To solve for the potential in the radiation dominated era, we choose Eq. (7.16). Dropping the matter source terms, we have

$$\Phi = \frac{6a^2 H^2}{k^2}\left[\Theta_{r,0} + \frac{3aH}{k}\Theta_{r,1}\right] \qquad (7.40)$$

since $H^2 = 8\pi G\rho_r/3$ in the radiation era. Also in the radiation era, $aH = 1/\eta$. Armed with this fact, we can use Einstein's equation (7.40) to eliminate $\Theta_{r,0}$ from the two radiation equations, (7.11) and (7.12). These become

$$-\frac{3}{k\eta}\dot{\Theta}_{r,1} + k\Theta_{r,1}\left[1 + \frac{3}{k^2\eta^2}\right] = -\dot{\Phi}\left[1 + \frac{k^2\eta^2}{6}\right] - \Phi\frac{k^2\eta}{3} \qquad (7.41)$$

$$\dot{\Theta}_{r,1} + \frac{1}{\eta}\Theta_{r,1} = \frac{-k}{3}\Phi\left[1 - \frac{k^2\eta^2}{6}\right] \tag{7.42}$$

We can turn these two first order-equations for Φ and $\Theta_{r,1}$ into one second-order equation for the potential. Use Eq. (7.42) to eliminate $\dot{\Theta}_{r,1}$ from the first equation, which then becomes

$$\dot{\Phi} + \frac{1}{\eta}\Phi = \frac{-6}{k\eta^2}\Theta_{r,1}. \tag{7.43}$$

We now have an expression for $\Theta_{r,1}$ solely in terms of the potential and its first derivative. To arrive at a second-order equation for Φ, we differentiate. When we do, we will encounter terms proportional to $\Theta_{r,1}$ and its derivative. Each of these can be eliminated with Eq. (7.42) and Eq. (7.43). The resulting second-order equation is

$$\ddot{\Phi} + \frac{4}{\eta}\dot{\Phi} + \frac{k^2}{3}\Phi = 0. \tag{7.44}$$

To determine the behavior of the potential in the radiation-dominated era, we must solve Eq. (7.44) subject to the initial conditions that Φ is constant. It can be solved analytically by defining $u \equiv \Phi\eta$. Then Eq. (7.44) becomes

$$\ddot{u} + \frac{2}{\eta}\dot{u} + \left(\frac{k^2}{3} - \frac{2}{\eta^2}\right)u = 0. \tag{7.45}$$

This is the spherical Bessel equation of order 1 (see Eq. (C.13)) with solutions $j_1(k\eta/\sqrt{3})$ — the spherical Bessel function — and $n_1(k\eta/\sqrt{3})$ — the spherical Neumann function. The latter blows up as η gets very small, so we discard it on the basis of the initial conditions. The spherical Bessel function of order 1 can be expressed in terms of trigonometric functions (Eq. (C.14)), so

$$\Phi = 3\Phi_{\rm p}\left(\frac{\sin(k\eta/\sqrt{3}) - (k\eta/\sqrt{3})\cos(k\eta/\sqrt{3})}{(k\eta/\sqrt{3})^3}\right) \tag{7.46}$$

where $\Phi_{\rm p}$ is the primordial value of Φ. The factor of 3 in front here arises because the $\eta \to 0$ limit of the expression in parentheses is $1/3$.

Equation (7.46) tells us that, as soon as a mode enters the horizon during the radiation-dominated era, its potential starts to decay. After decaying, the potential oscillates, as depicted in Figure 7.8. Qualitatively, we could have anticipated as much. From the qualitative discussion surrounding Eq. (7.1), we expected that when the pressure is large, as it is when radiation dominates, perturbations will oscillate in time. If perturbations to the dominant component (here radiation) do not grow, then the potential in an expanding universe will begin to decay simply due to the dilution of the zero-order density. This is evident in Eq. (7.40) which (neglecting the dipole well within the horizon) says that $\Phi \sim \Theta_0/\eta^2$. Since Θ_0 oscillates with fixed amplitude, the potential also ocillates, but with an amplitude decreasing as η^{-2}. Indeed, this is precisely the large $k\eta$ limit of Eq. (7.46). The decay and oscillation of the potential is shown in Figure 7.8, with both the analytic

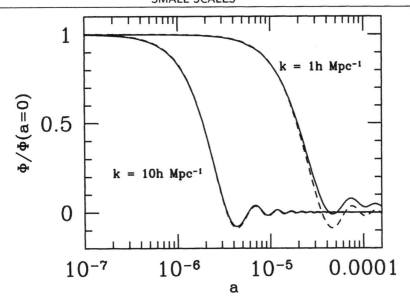

Figure 7.8. Evolution of the potential in the radiation-dominated era. For two small scale modes which enter the horizon well before equality, the exact (solid curve) solution is shown along with the approximate analytic solution (dashed curve) of Eq. (7.46).

expression of Eq. (7.46) and the numerical solution including matter perturbations. Note that the approximate description — in which the effect of matter on the potential is neglected — is valid only deep in the radiation era. The analytic solution for the $k = 1\,h$ Mpc^{-1} mode already begins to depart from the exact solution at $a \simeq 3 \times 10^{-5}$, well before equality (here, in the sCDM model I have taken for illustrative purposes, at $a \simeq 2 \times 10^{-4}$).

Armed with knowledge of the potential in the radiation dominated era, we can now determine the evolution of the matter perturbations, the second half of Figure 7.7. To do this, we turn the two matter evolution equations — (7.13) and (7.14) — into one second-order equation with the potentials serving as an external source. Differentiate Eq. (7.13) and use Eq. (7.14) to eliminate $\dot{v}$:

$$\ddot{\delta} + ik\left(-\frac{\dot{a}}{a}v + ik\Phi\right) = -3\ddot{\Phi}. \tag{7.47}$$

Now we can use Eq. (7.13) to eliminate v, leading to

$$\ddot{\delta} + \frac{1}{\eta}\dot{\delta} = S(k,\eta) \tag{7.48}$$

where the source term is

$$S(k,\eta) = -3\ddot{\Phi} + k^2\Phi - \frac{3}{\eta}\dot{\Phi}. \tag{7.49}$$

The two solutions to the homogeneous equation $(S = 0)$ associated with Eq. (7.48) are $\delta = $ constant and $\delta = \ln(a)$ (or, equivalently in the radiation-dominated era, $\ln[\eta]$). In general, the solution to a second-order equation is a linear combination of the two homogeneous solutions and a particular solution. In the absence of a revelation about the particular solution, one can construct it from scratch from the two homogeneous solutions (call them s_1 and s_2) and the source terms. It is the integral of the source term weighted by the Green's function $[s_1(\eta)s_2(\eta') - s_1(\eta')s_2(\eta)]/[\dot{s}_1(\eta')s_2(\eta') - s_1(\eta')\dot{s}_2(\eta')]$. So here, we have

$$\delta(k,\eta) = C_1 + C_2 \ln(\eta) - \int_0^\eta d\eta' S(k,\eta')\eta' \left(\ln[k\eta'] - \ln[k\eta]\right). \tag{7.50}$$

At very early times the integral is small, so our initial conditions (δ constant) dictate that the coefficient of $\ln(\eta)$, C_2, vanishes and $C_1 = \delta(k, \eta = 0) = 3\Phi_{\rm p}/2$. Now let us consider the integral in Eq. (7.50). The source function decays to zero along with the potential as the mode enters the horizon. Thus, the dominant contribution to the integral comes from the epochs during which $k\eta$ is of order 1. The integral over $S(\eta') \ln(k\eta')$ therefore will just asymptote to some constant, while the integral over $S(\eta') \ln(k\eta)$ will lead to a term proportional to $\ln(k\eta)$ with the constant of proportionality being just that, a constant. Thus, we expect that after the mode has entered into the horizon,

$$\delta(k,\eta) = A\Phi_{\rm p} \ln(Bk\eta), \tag{7.51}$$

i.e., a constant $(A\Phi_{\rm p} \ln[B])$ plus a logarithmic growing mode $(A\Phi_{\rm p} \ln[k\eta])$.

We can determine the constants A and B in Eq. (7.51) by referring to the relevant parts of Eq. (7.50). The constant term, $A\Phi_{\rm p} \ln(B)$, is equal to C_1 plus the integral over $\ln(\eta')$, or

$$A\Phi_{\rm p} \ln(B) = \frac{3}{2}\Phi_{\rm p} - \int_0^\infty d\eta' S(k,\eta')\eta' \ln(k\eta'), \tag{7.52}$$

while the coefficient of the $\ln(k\eta)$ term is set by the remaining integral

$$A\Phi_{\rm p} = \int_0^\infty d\eta' S(k,\eta')\eta'. \tag{7.53}$$

Note that in both integrals here, I have set the upper limit to infinity in accord with our expectation that the integrals asymptote to some constant value at large η. Using the expression for the source term, Eq. (7.49), and our analytic approximations to the potential, Eq. (7.46), we can evaluate the integrals here and determine A and B. I find $A = 9.0$ and $B = 0.62$. Hu and Sugiyama (1996), who introduced this method for following the dark matter evolution at early times, found that integrating the exact potentials (instead of the approximate ones of Eq. (7.46)) leads to slightly different values, $A = 9.6$ and $B = 0.44$.

Figure 7.9 shows the exact solution for δ in the radiation era along with the approximation of Eq. (7.51). Setting aside the details for a moment, we see that

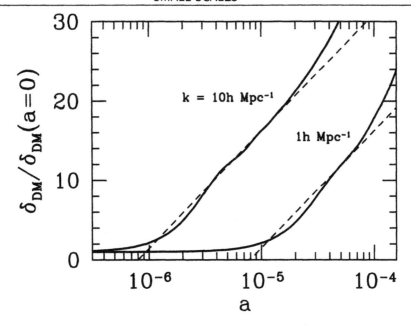

Figure 7.9. Matter perturbations in the radiation-dominated era. The two scales shown here both enter the horizon in the radiation era and lock onto the logarithmically growing mode after some oscillations. Heavy solid curves are the exact solutions, light dashed curves the logarithmic mode of Eq. (7.51). The perturbations have been artificially normalized by their values at early times: inflation actually predicts a larger initial amplitude (by a factor of $10^{3/2}$) for the larger scale mode.

matter perturbations do indeed grow even during the radiation era. The growth is not as prominent as during the matter era (when the constant potentials derived in Section 7.2 imply $\delta \propto a$) due to the pressure of the radiation, but it still exists. For both scales shown in Figure 7.9 the perturbations do indeed settle into the logarithmic growing mode once they enter the horizon. As the universe gets closer to matter domination, though, the pressure of the radiation becomes less important, and the perturbations begin to grow faster. Indeed, you might be worried that our approximation for the $k = 1\,h\ \text{Mpc}^{-1}$ mode is not very useful. Fortunately, we will be using these solutions only to set the initial conditions for growth in the sub-horizon epoch (next subsection), so the approximation need be valid only for a very limited range of times. As long as we choose the matching epoch appropriately, the logarithmic approximation will be extremely good.

7.3.2 Sub-horizon Evolution

We saw in the last subsection that radiation pressure causes the gravitational potentials to decay as modes enter the horizon during the radiation era. Although I did not focus on the radiation perturbations themselves (we will do this in the next chapter), you might expect that the pressure suppresses any growth in $\Theta_{r,0}$. This is correct, and it is in sharp contrast to the matter perturbations which, we

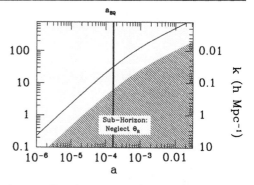

just saw, grow logarithmically. Although initially the potential is determined by the radiation (since the universe is radiation dominated), eventually the growth in the matter perturbations more than offsets the fact that there is more radiation than matter. That is, eventually $\rho_{\rm dm}\delta$ becomes larger than $\rho_r\Theta_{r,0}$ even if $\rho_{\rm dm}$ is smaller than ρ_r. Once this happens, the gravitational potential and the dark matter perturbations evolve together and do not care what happens to the radiation. In this subsection, we want to solve the set of equations governing the matter perturbations and the potential and then match on to the logarithmic solution (7.51) set up during the epoch in which the potential decays.

Once again our starting point is the set of equations governing dark matter evolution, (7.13) and (7.14), and the algebraic equation for the gravitational potential (7.16). And, once again, we want to reduce this set of three equations (two of which are first-order differential equations) to one second-order equation. We will want to follow the sub-horizon dark matter perturbations through the epoch of equality, so it proves convenient again to use y (Eq. (7.21))—the ratio of the scale factor to its value at equality—as the evolution variable. In terms of y, the three equations become

$$\delta' + \frac{ikv}{aHy} = -3\Phi' \tag{7.54}$$

$$v' + \frac{v}{y} = \frac{ik\Phi}{aHy} \tag{7.55}$$

$$k^2\Phi = \frac{3y}{2(y+1)}a^2H^2\delta. \tag{7.56}$$

Several comments are in order about this version of our fundamental equations. First, notice that the time derivatives in the first two equations have been replaced by derivatives with respect to y (indicated by primes), and this transformation leads to the factors of $\dot{y} = aHy$ in the denominators of the unprimed terms. Second, the gravitational potential is now expressed solely in terms of δ: there is no dependence on radiation perturbations because of our arguments above that these are subdominant, and there is no aHv/k dependence because the perturbations are well within the horizon and $aH/k \ll 1$. Finally, the coefficient of the δ source term is $4\pi G\rho_{\text{dm}}a^2 \to (3/2)a^2H^2y/(y+1)$ since we are interested in times early enough that any curvature or dark energy is negligible.

We now go through the familiar routine of turning Eqs. (7.54) and (7.55) into a second-order equation for δ: differentiate the first of these to get

$$\delta'' - \frac{ik(2+3y)v}{2aHy^2(1+y)} = -3\Phi'' + \frac{k^2\Phi}{a^2H^2y^2} \tag{7.57}$$

where v' has been eliminated using the velocity equation. Also I have used the fact that $d(1/aHy)/dy = -(1+y)^{-1}(2aHy)^{-1}$. The first term on the right is much smaller than the second, since the latter is multiplied by $(k/aH)^2$, and we are focusing on sub-horizon modes. Using Eq. (7.56), we recognize this second term as $3\delta/[2y(y+1)]$. We can rewrite the velocity on the left using Eq. (7.54) but neglecting the potential which on sub-horizon scales is much smaller than δ. Thus, the combination $ikv/(aHy)$ can be simply replaced by $-\delta'$ leaving

$$\delta'' + \frac{2+3y}{2y(y+1)}\delta' - \frac{3}{2y(y+1)}\delta = 0. \tag{7.58}$$

This is the *Meszaros equation* governing the evolution of sub-horizon cold, dark matter perturbations once radiation perturbations have become negligible.

To understand the growth of dark matter perturbations, we need to obtain the two independent solutions to the Meszaros equations and then match on to the logarithmic mode established in the previous subsection. To solve this differential equation, we can use our knowledge of the solution deep in the matter era. We know that sub-horizon perturbations in the matter era grow with the scale factor, so one of the solutions to Eq. (7.58) is a polynomial in y of order 1. Therefore, for one mode at least, δ'' vanishes. Therefore, the equation governing this first mode, the growing mode,[3] is $D_1'/D_1 = 3/(2+3y)$, the solution to which is

$$D_1(y) = y + 2/3. \tag{7.59}$$

To find the second solution, notice that the Meszaros equation tells us that $u \equiv \delta/D_1$ satisfies

$$(1+3y/2)u'' + \frac{u'}{y(y+1)}[(21/4)y^2 + 3y + 1] = 0. \tag{7.60}$$

Since there is no term proportional to u, Eq. (7.60) is actually a first-order equation[4] for u'. We can therefore integrate to obtain a solution for u' and then integrate again to get the second Meszaros solution. The first integral gives

---

[3]D_1 is the *growth function* mentioned in Section 7.1. Note though that in this section we are assuming that only matter, and not curvature or dark energy, dominates the landscape. Therefore, our expression for the growth function will be valid only when $a \lesssim 0.1$. For the generalization to later times, see Section 7.5.

[4]Indeed this is a general trick for obtaining the second solution to a differential equation once the first is known. We will use it again later on to obtain the growth factor.

$$u' \propto (y + 2/3)^{-2} y^{-1} (y + 1)^{-1/2}. \tag{7.61}$$

Integrating again leads to the second Meszaros solution

$$D_2(y) = D_1(y) \ln \left[\frac{\sqrt{1+y}+1}{\sqrt{1+y}-1} \right] - 2\sqrt{1+y}. \tag{7.62}$$

At late times ($y \gg 1$), the growing solution D_1 scales as y while the decaying mode D_2 falls off as $y^{-3/2}$.

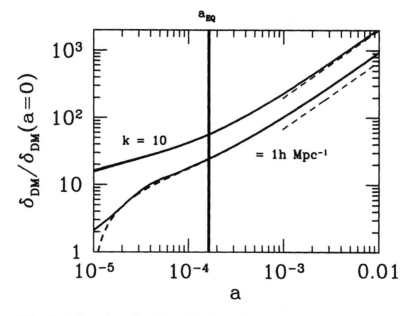

Figure 7.10. Evolution of small-scale, sub-horizon, dark matter perturbations. Solid curves are exact solutions; dashed curves (almost imperceptible because the goodness of fit in the $10\,h$ Mpc^{-1} case) the Meszaros solution with coefficients given by the matching condition, Eq. (7.64). The dashed straight lines at $a > 10^{-3}$ are the asymptotic solution of Eq. (7.67).

The general solution to the Meszaros equation is therefore

$$\delta(k, y) = C_1 D_1(y) + C_2 D_2(y) \qquad y \gg y_H \tag{7.63}$$

where y_H is the scale factor when the mode enters the horizon divided by the scale factor at equality (Exercise 6). To determine the constants C_1 and C_2 we can match on to the logarithmic solution of Eq. (7.51). That solution is valid within the horizon but before equality: $y_H \ll y \ll 1$. So we can hope to arrive at a reasonable approximation for the evolution of dark matter perturbations only for those modes that enter the horizon before equality. For those modes, we match the two solutions and their first derivatives

$$A \Phi_{\mathrm{p}} \ln(B y_m / y_H) = C_1 D_1(y_m) + C_2 D_2(y_m)$$

$$\frac{A\Phi_p}{y_m} = C_1 D_1'(y_m) + C_2 D_2'(y_m) \tag{7.64}$$

where the matching epoch y_m must satisfy $y_H \ll y_m \ll 1$. Note that I have replaced the argument of the log in Eq. (7.51) — $k\eta$ — with y/y_H, valid as long as the matching epoch is deep in the radiation era. Figure 7.10 shows the evolution of two modes along with the analytic solutions to the Meszaros equation with coefficients set by the matching conditions laid out in Eq. (7.64). Not suprisingly, for larger scale modes than the ones shown the approximation breaks down.

7.4 NUMERICAL RESULTS AND FITS

In Section 7.2 and Section 7.3, we derived analytic solutions following the dark matter perturbations deep into the matter era. Here, we assimilate these results and spline them together to form the transfer function. Also, I will present a well-known fitting function for the transfer function.

First, we need to transform our expression ((7.63) along with Eqs. (7.64)) for the small-scale matter density into an expression for the transfer function. The transfer function is determined by the behavior of δ well after equality when the decaying mode has long since vanished. We can extract an even simpler form for δ in this $a \gg a_{eq}$ limit. The key constant in that case is C_1, the coefficient of the growing mode. Multiplying the first matching condition in Eq. (7.64) by D_2' and the second by D_2 and then subtracting leads to

$$C_1 = \frac{D_2'(y_m)A\ln(By_m/y_H) - D_2(y_m)(A/y_m)}{D_1(y_m)D_2'(y_m) - D_1'(y_m)D_2(y_m)}\Phi_p. \tag{7.65}$$

The denominator $D_1 D_2' - D_1' D_2 = -(4/9)y_m^{-1}(y_m+1)^{-1/2}$, which is approximately equal to $-4/9y_m$ since $y_m \ll 1$. Similarly for small y_m, $D_2 \rightarrow (2/3)\ln(4/y) - 2$ and $D_2' \rightarrow -2/3y$. Therefore,

$$C_1 \rightarrow \frac{-9A\Phi_p}{4}\left[\frac{-2}{3}\ln(By_m/y_H) - (2/3)\ln(4/y_m) + 2\right], \tag{7.66}$$

which fortuitously does not depend on y_m. Therefore, at late times we have an approximate solution for the small-scale dark matter perturbations

$$\delta(\vec{k}, a) = \frac{3A\Phi_p(\vec{k})}{2}\ln\left[\frac{4Be^{-3}a_{eq}}{a_H}\right]D_1(a) \qquad a \gg a_{eq}. \tag{7.67}$$

On very small scales, the argument of the log simplifies because $a_{eq}/a_H = \sqrt{2}k/k_{eq}$ (Exercise 6). To turn Eq. (7.67) into a transfer function, we need to remember how δ is related to Φ_p. Comparing Eq. (7.8) with Eq. (7.67) leads to an analytic expression for the transfer function on small scales:

$$T(k) = \frac{5A\Omega_m H_0^2}{2k^2 a_{eq}}\ln\left[\frac{4Be^{-3}\sqrt{2}k}{k_{eq}}\right] \qquad k \gg k_{eq}. \tag{7.68}$$

Recall that the wavenumber entering the horizon at equality is defined as $k_{eq} \equiv a_{eq}H(a_{eq}) = \sqrt{2}H_0 a_{eq}^{-1/2}$, so the prefactor is also a function of k/k_{eq} only. Then, plugging in numbers leads to

$$T(k) = \frac{12k_{eq}^2}{k^2} \ln\left[\frac{k}{8k_{eq}}\right] \qquad k \gg k_{eq}. \qquad (7.69)$$

Figure 7.11 shows the power spectrum for a standard CDM model ($n = 1$; $h = 0.5$; but no baryons) matching the large-scale transfer function ($T = 1$) with the small-scale transfer function of Eq. (7.69). Also shown is the exact solution

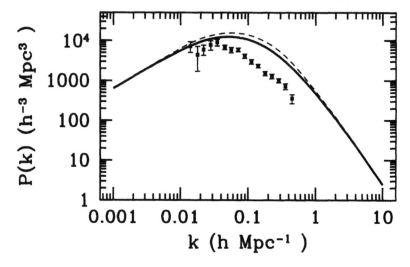

Figure 7.11. The power spectrum in a standard CDM model with a Harrison–Zel'dovich–Peebles spectrum. The thick solid curve uses the BBKS transfer function; the dashed curve interpolates between the analytic transfer function on large scales (equal to 1) and small scales (Eq. (7.68)). The data points are a compilation (and interpretation) by Peacock and Dodds (1994).

(again in the no-baryon limit), or equivalently, the fitting form of Bardeen, Bond, Kaiser, and Szalay (1986, BBKS),

$$T(x \equiv k/k_{eq}) = \frac{\ln[1 + 0.171x]}{(0.171x)}\left[1 + 0.284x + (1.18x)^2 + (0.399x)^3\right.$$

$$\left. + (0.490x)^4\right]^{-0.25}. \qquad (7.70)$$

Note that the BBKS form agrees very well with the analytic solution on small scales; i.e., both aymptote to $\ln(k)/k^2$ with the same coefficients. Since wavenumbers are measured in units of h Mpc^{-1}, the ratio k/k_{eq} depends on $\Omega_m h$. So defining $\Gamma \equiv \Omega_m h$, the BBKS transfer function can also be written as

$$T\left(q \equiv k/\Gamma h\,\mathrm{Mpc}^{-1}\right) = \frac{\ln[1 + 2.34q]}{(2.34q)}\left[1+3.89q+(16.2q)^2+(5.47q)^3+(6.71q)^4\right]^{-0.25}.$$

(7.71)

Several final comments are in order. First, our analytic work has enabled us to understand the origin of the asymptotic, small-scale behavior of the power spectrum. Had there been no logarithmic growth in the radiation era, the modes which entered very early on would have experienced no growth from horizon entry until the epoch of equality. Their amplitude relative to large-scale modes would then have been suppressed by a factor of order $(k_{eq}/k)^2$. The logarithmic growing mode in the radiation era somewhat ameliorates this suppression. Second, although our analytic expression and its BBKS counterpart are good approximations, it is important to be aware of some small effects which affect the transfer function in the real world. We have assumed no anisotropic stresses ($\Phi = -\Psi$). Dropping this assumption changes the factor of 9/10 by which the potential drops for large-scale modes to 0.86, resulting in a corresponding rise in the small-scale transfer function. Including a realistic amount of baryons leads to even more severe small-scale changes. We will address these in Section 7.6. Third, all of our work in this section has been on the transfer function, i.e., on the evolution of perturbations early on when the only components of the universe were matter and radiation. At very late times, the growth function depends on other hypothetical components, the most likely of which is dark energy. Finally, the theoretical power spectrum in Figure 7.11 has been normalized by fixing δ_H in Eq. (7.9) using the observations of CMB anisotropies on large scales (more on this in Chapter 8). We see that (i) the large-scale normalization is roughly correct and that (ii) the shape of the standard CDM power spectrum is wrong. The sCDM power spectrum turns over on relatively small scales, in distinct disagreement with the data. The universe as we observe it appears to have a smaller k_{eq} than sCDM. This observation motivates consideration of variations of sCDM; we will consider these in Section 7.6.

7.5 GROWTH FUNCTION

At late times ($z \lesssim 10$) all modes of interest have entered the horizon. You might think then, that the $y \gg 1$ limit of the Meszaros equation, which describes sub-horizon modes in the matter era, would apply. This is true if $\Omega_m = 1$. If the energy budget of the universe has another item at late times — either dark energy or curvature — then we must retrace the steps which led to the Meszaros equation. Before doing this, I want to point out that, no matter what constitutes the energy budget today, all modes will experience the same growth factor. We saw this in the previous section, where the Meszaros equation was independent of k. And we will soon see it again, when we generalize the Meszaros equation to account for other forms of energy. This uniform growth is a direct result of the fact that cold, dark matter has zero pressure. Therefore, once a mode enters the horizon, there is no way for pressure to smooth out the inhomogeneities and all modes evolve identically.

We want to derive an evolution equation analogous to the Meszaros equation, but allowing for the possibility of energy other than matter or radiation. We can take the $y \gg 1$ limit of Eqs. (7.54)–(7.56), but we must rethink the coefficient of the source term in the Poisson equation. Since radiation can be ignored, the coefficient multiplying δ in Eq. (7.56) is now $4\pi G\rho_{\mathrm{dm}} = (3/2)H_0^2\Omega_m a^{-3}$. Also when differentiating Eq. (7.54) previously, we set $(1/aHy)' = -(1+y)^{-1}(2aHy)^{-1}$; here we need to account for other contributions to H' so Eq. (7.57) becomes

$$\delta'' + ikv \left(\frac{d(aHy)^{-1}}{dy} - \frac{1}{aHy^2} \right) = \frac{3\Omega_m H_0^2}{2y^3 a^2 H^2 a_{\mathrm{eq}}} \delta. \tag{7.72}$$

Replacing the velocity term using the continuity equation as before leads to

$$\frac{d^2\delta}{da^2} + \left(\frac{d\ln(H)}{da} + \frac{3}{a} \right) \frac{d\delta}{da} - \frac{3\Omega_m H_0^2}{2a^5 H^2} \delta = 0. \tag{7.73}$$

Here I have divided by a_{eq}^2 and we will now use a as the variable instead of y. In this large y limit, all factors of a_{eq} disappear.

There are two solutions to Eq. (7.73). One solution is $\delta \propto H$. It is easy to check this if all the energy is nonrelativistic matter, so that the solution is proportional to $a^{-3/2}$. Then all three terms scale as $a^{-7/2}$; the coefficient of the first is $15/4$, the second $-9/4$, and the last $3/2$. The sum of these does indeed vanish. In Exercise 7, you will be asked to show that $\delta \propto H$ is a solution if there are other components of energy in the universe. This solution is pretty, but it is not the one we want since almost all current models of the universe have a nonincreasing Hubble rate. The modes we are interested in—those that remain long after horizon crossing—are the growing modes. So we are interested in the other solution of Eq. (7.73).

To obtain the growing mode, we try a solution of the form $u = \delta/H$. The evolution equation for u then becomes

$$\frac{d^2u}{da^2} + 3\left[\frac{d\ln(H)}{da} + \frac{1}{a} \right] \frac{du}{da} = 0. \tag{7.74}$$

This first-order equation for u' can be integrated to obtain

$$\frac{du}{da} \propto (aH)^{-3}. \tag{7.75}$$

Integrating again and remembering that the second solution, the growth factor, is uH leads to an expression for the growth factor

$$D_1(a) \propto H(a) \int^a \frac{da'}{(a'H(a'))^3}. \tag{7.76}$$

I have glossed over the proportionality constant. This is fixed by the definition of Eq. (7.4), which says that, early on when matter still dominates (say at $z \simeq 10$), D_1 should be equal to a. At those times, $H = H_0\Omega_m^{1/2}a^{-3/2}$ so the growth factor is

$$D_1(a) = \frac{5\Omega_m}{2} \frac{H(a)}{H_0} \int_0^a \frac{da'}{(a'H(a')/H_0)^3}. \tag{7.77}$$

The growth factor in an open universe without dark energy can be computed analytically (see Exercise 8).

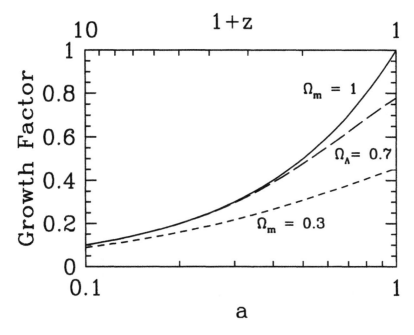

Figure 7.12. The growth factor in three cosmologies. Top two curves are for flat universes without and with a cosmological constant. Bottom curve is for an open universe.

Figure 7.12 shows the growth factor for three different cosmologies. As mentioned above, if the universe is flat and matter dominated, the growth factor is simply equal to the scale factor. In both open and dark energy cosmologies, though, growth is suppressed at late times. This leads to an important qualitative conclusion: structure in an open or dark energy universe developed much earlier than in a flat, matter-dominated universe. There has been relatively little evolution at recent times if the universe is open or dark energy-dominated. Therefore, whatever structure is observed today was likely in place at much earlier times. We will see some quantitative implications of this in Section 9.5.

7.6 BEYOND COLD DARK MATTER

There is more to the universe than just cold dark matter. Although CDM is the main component in most cosmological models, so that the transfer function we derived earlier is a good approximation to reality, there are trace amounts of other stuff. To be completely accurate we need to account for this other stuff. Here I

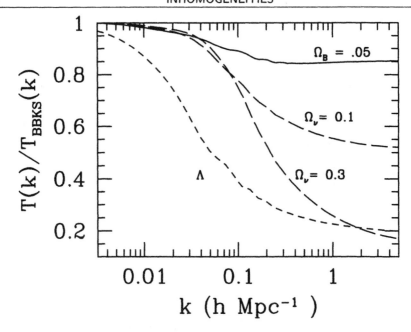

Figure 7.13. The ratio of the transfer function to the BBKS transfer function (Eq. (7.70)) which describes dark-matter-only (no baryons) perturbations. Top curve (and all other curves as well) has 5% baryons. Two middle curves show different values for a massive neutrino. Bottom curve has a cosmological constant $\Omega_\Lambda = 0.7$.

focus on three additional components. First, we consider the effect of the baryons, which constitute roughly 10% of the total matter in most models, on the transfer function. Then, we entertain the possibility that neutrinos have mass and examine the resultant effect on the transfer function. Finally, dark energy — one model for which is the cosmological constant — is considered.

Figure 7.13 shows the transfer functions accounting for these components. A realistic baryon fraction suppresses the transfer function on small scales. A massive neutrino does the same, with the nature and amplitude of the suppression depending on the neutrino mass. Dark energy, here in the form of a cosmological constant, moves the epoch of equality to later times, thereby reducing k_{eq}. The break in the transfer function therefore comes on much larger scales than in the standard CDM model, in apparent agreement with the data exhibited in Figure 7.11.

7.6.1 Baryons

Baryons account for about 4% of the total energy density in the universe. As such, their effect on the matter power spectrum is small. A careful examination of Figure 7.13 reveals two signatures of a nonzero baryon density. The first is that the power spectrum is suppressed on small scales. This is not surprising: at early times, before decoupling, baryons are tightly coupled to photons. Therefore, just as radiation perturbations decay when entering the horizon, so too do baryon overdensities. After decoupling, baryons are released from the relatively smooth radiation field and fall into the gravitational potentials set up by the dark matter. The depth of these wells is smaller than we estimated in Section 7.3, though, because only a fraction $\Omega_{\text{cdm}}/\Omega_m$ of the total matter was involved in the collapse.

The second effect of baryons is less noticable in Figure 7.13 and indeed may never get measured in real life either. Nonetheless, it is extremely important if only because it hints at a fundamental feature of the radiation field. In all the curves in Figure 7.13, except the $\Omega_\nu = 0.3$ case, you can see small oscillations in the transfer function centered around $k \simeq 0.1\,h$ Mpc^{-1}. These are not numerical artifacts. Rather, they are manifestations of the oscillations that the combined baryon/photon fluid experience before decoupling. We got a glimpse of these in Section 7.3.2 (e.g., Figure 7.8) when we considered the potential in the radiation-dominated era. Just as the potential oscillates in this era, the baryon/photon fluid also oscillates. It is the traces of these oscillations that are imprinted on the matter transfer function. They are barely (if at all) detectable because baryons are such a small fraction of the total matter. In the baryon-only model plotted in Figure 1.13, the oscillations were much more noticeable.[5] And these oscillations are also prominent in the spectrum of the radiation perturbations, as we will see in the next chapter.

7.6.2 Massive Neutrinos

Neutrinos are known to exist, and the standard Big Bang model gives a definite prediction for how many there are in the universe (Eq. (2.77)). Massive neutrinos may play an important role in structure formation. Conversely, an accurate measurement of the power spectrum may enable us to infer neutrino masses. For orientation, recall the difference between massless (Eq. (2.78)) and massive (Eq. (2.80)) neutrino energy densities. The best bet from experiments is that the most massive neutrino has a mass of order 0.05 eV, therefore contributing $\Omega_\nu \simeq 10^{-3}$. Even this trace amount might eventually be detectable if the power spectrum can be measured accurately enough. There is also the possibility that one or more neutrinos has a larger mass (see the footnote on Page 46). Current upper limits from structure formation hover around 2 eV (Elgaroy $et~al.$, 2002).

The reason why even a small admixture of massive neutrinos affects the power spectrum is that, especially if they are light, neutrinos can move fast (they are not $cold$ dark matter) and stream out of high-density regions. Perturbations on scales smaller than the free-streaming scale are therefore suppressed. Indeed, a long time ago, cosmologists considered the possibility that all the dark matter in the universe was in the form of neutrinos. If this were so, then there would be no power on small scales and structure would have to form from the "top down."

We can estimate the scale on which perturbations are damped by computing the comoving distance a massive neutrino can travel in one Hubble time at equality.

---

[5]Incidentally, we now also understand why the power in Figure 1.13 is so low in the baryon-only universe: there is no dark matter which can cluster before recombination.

This calculation is trivial, however, if the neutrino mass is in the eV range. For then, the average velocity, T_ν/m_ν, is of order unity at equality. So neutrinos can freestream out of horizon-scale perturbations at equality. This leads to a suppression in power on all scales smaller than k_{eq}.

Figure 7.13 shows this suppression. Note, though, that the effect is a little subtle. A lighter neutrino can free-stream out of larger scales, so the suppression begins at lower k for the $\Omega_\nu = 0.1$ mass than for the $\Omega_\nu = 0.3$ case. On the other hand, the more massive neutrino constitutes more of the total density so it suppresses small-scale power more than does the lighter neutrino.

7.6.3 Dark Energy

Cosmologists have recently accumulated tantalizing evidence for dark energy in the universe above and beyond the dark matter that we have spent so much time on in this chapter. If dark energy exists, how does it affect the matter perturbations?

The first effect of dark energy is indirect. Since theoretical prejudice and evidence both indicate that the universe is flat, $\Omega_{de} \simeq 0.6$-0.7 implies that the matter density, Ω_m, is less than 1. This has a huge impact on the power spectrum, because we have seen that the power spectrum turns over at k_{eq}, which is proportional to Ω_m. So dark energy leads to a turnover in the power spectrum on a scale much larger than predicted in standard CDM. In fact, as we saw in Figure 7.11, this is one of the pieces of evidence for dark energy. The turnover in the power spectrum does not appear on the scale predicted by standard CDM.

The second effect is again related to the smaller matter density in most models of dark energy. As a result of the Poisson equation (7.7), overdensities are inversely proportional to Ω_m for a fixed potential. Therefore, the amplitude of the power spectrum increases as the matter decreases, or equivalently in a flat universe as the dark energy content goes up. With a few caveats to be discussed in Chapter 8, large-angle CMB anisotropies fix the potential on large scales. When normalizing to these large-angle results, therefore, the power spectrum for a model with dark energy is normalized higher than one without.

The third effect of the dark energy on the density inhomogeneities is more direct and more model dependent. At late times, amplification of perturbations is controlled by the growth factor of Eq. (7.77). The evolution of the Hubble rate depends on the model of dark energy, so different models of dark energy predict different growth factors. If we parameterize the dark energy by its equation of state (2.84), then the Hubble rate in a flat universe evolves as

$$\frac{H(z)}{H_0} = \left[\frac{\Omega_m}{a^3} + \frac{\Omega_{de}}{a^{3[1+w]}} \right]^{1/2} \tag{7.78}$$

at late times. Using this time dependence, it is straightforward to perform the integral in Eq. (7.77) and find the growth factor for a given equation of state (see Exercise 11).

To sum up, dark energy affects the power spectrum by changing k_{eq} and the normalization (this depends only on Ω_{de}) and by changing the growth factor at late times (depends on both Ω_{de} and w). Careful observations of the matter spectrum therefore may enable us to learn about dark energy.

SUGGESTED READING

Once again *The Large Scale Structure of the Universe* (Peebles) is a useful reference. Since it was written before the implications of cold dark matter and inflation were explored, though, it does not contain a transfer function or power spectrum such as the ones we have derived (although Peebles himself was instrumental in computing these things several years after the book was published). A more up-to-date book, which is particularly strong on large-scale structure is *Structure Formation in the Universe* (Padmanabhan).

The first papers to work out the CDM transfer function are particularly instructive to read, not least because they also focus on some of the physical implications of the hierarchical theories. See Blumenthal *et al.* (1984) and Peebles (1982). The most important recent paper is Seljak and Zaldarriaga (1996), not so much because it contains a concise description of the set of coupled equations to be solved (although it does that), but because it makes available CMBFAST, a code which computes transfer functions and CMB anisotropy spectra. It is currently available at `http://physics.nyu.edu/matiasz/CMBFAST/cmbfast.html`. The treatment in this chapter follows most closely the small scale analytic solution of Hu and Sugiyama (1996), a paper which is extremely rich and well worth reading. A more recent paper by Eisenstein and Hu (1998) employs the analytic small-scale solution to derive accurate fitting formulae that move beyond those presented by Bardeen, Bond, Kaiser, and Szalay (1986, BBKS).

EXERCISES

Exercise 1. Derive Eqs. (7.11) and (7.12).
(a) First neglect the scattering term in Eq. (4.100), the one proportional to $\dot{\tau}$. Then the photon evolution equation is identical to the neutrino evolution equation (4.107). Show that this collisionless equation reduces to the two equations for the monopole and dipole. To get the monopole equation, multiply Eq. (4.107) by $(d\mu/2)\mathcal{P}_0(\mu) = d\mu/2$ and integrate from $\mu = -1$ to 1. To get the dipole, multiply by $(d\mu/2)\mathcal{P}_1(\mu)$ and integrate.
(b) Show that, in the limit of small baryon density, the scattering term in Eq. (4.100) can indeed be neglected. Neglect Π, since the quadrupole and polarization are very small. Then show that the scattering term is proportional to R, 3/4 times the baryon-to-photon ratio. You will want to use Eq. (4.106). It cannot be emphasized enough that this series of approximations is valid only for the purposes of this chapter, wherein we are interested in the matter distribution.

Exercise 2. Solve the set of five equations ((7.11)–(7.14) and (7.15)) numerically to obtain the transfer function for dark matter. Use the initial conditions derived in Chapter 6. The one numerical problem you may encounter using Eq. (7.15) occurs on small scales when you try to evolve all the way to the present. The photon moments then become difficult to track, and even a good differential equation solver will balk at late times. However, there are several simple solutions to this: (i) by the late times in question, the potential is constant so there is no need to evolve all the way to the present or (ii) stop following the photon moments after a certain time; they don't have any effect on the matter distribution at late times anyway. Plot the transfer function for sCDM (with Hubble constant $h = 0.5$) and ΛCDM (with $\Omega_\Lambda = 0.7$ and $h = 0.7$). Compare with the BBKS transfer function of Eq. (7.70).

Exercise 3. The four subsections in Sections 7.2 and 7.3 correspond to four different approximations to the full set of Einstein–Boltzmann equations. In the following table, fill in the regime of validity for each approximation:

	$a \ll a_{eq}$	$a \sim a_{eq}$	$a \gg a_{eq}$
$k\eta \ll 1$			
$k\eta \sim 1$			
$k\eta \gg 1$			

For example, the super-horizon solution of Section 7.2.1 is valid along the whole top row, since it sets $k\eta \to 0$. Note that time evolves from upper left to bottom right, so the fact that none of the approximations work in the center square means that only those scales that enter the horizon well before or well after equality will be subject to analytic techniques.

Exercise 4. Fill in some of the algebraic detail left out of Section 7.2.1.
(a) Show that Eq. (7.24) leads to Eq. (7.25) by carrying out the differentiation.
(b) Show that Eq. (7.25) is equivalent to Eq. (7.27) when the definition of u from Eq. (7.26) is used.
(c) Show that the integral in Eq. (7.31) can be done analytically with the result given in Eq. (7.32). One way to do the integral is to define a dummy variable $x \equiv \sqrt{1 + y}$.

Exercise 5. Find the wavenumber of the mode which equals the inverse comoving Hubble radius at equality. That is, define k_{eq} to be equal to $a_{eq} H(a_{eq})$. Show that this definition implies

$$k_{eq} = \sqrt{\frac{2\Omega_m H_0^2}{a_{eq}}}. \tag{7.79}$$

Then use Eq. (2.87) to show that k_{eq} is given by Eq. (7.39). Show that if you define k_{eq} by setting it to $1/\eta_{eq}$, you get a number 17% lower.

Exercise 6. Define a_H, the scale factor at which wavelength k equals the comoving Hubble radius, via $a_H H(a_H) \equiv k$. Express a_H/a_{eq} in terms of k and k_{eq}. Show that in the limit $k \gg k_{eq}$, this expression reduces to

$$\lim_{k \gg k_{\text{eq}}} \frac{a_H}{a_{\text{eq}}} = \frac{k_{\text{eq}}}{\sqrt{2k}}. \qquad (7.80)$$

Exercise 7. Show that $\delta \propto H$ is a solution to the evolution equation (7.73) if the universe is flat with a cosmological constant. You will need to use Eq. (1.2). Show also that the solution is valid if the universe has zero cosmological constant, but is open with $\Omega_m < 1$.

Exercise 8. Derive the growth factor for an open universe with $\Omega_m < 1$:

$$D_1(a, \Omega_m) = \frac{5\Omega_m}{2(1 - \Omega_m)} \left[3\frac{\sqrt{1 + x}}{x^{3/2}} \ln\left(\sqrt{1 + x} - \sqrt{x}\right) + 1 + \frac{3}{x} \right] \qquad (7.81)$$

where $x \equiv (1 - \Omega_m)a/\Omega_m$. There may be easier ways to do this (e.g., you might want to check *The Large Scale Structure of the Universe*, Section 11), but I found it easiest to define a dummy variable $y \equiv \Omega_m/a$; write the integral of Eq. (7.77) as

$$\int_{\Omega_m/a}^{\infty} \frac{dy}{y^2 (y + 1 - \Omega_m)^{3/2}} = 2 \left[\frac{d}{d\epsilon} \frac{d}{d\lambda} \int_{\Omega_m/a}^{\infty} \frac{dy}{(y + \epsilon)\sqrt{y + \lambda}} \right]_{\epsilon = 0, \lambda = 1 - \Omega_m} ; \qquad (7.82)$$

and then use 2.246 from Gradshteyn and Ryzhik.

Exercise 9. One popular way to characterize power on a particular scale is to compute the expected RMS overdensity in a sphere of radius R,

$$\sigma_R^2 \equiv \langle \delta_R^2(x) \rangle. \qquad (7.83)$$

Here

$$\delta_R(\vec{x}) \equiv \int d^3x' \delta(\vec{x}') W_R(\vec{x} - \vec{x}') \qquad (7.84)$$

where $W_R(x)$ is the *tophat* window function, equal to 1 for $x < R$ and 0 otherwise; the angular brackets denote the average over all space.

(a) By Fourier transforming, express σ_R in terms of an integral over the power spectrum.

(b) Use the BBKS transfer function to compute σ_8 ($R = 8\,h^{-1}$ Mpc) for a standard CDM model ($h = 0.5, n = 1, \Omega_m = 1$). We will see in Chapter 8 that COBE normalization for this model is

$$\delta_H = 1.9 \times 10^{-5}. \qquad (7.85)$$

The value of σ_8 you find is yet another sign of the sickness of the model. For galaxies, σ_8 is known to be unity (or less, depending on galaxy type). A model with $\sigma_8 > 1$ then requires galaxies to be less clustered than the dark matter. Present models of galaxy formation suggest that this is unlikely. There are even direct measures of σ_8 of the mass (e.g., Section 9.5); these too constrain σ_8 to be less than one.

(c) In the same model, plot σ_R as a function of R. Since σ_R monotically increases, small scales tend to go nonlinear before large scales, the signature of a hierarchical model.

Exercise 10. Rewrite σ from Exercise 9 as

$$\sigma_R^2 = \int_0^\infty \frac{dk}{k} \Delta^2(k) \tilde{W}_R^2(k), \tag{7.86}$$

where $\tilde{W}_R$ is the Fourier tranform of the tophat window function and $\Delta^2 = d\sigma^2/d\ln(k)$ is the contribution to the variance per $\ln(k)$. A useful transition point is the value of k at which Δ exceeds 1. Scales larger than this are linear, while smaller scales have gone nonlinear. Find k_{nl} defined in this way for the sCDM model described in Exercise 9.

Exercise 11. Compute the growth factors in a universe with $\Omega_{de} = 0.7, \Omega_m = 0.3$, and $w = -0.5$. Plot as a function of a. Compare with the cosmological constant model ($w = -1$) with the same Ω_{de}, Ω_m.

8

ANISOTROPIES

The primordial perturbations set up during inflation manifest themselves in the radiation as well as in the matter distribution. By understanding the evolution of the photon perturbations, we can make predictions about the expected anisotropy spectrum today. This evolution is again completely determined by the Einstein–Boltzmann equation we derived in Chapters 4 and 5, and one way to go would be to simply stick all the relevant equations in those chapters on a computer and solve them numerically. Historically, this is a pretty good caricature of what happened. Long before we developed deep insight into the physics of anisotropies, various groups had codes which determined the expected spectra from different models. Only much later did we come to understand both qualitatively and quantitatively why the spectra look like they do.[1] In this chapter, I will mangle the history and simply explain what we have learned about the physics of anisotropies.

Perturbations to the photons evolved completely differently before and after the epoch of recombination at $z \simeq 1100$. Before recombination, the photons were tightly coupled to the electrons and protons; all together they can be described as a single fluid (dubbed the "baryon–photon" fluid). After recombination, photons free-streamed from the "surface of last scattering" to us today. After an overview which qualitatively explains the anisotropy spectrum, Sections 8.2–8.4 work through the physics of the baryon-photon fluid before recombination. Then Sections 8.5–8.6 treat the post-recombination era, culminating in the predicted spectrum of anisotropies today. Finally Section 8.7 discusses how these spectra vary when the cosmological parameters change.

---

[1]Understanding the anisotropies actually helped make the codes much more efficient. The prime example of this is the popular code CMBFAST (Seljak and Zaldarriaga, 1996) which is based in part on the analytic solution presented in this chapter.

8.1 OVERVIEW

Let's begin as we did in the last chapter, by cheating and looking at the answers first. Figure 8.1 shows the evolution of the perturbations to the photons. Four Fourier modes corresponding to perturbations on four different scales are shown. Qualitatively, the most important feature of Figure 8.1 is that perturbations to the photons do not grow appreciably with time. This stands in stark contrast to the matter perturbations, which do grow. And this contrast is something we should have expected: the pressure of the photons is so large that it can withstand the tendency toward collapse. This means that the small perturbations set up during inflation stay small; they remain linear all the way up to the present.

Before going further and examining the evolution of the different modes in more detail, a technical note: I have plotted not simply the perturbation to the photons but rather the combination $k^{3/2}(\Theta_0 + \Psi)$. The $k^{3/2}$ factor balances the fact that the amplitude of the perturbations (in a simple inflationary model) scales as $k^{-3/2}$. I have added the gravitational potential Ψ because the photons we see today had to travel out of the potentials they were in at the time of recombination. As they emerged from these potential wells, their wavelengths were stretched (if the region was overdense and $\Psi < 0$), thereby decreasing their energy. Thus, the temperature we see today is actually Θ_0 at recombination plus Ψ.

The large-scale mode in Figure 8.1 evolves hardly at all. This is not surprising: no causal physics can affect perturbations with wavelengths larger than the horizon, so a super-horizon mode should exhibit little evolution. This means that when we observe large-scale anisotropies — which are sensitive to modes with wavelengths larger than the horizon at recombination — we are observing perturbations in their most pristine form, as they were set down at very early times, presumably during inflation.

Figure 8.1 shows that the smaller scale modes evolve in a more complicated way than the super-horizon modes. Consider the curve labeled "First Peak." As the mode enters the horizon, the perturbation begins to grow until it reaches an apparent maximum at the time of recombination. If we observe anisotropies on scales corresponding to this mode, we would expect to see large fluctuations. Hence the label: the anisotropy spectrum will have a peak at the angular scales corresponding to the mode which has just reached its peak at recombination.

The mode in Figure 8.1 which enters the horizon slightly earlier peaks earlier and then turns over so that its amplitude at recombination is zero. By recombination, it has undergone half of an oscillation; so we see our first clear signal of the *acoustic oscillations* due to the pressure of the relativistic photons. The phase of this mode is such that, at recombination its amplitude is zero. Therefore, when we observe anistropies today corresponding to these scales, we expect very small fluctuations. There will be a trough in the anisotropy spectrum on these angular scales.

And on it goes. The curve labeled "Second Peak" entered the horizon even earlier and has gone through one full oscillation by recombination. As such, this mode will have large fluctuations and lead to a second peak in the anisotropy spectrum. You

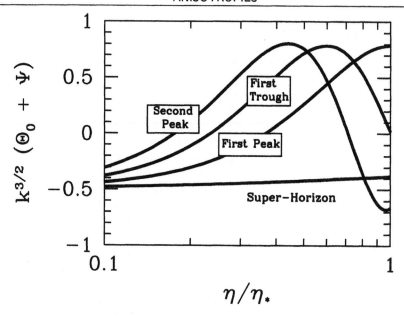

Figure 8.1. Evolution of photon perturbations of four different modes before recombination at η_*. Normalization is arbitrary, but the relative normalization of the 4 curves is appropriate for perturbations with a Harrison–Zel'dovich–Peebles ($n = 1$) spectrum. Model is standard CDM with $h = 0.5$, $\Omega_m = 1$, and $\Omega_b = 0.06$. Starting from the bottom left and moving upward, the wavenumbers for the modes are $k = (7 \times 10^{-4}, 0.022, 0.034, 0.045)$ Mpc^{-1} or $(8, 260, 400, 540)/\eta_0$.

might expect that there will be a never-ending series of peaks and troughs in the anisotropy spectrum corresponding to modes that entered the horizon earlier and earlier. And you would be right: this is exactly what happens.

We can see this more clearly by looking at the spectrum of perturbations at one time, the time of recombination. Figure 8.2 shows this spectrum for two different models, one with a very low baryon content. We do indeed see this pattern of peaks and troughs. There are two more quantitative features of these oscillations that are important. First, note that — at least in the higher baryon model — the heights of the peaks seem to alternate: the odd peaks seem higher than the even peaks. Second, and this is clearest in the low baryon model, perturbations on small scales $k\eta_0 \gtrsim 500$ are damped.

To understand the first of these features, we can write down a cartoon version of the equation governing perturbations. Very roughly, this equation is

$$\ddot{\Theta}_0 + k^2 c_s^2 \Theta_0 = F \tag{8.1}$$

where F is a driving force due to gravity and c_s is the sound speed of the combined baryon–photon fluid. This is the equation of a forced harmonic oscillator (see box on page 220). Qualitatively, it predicts the oscillations we have seen above. But it also

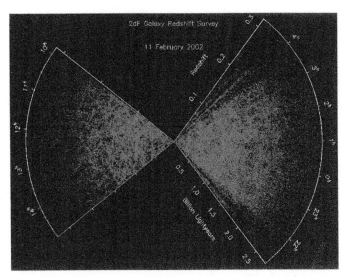

Plate 1.12. Distribution of galaxies in the Two Degree Field Galaxy Redshift Survey (2dF) (Colless *et al.*, 2001). By the end of the survey, redshifts for 250,000 galaxies will have been obtained. As shown here, they probe structure in the universe out to $z = 0.3$, corresponding to distances up to 1000 h^{-1} Mpc away from us (we are located at the center).

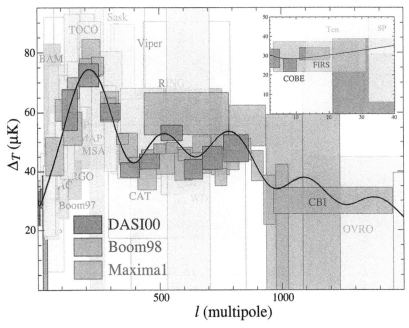

Plate 1.14. Anisotropies in the CMB predicted by the theory of inflation compared with observations. *x*-axis is multipole moment (e.g., $l = 1$ is the dipole, $l = 2$ the quadrupole) so that large angular scales correspond to low l; *y*-axis is the root mean square anisotropy (the square root of the two-point function) as a function of scale. The characteristic signature of inflation is the series of peaks and troughs, a signature which has been verified by experiment.

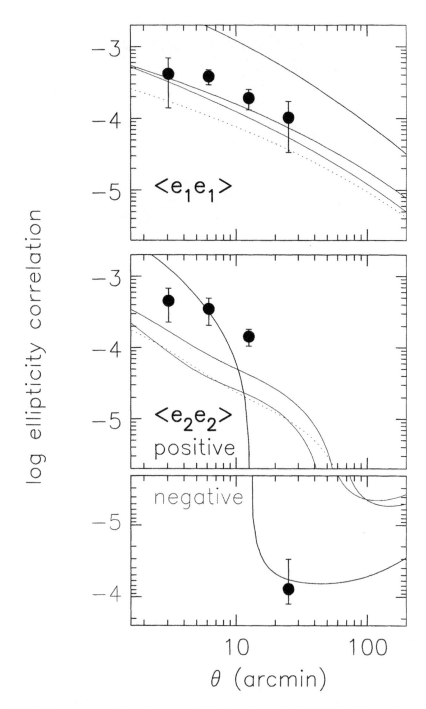

Plate 10.10. Measurement of the shear correlation functions using 145,000 background galaxies (Wittman *et al.*, 2000). Also shown are a variety of CDM models; topmost in top panel is standard CDM, ruled out here at many sigma. Note that $w_{\gamma_1} = \langle e_1 e_1 \rangle$ remains positive on all angular scales.

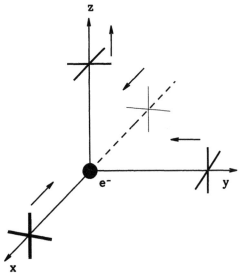

Plate 10.13. Incoming dipole radiation also produces no polarization. Heavy (thin) lines denote hot (cold) spots. Here the incoming radiation is hotter than average (average is medium thickness) from the $+\hat{x}$-direction, and colder than average from the $-\hat{x}$-direction. The two rays from the $\pm\hat{x}$-directions therefore produce the average intensity for the outgoing ray along the $\hat{y}$-direction. The outgoing intensity along the $\hat{x}$-direction is produced by the ray incident from the $\pm\hat{y}$-directions. Since these have the average intensity, the outgoing intensity is also the average along the $\hat{x}$-direction. The net result is outgoing unpolarized light.

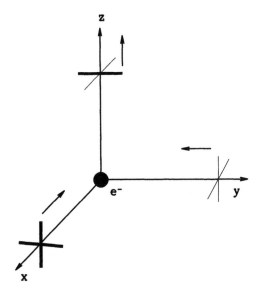

Plate 10.14. Incoming quadrupole radiation produces outgoing polarized light. The outgoing radiation has greater intensity along the y-axis than in the $\hat{x}$-direction. This is a direct result of the hotter radiation incident from the $\hat{x}$-direction.

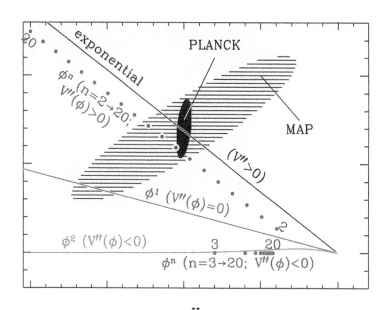

Plate 11.14. Expected 95% uncertainty on the inflationary parameters n and r from MAP and Planck (from Dodelson, Kinney, and Kolb, 1997). Three other parameters (normalization, Ω_B, and h) have been marginalized over. Every inflationary model gives a unique prediction somewhere in this plane; many such predictions are plotted.

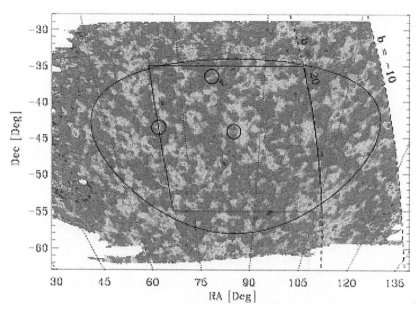

Plate 11.15. A map of the CMB temperature from observations by Boomerang (Netterfield *et al.*, 2001), a long-duration balloon flight at the South Pole. Hot and cold spots have amplitudes as large as $500\mu K$. Circles shows quasars identified in these radio observations. The large elliptical region delineates data analyzed to obtain bandpowers. The rectangular region is an earlier data set.

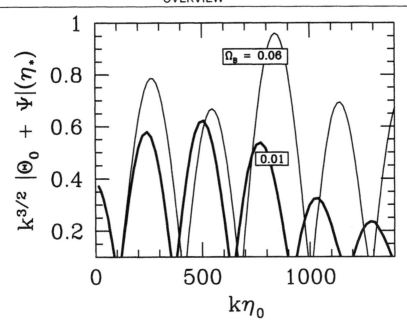

Figure 8.2. Perturbations to the photon distribution at recombination in two models. The larger damping length of the low-Ω_b model is clearly evident in the suppression of perturbations for modes with $k > 500/\eta_0$.

explains something about the heights of the peaks. As we add more baryons to the universe, the sound speed goes down (baryons are heavy so they reduce the speed). Thus the frequency of the oscillations goes down. The peaks at $n\pi/\omega$ are shifted to larger k (you really should read that box!), and the spacing between peaks gets correspondingly larger. Further, as the frequency goes down, the disparity between the heights of the odd and even peaks gets larger. We clearly see both of these features in Figure 8.2. Another way of understanding the alternating peak heights is to note that the perturbations for the first peak mode have been growing since they entered the horizon. By decreasing the pressure (or equivalently increasing the importance of gravity) these modes will grow even more. The second peak mode on the other hand, corresponds to an *underdensity* of photons in the potential wells. Decreasing the pressure makes it harder for photons to escape the well and therefore reduces the magnitude of the perturbation (makes it less underdense).

Consider a simple harmonic oscillator with mass m and force constant k. In addition to the restoring force, the oscillator is acted on by an external force F_0. Thus the full force is $F_0 - kx$ where x is the oscillator's position. The equation of motion is

$$\ddot{x} + \frac{k}{m}x = \frac{F_0}{m}. \tag{8.2}$$

The term on the right-hand side—representing the external force—is driving the oscillator to large values of x. The restoring force on the other hand tries to keep the oscillator as close to the origin as possible. The solution therefore will be that oscillations will be set up around a new zero point, at positive x.

The solution to Eq. (8.2) is the sum of the general solution to the homogeneous equation (with the right-hand side set to zero) and a particular solution. The general solution has two modes, best expressed as a sine and cosine with arguments ωt, with the frequency ω defined as $\omega \equiv \sqrt{\frac{k}{m}}$. A particular solution to Eq. (8.2) is constant $x = F_0/m\omega^2$, so the full solution is the sum of the sine and cosine modes plus this constant. Let us assume that the oscillator is initially at rest. Then, since $\dot{x}(0)$ is proportional to the coefficient of the sine mode, this coefficient must vanish, leaving

$$x = A\cos(\omega t) + \frac{F_0}{m\omega^2}. \tag{8.3}$$

This solution is shown in the figure at right. The solid line is the unforced solution: oscillations about the origin. The

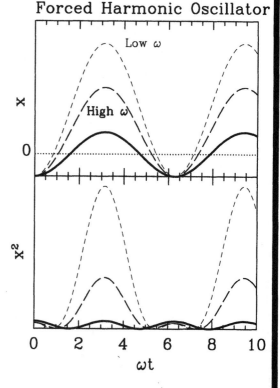

Forced Harmonic Oscillator

dashed curves are the forced solutions for two different choices of frequencies. In both cases, the oscillations are not around $x = 0$ as they would be if the system was unforced. Once an external force is introduced, the zero point of the oscillations shifts in the direction of the force. Two curves are drawn to show that this shift is more dramatic for lower frequencies. The bottom panel shows the square of the oscillator position as a function of time. All three oscillators experience a series of peaks at $t = n\pi/\omega$ corresponding to the minima/maxima of the cosine mode. (Note that if only the sine mode was present these peaks would be at $t = (2n + 1)\pi/\omega$.) The heights of these peaks are identical in the case of the unforced oscillator and equal to the height at $t = 0$. In the forced case, though, the height of the odd peaks—those at $t = (2n+1)\pi/\omega$—is greater than that of the even peaks. The effect is most dramatic for low frequencies. If the frequency is low, the force has a greater effect, producing the greater zero-point offset, and hence the greater odd/even disparity. The other feature of this example is that the even peaks correspond to negative positions of the oscillator: points at which it is farthest from where the force wants it to go.

Photon Diffusion

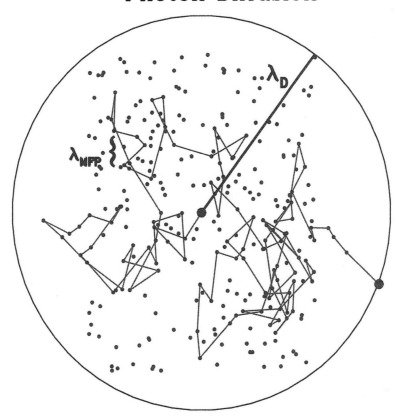

Figure 8.3. Photon diffusion through the electron gas. Electrons are denoted as points. Shown is a typical photon path as it scatters off electrons. The mean free path is λ_{MFP}. After a Hubble time, the photon has scattered many times, so that it has moved a distance of order λ_D.

To understand the damping evident in Figure 8.2, we need to remember that the approximation of the photons and electrons and baryons moving together as a single fluid is just that, an approximation. It is valid only if the scattering rate of photons off of electrons is infinite. Of course this condition is not met: photons travel a finite distance in between scatters. Consider Figure 8.3, which depicts the path of a single photon as it scatters off a sea of electrons. It travels a mean distance λ_{MFP} in between each scatter. In our case this distance is $(n_e \sigma_T)^{-1}$. If the density of electrons is very large, then the mean free path is correspondingly small. In the course of a Hubble time, H^{-1}, a photon scatters of order $n_e \sigma_T H^{-1}$ times (simply the product of the rate and the time). As depicted in Figure 8.3, each scatter contributes to the random walk of the photon. We know that the total distance traveled in the course of a random walk is the mean free path times the square root of the total number of steps. Therefore, a cosmological photon moves a mean distance

$$\lambda_D \sim \lambda_{\mathrm{MFP}} \sqrt{n_e \sigma_T H^{-1}}$$

$$= \frac{1}{\sqrt{n_e \sigma_T H}} \tag{8.4}$$

in a Hubble time. Any perturbation on scales smaller than λ_D can be expected to be washed out. In Fourier space this will correspond to damping of all high k-modes. Note that this crude estimate gets the Ω_b dependence right. Models with small baryon density have a larger λ_D (since n_e is proportional to Ω_b when the universe is ionized). Therefore, the damping sets in at larger scales, or smaller k. This is precisely what we saw in Figure 8.2.

The final step is to relate the perturbations at recombination, as depicted in Figure 8.2, to the anisotropies we observe today. The math of this is a little complicated, but the physics is perfectly straightforward. Consider one Fourier mode, a plane-wave perturbation. Figure 8.4 shows the temperature variations for one mode at recombination. Photons from hot and cold spots separated by a typical (comoving) distance k^{-1} travel to us coming from an angular separation $\theta \simeq k^{-1}/(\eta_0 - \eta_*)$ where $\eta_0 - \eta_*$ is the (comoving) distance between us and the surface of last scattering.[2] If we decompose the temperature field into multipole moments, then an angular scale θ roughly corresponds to $1/l$. So, using the fact that $\eta_* \ll \eta_0$, we project inhomogeneities on scales k onto anisotropies on angular scales $l \simeq k\eta_0$.

There is one final caveat to this picture of free-streaming. We have been implicitly assuming that nothing happens to the photons on their journey from the last scattering surface to Earth. In fact, if the universe was flat and matter dominated through this whole time, then gravitational potentials remain constant, and this assumption is correct. However, recombination takes places not too much later than the epoch of equality, so the remnant radiation density means potentials are not exactly constant right after recombination. Also, at late times, dark energy does not behave like matter and leads to potential decay. You can imagine other disruptions to matter domination. All of these so-called *integrated Sachs–Wolfe effects* produce new perturbations to the photons, leading to changes typically of order 10%.

And that's it; we now understand how primordial perturbations are processed to form the present-day anisotropy spectrum. Let's work through it again quantitatively.

8.2 LARGE-SCALE ANISOTROPIES

To find the large-scale solution for the photon perturbation, we make use of the super-horizon equation, (7.17). This immediately tells us that $\Theta_0 = -\Phi$ plus a

---

[2]This is true only in a flat universe. In an open universe, the distance to the last scattering surface is larger, so the same physical scale is projected onto a smaller angular scale.

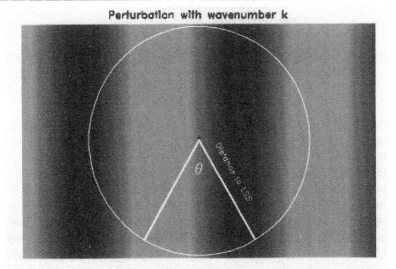

Figure 8.4. Free-streaming. Perturbations in the temperature at recombination from one plave wave with wavenumber k. Hot and cold spots are shaded light and dark. After recombination, photons from the hot and cold spots travel freely to us, here denoted by the star at the center. This mode contributes anisotropy on a scale $\theta \sim k^{-1}/(\text{Distance to last scattering surface})$.

constant. The initial conditions are such that $\Theta_0 = \Phi/2$, so the constant is $3\Phi_{\mathrm{p}}/2$, where Φ_{p} is the primordial potential set up during inflation. We have an exact expression for the large-scale evolution of Φ, Eq. (7.32). If recombination takes place long after the epoch of equality, then we can take the $y \gg 1$ limit of this expression, $\Phi \to 9\Phi_{\mathrm{p}}/10$. Therefore, at recombination, large-scale photon perturbations satisfy

$$\Theta_0(k, \eta_*) = -\Phi(k, \eta_*) + \frac{3\Phi_{\mathrm{p}}(k)}{2}$$

$$= \frac{2\Phi(k, \eta_*)}{3}. \tag{8.5}$$

The observed anisotropy is $\Theta_0 + \Psi$, which to a good approximation is $\Theta_0 - \Phi$ since $\Psi \simeq -\Phi$. Therefore,

$$(\Theta_0 + \Psi)(k, \eta_*) = \frac{1}{3}\Psi(k, \eta_*). \tag{8.6}$$

Another useful way of expressing the large-scale perturbations at recombination is in terms of the density field. The initial conditions derived in Chapter 6 were that $\delta = 3\Phi/2$. Integrating the large-scale evolution equation, $\dot{\delta} = -3\dot{\Phi}$, leads to

$$\delta(\eta_*) = \frac{3}{2}\Phi_{\mathrm{p}} - 3[\Phi(\eta_*) - \Phi_{\mathrm{p}}]$$

$$= 2\Phi(\eta_*). \tag{8.7}$$

So the observed anisotropy expressed in terms of the dark matter overdensity is

$$(\Theta_0 + \Psi)(k, \eta_*) = -\frac{1}{6}\delta(\eta_*). \tag{8.8}$$

Equations (8.6) and (8.8) will be useful to us when we compute the large-scale anisotropy spectrum. However, even now, they contain a fascinating piece of information. From the Fourier transform of Eq. (8.8), we see that the observed anisotropy of an overdense region will be *negative*. This is such a surprising result that it is worth repeating. For large-scale perturbations, overdense regions do indeed contain hotter photons at recombination than do underdense regions: i.e., $\Theta_0 > 0$ when $\Psi < 0$. However, to get to us today, these photons must travel out of their potential wells. In so doing they lose energy, and this energy loss more than compensates for the fact that the photons were initially hotter than average: i.e., $\Theta_0 + \Psi$ is negative when $\Psi < 0$. To sum up, when we observe large-scale hot spots on the sky today, we are actually observing regions that were underdense at the time of recombination.

The other important feature of Eq. (8.8) is the coefficient $1/6$. It enables us to relate "$\delta T/T$" (the left-hand side) to "$\delta\rho/\rho$" (the right). Very roughly speaking, an anisotropy of order 10^{-5} corresponds to an overdensity of 6×10^{-5}. One of the important questions which must be addressed by the picture of gravitational instability is whether the observed anisotropy is consistent with the overdensities needed to form structure by today. This factor of 6 is a huge help. In almost all models of structure formation other than inflation, this factor of 6 is replaced by a number much closer to unity (see Exercise 1 for a specific example). Therefore, they struggle with the fact that the observed level of anisotropy is too small to account for the clustering of matter in the universe. Equivalently, when normalized to large-angle anisotropies, the matter power spectrum is too small.[3]

8.3 ACOUSTIC OSCILLATIONS

8.3.1 Tightly Coupled Limit of the Boltzmann Equations

When all electrons were ionized, before η_*, the mean free path for a photon was much smaller than the horizon of the universe. Compton scattering caused the electron–proton fluid to be tightly coupled with the photons. We now proceed to explore this regime quantitatively using the Boltzmann equations.

The tightly coupled limit corresponds to the scattering rate being much larger than the expansion rate: $\tau \gg 1$, where τ is the optical depth defined in Eq. (4.61). I

---

[3]This realization has been most important in theories in which structure is generated by cosmic strings. Several papers which pointed out the problem in the aftermath of the COBE detection include: Albrecht and Stebbins (1992), Perivolaropoulos and Vachaspati (1994), and Pen and Spergel (1995).

want to argue that in the $\tau \gg 1$ limit, the only nonnegligible moments, Θ_l, are the monopole ($l = 0$) and the dipole ($l = 1$). All others are suppressed. In this sense, photons behave just like a fluid, which can be described with only two variables: the density ρ and the velocity $\vec{v}$. In order to show this, let's go back to the Boltzmann equation (4.100) for photons. We want to turn this differential equation for $\Theta(\eta, \mu)$ into an infinite set of coupled equations for $\Theta_l(\eta)$. The advantage is that — as we will see — the higher moments are small and so can be neglected. The strategy is to multiply by $\mathcal{P}_l(\mu)$ and then integrate over μ. Using Eq. (4.99), the Boltzmann equation for $l > 2$ becomes

$$\dot{\Theta}_l + \frac{k}{(-i)^{l+1}} \int_{-1}^{1} \frac{d\mu}{2} \mu \mathcal{P}_l(\mu) \Theta(\mu) = \dot{\tau} \Theta_l. \tag{8.9}$$

Note that all other terms (e.g., $-\dot{\Phi}$) have simple μ dependence (scale as μ^0 or μ^1) so all $l > 2$ moments vanish for them. To do the integral, we make use of the recurrence relation for Legendre polynomials, Eq. (C.3), to get

$$\dot{\Theta}_l - \frac{kl}{2l+1} \Theta_{l-1} + \frac{k(l+1)}{2l+1} \Theta_{l+1} = \dot{\tau} \Theta_l. \tag{8.10}$$

Let us consider the order of magnitude of the terms in Eq. (8.10). The first term on the left is of order Θ_l/η which is much smaller than the term on the right which is of order $\tau \Theta_l/\eta$. Neglecting the Θ_{l+1} term for the moment, this tells us that in the tightly coupled regime

$$\Theta_l \sim \frac{k\eta}{2\tau} \Theta_{l-1}. \tag{8.11}$$

For horizon size modes $k\eta \sim 1$, this means that $\Theta_l \ll \Theta_{l-1}$. (By the way, this is justification for throwing out the Θ_{l+1} term in making our estimate.) This estimate is valid for all modes higher than the dipole, so all such modes are very small compared to the monopole and dipole.

Before making use of this fact and deriving the tightly coupled equations in the limit in which only the monopole and dipole are nonzero (the fluid approximation), I want to explain *why* higher moments are damped in a tightly coupled environment. Indeed this observation is extremely important not only in cosmology but in all settings in which the fluid approximation is used. To understand the fluid approximation, consider one plane-wave perturbation as depicted in Figure 8.5. An observer sitting at the center of the perturbation sees photons arriving from a distance of order the mean free path, η/τ. A wavelength of order the horizon η is much larger than this distance, so the photons arriving at the observer all have the same temperature. There is very little anisotropy. You might think that a perturbation with a very small wavelength (with $k\eta \sim \tau$) would lead to anisotropy. In fact, though, such a mode has a wavelength much smaller than the damping scale. So all perturbations on such small scales are smoothed out, again leading to no anisotropy. The bottom line is there is essentially no anisotropy beyond the monopole and the dipole in the tightly coupled regime.

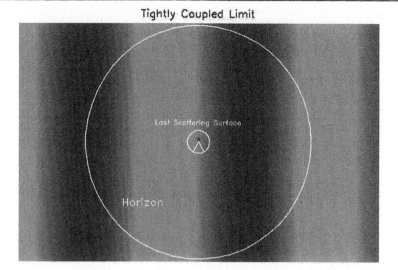

Figure 8.5. Anisotropies in the tightly coupled era. Perturbations on the scale of the horizon cannot be observed by an observer [denoted by the star here], for the photons observed come from the last scattering surface a distance η/τ away. This last scattering surface is so close that photons arriving from all angles have virtually identical temperatures.

Armed with this knowledge, we can now turn to the equations for the first two moments, which — after disposing of Θ_2 — read:

$$\dot{\Theta}_0 + k\Theta_1 = -\dot{\Phi} \tag{8.12}$$

$$\dot{\Theta}_1 - \frac{k\Theta_0}{3} = \frac{k\Psi}{3} + \dot{\tau}\left[\Theta_1 - \frac{iv_b}{3}\right]. \tag{8.13}$$

These follow by mutiplying Eq. (4.100) by $\mathcal{P}_0(\mu)$ and $\mathcal{P}_1(\mu)$ and integrating over μ. They are supplemented by the equations for the electron–baryon fluid, Eqs. (4.105) and (4.106). Let us first rewrite the velocity equation, (4.106), as

$$v_b = -3i\Theta_1 + \frac{R}{\dot{\tau}}\left[\dot{v}_b + \frac{\dot{a}}{a}v_b + ik\Psi\right]. \tag{8.14}$$

The second term on the right here is much smaller than the first since it is suppressed by a relative factor of order τ^{-1}. Thus, to lowest order, $v_b = -3i\Theta_1$. A systematic way to expand, then, is to use this lowest order expression everywhere in the second term, leading to

$$v_b \simeq -3i\Theta_1 + \frac{R}{\dot{\tau}}\left[-3i\dot{\Theta}_1 - 3i\frac{\dot{a}}{a}\Theta_1 + ik\Psi\right]. \tag{8.15}$$

Now let us insert this expression into Eq. (8.13), eliminating v_b. After rearranging terms, we find

$$\dot{\Theta}_1 + \frac{\dot{a}}{a}\frac{R}{1+R}\Theta_1 - \frac{k}{3[1+R]}\Theta_0 = \frac{k\Psi}{3}. \tag{8.16}$$

We now have two first-order coupled equations for the first two photon moments, Eqs. (8.12) and (8.16). We can turn these into one second-order equation by differentiating Eq. (8.12) and using Eq. (8.16) to eliminate $\dot{\Theta}_1$:

$$\ddot{\Theta}_0 + k\left[\frac{k\Psi}{3} - \frac{\dot{a}}{a}\frac{R}{1+R}\Theta_1 + \frac{k}{3[1+R]}\Theta_0\right] = -\ddot{\Phi}. \tag{8.17}$$

Finally, we use Eq. (8.12) to eliminate Θ_1 here. This leaves

$$\ddot{\Theta}_0 + \frac{\dot{a}}{a}\frac{R}{1+R}\dot{\Theta}_0 + k^2 c_s^2\Theta_0 = -\frac{k^2}{3}\Psi - \frac{\dot{a}}{a}\frac{R}{1+R}\dot{\Phi} - \ddot{\Phi} \equiv F(k,\eta) \tag{8.18}$$

where I have defined the forcing function on the right as F and the sound speed of the fluid as

$$c_s \equiv \sqrt{\frac{1}{3(1+R)}}. \tag{8.19}$$

The sound speed depends on the baryon density in the universe. In the absence of baryons, it has the standard value for a relativistic fluid, $c_s = 1/\sqrt{3}$. The presence of baryons, though, makes the fluid heavier, thereby lowering the sound speed. We will see shortly that the fluid oscillates in both space and time, with a period which is determined by the sound speed, and hence by the baryon density. Note that Eq. (8.18) is the "grown-up" version of Eq. (8.1); it differs only through the $\dot{\Theta}_0$ damping[4] term (see Exercise 2). The presence of this term does not change any of the qualitiative conlcusions we reached in Section 8.1. Finally, note that Φ enters on the right in a very similar way as Θ_0 does on the left. An alternate version of Eq. (8.18) takes adavantage of this:

$$\left\{\frac{d^2}{d\eta^2} + \frac{\dot{R}}{1+R}\frac{d}{d\eta} + k^2 c_s^2\right\}[\Theta_0 + \Phi] = \frac{k^2}{3}\left[\frac{1}{1+R}\Phi - \Psi\right]. \tag{8.20}$$

8.3.2 Tightly Coupled Solutions

The equation we have derived governing acoustic oscillations of the photon–baryon fluid, (8.20), is a second-order ordinary differential equation. To solve it, we will again (as in Section 7.3.1) use Green's method to find the full solution. First we find the two solutions to the homogeneous equation. Then we use these to construct the particular solution.

In prnciple, to obtain the homogeneous solutions, we must solve the damped, harmonic oscillator equation, (8.20) with the right-hand side equal to zero. In practice, the damping term is of order $R(\Theta_0 + \Phi)/\eta^2$ while the pressure term is much

---

[4]This "damping" term is not to be confused with the damping of perturbations on small scales treated in the next section. They are completely different effects.

larger, of order $k^2 c_s^2 (\Theta_0 + \Phi)$ (at least it's larger when modes are within the horizon or when R is small). Physically we expect pressure to induce oscillations in the photon temperature; the time scale for these oscillations is much shorter than the damping introduced by the expansion of the universe. To a first approximation, then, let us neglect the damping term and simply obtain the oscillating solutions.[5] In this limit, the two homogeneous solutions are

$$S_1(k, \eta) = \sin [k r_s(\eta)] \quad ; \quad S_2(k, \eta) = \cos [k r_s(\eta)] \qquad (8.21)$$

where I have defined the *sound horizon* as

$$r_s(\eta) \equiv \int_0^\eta d\eta' c_s(\eta'). \qquad (8.22)$$

Since c_s is the sound speed, the sound horizon is the comoving distance traveled by a sound wave by time η.

The tightly coupled solution for the photon temperature can be constructed from the homogeneous solutions of Eq. (8.21):

$$\Theta_0(\eta) + \Phi(\eta) = C_1 S_1(\eta) + C_2 S_2(\eta)$$

$$+ \frac{k^2}{3} \int_0^\eta d\eta' \left[\Phi(\eta') - \Psi(\eta') \right] \frac{S_1(\eta') S_2(\eta) - S_1(\eta) S_2(\eta')}{S_1(\eta') \dot{S}_2(\eta') - \dot{S}_1(\eta') S_2(\eta')}. \qquad (8.23)$$

Here again, I have dropped all occurences of R except in the arguments of the rapidly varying sines and cosines. That is, the argument of S_1, for example, is still taken to be $k r_s$ with its nonzero value of R. We can fix the constants C_1 and C_2 in Eq. (8.23) by appealing to the initial conditions, when both Θ_0 and Φ are constants. The coefficient of the sine term therefore, C_1, must vanish, and $C_2 = \Theta_0(0) + \Phi(0)$. The denominator in the integrand reduces to $-k c_s(\eta') \to -k/\sqrt{3}$ in the limit in which we are working. Finally, the difference of the products in the numerator of the integrand is simply $- \sin[k(r_s - r_s')]$, so

$$\Theta_0(\eta) + \Phi(\eta) = [\Theta_0(0) + \Phi(0)] \cos(k r_s)$$

$$+ \frac{k}{\sqrt{3}} \int_0^\eta d\eta' \left[\Phi(\eta') - \Psi(\eta') \right] \sin \left[k(r_s(\eta) - r_s(\eta')) \right]. \qquad (8.24)$$

Equation (8.24) is an expression for the anisotropy in the tightly coupled limit, first derived by Hu and Sugiyama in 1995. If you are not impressed with this solution since it still involves an integral over the gravitational potentials, I urge you to reconsider. First, look at Figure 8.6, which compares the solution of Eq. (8.24) with exact results obtained by integrating the full set of coupled Einstein–Boltzmann equations. The approximate solution gets the peak locations dead on, and it does fairly well with the heights as well. The later peaks — those at $k \eta_0 > 500$ — are

---

[5]You can rectify this by applying the WKB approximation in Exercise 5.

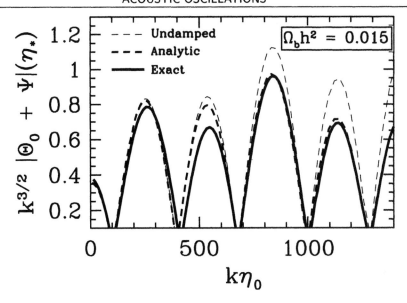

Figure 8.6. The monopole at recombination in a standard CDM model. The exact solution is the heavily weighted solid line. The light dashed line is the undamped solution of Section 8.3, Eq. (8.24); the heavier curve in the middle accounts for damping using the treatment of Section 8.4.

clearly overestimated by our solution, but we will shortly rectify this when we include damping due to diffusion in the next section. A second reason to respect the approximate solution is that it divides the problem neatly into first (i) a calculation of the external gravitional potentials generated by the dark matter and then (ii) the effect of these potentials on the anisotropies. Third, the solution clearly illustrates that the cosine mode is the one excited by inflationary models. This is important, because it is very hard to imagine this mode excited by any other mechanism. If causality is respected, then there should be no perturbations with $k\eta \ll 1$ early on. We know that inflation evades this constraint by changing the true horizon; it is tempting to say that if this mode is observed, we are seeing evidence for inflation. Fourth, we now have a more accurate expression for the frequency of oscillations and therefore for the locations of the acoustic peaks. In the limit that the first term in Eq. (8.24) dominates, the peaks should appear at the extrema of $\cos(kr_s)$, e.g., at

$$k_\mathrm{p} = n\pi/r_s \qquad n = 1, 2, \ldots \qquad . \qquad (8.25)$$

And the final reason Eq. (8.24) is impressive is that the full set of Einstein–Boltzmann equations involve literally thousands of coupled variables (e.g., the Θ_l's). Reducing those thousands of differential equations to just one is a huge leap in knowledge.

In addition to the monopole, the photon distribution has a nonnegligible dipole at recombination. Using Eq. (8.12), we can obtain an analytic solution for the dipole by differentiating Eq. (8.24):

$$\Theta_1(\eta) = \frac{1}{\sqrt{3}} \left[\Theta_0(0) + \Phi(0)\right] \sin(kr_s)$$

$$- \frac{k}{3} \int_0^\eta d\eta' \left[\Phi(\eta') - \Psi(\eta')\right] \cos\left[k(r_s(\eta) - r_s(\eta'))\right]. \tag{8.26}$$

The first term is completely out of phase with the monopole ($\sin(kr_s)$ versus $\cos(kr_s)$). Figure 8.7 shows that this feature remains even after accounting for the integral term. This mismatch of phase will have important implications for the final anisotropy spectrum.

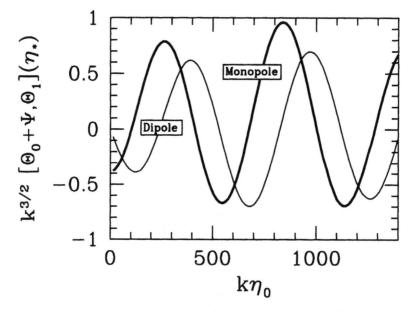

Figure 8.7. The monopole and dipole at recombination in a standard CDM model. The dipole vanishes for the longest wavelength modes that have not entered the horizon by recombination. It is completely out of phase with the monopole.

8.4 DIFFUSION DAMPING

Figure 8.6 makes it clear that we must account for diffusion to get accurate CMB spectra. To analyze diffusion quantitatively, we must return to the equations for the moments of the photon distribution, Eqs. (8.12), (8.13) and (8.10). Until now, we have neglected Θ_2 and all higher moments. Diffusion is characterized by a small but nonnegligible quadrupole.

We must therefore supplement the set of equations we wrote down in the last section with an equation for the quadrupole, Θ_2. Our task is somewhat simplified by the fact that we will be interested in phenomena occurring only on small scales. On these scales, recall from Chapter 7 (e.g., Figure 7.8) that the potentials are very small because of radiation pressure, so we can drop Φ and Ψ everywhere. Also, we will see that diffusion manifests itself in the moments by making each successive moment proportional to a higher power of $1/\dot{\tau}$. Thus we will need to keep only the $l = 2$ mode; all higher ones can be neglected. With these approximations, we have

$$\dot{\Theta}_0 + k\Theta_1 = 0 \tag{8.27}$$

$$\dot{\Theta}_1 + k\left(\frac{2}{3}\Theta_2 - \frac{1}{3}\Theta_0\right) = \dot{\tau}\left(\Theta_1 - \frac{iv_b}{3}\right) \tag{8.28}$$

$$\dot{\Theta}_2 - \frac{2k}{5}\Theta_1 = \frac{9}{10}\dot{\tau}\Theta_2. \tag{8.29}$$

These three equations need to be supplemented by an equation for v_b. This is best expressed as a slight rewriting of Eq. (8.14):

$$3i\Theta_1 + v_b = \frac{R}{\dot{\tau}}\left[\dot{v}_b + \frac{\dot{a}}{a}v_b\right], \tag{8.30}$$

where again I have dropped the gravitational potential.

To solve this set of equations, we appeal to the high-frequency nature of damping. Let us write the time dependence of the velocity as

$$v_b \propto e^{i\int \omega d\eta} \tag{8.31}$$

and similarly for all other variables. We already know that $\omega \simeq kc_s$ in the tightly coupled limit. Now we are searching for damping, an imaginary part to ω. Since damping occurs on small scales, or high frequencies,

$$\dot{v}_b = i\omega v_b \gg \frac{\dot{a}}{a}v_b; \tag{8.32}$$

$\dot{a}/a$ is of order η^{-1} while ω is of order k. So we can drop the second term on the right in Eq. (8.30) and the velocity equation then becomes

$$v_b = -3i\Theta_1\left[1 - \frac{i\omega R}{\dot{\tau}}\right]^{-1}$$

$$\simeq -3i\Theta_1\left[1 + \frac{i\omega R}{\dot{\tau}} - \left(\frac{\omega R}{\dot{\tau}}\right)^2\right] \tag{8.33}$$

where I have expanded out to $\dot{\tau}^{-2}$ because $v_b + 3i\Theta_1$ is multiplied by $\dot{\tau}$ in Eq. (8.28).

The equation for the second moment of the photon field, (8.29), can be reduced similarly. First we can drop the $\dot{\Theta}_2$ term since it is much smaller than $\dot{\tau}\Theta_2$. This leaves simply

$$\Theta_2 = -\frac{4k}{9\dot{\tau}}\Theta_1 \tag{8.34}$$

which shows that our approximation scheme is controlled: higher moments are suppressed by additional powers of $k/\dot{\tau}$. The equation for the zeroth moment becomes

$$i\omega\Theta_0 = -k\Theta_1. \tag{8.35}$$

Inserting all of these into Eq. (8.28) gives the dispersion relation

$$i\omega - \frac{8k^2}{27\dot{\tau}} + (k^2/3i\omega) = \dot{\tau}\left(1 - \left[1 + \frac{i\omega R}{\dot{\tau}} - \left(\frac{\omega R}{\dot{\tau}}\right)^2\right]\right). \tag{8.36}$$

Collecting terms we get

$$\omega^2(1+R) - \frac{k^2}{3} + \frac{i\omega}{\dot{\tau}}\left[\omega^2 R^2 + \frac{8k^2}{27}\right] = 0. \tag{8.37}$$

The first two terms on the left, the leading ones in the expansion of $1/\dot{\tau}$, recover the result of the previous section, that the frequency is the wavenumber times the speed of sound. We can write the frequency as this zero-order piece plus a first-order correction, $\delta\omega$. Then, inserting the zero-order part into the terms inversely proportional to $\dot{\tau}$ leads to

$$\delta\omega = -\frac{ik^2}{2(1+R)\dot{\tau}}\left[c_s^2 R^2 + \frac{8}{27}\right]. \tag{8.38}$$

Therefore, the time dependence of the perturbations is

$$\Theta_0, \Theta_1 \sim \exp\left\{ik\int d\eta c_s\right\}\exp\left\{-\frac{k^2}{k_D^2}\right\} \tag{8.39}$$

where the damping wavenumber is defined via

$$\frac{1}{k_D^2(\eta)} \equiv \int_0^\eta \frac{d\eta'}{6(1+R)n_e\sigma_T a(\eta')}\left[\frac{R^2}{(1+R)} + \frac{8}{9}\right]. \tag{8.40}$$

Putting aside factors of order unity, this equation says that $1/k_D \sim [\eta/n_e\sigma_T a]^{1/2}$, which agrees with our heuristic estimate at the beginning of this chapter.

As a first estimate of the damping scale, we can work in the prerecombination limit, in which all electrons (except those in helium) are free. In Chapter 3 we estimated the optical depth in this limit, but ignored helium. The mass fraction of helium is usually denoted as Y_p and is approximately 0.24. Since each helium nucleus contains four nucleons, the ratio of helium to the total number of nuclei is $Y_p/4$. Each of these absorbs two electrons (one for each proton), so when counting the number of free electrons before hydrogen recombination, we must multiply our estimate of Eq. (3.46) by $1 - Y_p/2$. Using the fact that $H_0 = 3.33 \times 10^{-4}\, h$ Mpc^{-1}, we have, in the prerecombination limit,

$$n_e \sigma_T a = 2.3 \times 10^{-5} \text{Mpc}^{-1} \Omega_b h^2 a^{-2} \left(1 - \frac{Y_p}{2}\right). \qquad (8.41)$$

Using this, you can show (Exercise 8) that an approximation for the damping scale is

$$k_D^{-2} = 3.1 \times 10^6 \text{ Mpc}^2 a^{5/2} f_D(a/a_{\text{eq}}) \left(\Omega_b h^2\right)^{-1} \left(1 - \frac{Y_p}{2}\right)^{-1} \left(\Omega_m h^2\right)^{-1/2} \qquad (8.42)$$

where f_D, defined in Eq. (8.88), goes to 1 as a/a_{eq} gets large.

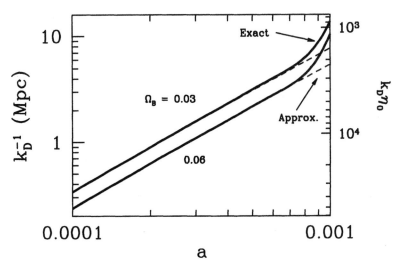

Figure 8.8. Damping scale as a function of the scale factor for two different values of Ω_b (with $h = 0.5$). Heavy curves (exact) numerically integrate over the standard recombination history, while light curves use the approximation of Eq. (8.42) which assumes electrons remain ionized. Right axis shows the equivalent $k_D \eta_0$; damping occurs on angular scales $l > k_D \eta_0$.

Figure 8.8 shows the evolution of the damping scale before recombination. Neglecting recombination is a good approximation at early times but, as expected, leads to quantitative errors right near η_*, when using Eq. (8.41) for the free electron density does not accurately account for the electrons swept up into neutral hydrogen. In the absence of recombination, k_D scales as $\Omega_b^{1/2}$. Note from the late time behavior in Figure 8.8 that the messy details of recombination change this simple scaling: k_D for the $\Omega_b = 0.06$ case is less than $2^{0.5}$ as big as the $\Omega_b = 0.03$ case.

Figure 8.8 requires one final comment. The damping of anisotropies due to photon diffusion is sometimes referred to as being caused by the "finite thickness of the last scattering surface." That is, it is argued that if recombination took place instantaneously at η_*, then there would be no damping. Figure 8.8 shows that this is patently false. Even if recombination had occurred in this way, the universe before recombination would *not* have been inifinitely optically thick. Photons would still stream a reasonable distance and hence damp anisotropies. In the examples shown, the damping scale would have been smaller (larger l) by less than a factor of 2 if recombination had occurred instantaneously. On the other hand, we will see in the next section that the anisotropies today are determined by integrating over the *visibility function*, essentially a filter centered at the epoch of recombination but broadened due to the finite thickness of the last scattering surface. When incorporating the effects of damping (Seljak, 1994; Hu and Sugiyama, 1995), one must account for this finite thickness by integrating the damping function e^{-k^2/k_D^2} weighted by the visibility function. Thus the finite thickness of the last scattering surface has both qualitative and quantitative effects on the final anisotropy spectrum.

8.5 INHOMOGENEITIES TO ANISOTROPIES

We now have a good handle on the perturbations to the photons at recombination. It is time to transform this understanding into predictions for the anisotropy spectrum today. First, we will solve for the moments Θ_l today in the next subsection. Then we will spend a bit of time relating these moments to the observables. Thus the main purpose of the following subsections is to derive Eq. (8.56), which relates the moments today to the monopole and dipole at recombination, and Eq. (8.68), which expresses the CMB power spectrum in terms of the Fourier moments today.

8.5.1 Free Streaming

We want to derive a formal solution for the photon moments today $\Theta_l(\eta_0)$ in terms of the monopole and dipole at recombination. A formal solution can be obtained by returning to Eq. (4.100). Subtracting $\dot{\tau}\Theta$ from both sides leads to

$$\dot{\Theta} + (ik\mu - \dot{\tau})\Theta = e^{-ik\mu\eta+\tau}\frac{d}{d\eta}\left[\Theta e^{ik\mu\eta-\tau}\right] = \tilde{S} \qquad (8.43)$$

where the source function is defined as

$$\tilde{S} \equiv -\dot{\Phi} - ik\mu\Psi - \dot{\tau}\left[\Theta_0 + \mu v_b - \frac{1}{2}\mathcal{P}_2(\mu)\Pi\right]. \qquad (8.44)$$

Hold your curiosity about the ~ in the definition. Multplying both sides of Eq. (8.43) by the exponential and then integrating over η leads directly to

$$\Theta(\eta_0) = \Theta(\eta_{\text{init}})e^{ik\mu(\eta_{\text{init}}-\eta_0)}e^{-\tau(\eta_{\text{init}})} + \int_{\eta_{\text{init}}}^{\eta_0} d\eta \tilde{S}(\eta)e^{ik\mu(\eta-\eta_0)-\tau(\eta)} \qquad (8.45)$$

where I have used the fact that $\tau(\eta = \eta_0) = 0$ since τ is defined as the scattering rate integrated from η up to η_0. We also know that, if the initial time η_{init} is early enough, then the optical depth $\tau(\eta_{\text{init}})$ will be extremely large. Therefore, the first term on the right side of Eq. (8.45) vanishes. This corresponds to the fact that any initial anisotropy is completely erased by Compton scattering. By the same reasoning, we can set the lower limit on the integral to zero: any contribution to the integrand from $\eta < \eta_{\text{init}}$ is completely negligible. Thus, the solution for the perturbations is

$$\Theta(k,\mu,\eta_0) = \int_0^{\eta_0} d\eta \tilde{S}(k,\mu,\eta)e^{ik\mu(\eta-\eta_0)-\tau(\eta)}. \tag{8.46}$$

Equation (8.46) looks simple, but of course all of the complication is hidden in the source function $\tilde{S}$. Notice that $\tilde{S}$ depends somewhat on the angle μ. If it did *not* depend on μ, we could immediately turn Eq. (8.46) into an equation for each of the Θ_l's. For, we could multiply each side by the Legendre polynomial $\mathcal{P}_l(\mu)$ and then integrate over all μ. By Eq. (4.99), the left side would give $(-i)^l\Theta_l$ and the right would contain the integral

$$\int_{-1}^1 \frac{d\mu}{2}\mathcal{P}_l(\mu)e^{ik\mu(\eta-\eta_0)} = \frac{1}{(-i)^l}j_l\left[k(\eta-\eta_0)\right] \tag{8.47}$$

where j_l is the spherical Bessel function. This approach looks so promising that we should pursue it to its end, again forgetting for the moment that $\tilde{S}$ really does have some μ dependence. The expression for Θ_l would be

$$\Theta_l(k,\eta_0) = (-1)^l \int_0^{\eta_0} d\eta \tilde{S}(k,\eta)e^{-\tau(\eta)}j_l\left[k(\eta-\eta_0)\right]. \tag{8.48}$$

What about the μ dependence in $\tilde{S}$? We can account for this by noting that $\tilde{S}$ multiplies the exponential $e^{ik\mu(\eta-\eta_0)}$ in Eq. (8.46). Thus, everywhere we encounter a factor of μ in $\tilde{S}$ we can replace it with a time derivative:

$$\mu \rightarrow \frac{1}{ik}\frac{d}{d\eta}. \tag{8.49}$$

Let me demonstrate this explicitly with the $-ik\mu\Psi$ term in $\tilde{S}$. The integral is

$$-ik\int_0^{\eta_0} d\eta\ \mu\Psi e^{ik\mu(\eta-\eta_0)-\tau(\eta)} = -\int_0^{\eta_0} d\eta\Psi e^{-\tau(\eta)}\frac{d}{d\eta}e^{ik\mu(\eta-\eta_0)}$$

$$= \int_0^{\eta_0} d\eta e^{ik\mu(\eta-\eta_0)}\frac{d}{d\eta}\left[\Psi e^{-\tau(\eta)}\right] \tag{8.50}$$

where the last line follows by integration by parts. Note that the surface terms can be dropped: at $\eta = 0$ they are damped by the $e^{-\tau(0)}$ factor. The terms at $\eta = \eta_0$ are not small, but they are irrelevant since they have no angular dependence. They alter the monopole, an alteration which we cannot detect. Thus, accounting for the integration by parts changes the substitution rule of Eq. (8.49) by a minus sign, with the understanding that the derivative does *not* act on the oscillating part of the exponential, $e^{ik\mu(\eta-\eta_0)}$. The solution in Eq. (8.48) therefore becomes

$$\Theta_l(k,\eta_0) = \int_0^{\eta_0} d\eta S(k,\eta) j_l \left[k(\eta_0 - \eta) \right] \tag{8.51}$$

with the source function now defined as

$$S(k,\eta) \equiv e^{-\tau} \left[-\dot{\Phi} - \dot{\tau}(\Theta_0 + \frac{1}{4}\Pi) \right]$$

$$+ \frac{d}{d\eta} \left[e^{-\tau} \left(\Psi - \frac{iv_{\mathrm{b}}\dot{\tau}}{k} \right) \right] - \frac{3}{4k^2} \frac{d^2}{d\eta^2} \left[e^{-\tau}\dot{\tau}\Pi \right]. \tag{8.52}$$

In Eq. (8.51), I have also used the property of spherical Bessel functions: $j_l(x) = (-1)^l j_l(-x)$.

At this stage, it is useful to introduce the *visibility function*

$$g(\eta) \equiv -\dot{\tau} e^{-\tau}. \tag{8.53}$$

The visibility function has some interesting properties. The integral $\int_0^{\eta_0} d\eta g(\eta) = 1$, so we can think of it as a probability density. It is the probability that a photon last scattered at η. In the standard recombination, since τ is so large early on, this probability is essentially zero for η earlier than the time of recombination. It also declines rapidly after recombination, because the prefactor $-\dot{\tau}$, which is the scattering rate, is quite small. Figure 8.9 shows the visibility function for two values of the baryon density.

The source function in Eq. (8.52) can now be expressed in terms of the visibility function. If we drop the polarization tensor Π in the source since it is very small, then the source function becomes

$$S(k,\eta) \simeq g(\eta) \left[\Theta_0(k,\eta) + \Psi(k,\eta) \right]$$

$$+ \frac{d}{d\eta} \left(\frac{iv_{\mathrm{b}}(k,\eta)g(\eta)}{k} \right)$$

$$+ e^{-\tau} \left[\dot{\Psi}(k,\eta) - \dot{\Phi}(k,\eta) \right]. \tag{8.54}$$

We can take our analytic solution one step further by performing the time integral in Eq. (8.51). The source term proportional to v_{b} is best treated by integrating by parts. Then,

$$\Theta_l(k,\eta_0) = \int_0^{\eta_0} d\eta \, g(\eta) \, \left[\Theta_0(k,\eta) + \Psi(k,\eta) \right] \, j_l \left[k(\eta_0 - \eta) \right]$$

$$- \int_0^{\eta_0} d\eta \, g(\eta) \, \frac{iv_{\mathrm{b}}(k,\eta)}{k} \, \frac{d}{d\eta} j_l \left[k(\eta_0 - \eta) \right]$$

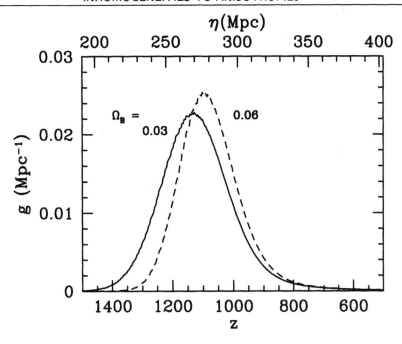

Figure 8.9. The visibility function. Most electrons last scatter at around $z \simeq 1100$ with little dependence on the baryon density. Note that the integral of g over conformal time is 1. Here $h = 0.5$.

$$+ \int_0^{\eta_0} d\eta \; e^{-\tau} \left[\dot{\Psi}(k, \eta) - \dot{\Phi}(k, \eta) \right] \; j_l \left[k(\eta_0 - \eta) \right]. \qquad (8.55)$$

There are two types of terms in Eq. (8.55). First, there are those wherein the integral is weighted by $e^{-\tau}$. These contribute as long as $\tau \lesssim 1$, that is, at all times after recombination. Note that these are also proportional to derivatives of the potentials. If the potentials are constant after recombination, these terms vanish. In many theories, as we saw in Chapter 7, this is precisely what happens: the universe is purely matter dominated after recombination and in such an environment, the potentials generally remain constant. The corrections due to changing potentials are therefore important to get things right quantitatively, but do not affect the qualitative structure of the anisotropy spectrum. Rather, the dominant terms in Eq. (8.55) are the second types of terms, the ones with integrals weighted by the visibility function.

Since the visibility function is so sharply peaked, the integrals in the first two terms become very simple. To see why, consider Figure 8.10 which shows the three parts of the integrand of the first term (the monopole) in Eq. (8.55). Since the visibility function changes rapidly compared with the other two functions, we can evaluate those other functions at the peak of the visibility function, i.e., at $\eta = \eta_*$, and remove them from the integral. But then, the integral is simply $\int d\eta g(\eta) = 1$. Thus, we are left with

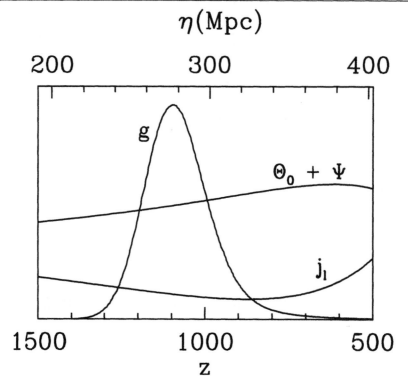

Figure 8.10. The three components of the integrand in the monopole term of Eq. (8.55). The visibility function is sharply peaked, so it changes rapidly compared with the monopole $\Theta_0 + \Psi$ and the Bessel function $j_l(k[\eta - \eta_0])$. Figure is for $l = 100, k = 0.013\,h\ \mathrm{Mpc}^{-1}$.

$$\Theta_l(k, \eta_0) \simeq [\Theta_0(k, \eta_*) + \Psi(k, \eta_*)]\ j_l\left[k(\eta_0 - \eta_*)\right]$$

$$+ 3\Theta_1(k, \eta_*)\left(j_{l-1}\left[k(\eta_0 - \eta_*)\right] - \frac{(l+1)j_l\left[k(\eta_0 - \eta_*)\right]}{k(\eta_0 - \eta_*)}\right)$$

$$+ \int_0^{\eta_0} d\eta\ e^{-\tau}\left[\dot{\Psi}(k, \eta) - \dot{\Phi}(k, \eta)\right]\ j_l\left[k(\eta_0 - \eta)\right]. \qquad (8.56)$$

Here I have used the spherical Bessel function identity of Eq. (C.18) to rewrite the Bessel function derivative in the velocity term and also the fact that $v_{\mathrm{b}} \simeq -3i\Theta_1$ at η_*. On scales much smaller than the one shown in Figure 8.10, $\Theta_0 + \Psi$ changes more rapidly because of the rapid change in the damping scale around recombination. However, this effect can be incorporated by modifying the damping function from

$$e^{-k^2/k_D(\eta_*)^2} \rightarrow \int d\eta g(\eta) e^{-k^2/k_D(\eta)^2}. \qquad (8.57)$$

Equation (8.56) is the basis for semianalytic calculations (Seljak, 1994; Hu and Sugiyama, 1995) of C_l spectra which agree with the exact (numerical) solutions to within 10%. From Eq. (8.56), we see that, to solve for the anisotropies today, we must know the monopole (Θ_0), dipole (Θ_1), and potential (Ψ) at the time of recombination. Further, there will be small but noticeable corrections if the potentials are time dependent. These corrections, encoded in the last line of Eq. (8.56), are often called *integrated Sachs–Wolfe* (ISW) terms.

The monopole term — the first in Eq. (8.56) — is precisely what we expected from the rough arguments of Section 8.1. In particular, the spherical Bessel function, $j_l[k(\eta_0-\eta_*)]$, determines how much anisotropy on an angular scale l^{-1} is contributed by a plane wave with wavenumber k. On very small angular scales,

$$\lim_{l\to\infty} j_l(x) = \frac{1}{l} \left(\frac{x}{l}\right)^{l-1/2}. \tag{8.58}$$

That is, $j_l(x)$ is extremely small for large l when $x < l$. In our case, this means that $\Theta_l(k,\eta_0)$ is very close to zero for $l > k\eta_0$. This makes sense physically. Returning to Figure 8.4, we see that very small angular scales will see little anisotropy from a perturbation with a large wavelength. The converse is also true: angular scales larger than $1/(k\eta_0)$ get little contribution from such a perturbation. To sum up, a perturbation with wavenumber k contributes predominantly on angular scales of order $l \sim k\eta_0$. One last comment about the monopole term: the final anisotropy today depends on not just Θ_0, but rather $\Theta_0 + \Psi$, again something we anticipated since photons must climb out of their potential wells to reach us today.

8.5.2 The C_l's

How is the observed anisotropy pattern today related to the rather abstract $\Theta_l(k,\eta_0)$? To answer this question, we must first describe the way in which the temperature field is characterized today and then relate this characterization to Θ_l.

Recall that in Eq. (4.34), we wrote the temperature field in the universe as

$$T(\vec{x},\hat{p},\eta) = T(\eta)\left[1 + \Theta(\vec{x},\hat{p},\eta)\right]. \tag{8.59}$$

Although this field is defined at every point in space and time, we can observe it only here (at $\vec{x}_0$) and now (at η_0).[6] Our only handle on the anisotropies is their dependence on the direction of the incoming photons, $\hat{p}$. So all the richness we observe comes from the changes in the temperature as the direction vector $\hat{p}$ changes. Observers typically makes maps, wherein the temperature is reported at a number of incoming directions, or "spots on the sky." These spots are labeled not by the $\hat{p}_x,\hat{p}_y,\hat{p}_z$ components of $\hat{p}$, but rather by polar coordinates θ,ϕ. However, it

---

[6]We do make small excursions from this point in space-time. For example, satellites are not located on Earth and anisotropy measurements have been made over the past 30 years. These are completely insignificant on scales over which the temperature is varying, which are of order the Hubble time (or distance).

is a simple matter to move back and forth between the 3D unit vector $\hat{p}$ and polar coordinates.[7] I'll stick with $\hat{p}$ in the ensuing derivation.

We now expand the field in terms of spherical harmonics. That is, we write

$$\Theta(\vec{x}, \hat{p}, \eta) = \sum_{l=1}^{\infty} \sum_{m=-l}^{l} a_{lm}(\vec{x}, \eta) Y_{lm}(\hat{p}). \qquad (8.60)$$

The subscripts l, m are conjugate to the real space unit vector $\hat{p}$, just as the variable $\vec{k}$ is conjugate to the Fourier transform variable $\vec{x}$. We are all familiar with Fourier transforms, so it is useful to think of the expansion in terms of spherical harmonics as a kind of generalized Fourier transform. Whereas the complete set of eigenfunctions for the Fourier transform are $e^{i\vec{k}\cdot\vec{x}}$, here the complete set of eigenfunctions for expansion on the surface of a sphere are $Y_{lm}(\hat{p})$. All of the information contained in the temperature field T is also contained in the space-time dependent amplitudes a_{lm}. As an example of this, consider an experiment which maps the full sky with an angular resolution of $7°$. The full sky has 4π radians$^2 \simeq 41,000$ degrees^2, so there are 840 pixels with area of $(7°)^2$. Thus, such an experiment would have 840 independent pieces of information. Were we to characterize this information with a_{lm}'s instead of temperatures in pixels, there would be some $l_{\max}$ above which there is no information. One way to determine this $l_{\max}$ is to set the total number of recoverable a_{lm}'s as $\sum_{l=0}^{l_{\max}}(2l+1) = (l_{\max}+1)^2 = 840$. So the information could be equally well characterized by specifying all the a_{lm}'s up to $l_{\max} = 28$. Incidentally, this is a fairly good caricature of the COBE experiment (Smoot et al., 1992; Bennett et al., 1996). They presented temperature data over many more pixels, but many of these pixels were overlapping. So, the independent information was contained in multipoles up to $l \sim 30$. Experiments currently under way or well along in the planning stage are capable of measuring the moments all the way up to $l \sim 10^4$.

We want to relate the observables, the a_{lm}'s, to the Θ_l we have been dealing with. To do this, we can use the orthogonality property of the spherical harmonics. The Y_{lm}'s are normalized via Eq. (C.11),

$$\int d\Omega Y_{lm}(\hat{p}) Y_{l'm'}^*(\hat{p}) = \delta_{ll'}\delta_{mm'}, \qquad (8.61)$$

where Ω is the solid angle spanned by $\hat{p}$. Therefore the expansion of Θ in terms of spherical harmonics, Eq. (8.60), can be inverted by multiplying both sides by $Y_{lm}^*(\hat{p})$ and integrating:

$$a_{lm}(\vec{x}, \eta) = \int \frac{d^3k}{(2\pi)^3} e^{i\vec{k}\cdot\vec{x}} \int d\Omega Y_{lm}^*(\hat{p}) \Theta(\vec{k}, \hat{p}, \eta). \qquad (8.62)$$

Here I have written the right-hand side in terms of the Fourier transform ($\Theta(\vec{k})$) instead of $\Theta(\vec{x})$), since that is the quantity for which we obtained solutions.

---

[7] $\hat{p}_z = \cos\theta$, $\hat{p}_x = \sin\theta\cos\phi$, and $\hat{p}_y = \sin\theta\sin\phi$.

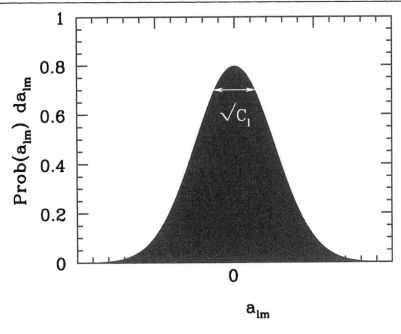

Figure 8.11. The distribution from which the a_{lm}'s are drawn. The distribution has expectation equal to zero and a width of $C_l^{1/2}$.

As with the density perturbations, we cannot make predictions about any particular a_{lm}, just about the distribution from which they are drawn, a distribution which traces its origin to the quantum fluctuations first laid down during inflation. Figure 8.11 illustrates this distribution. The mean value of all the a_{lm}'s is zero, but they will have some nonzero variance. The variance of the a_{lm}'s is called C_l. Thus,

$$\langle a_{lm} \rangle = 0 \qquad ; \qquad \langle a_{lm} a_{l'm'}^* \rangle = \delta_{ll'} \delta_{mm'} C_l. \tag{8.63}$$

It is very important to note that, for a given l, each a_{lm} has the same variance. For $l = 100$, say, all 201 $a_{100,m}$'s are drawn from the same distribution. When we measure these 201 coefficients, we are sampling the distribution. This much information will give us a good handle on the underlying variance of the distribution. On the other hand, if we measure the five components of the quadrupole ($l = 2$), we do not get very much information about the underlying variance, C_2. Thus, *there is a fundamental uncertainty in the knowledge we may get about the C_l's.* This uncertainty, which is most pronounced at low l, is called *cosmic variance.* Quantitatively, the uncertainty scales simply as the inverse of the square root of the number of possible samples, or

$$\left(\frac{\Delta C_l}{C_l} \right)_{\text{cosmic variance}} = \sqrt{\frac{2}{2l+1}}. \tag{8.64}$$

We can now obtain an expression for C_l in terms of $\Theta_l(k)$. First we square a_{lm} in Eq. (8.62) and take the expectation value of the distribution. For this we need $\langle \Theta(\vec{k}, \hat{p})\Theta^*(\vec{k}', \hat{p}') \rangle$, where from now on we will keep the η dependence implicit. This expectation value is complicated because it depends on two separate phenomena: (i) the initial amplitude and phase of the perturbation is chosen during inflation from a Gaussian distribution and (ii) the evolution we have studied in this chapter turns this initial perturbation into anisotropies, i.e. produces the dependence on $\hat{p}$. To simplify then, it makes sense to separate these two phenomena and write the photon distribution as $\delta \times (\Theta/\delta)$, where the dark matter overdensity δ does not depend on any direction vector. The ratio Θ/δ is precisely what we have solved for in the last two chapters: given the initial amplitude of a mode, we have learned how to evolve forward in time. The ratio does *not* depend on the initial amplitude, so it can be removed from the averaging over the distribution. Therefore,

$$\langle \Theta(\vec{k}, \hat{p})\Theta(\vec{k}', \hat{p}') \rangle = \langle \delta(\vec{k})\delta^*(\vec{k}') \rangle \frac{\Theta(\vec{k}, \hat{p})}{\delta(\vec{k})} \frac{\Theta^*(\vec{k}', \hat{p})}{\delta^*(\vec{k}')}$$

$$= (2\pi)^3 \delta^3(\vec{k} - \vec{k}')P(k) \frac{\Theta(k, \hat{k} \cdot \hat{p})}{\delta(k)} \frac{\Theta^*(k, \hat{k} \cdot \hat{p}')}{\delta^*(k)}, \quad (8.65)$$

where the second equality uses the definition of the matter power spectrum $P(k)$, but also contains a subtlety in the ratio Θ/δ. This ratio, which is determined solely by the evolution of both δ and Θ, depends only on the magnitude of $\vec{k}$ and the dot product $\hat{k} \cdot \hat{p}$. Two modes with the same k and $\hat{k} \cdot \hat{p}$ evolve identically even though their initial amplitudes and phases are different.

After squaring Eq. (8.62), we see that the anisotropy spectrum is

$$C_l = \int \frac{d^3k}{(2\pi)^3} P(k) \int d\Omega Y_{lm}^*(\hat{p}) \frac{\Theta(k, \hat{k} \cdot \hat{p})}{\delta(k)} \int d\Omega' Y_{lm}(\hat{p}') \frac{\Theta^*(k, \hat{k} \cdot \hat{p}')}{\delta^*(k)}. \quad (8.66)$$

Now we can expand $\Theta(k, \hat{k} \cdot \hat{p})$ and $\Theta(k, \hat{k} \cdot \hat{p}')$ in spherical harmonics using the inverse of Eq. (4.99), $\Theta(k, \hat{k} \cdot \hat{p}) = \sum_l (-i)^l (2l + 1)\mathcal{P}_l(\hat{k} \cdot \hat{p})\Theta_l(k)$. This leaves

$$C_l = \int \frac{d^3k}{(2\pi)^3} P(k) \sum_{l'l''} (-i)^{l'} (i)^{l''} (2l' + 1)(2l'' + 1) \frac{\Theta_{l'}(k)\Theta_{l''}^*(k)}{|\delta(k)|^2}$$

$$\times \int d\Omega \mathcal{P}_{l'}(\hat{k} \cdot \hat{p}) Y_{lm}^*(\hat{p}) \int d\Omega' \mathcal{P}_{l''}(\hat{k} \cdot \hat{p}') Y_{lm}(\hat{p}'). \quad (8.67)$$

The two angular integrals here (Exercise 9) are identical. They are nonzero only if $l' = l$ and $l'' = l$, in which case they are equal to $4\pi Y_{lm}(\hat{k})/(2l + 1)$ (or the complex conjugate). The angular part of the d^3k integral then becomes an integral over $|Y_{lm}|^2$, which is just equal to 1, leaving

$$C_l = \frac{2}{\pi} \int_0^\infty dk \, k^2 \, P(k) \left| \frac{\Theta_l(k)}{\delta(k)} \right|^2. \quad (8.68)$$

For a given l, then, the variance of a_{lm}, C_l, is an integral over all Fourier modes of the variance of $\Theta_l(\vec{k})$. We can now use Eqs. (8.56) and (8.68) to plot the anisotropy spectrum today.

8.6 THE ANISOTROPY SPECTRUM TODAY

8.6.1 Sachs–Wolfe Effect

Large-angle anisotropies are not affected by any microphysics: at the time of recombination, the perturbations responsible for these anisotropies were on scales far larger than could be connected via causal processes. On these largest of scales, only the monopole contributes to the anisotropy; this is the first term in Eq. (8.56). So the large-angle anisotropy is determined by $\Theta_0 + \Psi$ evaluated at recombination. The large-scale solution we found in Eq. (8.6) was that this combination is equal to $\Psi(\eta_*)/3$. In most cosmological models, recombination occurs far enough after matter/radiation equality that we can approximate the potential back then to be equal to the potential today modulo the growth factor, so

$$\Theta_0(\eta_*) + \Psi(\eta_*) \simeq \frac{1}{3D_1(a=1)}\Psi(\eta_0) = -\frac{1}{3D_1(a=1)}\Phi(\eta_0). \qquad (8.69)$$

The last equality holds here because at very late times, there are no appreciable anisotropic stresses, and $\Phi = -\Psi$.

We may use Eq. (7.7) to express the potential Φ today in terms of the dark matter distribution, so that

$$\Theta_0(\eta_*) + \Psi(\eta_*) \simeq -\frac{\Omega_m H_0^2}{2k^2 D_1(a=1)}\delta(\eta_0). \qquad (8.70)$$

This gives us what we need: an expression for the sum of $\Theta_0 + \Psi$ at recombination that we can plug into the monopole term in Eq. (8.56). To get the anisotropy spectrum today, we then integrate as in Eq. (8.68), leaving

$$C_l^{\mathrm{SW}} \simeq \frac{\Omega_m^2 H_0^4}{2\pi D_1^2(a=1)}\int_0^\infty \frac{dk}{k^2}j_l^2\left[k(\eta_0 - \eta_*)\right]P(k) \qquad (8.71)$$

where the superscript denotes *Sachs-Wolfe*, in honor of the first people to compute the large-angle anisotropy (Sachs and Wolfe, 1967). The power spectrum is given by Eq. (7.9) with the transfer function set to 1 (since we're considering very large scales). Therefore,

$$C_l^{\mathrm{SW}} \simeq \pi H_0^{1-n}\left(\frac{\Omega_m}{D_1(a=1)}\right)^2 \delta_H^2 \int_0^\infty \frac{dk}{k^{2-n}}j_l^2\left[k(\eta_0 - \eta_*)\right]. \qquad (8.72)$$

The large-scale anisotropies in Eq. (8.72) can be computed analytically. First, we will use the fact that $\eta_* \ll \eta_0$ and define the dummy variable $x \equiv k\eta_0$. Then the spectrum can be rewritten as

$$C_l^{\text{SW}} \simeq \pi(\eta_0 H_0)^{1-n} \left(\frac{\Omega_m}{D_1(a=1)}\right)^2 \delta_H^2 \int_0^\infty \frac{dx}{x^{2-n}} j_l^2(x). \qquad (8.73)$$

The integral over the spherical Bessel functions can be analytically expressed (Eq. (C.17) from Gradshteyn and Ryzhik, 6.574.2) in terms of gamma functions, leaving

$$C_l^{\text{SW}} \simeq 2^{n-4}\pi^2(\eta_0 H_0)^{1-n} \left(\frac{\Omega_m}{D_1(a=1)}\right)^2 \delta_H^2 \frac{\Gamma\left(l+\frac{n}{2}-\frac{1}{2}\right)}{\Gamma\left(l+\frac{5}{2}-\frac{n}{2}\right)} \frac{\Gamma(3-n)}{\Gamma^2\left(2-\frac{n}{2}\right)}. \qquad (8.74)$$

If the spectrum is Harrison–Zel'dovich–Peebles, $n = 1$, then the first ratio of the gamma functions $\Gamma(l)/\Gamma(l+2)$ is equal to $[l(l+1)]^{-1}$ using Eq. (C.24). The remaining ratio of gamma functions $\Gamma(2)/\Gamma^2(3/2) = 4/\pi$ using Eq. (C.25), so

$$l(l+1)C_l^{\text{SW}} = \frac{\pi}{2} \left(\frac{\Omega_m}{D_1(a=1)}\right)^2 \delta_H^2, \qquad (8.75)$$

a constant. Indeed, this is the reason why workers in the field typically plot $l(l+1)C_l$: at low l, where the Sachs–Wolfe approximation is a good one, we expect a plateau.

Figure 8.12 shows the COBE measurements of the large-angular-scale anisotropies along with the Boltzmann solutions of three CDM models. Note that, even for $n = 1$, the true spectrum is *not* completely flat as suggested by Eq. (8.75). The dipole at recombination (neglected in Eq. (8.74)) contributes slightly. The integrated Sachs–Wolfe effect also is not completely negligible, especially in the Λ model, wherein the potential starts to decay once the universe becomes Λ-dominated at late times. For an $n = 1$ spectrum, the best fit values of δ_H from COBE are

$$\delta_H = 1.9 \times 10^{-5} \qquad \Omega_m = 1$$

$$\delta_H = 4.6 \times 10^{-5} \qquad \Omega_m = 0.3\,;\ \Omega_\Lambda = 0.7. \qquad (8.76)$$

Also shown in Figure 8.12 is a *tilted* model, one in which the primordial spectral index n is not equal to 1. In such models, the anisotropy should scale as l^{n-1} compared with the Harrison–Zel'dovich–Peebles $n = 1$ spectrum. You can see this scaling from Eq. (8.74) or more directly from the integral in Eq. (8.73). The integrand peaks at $x \sim l$, so very roughly every appearence of x there can be replaced by l. The generalization of the integrand from x^{-1} to x^{n-2} therefore leads to a change in the spectrum that scales as l^{n-1}. As indicated in Figure 8.12, the COBE data have the greatest weight at $l \sim 10$, but cover a range of l spanning an order of magnitude. Extreme values of tilt are therefore ruled out by COBE. To get much better constraints on the tilt, though, measurements spanning a larger range of l are necessary.

8.6.2 Small Scales

The small-scale anisotropy spectrum depends not only on the monopole, but also on the dipole and the integrated Sachs–Wolfe effect. Figure 8.13 shows all these contributions to the spectrum. Let's consider each in turn.

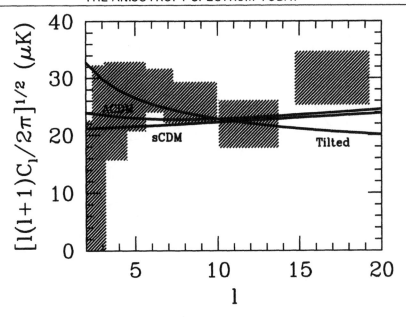

Figure 8.12. Large-scale anisotropies. Hatched boxes show measurements by COBE satellite (Bennett *et al.*, 1996). Curves show the spectra for standard CDM and ΛCDM (both with $n = 1$). The *tilted* model is identical to standard CDM, except $n = 0.5$. The late time integrated Sachs–Wolfe effect enhances anisotropy on the largest scales in ΛCDM. Note that here, and in subsequent C_l figures, the root mean square anisotropy is plotted, proportional to $C_l^{1/2}$. C_l is dimensionless so the units of μK come from multiplying by the present background temperature, $T = 2.73$K.

The monopole at recombination $(\Theta_0 + \Psi)(k, \eta_*)$ free-streams to us today, creating anisotropies on angular scales $l \sim k\eta_0$. This is what we expected back in Figure 8.4, showed to be true in Eq. (8.56), and can now see directly in the top panel of Figure 8.13. There are two interesting features of the quantitative aspect of the free-streaming process. First, note that the "zeroes" in the monopole spectrum, here at $400, 650$, and 970, are smoothed out because many modes contribute to anisotropy on a given angular scale. If only the $k = 400/\eta_0$ modes contributed to the anisotropy at $l = 400$, then C_{400} would really be zero. But many nonzero modes, with wavenumber greater than $400/\eta_0$, contribute. These change the zero to a trough in the C_l spectrum.

The second feature of free-streaming worth noticing is that our initial estimate tof the peak positions is not exactly right. Inhomogeneity on scale k does *not* show up as anisotropy precisely on angular scale $l = k\eta_0$. Rather, There is a noticable shift in the top panel, suggesting that a given k-mode contributes to slightly smaller l than we anticipated. This shift arises from the spherical Bessel function in Eq. (8.56). As indicated in Figure 8.14, the peak in the Bessel function comes not when $l = k\eta_0$, but rather at slightly smaller values of l. In addition, our initial

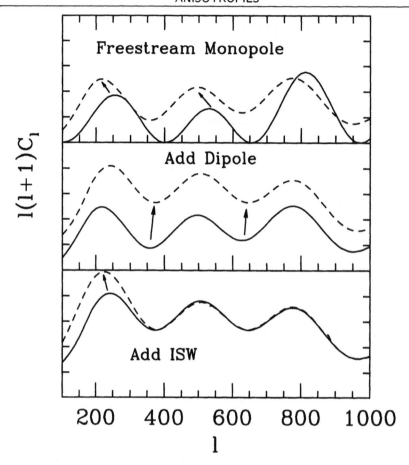

Figure 8.13. Small-scale anisotropy. *Top panel:* The monopole at recombination $(\Theta_0 + \Psi)(k = l/\eta_0, \eta_*)$ contains most of the structure of the final anisotropy spectrum. When free-streamed via the integral in Eq. (8.56), the spectrum shifts slightly to lower l. *Middle panel:* Accounting for the dipole raises the anisotropy spectrum. Since the dipole is out of phase with the monopole, the troughs become less pronounced. *Bottom panel:* The integrated Sachs–Wolfe effect enhances the anisotropy on scales comparable to the horizon. In this case, the potential changes near recombination since the universe is not purely matter dominated then. Thus the first peak gets most of the excess power. Throughout, $h = 0.5, \Omega_b = 0.06, \Omega_m = 1$.

estimate for the location of the peaks in k-space, Eq. (8.25), is also slightly high. For example, the expected position of the first peak, $\pi\eta_0/r_s$, for the model depicted in Figure 8.13 is a little over 280. The first peak in the monopole in k-space, however, appears at $k\eta_0 \sim 260$. These two effects—fixed k projects to slightly smaller l and peaks on slightly larger scales than expected from Eq. (8.25)—serve to move the predicted positions of the peaks to lower l. A better approximation for the first

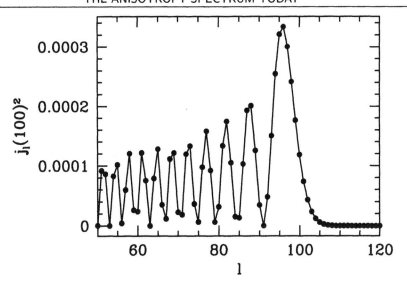

Figure 8.14. The spherical Bessel function, $j_l(100)$. Note that the peak occurs at $l \simeq 90$, slightly smaller than the argument.

peak position is $l_p \simeq 0.75\pi\eta_0/r_s$.

The dipole at recombination is smaller than the monopole and out of phase with it. The middle panel in Figure 8.13 shows that adding in the dipole raises the overall anisotropy level, but particularly fills in the troughs. Without the dipole (in this model) the ratio of the height of the first peak (at $l \sim 200$) to the height of the first trough (at $l \sim 400$) is about 2.5:1; the dipole lowers this ratio to 1.5:1. This is a direct manifestation of the dipole and monopole being out of phase with one another. That is, at the places where the monopole contributes least to the anisotropies, at its troughs, the dipole contributes the most. One other comment about the relation between the monopole and the dipole: they add incoherently. By incoherently, I mean that the cross term of Θ_l from the monopole multiplied by Θ_l from the dipole vanishes when integrating over all k-modes to get the C_l's. This can be seen mathematically from the properties of the spherical Bessel function (Exercise 12). Incoherence implies that the dipole is not as important in the power spectrum as one might naively think. If the amplitude of the dipole is 30% of that of the monopole at recombination, the dipole's contribution to the C_l's is only 10% ($1^2 + 0.3^2 = 1.1$).

The integrated Sachs–Wolfe effect is also important if the potential changes after recombination. To see which scales are affected by the ISW effect, consider the integral in Eq. (8.56). Suppose the potential changes at time η_c, with all sub-horizon scales ($k\eta_c > 1$) being affected. The Bessel function peaks at $l \sim k(\eta_0 - \eta_c)$; so all angular scales $l > (\eta_0 - \eta_c)/\eta_c$ are affected. The largest effect is typically at the horizon.

The best, and most prevalent, example of the ISW effect is that due to residual radiation at recombination. If the universe were purely matter dominated, there would be no such effect. But, the transition to pure matter domination is not abrupt, and even for $a_{eq} \sim 10^{-4}$, an ISW effect occurs right after recombination. This early ISW effect is particularly important because it adds coherently with the monopole. To see this, integrate the last term in Eq. (8.56) by parts. Then, the dominant contribution comes from $\eta \simeq \eta_*$, so the Bessel function can be evaluated there, leaving the trivial integral which gives

$$\Theta_l(k, \eta_0)^{\text{early ISW}} = [\Psi(k, \eta_0) - \Psi(k, \eta_*) - \Phi(k, \eta_0) + \Phi(k, \eta_*)] \, j_l \, [k(\eta_0 - \eta_*)].$$
$$(8.77)$$

This adds exactly in phase with the monopole (which is proportional to the same Bessel function) so even though the magnitude of the effect on Θ_l is much smaller than is the dipole, the effect on the anisotropy spectrum is disproportionate. A 30% dipole leads to a 10% shift in the C_l's, while a 5% ISW effect leads to the same 10% shift in the C_l's. The bottom panel shows that the large scales, those with $l \sim \eta_0/\eta_*$, get a big boost from this early ISW effect.

8.7 COSMOLOGICAL PARAMETERS

The anisotropy spectrum depends on cosmological parameters. This fundamental realization initially caused great consternation ("We will never be able to measure any one parameter because there is too much degeneracy"). As more quantitative studies were carried out, the pendulum swung to the other side ("We will be able to disentangle the degeneracies and measure cosmological parameters to percent accuracy"). More recently, the community has settled into a state of cautious optimism. Indeed, just a decade after the initial discovery of large scale anisotropies by COBE, there were a host of experiments which together seemed to pin down one parameter (the total energy density) by measuring the location of the first peak. Several of these had measured the subsequent two peaks, allowing an inference of the baryon density, the parameter which most affects the heights and locations of these peaks.

We now have developed the theoretical tools needed to participate in the parameter determination discussion. In this section, we apply these tools to understand how the anisotropy spectrum varies as cosmological parameters vary.

One very important decision that must be made is which parameters will be allowed to vary. I will consider eight parameters:

- Curvature density, $\Omega_k \equiv 1 - \Omega_m - \Omega_\Lambda$
- Normalization, C_{10}
- Primordial tilt, n
- Tensor modes, r (for a precise definition, see Exercise 18)
- Reionization, parametrized by τ back to recombination
- Baryon density, $\Omega_b h^2$
- Matter density, $\Omega_m h^2$
- Cosmological constant energy density, Ω_Λ

There are two aspects of this list worth stressing. The first is that obviously it does not include all possible cosmological parameters. Some favorites missing are a neutrino mass (I will set all masses to zero in the following), the equation of state for dark energy w (will be fixed at -1 corresponding to a cosmological constant), and tensor tilt n_T (fixed at zero). The second important point is that I have deliberately chosen very specific combinations of these parameters, e.g., $\Omega_b h^2$, not Ω_b and h separately. While there is good reason for this (e.g., the alternating peaks effect depends on $\Omega_b h^2$), it also is a source of confusion. A common complaint is that, within the context of a flat universe (the first parameter, the curvature density, equal to zero), why should both the cosmological constant and the matter density be allowed to vary? Mustn't their sum equal 1? It is true that $\Omega_m + \Omega_\Lambda$ must equal 1 in a flat universe. But that does not preclude us from varying both $\Omega_m h^2$ and Ω_Λ, since h can change while the sum of the two densities is 1.

To harp on this point, consider two analysts. Analyst A works in the context of a flat universe and uses $\Omega_m h^2$ and Ω_Λ as her two free parameters. Analyst B also assumes the universe is flat, but takes h and Ω_Λ as his two parameters. When A raises Ω_Λ, the matter density ($\Omega_m h^2$) is kept fixed, so the epoch of equality is kept fixed. However, when analyst B raises his Ω_Λ, to keep the universe flat, he must lower Ω_m. He is therefore also lowering the matter density (since h is kept fixed), thereby moving $a_{\rm eq}$ closer to today. That change in $a_{\rm eq}$ will lead to an enhanced ISW effect, and therefore a larger first peak. Analyst A, who had the foresight to separate out this effect by choosing $\Omega_m h^2$ as one of her parameters, sees no such enhancement. And, indeed the enhancement is caused only indirectly by Ω_Λ: rather it is the direct result of a smaller $\Omega_m h^2$.

Let's now consider the effect of each parameter in turn.

8.7.1 Curvature

If the universe is not flat, then the simple picture of Figure 8.4 is no longer accurate since the geodesics of massless particles are such that photons starting out parallel to each other slowly diverge. Consider the implication of this divergence for anisotropies. Suppose the identical pattern of inhomogeneities was in place at recombination in both a flat and open universe. As shown in Figure 8.15, the physical scale with maximal anisotropy (the first peak) gets projected onto a much smaller angular scale in an open universe. The peaks should therefore be shifted to higher l. As shown in Figure 8.16, this is precisely what happens.

The magnitude of this effect is determined by the comoving angular diameter distance to the last scattering surface, in a flat universe simply equal to $\eta_0 - \eta_*$, and in a universe with curvature given by Eq. (2.46) out to z_*. Figure 8.17 shows this distance as a function of the curvature density with all other parameters held fixed. The angular diameter distance scales as $(1 - \Omega_k)^{-0.45}$, so that it is a factor of 1.7 larger in an open $\Omega_k = 0.7$ universe than in a flat universe. Notice from Figure 8.16 that this is precisely the factor by which the first peak shifts from one model to

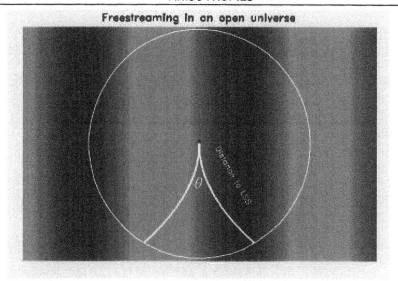

Figure 8.15. Photon trajectories in an open universe diverge. Perturbations at last scattering turn up on smaller scales in an open universe than they do in a flat universe.

the other. Of all the parameters under consideration, curvature by far causes the largest shift in the location of the peaks.

Figure 8.16 also shows data circa 2002. There is a clear rise up to a first peak at $l \sim 200$ and an equally clear fall past this first peak. When the data first started coming in (around 1998), a skeptic could plausibly claim that no one data set spanned the whole peak, and it is difficult to combine data sets. Within a year or two, though, this objection vanished as larger data sets such as TOCO (Miller *et al.*, 1999), Boomerang (de Bernardis *et al.*, 2000), and Maxima (Hanany *et al.*, 2000) all contained enough information by themselves to rule out an open universe. The DASI detection (Halverson *et al.*, 2002), together with the reanalyzed Boomerang and Maxima data, cemented the case for a flat universe.

Of course, a truly flat universe is only one point in parameter space, the point at which the sum of the energy densities exactly equals the critical density, and no data will ever rule out all values except for this one point. Rather, the data now suggest that the total density is equal to the critical density with an error of about 5%. The classic open universe once favored by astronomers had 30% of the critical density, and so is ruled out with very high confidence.

8.7.2 Degenerate Parameters

Figure 8.18 shows the results of varying four parameters. Before considering each in turn, it is important to state the obvious. All of these parameters change the spectrum in very similar ways. The shape of the spectrum varies hardly at all; rather, these parameters simply move the spectrum up and down.

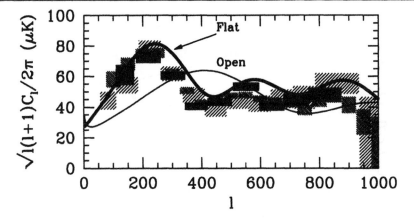

Figure 8.16. The anisotropy spectrum in flat versus open universe. Also shown are data from three small-scale experiments: DASI (darkest; Halverson *et al.*, 2002), Boomerang (medium; Netterfield *et al.*, 2002), and Maxima (lightest; Lee *et al.*, 2001). The pattern of peaks and troughs persists in the open universe but is shifted to smaller scales. The data clearly favor the flat case. Both curves have identical parameters $n = 1, \Omega_m h^2 = 0.15, \Omega_b h^2 = 0.02$ with no reionization, tensors, or cosmological constant. *Open* curve has $\Omega_k = 1 - \Omega_m = 0.7$; *flat* has the same parameters except $\Omega_k = 0$.

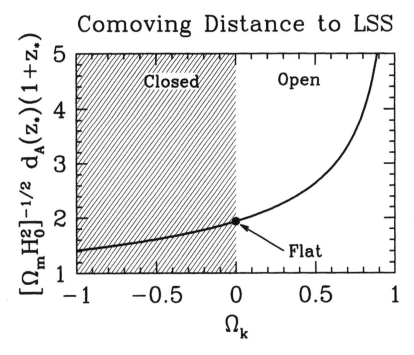

Figure 8.17. Comoving angular diameter distance back to the last scattering surface at $z_* \simeq 1100$ as a function of curvature. The distance is larger in an open universe than in a closed universe.

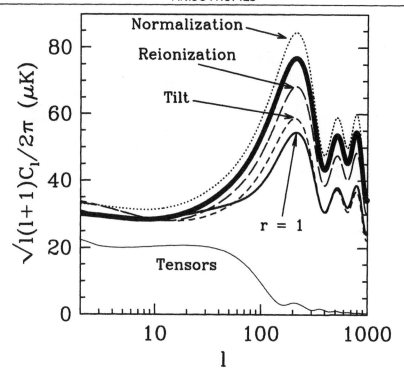

Figure 8.18. Changes in the anisotropy spectrum as C_{10}, τ, r, and n vary. The base model (thick curve) is a flat universe with no reionization or tensors, $n = 1$, $\Omega_m h^2 = 0.16$, $\Omega_b h^2 = 0.021$, and $\Omega_\Lambda = 0.7$. The thin curves vary one parameter each. *Reionization* corresponds to letting the optical depth back to the last scattering surface equal 0.2 instead of zero; *tilt* has a primordial spectum with $n = 0.8$; $r = 1$ has an equal contribution of scalars and tensors to the quadrupole; and *normalization* has C_{10} 10% higher than the base model. The curve labeled *tensors* is the contribution to the anisotropy from tensors only. Only the $r = 1$ curve includes this contribution; all others assume no anisotropy from tensors.

Normalization. The parameter C_{10} trivially moves the spectrum up or down. Note that, of the four parameters varied in Figure 8.18, it is the only one which can raise the amplitude of the spectrum.

Tilt. We have already considered the large-angle effects of a tilted ($n \neq 1$) spectrum. If $n < 1$, then the small-scale anisotropies are smaller than in the $n = 1$ model. Figure 8.18 shows that, as smaller and smaller scales are probed, the effect becomes more pronounced. So of the four parameters considered here, tilt has the most distinctive shape—it is not a simple up–down shift—and perhaps will be most easily extracted. Quantitatively, the spectrum scales as

$$\frac{C_l(n)}{C_l(n=1)} \simeq \left(\frac{l}{l_{\text{pivot}}}\right)^{n-1} \tag{8.78}$$

where here $l_{\text{pivot}} = 10$ since we are fixing C_{10}. Accounting for the fact that $\sqrt{C_l}$ is plotted in Figure 8.18, we see from the point at $l = 1000$ that this scaling works extremely well.

Reionization. The universe was almost certainly reionized at late times. We see this in the absorption spectra of high-redshift quasars, where no evidence is seen of a uniform background of neutral hydrogen until we go back as least as far as $z \sim 6$ (Becker *et al.*, 2001; Fan *et al.*, 2002). Reionization brings the CMB back in contact with electrons. If enough scattering takes place, that is, if the optical depth back to the last scattering surface is high enough, isotropy is restored; equivalently, primordial anisotropies are washed out.

There are several ways to see the effect of reionization quantitatively. One is to imagine a photon traveling in our direction with temperature $T[1 + \Theta]$, where T is the background temperature and Θ is the perturbation for which we have solved. If these photons hit a region with optical depth τ, only a fraction $e^{-\tau}$ will escape and continue on their way to us. In addition to these, we will also get a fraction $1 - e^{-\tau}$ from the ionized region. All of these have the equilibrated temperature, T. So the temperature we see today is

$$T[1 + \Theta]e^{-\tau} + T\left(1 - e^{-\tau}\right) = T\left[1 + \Theta e^{-\tau}\right]. \tag{8.79}$$

Subtracting from this the mean temperature T tells us that the fractional anisotropy will be Θ, the primordial one set up at $z \simeq 1100$, multiplied by $e^{-\tau}$. Of course this argument can affect only those scales within the horizon at the time of reionization, so multipoles l larger than $\eta_0/\eta_{\text{reion}}$ will be suppressed by $e^{-\tau}$; small l will be unaffected. This is seen in Figure 8.18, where the reionization curve falls on top of the base model on large scales but is uniformly suppressed on small scales.

Tensors. We saw in Chapter 5 that once they enter the horizon, the amplitude of gravitational waves dies away. Therefore, gravity waves affect the anisotropy spectrum only on scales larger than the horizon at recombination. Typically, this translates into angular scales $l < 100$. Indeed the *tensors* curve in Figure 8.18 shows that tensors die out after $l > 100$. We can observe only the sum of anisotropies due to tensors and scalars. So if tensor perturbations were produced during inflation, and if the total (scalar plus tensor) anisotropy spectrum is fit to the large-scale (COBE) data, then the small-scale scalar amplitude is smaller than it would otherwise be. Therefore, on scales $l > 100$ where only scalars remain, the anisotropy spectrum is identical to the base model in Figure 8.18, but with a lower amplitude.

8.7.3 Distinct Imprints

The final variations we will consider are changes in the baryon density $\Omega_b h^2$, the matter density $\Omega_m h^2$, and the cosmological constant. As can be seen from Figure 8.19, these changes lead to richer variations in the anisotropy spectrum; as such they are somewhat harder to understand (but easier to extract from the data!) than the parameters in the previous subsection.

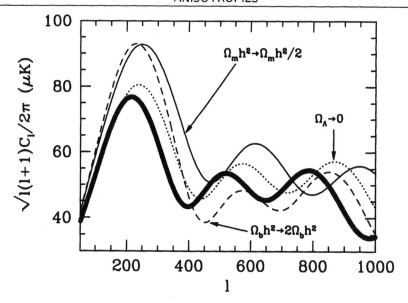

Figure 8.19. Changes in the anisotropy spectrum as baryon density, matter density, and cosmological constant vary. Same base model as Figure 8.18.

Each of these parameters induces a small shift in the locations of the peaks and troughs in the spectrum. To understand these shifts, it is important to recall that since inhomogenities on scales k show up at $l = k\eta_0$ in a flat universe, the peaks in a flat universe will show up at $l_p \simeq k_p\eta_0 \simeq n\pi\eta_0/r_s(\eta_*)$ (Eq. (8.25), but also see the discussion on page 247 that argues that the actual value of l_p is $\sim 25\%$ lower). Figure 8.20 shows this ratio as a function of matter and baryon density. It is more sensitive to the matter density, so the peak spacing increases as the matter density goes down. But there is also a little sensitivity to the baryon density. With the densities fixed, introducing a cosmological constant does not change the sound horizon, but it does slightly affect η_0, so the peaks shift in that cases as well.

Baryon density. In addition to the lateral shift in the spectrum due to the change in the sound horizon, changes in the baryon density affect the heights of the peaks as well. We have already touched on the ways in which the anisotropy spectrum depends on the baryon density. The foremost, clearly visible in Figure 8.19, is that odd peaks (first and third in the figure) are higher than the even peaks when the baryon density is large. This is a direct ramification of the lower frequency of oscillations due to the massive baryons. This change is virtually unique, making the baryon density one of the easiest parameters to extract from the CMB. Observations as of 2001 (e.g., Pryke *et al.*, 2001) pin down $\Omega_b h^2 = 0.022\pm0.04$, and this constraint will undoubtedly get tighter with data from the Map and Planck satellites. The second change due to $\Omega_b h^2$ is that an increased baryon density reduces the diffusion length. Therefore, a larger baryon density means damping moves to smaller angular scales, so the anisotropy spectrum on scales $l > 1000$ is larger in a high-$\Omega_b h^2$ model.

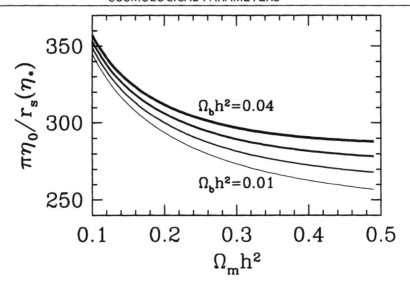

Figure 8.20. The inverse sound horizon at recombination. In a flat universe, the spacing between the acoustic peaks in the CMB is equal to $\pi\eta_0/r_s(\eta_*)$.

Cosmological constant. The cosmological constant is a late-time phenomenon. It was not around at recombination, and therefore could not have affected perturbations then. Therefore, the only possible effects of a cosmological constant are on free-streaming and on the largest angular scales just entering the horizon at recent times. The change due to free-streaming is evident in Figure 8.19. The spectrum is shifted to smaller angular scales if there is no cosmological constant. You will show in Exercise 13 that this small shift can be readily explained by comparing the conformal times in a Λ universe and a matter-dominated universe. Figure 8.19 also shows that the anisotropy spectrum is slightly lower on small scales in a Λ universe. This is a direct result of the large-angle normalization. In a Λ universe, there is a late-time ISW effect, which enhances the anisotropies on large angles. If we normalize on these scales, then the small-scale anisotropy gets correspondingly smaller.

Matter density. If the matter density is low, the epoch of equality occurred closer to recombination, so that the radiation density must be accounted for in computing the inhomogeneities at recombination. In particular, the decaying potential due to the inability of the radiation to cluster provides a strong driving force for the oscillations. Therefore, $\Theta_0(\eta_*)$ is larger than in a purely matter-dominated universe. Further, after recombination, since the potential is not constant, the ISW effect also contributes significantly to the final anisotropy spectrum. Therefore, the small-scale anisotropies increase if the matter density is low. This effect too has apparently been detected, with measurements (Pryke *et al.*, 2001) implying $\Omega_m h^2 = 0.16 \pm 0.04$.

SUGGESTED READING

The large-scale Sachs–Wolfe effect was first predicted by Sachs and Wolfe (1967), just several years after the discovery of the CMB. Several groups initiated the study of anisotropies in the tightly coupled limit: Doroshkevich, Zel'dovich, and Sunyaev (1978), Atrio-Barandela and Doroshkevich (1994), and Jorgenson *et al.* (1995). The approach was perfected by Seljak (1994) and Hu and Sugiyama (1995), the latter of which is the basis for the semianalytic treatment of this chapter. Again, CMBFAST described in Seljak and Zaldarriaga (1996) is a crucial tool for fast, accurate numerical work. Diffusion damping is sometimes called *Silk* damping because of the Silk (1968) paper recognizing its importance. Two other papers of interest are Zaldarriaga and Harari (1995) which discusses the effect of polarization on the damping scale (see Exercise 7) and Hu and White (1997a) which, among other things, gives fits to the damping scale valid for a wide range of parameters.

The question of how the anisotropy spectrum depends on cosmological parameters has been explored in literally hundreds of papers over the past decade. I remember Dick Bond, one of the pioneers in the field, giving a talk in 1992 at a conference about the early COBE data waving his hands through an invisible multidimensional parameter space, explaining that our goal now was to navigate through this space. Among the most important realizations were the dependence on curvature (Kamionkowski, Spergel, and Sugiyama 1994), the degeneracy of the height of the first peak (Bond *et al.*, 1994), and breaking of this degeneracy by smaller scale information (Jungman *et al.*, 1995). More recently, Hu *et al.* (2000) is a good reference.

I have given short shrift (or no shrift) to some important parameters. The effect of dark energy on the CMB has now been well studied: first by Coble, Dodelson, and Frieman (1997) and then more generally by Caldwell, Dave, and Steinhardt (1998). Massive neutrinos affect the anisotropy spectrum at the 5-10% level (Ma and Bertschinger, 1995 and Dodelson, Gates and Stebbins, 1996). The anisotropies due to tensors became a hot topic after the COBE discovery. For a semianalytic treatment and references to the dozens of papers relating the tensor anisotropy to parameters in the potential, see Turner, White and Lidsey (1993). Although the effect of reionization on the primary anisotropies generated before recombination is well understood, a hot topic now is *secondary* anisotropies, those generated after reionization. These will likely be probed by the next generation of experiments.

3K: The Cosmic Microwave Background (Partridge) is a good introduction to some of the experimental issues I have neglected in this book. The COBE discovery paper is Smoot *et al.* (1992) with the 4-year observations presented in Bennett *et al.* (1996). There were many good analyses papers written on the COBE data; I've relied on Bunn and White (1997), which is especially good for using COBE to normalize the matter power spectrum, and Tegmark (1997), from which the points in Figure 8.12 are taken. The two satellite experiments are Map (http://map.gsfc.nasa.gov) and Planck (http://sci.esa.int/planck/). Map was launched in 2001, and Planck is scheduled to be launched in 2007.

EXERCISES

Exercise 1. Most of this book is devoted to understanding adiabatic perturbations with the initial conditions derived in Chapter 6. Another class of perturbations are *isocurvature* perturbations with initial conditions $\Theta_0 = \Psi = \Phi = 0$. Show that these initial conditions imply that

$$\Theta_0(\eta_*) + \Psi(\eta_*) = 2\Psi(\eta_*). \tag{8.80}$$

Exercise 2. The equation for a damped harmonic oscillator is

$$m\ddot{x} + b\dot{x} + kx = 0. \tag{8.81}$$

Find the solutions to this equation if $k/m > (b/2m)^2$. What is the frequency of oscillations? How does this differ from the undamped ($b = 0$) solution? What is the other effect of nonzero b besides the change in frequency?

Exercise 3. Determine $R(\eta_*)$ when $\Omega_b h^2 = 0.01, 0.02$. Plot the sound speed as a function of the scale factor for these two values of $\Omega_b h^2$.

Exercise 4. Show that the sound horizon can be expressed in terms of the conformal time as

$$r_s(\eta) = \frac{2}{3k_{eq}} \sqrt{\frac{6}{R(\eta_{eq})}} \ln \left\{ \frac{\sqrt{1+R} + \sqrt{R + R(\eta_{eq})}}{1 + \sqrt{R(\eta_{eq})}} \right\}, \tag{8.82}$$

where k_{eq} is given in Eq. (7.39).

Exercise 5. Obtain the WKB solution to Eq. (8.18). Write

$$\Theta_0 = Ae^{iB} \tag{8.83}$$

with A and B real. Show that the homogeneous part of Eq. (8.18) breaks up into two equations, coming from the real and imaginary part:

$$\text{Real}: -(\dot{B})^2 + \frac{\ddot{A}}{A} + \frac{\dot{R}}{1+R}\frac{\dot{A}}{A} + k^2 c_s^2 = 0 \tag{8.84}$$

$$\text{Imaginary}: 2\dot{B}\frac{\dot{A}}{A} + \ddot{B} + \frac{\dot{R}}{1+R}\dot{B} = 0. \tag{8.85}$$

Find B using the real part and the fact that B changes much more rapidly than A. Then, use the imaginary equation to determine A. Show that the homogeneous solutions obtained in this way differ from the simple oscillatory solutions of Eq. (8.21) by a factor of $(1+R)^{1/4}$.

Exercise 6. Obtain a semianalytic solution for $\Theta_0 + \Psi$ and Θ_1 at recombination by carrying out the integrals in Eqs. (8.24) and (8.26). To do this you will need expressions for the gravitational potentials. Hu and Sugiyama (1995) provided the following convenient fits:

$$\Phi(k,y) = \bar{\Phi}(k,y)\left\{[1 - T(k)]\exp[-0.11(ky/k_{\rm eq})^{1.6}] + T(k)\right\}$$

$$\Psi(k,y) = \bar{\Psi}(k,y)\left\{[1 - T(k)]\exp[-0.097(ky/k_{\rm eq})^{1.6}] + T(k)\right\}$$

where $y \equiv a/a_{\rm eq}$, $T(k)$ is the BBKS transfer function and the large-scale potentials are

$$\bar{\Phi}(k,y) = \frac{3}{4}\left(\frac{k_{\rm eq}}{k}\right)^2 \frac{y+1}{y^2}\bar{\Delta}_T(y)$$

$$\bar{\Psi}(k,y) = -\frac{3}{4}\left(\frac{k_{\rm eq}}{k}\right)^2 \frac{y+1}{y^2}\left(\Delta_T(y) + 0.65 N_2/(1+y)\right). \tag{8.86}$$

Finally the two functions N_2 and Δ_T are

$$N_2(y) = -0.1\frac{20y + 19}{3y + 4}\frac{y^2}{y+1}\Phi_{ls} - \frac{8}{3}\frac{y}{3y+4} + \frac{8}{9}\ln[3y/4 + 1]$$

$$\Delta_T = \left[1.16 - \frac{0.48y}{y+1}\right]\Phi_{ls}\frac{y^2}{y+1}. \tag{8.87}$$

Here Φ_{ls} is the large-scale solution of Eq. (7.32).

Exercise 7. Our treatment of diffusion damping neglected the effect of polarization. Go through the same expansion in $\dot{\tau}^{-1}$ that we carried out in Section 8.4 this time accounting for polarization. Show that this changes the factor of 8/9 in Eq. (8.40) to 16/15. This beautiful result was obtained by Zaldarriaga and Harari (1996) when the first author was an undergraduate!

Exercise 8. Assume that all electrons associated with hydrogen stay ionized and set $R = 0$. Evaluate the damping scale, k_D, defined in Eq. (8.40). Show that in this limit, the damping scale is given by Eq. (8.42), where

$$f_D(y) = 5\sqrt{1 + 1/y} - \frac{20}{3}\left(1 + 1/y\right)^{3/2} + \frac{8}{3}\left[\left(1 + 1/y\right)^{5/2} - 1/y^{5/2}\right]. \tag{8.88}$$

Exercise 9. Show that

$$\int d\Omega\, Y_{lm}(\hat{p})\mathcal{P}_l(\hat{p}\cdot\hat{k}) = \frac{4\pi}{2l + 1}. \tag{8.89}$$

Exercise 10. There is a different way to go from the inhomogeneous temperature field at recombination, $\Theta_0(\vec{x}, \eta_*)$ or $\Theta_0(\vec{k}, \eta_*)$, to the anisotropy pattern today, a_{lm}, than that given in the text.
(a) Assume that the photons we see today from direction $\hat{p}$ come from the surface of last scattering: $\Theta(\vec{x}_0, \hat{p}, \eta_0) = (\Theta_0 + \Psi)(\vec{x} = \chi_*\hat{p}, \eta_*)$ where x_0 is our position. Fourier transform the right-hand side and expand the left in terms of spherical harmonics to get

$$\sum_{lm} a_{lm} Y_{lm}(\hat{p}) = \int \frac{d^3k}{(2\pi)^3} \, e^{i\vec{k}\cdot\hat{p}\chi_*} (\tilde{\Theta} + \tilde{\Psi})(\vec{k}, \eta_*). \tag{8.90}$$

Now expand the exponential using Eq. (C.16) and then expand the resulting Legendre polynomial using Eq. (C.12). Equate the coefficients of $Y_{lm}(\hat{p})$ to get an expression for a_{lm}.
(b) Square the a_{lm} you got in **(a)** and take the expectation value to get an expression for C_l. You should recapture the expression in Eq. (8.68) with Θ_l given by the first term in Eq. (8.56).

Exercise 11. A simple way to estimate the COBE normalization of δ_H is to fix C_{10}. From Figure 8.12, estimate C_{10}. Use this and the Sachs–Wolfe formula, Eq. (8.75), to estimate δ_H for a flat, matter-dominated universe. Compare with the number given in Eq. (8.76).

Exercise 12. Show that the cross-terms from the monopole and dipole vanish when summing over all modes. The monpole is proportional to $j_l(k\eta_0)$ while the dipole is proportional to $j_l'(k\eta_0)$. Compute the three possible integrals

$$\int_0^\infty dx j_l j_l \quad ; \quad \int_0^\infty dx j_l j_l' \quad ; \quad \int_0^\infty dx j_l' j_l'. \tag{8.91}$$

Show that the integrals of the squares (j_l^2 and $(j_l')^2$) are much larger than the integral of the cross-term $j_l j_l'$. Do the integrals for $l = 10$ up to $l = 50$.

Exercise 13. Determine the shift in the locations of the peaks and troughs in the CMB anisotropy spectrum if the universe is flat with a cosmological constant as opposed to flat and matter dominated. Keep the sound horizon fixed in this calculation by fixing $\Omega_m h^2 = 0.15$. The peak positions then depend only on the distance to the last scattering surface, $\eta_0 - \eta_*$. Consider two flat models: (i) $\Omega_\Lambda = 0$ (so that $\Omega_m = 1$) and (ii) $\Omega_\Lambda = 0.7$ (so that $\Omega_m = 0.3$). What value of h is needed in the two cases to keep $\Omega_m h^2$ fixed? Determine $\eta_0 - \eta_*$ in each case (in the the cosmological constant case, you will have to do the integral numerically). Compare your result with the fitting formula: $\eta_0 \propto 1 + \ln(\Omega_m^{0.085})$ and with the shift in Figure 8.19.

Exercise 14. Compute the conformal time today in a flat model with dark energy $\Omega_{de} = 0.7$ today with $w = -0.5$. Compare the expected shift in the anisotropy spectrum with the cosmological constant model of the previous problem.

Exercise 15. Determine the effects of reionization using the Boltzmann equation. Neglect the gravitational potentials, the velocity, and Θ_0 in the Boltzmann equation for photons. Start with a spectrum $\Theta_l(\eta)$ and evolve till today. Show that the moments are indeed suppressed by $e^{-\tau}$.

Exercise 16. Assume that recombination took place instantaneously. Show that the solution for the lth moment due to tensor perturbations (Eq. (4.116)) is

$$\Theta_l^T = -\frac{1}{2} \int_{\eta_*}^{\eta_0} d\eta \; \dot{h} \; j_l[k(\eta_0 - \eta)]. \tag{8.92}$$

Exercise 17. Using the decomposition for tensor modes given in Eq. (4.115), find the contribution to the C_l's from $\Theta_l^T(k)$. That is, show that the analogue of Eq. (8.68) for tensors is

$$C_{l,i}^T = \frac{(l-1)l(l+1)(l+2)}{\pi} \int_0^\infty dk \; k^2$$

$$\times \left| \frac{\Theta_{l-2,i}^T}{(2l-1)(2l+1)} + 2\frac{\Theta_{l,i}^T}{(2l-1)(2l+3)} + \frac{\Theta_{l+2,i}^T}{(2l+1)(2l+3)} \right|^2, \tag{8.93}$$

where i denotes the two different components $+$ and $\times$.

Exercise 18. Determine the spectrum of anisotropies due to gravity waves produced during inflation.
(a) Combine the results of the previous two problems, your solution to Exercise 5.12, and the primordial amplitude of gravity waves in Eq. (6.100) to find the large-angle C_l's due to inflation-produced gravity waves.
(b) Tensor anisotropies are often parametrized by

$$r \equiv \frac{C_2^T}{C_2^S} \tag{8.94}$$

where C_2^T is the variance of the quadrupole due to tensors and C^S is the same due to scalars.[8] We already derived an expression for the scalar C_2 in Eq. (8.75). Find C_2^T and compute r to first order in the slow roll parameter ϵ.
(c) The results of part (b) and Eq. (6.104) imply a *consistency relation* – a robust prediction of inflation – between the two observables n_T and r. What is the consistency relation?

---

[8]Note that this convention is not universal; r is sometimes defined to the (more precisely constrained) ratio at $l \sim 10$.

9

PROBES OF INHOMOGENEITIES

The power spectra we have explored in the previous two chapters — $P(k)$ of the density field and C_l of the anisotropies — are the most obvious first tests of any cosmological model. The most direct way of measuring $P(k)$ is to do a redshift survey, wherein the angular positions and the redshifts (which are a measure of radial distance) of galaxies are recorded. There are, however, a number of problems with redshift surveys and their interpretation. The first is the simple fact that taking redshifts is time consuming: it is much easier to get the angular positions of galaxies than it is to also measure redshifts. In the same time that a $10,000$ galaxy redshift survey, say, could be completed, a million angular positions of galaxies could be obtained. With the much greater statistics, angular surveys often compensate for the lack of radial information. Indeed, some claim that the best information we have on large-scale clustering comes from angular surveys. Clearly, then, one skill we must acquire is the ability to make predictions about the angular correlation function $w(\theta)$. In Section 9.1 we will see that the angular correlation function is an integral over the 3D power spectrum.

Redshift surveys suffer from another, more profound problem than the fact that they are time consuming. While is true that the redshift gives a reasonable estimate of radial distance (by *radial* distance I simply mean distance from us), it is not true that this estimate is completely accurate. Recall that a galaxy's velocity is determined solely by the Hubble expansion (and hence redshift is a perfect indicator of distance) only if the galaxy is stationary on the comoving grid. Most galaxies have nonnegligible *peculiar velocities*; that is, they are moving on this grid. A galaxy's total velocity, which is measurable, is

$$\vec{v} = \vec{v}_{\text{pec}} + \hat{x} v_{\text{H}} \qquad (9.1)$$

where the Hubble velocity $v_{\text{H}} \equiv \chi da/dt = \chi aH$. Recall that χ is the comoving distance between us and a distant galaxy, so its physical distance from us is $a\chi$. In the absence of peculiar velocities, $\hat{x} \cdot \vec{v}/H$ is a perfect distance indicator. In the real world, though, where peculiar velocities do not vanish, even an accurate measurement of a galaxy's recession velocity does *not* translate into an unambiguous measurement of its radial distance away from us.

The ambiguity introduced by peculiar velocities offers an opportunity. In linear theory, peculiar velocities are determined by the surrounding density field, so we can correct for the distortions induced by working with redshifts. Indeed, we can go even further and use these distortions to learn about the growth of perturbations, for the precise way in which velocities are related to the density field is determined by the rate at which perturbations grow. Since this rate is determined in cosmology by Ω_m, studying the distribution of galaxies in redshift space is one of the more promising ways to measure Ω_m.

Finally, another way of gleaning information about the underlying mass density is by studying clusters of galaxies. Although strictly speaking, this topic falls in the realm of nonlinear evolution of the density field, and therefore beyond the boundary of this book, the Press–Schechter method of approximating cluster abundances is only a very small step away from linear theory and has been shown to be quite accurate in its predictions. Further, the study of clusters is advancing at an extraordinarily rapid rate, since clusters can now be probed with many different astronomical techniques. Section 9.5 introduces the basic predictions of the Press–Schechter theory and the implications for cold dark matter models.

9.1 ANGULAR CORRELATIONS

Figure 9.1 shows the angular positions of over a million galaxies from the Automated Plate Measuring (APM) Survey. In an angular survey such as this, what statistic can be computed that can be compared with theory? The simplest statistic is the two-point function: in real space it is $w(\theta)$ the *angular correlation function*. In Fourier space, the relevant function is the Fourier transform of w, $P_2(l)$, the two-dimensional power spectrum. In this section we compute both of these very important functions, relating them to the three-dimensional power spectrum.

First let me introduce some notation. Figure 9.2 shows the geometry: a given galaxy is at comoving distance $\chi(z)$ (Eq. (2.42)) away from us. The z-axis is typically chosen so that it points to the center of the distribution of galaxies. In the plane perpendicular to this axis, a galaxy's position is determined by the two-dimensional vector $\vec{\theta} = (\theta_1, \theta_2)$. Therefore, the three-dimensional position vector $\vec{x}$ has components

$$\vec{x}(\chi(z), \vec{\theta}) = \chi(z)(\theta_1, \theta_2, 1). \tag{9.2}$$

The assumption that all galaxies are located near the z-axis clearly breaks down if the survey measures structure on very large angular scales. As an example the APM survey in Figure 9.1 covers roughly 50 by 100 square degrees. From this data, one can measure the correlation function accurately out to about $10°$ or 0.17 radians, safely smaller than unity.

We measure all galaxies along the line of sight, effectively integrating over $\chi(z)$. Therefore an overdensity at angular position $\vec{\theta}$ is

Figure 9.1. The distribution of galaxies in the APM survey. Blacked-out regions were not observed during the survey.

$$\delta_2(\vec{\theta}) = \int_0^{\chi_\infty} d\chi W(\chi)\delta\left(\vec{x}(\chi,\vec{\theta})\right) \tag{9.3}$$

where the subscript $_2$ denotes the fact that δ on the left is the angular — or two-dimensional — overdensity, while δ on the right is the full three-dimensional overdensity. (And I will stick with this convention: for example P_2 is the 2D power spectrum while P denotes the 3D spectrum.) The upper limit on the χ integral corresponds to $z \to \infty$, equal to $\chi_\infty = 2/H_0$ in a flat, matter-dominated universe. In practice, the magnitude-limited surveys that have yielded the most cosmological information to date have probed $z \lesssim 0.5$. The selection function $W(\chi)$ encodes this information: it is the probability of observing a galaxy a comoving distance χ from us. Galaxies at large distances are too faint to be included in a survey, whereas there are relatively few galaxies at very low redshift simply because the volume is small. Since it is a probability, the selection function is normalized so that $\int_0^{\chi_\infty} d\chi W(\chi) = 1$.

The 2D vector conjugate to $\vec{\theta}$ will be $\vec{l}$, so that the Fourier transform of $\delta_2(\vec{\theta})$ is

$$\tilde{\delta}_2(\vec{l}) = \int d^2\theta e^{-i\vec{l}\cdot\vec{\theta}}\delta_2(\vec{\theta}). \tag{9.4}$$

The two-dimensional power spectrum is defined just as was the 3D:

$$< \tilde{\delta}_2(\vec{l})\tilde{\delta}_2^*(\vec{l}') >= (2\pi)^2\delta^2(\vec{l}-\vec{l}')P_2(l); \tag{9.5}$$

here $\delta^2()$ is the 2D Dirac delta function, not to be confused with the overdensities $\delta_2(\vec{\theta})$. Integrating, we can therefore write the 2D power spectrum as

$$P_2(l) = \frac{1}{(2\pi)^2}\int d^2l' \left\langle \delta_2(\vec{l})\delta_2^*(\vec{l}') \right\rangle$$

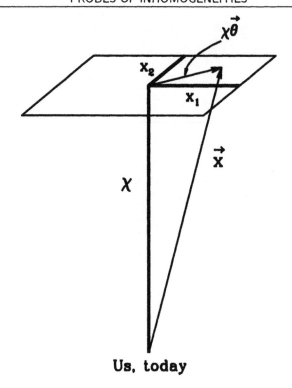

Figure 9.2. A distant galaxy is located at position $\vec{x}$ with respect to us at the origin today. This position can also be expressed in terms of $\chi(z)$, the comoving distance out to the redshift of the galaxy. The $x_1 - x_2$ plane is perpendicular to a suitably chosen x_3-axis. In this plane, a galaxy's position is given by the two-dimensional vector $\chi\vec{\theta}$, so $\vec{x} \simeq \chi(z)(\theta_1, \theta_2, 1)$.

$$= \frac{1}{(2\pi)^2} \int d^2l' \int d^2\theta \int d^2\theta' e^{-i\vec{l}\cdot\vec{\theta}} e^{i\vec{l}'\cdot\vec{\theta}'}$$

$$\times \int_0^{\chi_\infty} d\chi W(\chi) \int_0^{\chi_\infty} d\chi' W(\chi') \left\langle \delta\left(\vec{x}(\chi,\vec{\theta})\right) \delta\left(\vec{x}'(\chi',\vec{\theta}')\right) \right\rangle. \quad (9.6)$$

The integral over $\vec{l}'$ gives $(2\pi)^2$ times a Dirac delta function in $\vec{\theta}'$ and the brackets give the 3D correlation function,

$$\xi(\vec{x} - \vec{x}') \equiv \langle \delta(\vec{x})\delta(\vec{x}') \rangle$$

$$= \int \frac{d^3k}{(2\pi)^3} P(k) e^{i\vec{k}\cdot(\vec{x}-\vec{x}')}. \quad (9.7)$$

The average here $\langle \ldots \rangle$ is over all realizations of the density field. At very small distance, we expect galaxies to be clustered strongly as a result of gravity, so ξ is positive. As the distance gets larger, correlations die off and ξ gets smaller and eventually goes negative. The second line follows since the correlation function is the Fourier transform of the power spectrum (Exercise 1).

Performing the integral over $\vec{\theta}'$ in Eq. (9.6) then leads to

$$P_2(l) = \int d^2\theta e^{-i\vec{l}\cdot\vec{\theta}} \int_0^{\chi_\infty} d\chi W(\chi) \int_0^{\chi_\infty} d\chi' W(\chi') \int \frac{d^3k}{(2\pi)^3} P(k) e^{i\vec{k}\cdot[\vec{x}(\chi,\vec{\theta})-\vec{x}(\chi',0)]}.$$
(9.8)

The argument of the exponential at the end here is $i[k_1\chi\theta_1 + k_2\chi\theta_2 + k_3(\chi - \chi')]$, so the integral over angles $\vec{\theta}$ gives Dirac delta functions setting $l_1 = \chi k_1$ and $l_2 = \chi k_2$. We can use these delta functions to do the k_1 and k_2 parts of the d^3k integration, remembering to divide by the derivative of the argument, in this case putting a factor of χ^2 in the denominator. We are thus left with

$$P_2(l) = \int_0^{\chi_\infty} d\chi \frac{W(\chi)}{\chi^2} \int_0^{\chi_\infty} d\chi' W(\chi') \int_{-\infty}^{\infty} \frac{dk_3}{(2\pi)} P\left(\sqrt{k_3^2 + l^2/\chi^2}\right) e^{ik_3[\chi-\chi']}.$$
(9.9)

Until now, we have been doing math; to complete the calculation of the power spectrum we need to introduce some physics. I claim that the only 3D Fourier modes that contribute to the integral are those with k_3 very small, much smaller than l/χ. To see why, we first need to estimate l, the variable conjugate to θ. Roughly, l^{-1} is of order the angular scales probed by the survey[1]. Since we are working in the small angle approximation, $l/\chi \sim 1/(\chi\theta) \gg 1/\chi$. Now let's consider Figure 9.3. There

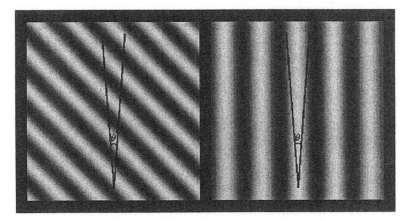

Figure 9.3. Two plane-wave perturbations and their contributions to the 2D power spectrum. Left panel shows a perturbation with longitudinal wavenumber $k_3 \gg \chi^{-1}$ (the $\hat{z}$ direction is vertical). Right panel shows a mode with $k_3 \sim \chi^{-1}$. Angular correlations due to the large k_3 mode (left panel) are negligible since there are many cancellations along the line of sight.

[1] In this sense, l is very similar to the degree of the Legendre polynomials introduced in Chapter 8 to study anisotropies. In the same sense, $P_2(l)$ is very similar to C_l; indeed in Exercise 5 you will show that they are identical on small scales.

we see that modes with longitudinal wavenumber k_3 much greater than χ^{-1} do not give rise to angular correlations because of cancellations along the line of sight. Only modes with k_3 smaller than χ^{-1} lead to angular correlations. Therefore, the relevant transverse wavenumbers l/χ are much larger than the relevant longitudinal wavenumbers, and we can safely set the argument of the 3D power spectrum to l/χ (see Exercise 2 for a more systematic justification). With this approximation, the k_3 integral gives a Dirac delta function in $\chi - \chi'$ so

$$P_2(l) = \int_0^{\chi_\infty} d\chi \frac{W^2(\chi)}{\chi^2} P(l/\chi). \tag{9.10}$$

This is an expression for the 2D power spectrum as an integral over the line of sight. We can change dummy variables from $\chi \rightarrow k \equiv l/\chi$ to rewrite the integral as

$$P_2(l) = \frac{1}{l} \int_0^{\infty} dk \, P(k) W^2(l/k). \tag{9.11}$$

The angular correlation function is the Fourier transform of the 2D power spectrum, so

$$w(\theta) = \int \frac{d^2 l}{(2\pi)^2} e^{i\vec{l}\cdot\vec{\theta}} P_2(l). \tag{9.12}$$

Since P_2 depends only on the magnitude of $\vec{l}$, the angular part of the integration over l is $\int_0^{2\pi} d\phi \, e^{il\theta \cos\phi}$, which is proportional to $J_0(l\theta)$, the Bessel function of order zero (Eq. (C.21)). Therefore,

$$w(\theta) = \int_0^{\infty} \frac{dl}{2\pi} l P_2(l) J_0(l\theta)$$

$$= \int_0^{\infty} dk \, k \, P(k) F(k, \theta), \tag{9.13}$$

where the second line follows from changing the order of integration. Here the *kernel* for the angular correlation function is

$$F(k, \theta) \equiv \frac{1}{k} \int_0^{\infty} \frac{dl}{2\pi} J_0(l\theta) W^2(l/k). \tag{9.14}$$

The kernel is plotted in Figure 9.4 for two surveys. Note that it is a function of $k\theta$ (see Exercise 3). The kernel is constant at small $k\theta$ and then begins damped oscillations. The contribution from small k though is suppressed because the integral in Eq. (9.13) is over the kernel weighted by $kP(k)$, and the latter goes to zero as $k \rightarrow 0$. Therefore, the modes that contribute most to $w(\theta)$ are typically those with wavenumbers of order the first turnover in the kernel, $k\theta \sim 0.2\,h$ Mpc^{-1} degrees for APM and a factor of 3 smaller for the deeper Sloan Digital Sky Survey (SDSS). This means that the angular correlation function at $5°$ in APM is most sensitive to power at $k = 0.04\,h$ Mpc^{-1}. The wavenumbers contributing to $w(\theta)$ in a deeper

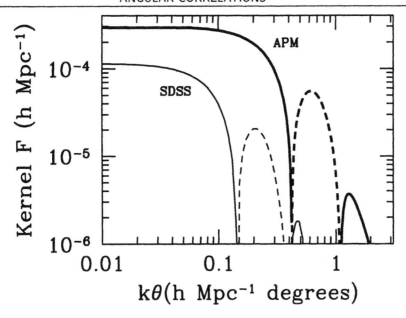

Figure 9.4. The kernel relating the angular correlation function to the 3D power spectrum in two surveys. Kernel is negative when line is dashed, positive otherwise. APM Survey probes galaxies brighter than apparent magnitude $m = 20$, while the Sloan Digital Sky Survey (SDSS) will go much deeper, potentially sensitive to galaxies brighter than $m = 23$.

angular survey are smaller. This makes sense: the same angle probes larger physical scales in a deeper survey.

Figure 9.5 shows the measurements of the angular correlation function from the APM survey. The most important conclusion from the data is that standard Cold Dark Matter — with $\Omega_m = 1$ and $h = 0.5$ — is a bad fit. To quote from the abstract of the Maddox *et al.* (1990) paper which measured the correlation function, "more large-scale clustering than predicted by popular versions of the Cold Dark Matter cosmogony is implied." Although sCDM has died many deaths since its inception in the early 1980s, the death from APM was perhaps its most celebrated. Despite long, hard work by many people trying to find systematic problems with this and other surveys, nothing significant has changed over the past decade to alter the conclusion that the standard Cold Dark Matter model of structure formation fails to predict accurately the observed pattern of large-scale clustering. Having said that, I want to emphasize that the situation is not quite as severe as you might imagine from a cursory examination of Figure 9.5. Consider the prediction from a ΛCDM model, also shown in Figure 9.5. Although the agreement is much better than with sCDM, there are clear discrepancies on both large and small scales. The small-scale discrepancies are completely illusory, though, because I have used the linear power spectrum to compute $w(\theta)$. Nonlinearites become important on scales of order $k \sim 0.2\,h\,\mathrm{Mpc}^{-1}$ as we saw in Chapter 7 (Exercise 10). These scales

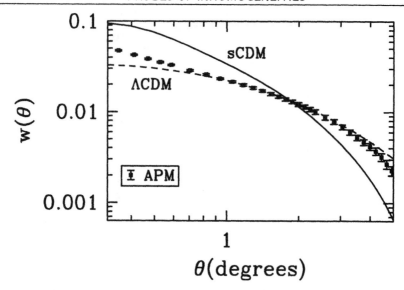

Figure 9.5. Angular correlation function in the APM survey and two theoretical models. Standard CDM (solid curve) is a bad fit to the data, while a model with a cosmological constant (here $\Omega_\Lambda = 0.7$) fits well. The apparent disagreements between the data and ΛCDM on small and large scales are illusory; see text.

contribute to $w(\theta)$ in APM when θ is of order $1°$. So to compare to the data fairly, we really need to account for nonlinearities; I have not done this, so we cannot take the small-angle discrepancy seriously. On large angles, people have begun to realize that the data have been overinterpreted. The basic problem is that the points on large angles are highly correlated. So the slight discrepancy between the data and ΛCDM on large angles looks worse than it really is.

The angular correlation function can be inverted to obtain the 3D power spectrum: the results from the APM Survey are shown in the top panel of Figure 9.6. We will discuss inversion techniques in Chapter 11, but you should be aware that the points in Figure 9.6 are the result of a long process (i.e., many papers), in the midst of which error bars were often vastly underestimated. Using only the results on large scales (where nonlinear effects are irrelevant and the relation between mass overdensity and galaxy overdensity is expected to be simple), Efstathiou and Moody (2001) placed constraints on CDM models, shown in the bottom panel of Figure 9.6. We found in Chapter 7 that the CDM transfer function depends only on k/k_{eq}. Since k_{eq} scales as $\Omega_m h^2$ and since surveys measure the wavenumber k in units of $h\,\mathrm{Mpc}^{-1}$, the combination $\Omega_m h$ determines the shape of the power spectrum. For this reason, fits to large-scale structure data are often given in terms of the shape parameter $\Gamma \equiv \Omega_m h$ (sometimes modified to account for baryons: see Exercise 8). Standard CDM with $h = 0.5$ and $\Omega_m = 1$ corresponds to $\Gamma = 0.5$; if the index of the primordial spectrum is $n = 1$, then the bottom right panel of Figure 9.6 shows

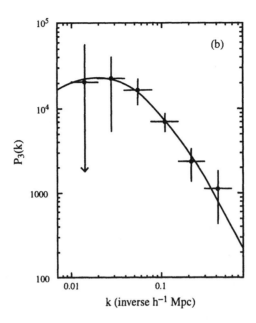

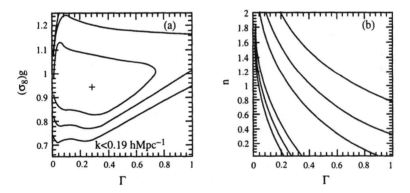

Figure 9.6. Results from the APM Survey (Efstathiou and Moody 2001). *Top panel.* The 3D power spectrum inferred from the angular correlation function. The curve is a CDM model with $\Gamma(\simeq \Omega_m h) = 0.2$, normalized to fit the data. *Bottom Panel.* Constraints on CDM models from the power spectrum. Three parameters—the shape parameter Γ, the amplitude σ_8, and the index of the primordial power spectrum n—were varied. Contours delineate one-, two-, and three-sigma regions. Left panel shows constraints after integrating over n; right after integrating over the amplitude. Standard CDM has $n = 1$; $\Gamma = 0.5$; and (COBE-normalized) $\sigma_8 = 1.15$.

that sCDM is ruled out at the 2-sigma level. ΛCDM has $\Gamma \simeq 0.2$ and is indeed a better fit to the data if n is close to 1.

9.2 PECULIAR VELOCITIES

In linear theory, velocities are related in a simple way to nearby overdensities. We first derive this relation in this section and then consider some of its ramifications. In particular, we will see that by measuring both the peculiar velocity field and the density field, one can infer the present value of the matter density, Ω_m.

In linear theory, we have already derived the equation which determines the velocity field. On scales well within the horizon, the continuity equation (4.103) reduces to

$$\dot{\delta} + ikv = 0. \tag{9.15}$$

At late times, though, we have solved for the evolution of δ: we know that it scales as the growth factor D_1, so

$$v(k, \eta) = \frac{i}{k} \frac{d}{d\eta} \left[\frac{\delta}{D_1} D_1 \right] = \frac{i\delta(k, \eta)}{kD_1} \frac{dD_1}{d\eta}. \tag{9.16}$$

A function commonly employed to relate the velocity to the density is the dimensionless linear growth rate,

$$f \equiv \frac{a}{D_1} \frac{dD_1}{da}. \tag{9.17}$$

Since $d/d\eta = a^2 H d/da$, the velocity is related to the density via

$$v(k, a) = \frac{ifaH\delta(k, a)}{k}. \tag{9.18}$$

The linear growth rate can be computed from Eq. (7.77). Most probes of the velocity field to date have been limited to relatively nearby objects, $z \lesssim 0.1$, so it is a reasonable approximation to evaluate f today and neglect its evolution. Figure 9.7 shows the linear growth rate today as a function of the matter density. For small Ω_m there is less growth: mass collapsing into overdense regions has lower velocity than if there was a critical density of matter. This makes sense since an overdensity in a lower density universe has less mass and therefore exerts a weaker gravitational pull on infalling matter. Figure 9.7 shows that, for all practical purposes, the growth rate depends only on the matter density and not, say, on the cosmological constant. Also, the approximation

$$f = \Omega_m^{0.6} \tag{9.19}$$

is seen to work extremely well (see Exercise 7 for a slightly better fit to a flat universe).

There are two important points about the relation between the density and the velocity in Eq. (9.18) that should be emphasized. First, we need to remember that

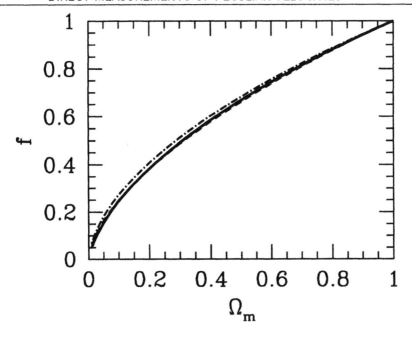

Figure 9.7. The linear growth rate as a function of matter density. There are three curves here, nearly indistinguishable. One (solid) is for an open universe with $\Omega_\Lambda = 0$; another (dot-dashed) for a flat universe with $\Omega_\Lambda = 1 - \Omega_m$; and the last (dashed) is $\Omega_m^{0.6}$.

the velocity is of course a vector and v in Eq. (9.18) is the Fourier component of the velocity parallel to $\vec{k}$. Explicitly, at low redshift we have

$$\vec{v}(\vec{k}) = if H_0 \delta(\vec{k}) \frac{\vec{k}}{k^2}. \tag{9.20}$$

The second point about the relation between the velocity and the density is that it holds only in linear theory. This has turned out to be a big problem for those who have tried to extract information from velocity studies. Velocities are easiest to obtain on small scales, but easiest to compare with theory on large scales.

9.3 DIRECT MEASUREMENTS OF PECULIAR VELOCITIES

A number of surveys have directly measured peculiar velocities. Measuring radial velocities is relatively easy: one just looks for shifted features in the spectrum of the galaxy. The hard part is breaking the radial velocity into the part due to the Hubble expansion and the remainder, the peculiar velocity. Subtracting off the Hubble expansion requires independent (i.e., other than the redshift) knowledge of the galaxy's distance from us. How accurate must a distance indicator be to be useful? Very roughly, typical peculiar velocities are of order 500 km sec^{-1}, while the Hubble velocity is $H_0 \chi = 100\,\text{km sec}^{-1} h\,(\chi/\text{Mpc})$ for a galaxy a distance χ

away. So a galaxy 50 Mpc away has a Hubble velocity roughly 10 times as large as its peculiar velocity. To be useful, therefore, a distance indicator for such galaxies much have an accuracy of order 10%. Indeed this is roughly the best one can hope for, so 50 Mpc is roughly the farthest one can hope to go in a velocity survey.

There are a number of ways of extracting cosmologically useful information from a velocity survey, but I will focus on just one of these: the two point function. With enough velocities, a survey can hope to measure the correlation function

$$\xi_v(\vec{x}_1, \vec{x}_2) \equiv \langle \vec{v}(\vec{x}_1) \cdot \hat{x}_1 \ \vec{v}(\vec{x}_2) \cdot \hat{x}_2 \rangle, \tag{9.21}$$

where the radial components $\vec{v} \cdot \vec{x}$ appear because these are all that can be measured using redshifts. Let us compute this correlation function using linear theory. We will see that it is an integral over the power spectrum, so — just like the angular correlation function of galaxies we considered in Section 9.1 — the radial velocity correlation function is a probe of the power spectrum. The observational obstacles involved in obtaining accurate peculiar velocities are daunting. However, the promise of measuring the *matter* power spectrum (the velocities are due to the matter, which may or may not be aligned with the galaxies) as opposed to the *galaxy* power spectrum sampled by $w(\theta)$ is so alluring that it is likely that peculiar velocity surveys will continue to play an important role in cosmology.

To evaluate the velocity correlation function, we can Fourier transform the velocities appearing in Eq. (9.21) so that

$$\xi_v(\vec{x}_1, \vec{x}_2) = \int \frac{d^3k}{(2\pi)^3} e^{i\vec{k}\cdot\vec{x}_1} \int \frac{d^3k'}{(2\pi)^3} e^{-i\vec{k}'\cdot\vec{x}_2} \langle \vec{v}(\vec{k}) \cdot \hat{x}_1 \vec{v}^*(\vec{k}') \cdot \hat{x}_2 \rangle. \tag{9.22}$$

Using linear theory for $\vec{v}$ (9.20) and the fact that $\langle \delta(\vec{k})\delta^*(\vec{k}') \rangle = (2\pi)^3 \delta^3(\vec{k} - \vec{k}') P(k)$ leads to

$$\xi_v(\vec{x}_1, \vec{x}_2) = f^2 H_0^2 \int_0^\infty \frac{dk\ k^2}{(2\pi)^3} P(k) \int d\Omega_k e^{i\vec{k}\cdot(\vec{x}_1 - \vec{x}_2)} \frac{\vec{k} \cdot \hat{x}_1 \vec{k} \cdot \hat{x}_2}{k^4}. \tag{9.23}$$

One way to do the angular integral here is to write the occurrences of $\vec{k}$ in the integrand as

$$\vec{k} e^{i\vec{k}\cdot\vec{x}} = \frac{1}{i} \frac{\partial}{\partial\vec{x}} e^{i\vec{k}\cdot\vec{x}} \tag{9.24}$$

where

$$\vec{x} \equiv \vec{x}_1 - \vec{x}_2. \tag{9.25}$$

Then taking the derivatives outside leads to an angular integral over the exponential, which has no azimuthal dependence. The integral of $e^{ikx\mu}$ over μ is is $4\pi j_0(kx)$ (Eq. (C.15)). So we have

$$\xi_v(\vec{x}_1, \vec{x}_2) = -f^2 H_0^2 \hat{x}_{1,i} \hat{x}_{2,j} \int_0^\infty \frac{dk}{2\pi^2 k^2} P(k) \frac{\partial^2}{\partial x_i \partial x_j} j_0(kx). \tag{9.26}$$

The first partial derivative here is

$$\frac{\partial j_0(kx)}{\partial x_j} = \frac{\partial(kx)}{\partial x_j} j_0'(kx) = k\frac{x_j}{x} j_0'(kx) \qquad (9.27)$$

where the prime here is derivative with respect to the argument kx. The second derivative then gives

$$\frac{\partial}{\partial x_i}\left[k\frac{x_j}{x}j_0'(kx)\right] = k^2 \left[\left\{\delta_{ij} - \hat{x}_i\hat{x}_j\right\}\frac{j_0'(kx)}{kx} + \hat{x}_i\hat{x}_j j_0''(kx)\right]. \qquad (9.28)$$

Then, the velocity correlation function is

$$\xi_v(\vec{x}_1, \vec{x}_2) = -f^2 H_0^2 \hat{x}_{1,i}\hat{x}_{2,j} \int_0^\infty \frac{dk}{2\pi^2} P(k) \left[\left\{\delta_{ij} - \hat{x}_i\hat{x}_j\right\}\frac{j_0'(kx)}{kx} + \hat{x}_i\hat{x}_j j_0''(kx)\right]$$

$$\equiv \hat{x}_{1,i}\hat{x}_{2,j}\left\{\delta_{ij} - \hat{x}_i\hat{x}_j\right\}\xi_{v,\perp} + \hat{x}_{1,i}\hat{x}_{2,j}\hat{x}_i\hat{x}_j\xi_{v,\parallel}. \qquad (9.29)$$

The definitions here reflect the fact that the first term is sensitive to the component of velocity perpendicular to the line connecting two galaxies, while the second probes the velocity parallel to this line. Finally let's define the angles

$$\cos\theta_1 \equiv \hat{x}_1 \cdot \hat{x} \qquad ; \qquad \cos\theta_2 \equiv \hat{x}_2 \cdot \hat{x}. \qquad (9.30)$$

With the aid of Figure 9.8 we see that $\hat{x}_1 \cdot \hat{x}_2$ is equal to $\cos(\theta_1 - \theta_2)$, so performing the sums over i, j leads to

$$\xi_v(\vec{x}_1, \vec{x}_2) = \sin\theta_1 \sin\theta_2 \xi_{v,\perp} + \cos\theta_1 \cos\theta_2 \xi_{v,\parallel}. \qquad (9.31)$$

Both components of ξ_v are integrals over the power spectrum. We can write

$$\begin{pmatrix} \xi(x) \\ \xi_{v,\perp}(x) \\ \xi_{v,\parallel}(x) \end{pmatrix} = \int_0^\infty \frac{dk}{2\pi^2 k} P(k) \begin{pmatrix} k^3 j_0(kx) \\ -f^2 H_0^2 j_0'(kx)/x \\ -f^2 H_0^2 k j_0''(kx) \end{pmatrix}. \qquad (9.32)$$

The weighted kernels are shown in Figure 9.9 and compared with the correlation function of the density. The key feature of Figure 9.9 is that, at fixed distance x, the density correlation function probes power on *smaller* scales than do the velocity correlation functions. Put another way, velocity surveys may be limited in how far out they go, but they pack an extra punch since they are sensitive to long wavelength modes. This is a direct result of the fact that $v \propto \delta/k$. The extra factor of $1/k$ gives additional weight to large scales.

Before going further and looking at some results from a velocity survey, I must digress to make note of an important feature in Figure 9.9. The density correlation function, as opposed to the velocity correlation functions, gets its contribution at a fixed distance from many Fourier modes. The figure shows the contributions to $\xi(50\,h^{-1}\,\text{Mpc})$. Naively, we would expect the main contribution to $\xi(50\,h^{-1}\,\text{Mpc})$ to come from Fourier modes with $k \sim x^{-1} = 0.02\,h\,\text{Mpc}^{-1}$. Since overdensities are small, i.e., still in the linear regime, on these scales, the naive expectation is

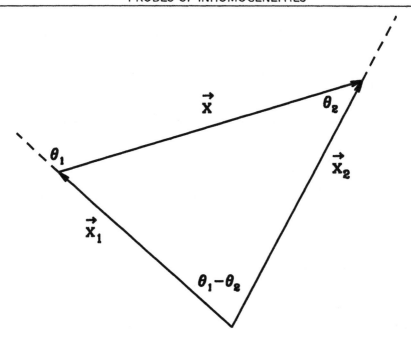

Figure 9.8. The vectors and angles of the the velocity correlation function. $\vec{x}$ is the difference vector; since the angle complementary to θ_1 is $\pi - \theta_1$, and since the three angles in the triangle must sum to π, the angle between the two galaxies is $\theta_1 - \theta_2$.

that $\xi(50\,h^{-1}\,\mathrm{Mpc})$ probes the linear power spectrum. Figure 9.9 shows that this expectation is incorrect. Modes with k as small as $0.02\,h\,\mathrm{Mpc}^{-1}$ do contribute to the correlation function, but contributions extend out to $k \sim 0.3\,h\,\mathrm{Mpc}^{-1}$ and beyond. This means that even on scales you would think would be safely linear, the correlation function depends on the small-scale power. Ultimately we want to compare theory with observations, and we are most confident doing so for modes that are still linear. The correlation function mixes up linear and nonlinear modes, so makes it difficult to compare theory with observations. For this reason, the power spectrum has gained preeminence as the statistic of choice for large-scale structure.

Returning to the velocity correlations, let's consider Figure 9.10. It shows one attempt (Freudling *et al.*1999) to extract cosmological information from a velocity survey, using 1300 velocities in the all-sky SFI catalogue (Haynes *et al.*1999), which goes out to $70\,h^{-1}\,\mathrm{Mpc}$. The power spectrum was parameterized by an amplitude A and the Γ parameter in the BBKS transfer function. For the analysis shown in Figure 9.10, the primordial spectral index was set to 1. The standard COBE-normalized CDM model, with critical matter density and $h = 0.5$, has an amplitude of $A = 0.29A_0$ (Eq. (7.9)) and shape parameter $\Gamma = 0.5$, so the SFI survey rules out this model at many sigma.

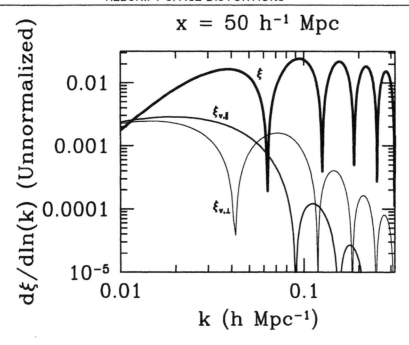

Figure 9.9. Contribution to various correlation functions from wavenumber k. Note that the velocity correlation functions get most of their contribution from smaller k than does the ordinary density correlation function.

9.4 REDSHIFT SPACE DISTORTIONS

Redshift surveys supplement the angular information about galaxies with an estimator for the radial position, the redshift. The simplest guess about the radial position of a galaxy with redshift z is that it lies a distance

$$\chi_s(z) = \frac{z}{H_0} \tag{9.33}$$

away from us, where the subscript $_s$ denotes *redshift space*. Redshift space then corresponds to assigning Cartesian coordinates to a galaxy equal to

$$\vec{x}_s = \frac{z}{H_0} \left(\sin\theta\cos\phi, \sin\theta\sin\phi, \cos\theta \right). \tag{9.34}$$

This assignment neglects several unpleasant realities. First, the comoving distance out to a galaxy at redshift z is equal to z/H_0 only at relatively low redshifts. A glance back at Figure 2.3 should convince you that this approximation is off by as much as 50% at $z = 1$. Fortunately, this first problem has not yet been much of a problem since most redshift surveys to date have probed $z \lesssim 0.1$.

A second, more pernicious, problem with redshift space is that the estimate for the distance in Eq. (9.33) neglects peculiar velocities. Figure 9.11 illustrates the

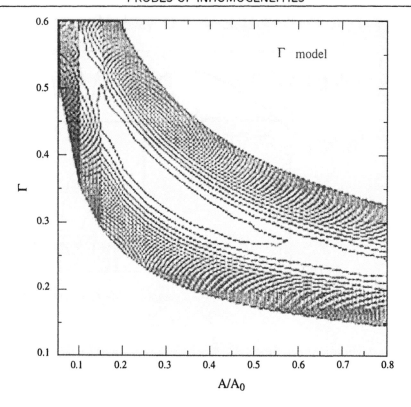

Figure 9.10. Likelihood contours on the amplitude A of the power spectrum and the shape parameter $\Gamma \equiv \Omega_m h$ from the SFI peculiar velocity survey. The amplitude is given in units of $A_0 = 2 \times 10^6 (h^{-1}\ \text{Mpc})^4$, and the contours indicate shifts in the likelihood function by successive factors of $1/e$. The allowed region — delineated by banana-shaped contour in the center — is consistent with ΛCDM, but strongly disfavors standard CDM, which has $\Gamma = 0.5$ and $A = 0.29 A_0$. From Freudling *et al.* (1999).

distortions that appear in redshift space. A slightly overdense region which is just beginning to collapse appears squashed in redshift space: the galaxies closest to us are moving toward the center of the overdense region and hence away from us, so they appear farther from us (and closer to the center of the overdense region) than they actually are. Similarly, galaxies on the "other side" of the perturbation are moving toward us, so they appear closer to us than they actually are. The overall effect is to induce an apparent quadrupole moment in an otherwise circular overdensity.

As a region becomes more overdense, the nature of the redshift space distortion changes. The bottom part of Figure 9.11 shows that a more collapsed object gets distorted in a much different way. It is elongated along the line of sight. More quantitatively, its quadrupole moment has the opposite sign as does a linear overdensity. It is clear then that accounting for redshift space distortions will be a tricky busi-

Real Space   Redshift Space

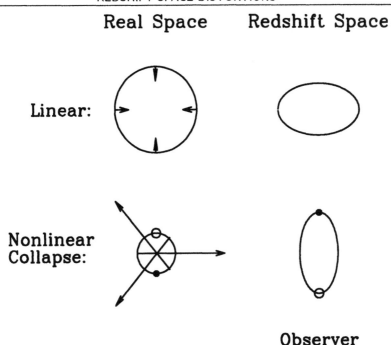

Linear:

Nonlinear
Collapse:

Observer

Figure 9.11. Redshift space distortions. In each case, a contour of constant density (circular in real space) is distorted in redshift space so that it looks asymmetric. Arrows denote direction and magnitude of velocity. In the case of nonlinear collapse, the velocities are so large that a point on "our side" (the bottom) of the center is mapped onto a point on the opposite side (compare the position of the solid dot on the bottom left and right).

ness, requiring careful treatment not only of linear overdensities, but also of the much more complicated effects of nonlinearities. We will content ourselves with a quantitative treatment of linear distortions, since this applies on large scales and is the starting point for all further work.

Suppose we measure the power spectrum in redshift space. How is this distorted power spectrum related to the underlying true spectrum in real space? Kaiser (1987) first solved this problem, working within the context of linear theory. The starting point is the realization that the number of galaxies in a particular region is the same, whether we use redshift-space or real-space coordinates. Therefore,

$$n_s(\vec{x}_s)d^3x_s = n(\vec{x})d^3x \tag{9.35}$$

where n is the density of galaxies at $\vec{x}$ in real space, and n_s is the density in redshift space. The infinitesimal volume around a point in redshift space is $d^3x_s = dx_s x_s^2 \sin\theta d\theta d\phi$, while the volume around a point in real space is $d^3x = dx x^2 \sin\theta d\theta d\phi$. The angular volume elements are identical, so

$$n_s(\vec{x}_s) = n(\vec{x})J \tag{9.36}$$

where the Jacobian J is given by

$$J \equiv \left| \frac{d^3 x}{d^3 x_s} \right| = \frac{dx}{dx_s} \frac{x^2}{x_s^2}. \tag{9.37}$$

To compute the Jacobian, we use the fact that the observed redshift is the sum of two terms:

$$z = H_0 x + \vec{v} \cdot \hat{x}. \tag{9.38}$$

The first term is the standard Hubble law, which says that redshift is proportional to distance; the second is the velocity along the line of sight. Recalling that redshift space corresponds to equating a galaxy's redshift with its distance from us, we see that, after dividing by H_0, this equation becomes

$$x_s = x + \frac{\vec{v} \cdot \hat{x}}{H_0}. \tag{9.39}$$

The Jacobian can be now be read off as

$$J = \left(1 + \frac{\partial}{\partial x} \left[\frac{\vec{v} \cdot \hat{x}}{H_0} \right] \right)^{-1} \left(1 + \frac{\vec{v} \cdot \hat{x}}{H_0 x} \right)^{-2}. \tag{9.40}$$

Kaiser realized that the correction term due to the derivative of the velocity is much more important than the $\vec{v} \cdot \hat{x}/H_0 x$ term. The argument goes as follows. For a plane wave perturbation, the term with the derivative of the velocity is of order $v'/H_0 \sim kv/H_0$, while the other correction is of order $v/H_0 x$. That is, the first correction term is larger than the second by a factor of order kx. Why do I say it's *larger*? That is, why is kx larger than unity? Kaiser's argument is that x is of order the size of the survey, while k is of order the Fourier modes we can hope to measure in the survey. Perturbations on the largest scale probed by the survey $k \sim x^{-1}$ are very poorly determined, since there are only a handful of Fourier modes with wavelength of order the survey size. Modes with smaller wavelength are much easier to measure since there are many such modes, and we effectively average over all of them to get an estimate of the power spectrum. Therefore, we are really interested only in modes with $kx \gg 1$. Expanding the remaining denominator about $v = 0$, we see that

$$J \simeq \left(1 - \frac{\partial}{\partial x} \left[\frac{\vec{v} \cdot \hat{x}}{H_0} \right] \right). \tag{9.41}$$

The number densities in real and redshift space are $n = \bar{n}(1 + \delta)$ and $n_s = \bar{n}(1 + \delta_s)$, respectively, with $\bar{n}$ the average number density. In light of Eq. (9.36), the overdensity in redshift space is

$$1 + \delta_s = [1 + \delta] \left(1 - \frac{\partial}{\partial x} \left[\frac{\vec{v} \cdot \hat{x}}{H_0} \right] \right). \tag{9.42}$$

Expanding to first order, we see that the overdensity in redshift space is actually a sum of the overdensity in real space and a correction due to peculiar velocity,

$$\delta_s(\vec{x}) = \delta(\vec{x}) - \frac{\partial}{\partial x}\left[\frac{\vec{v}(\vec{x}) \cdot \hat{x}}{H_0}\right]. \tag{9.43}$$

Now I want to introduce the *distant observer* approximation. The idea is that, in most cases of interest, the direction vector $\vec{x}$ is fixed, varying little from galaxy to galaxy. To see this, go back to Figure 9.2: $\vec{x}$ is mostly radial, with only small components in the transverse direction (proportional to the θ_1 and θ_2). As long as the galaxies are relatively close to each other in this plane, we can approximate $\hat{x} \cdot \vec{v} \rightarrow \hat{z} \cdot \vec{v}$, where $\hat{z}$ is a radial vector pointing to the center of the galaxies of interest.

In the distant observer approximation, we can compute the Fourier transform of the redshift space overdensity (here denoting the Fourier-transformed density by $\tilde{\delta}$ to avoid confusion),

$$\tilde{\delta}_s(\vec{k}) = \int d^3x \, e^{-i\vec{k}\cdot\vec{x}}\left[\delta(\vec{x}) - \frac{\partial}{\partial x}\left[\frac{\vec{v}(\vec{x}) \cdot \hat{z}}{H_0}\right]\right]$$

$$= \tilde{\delta}(\vec{k}) - if \int d^3x \, e^{-i\vec{k}\cdot\vec{x}}\frac{\partial}{\partial x}\left[\int \frac{d^3k'}{(2\pi)^3}e^{i\vec{k}'\cdot\vec{x}}\tilde{\delta}(\vec{k}')\frac{\vec{k}'}{k'^2}\cdot\hat{z}\right], \tag{9.44}$$

the first equality following from our Fourier convention and Eq. (9.43) and the second from Eq. (9.20) for linear velocities. The derivative with respect to the length x acts on the exponential, bringing down a factor of $i\vec{k}'\cdot\hat{x}$, which we again set to $i\vec{k}'\cdot\hat{z}$, so

$$\tilde{\delta}_s(\vec{k}) = \tilde{\delta}(\vec{k}) + \int \frac{d^3k'}{(2\pi)^3}\ \tilde{\delta}(\vec{k}')\left[f\left(\hat{k}'\cdot\hat{z}\right)^2\right]\int d^3x\, e^{i(\vec{k}'-\vec{k})\cdot\vec{x}}. \tag{9.45}$$

The $\vec{x}$ integral gives a Dirac delta function, equating $\vec{k}'$ with $\vec{k}$. Therefore, in the distant observer approximation,

$$\tilde{\delta}_s(\vec{k}) = \left[1 + f\mu_{\mathbf{k}}^2\right]\tilde{\delta}(\vec{k}). \tag{9.46}$$

Here $\mu_{\mathbf{k}}$ is defined to be $\hat{z}\cdot\hat{k}$, the cosine of the angle between the line of sight and the wavevector $\hat{k}$. Equation (9.46) quantifies what we should have anticipated about (linear) redshift space distortions. First of all, since $f\mu_{\mathbf{k}}^2 \geq 0$, the apparent overdensity in redshift space is *larger* than in real space. This is clear from Figure 9.12. The central region of the galaxies is clearly more overdense in redshift space than in real space, the enhancement due to the illusion that infalling galaxies are located close to the center. The second feature of Eq. (9.46) worth noting is that the enhancement is for waves with wavevector parallel to the line of sight. A plane wave perturbation with $\vec{k}$ perpendicular to the line of sight — one in which the density along the line of sight is constant — experiences no redshift space distortion.

The power spectrum in redshift space depends not only on the magnitude of $\vec{k}$ but also on its direction, which we are parameterizing with μ. It follows immediately from Eq. (9.46) that

Real Space

Redshift Space

Figure 9.12. A hundred galaxies in real space squashed in redshift space due to linear velocities. The apparent overdensity in redshift space is much larger near the center than it is in real space. We, the observers, are sitting at the bottom of the page.

$$P_s(\vec{k}) = P(k)\left[1 + \beta\mu_{\mathbf{k}}^2\right]^2. \tag{9.47}$$

Here I have introduced the parameter β, which you might think is simply equal to f, the linear growth rate. There is an additional factor in β, though, due to the fact that the mass overdensity δ is not necessarily equal to the overdensity in galaxies, δ_g. The velocity field samples the mass overdensity. So if we define the *bias*

$$b \equiv \frac{\delta_g}{\delta} \tag{9.48}$$

then $\vec{v} \propto \delta \propto \delta_g/b$. Therefore, the correction due to redshift space distortions in Eq. (9.47) is proportional to

$$\beta = \frac{f}{b} \simeq \frac{\Omega_m^{0.6}}{b}. \tag{9.49}$$

The redshift space distortion in the power spectrum, encoded in Eq. (9.47), is both good news and bad news. The good news is that by measuring the distortion in the redshift space power spectrum, we can hope to measure β, a combination of the density and bias. A quantitative way to do this is to measure the ratio of the quadrupole to the monopole of the power spectrum:

$$\frac{P_s^{(2)}(k)}{P_s^{(0)}(k)} \equiv \frac{5\int_{-1}^{1}\frac{d\mu_{\mathbf{k}}}{2}\mathcal{P}_2(\mu_{\mathbf{k}})P_s(\vec{k})}{\int_{-1}^{1}\frac{d\mu_{\mathbf{k}}}{2}\,\mathcal{P}_0(\mu_{\mathbf{k}})\,P_s(\vec{k})}. \tag{9.50}$$

Recall that $\mathcal{P}_l$ is the Legendre polynomial of order l, while P_s is the redshift-space power spectrum. Since (Exercise 11)

$$(1+\beta\mu_{\mathbf{k}}^2)^2 = \left[1 + \frac{2}{3}\beta + \frac{1}{5}\beta^2\right]\mathcal{P}_0(\mu_{\mathbf{k}}) + \left[\frac{4}{3}\beta + \frac{4}{7}\beta^2\right]\mathcal{P}_2(\mu_{\mathbf{k}}) + \frac{8}{35}\beta^2\mathcal{P}_4(\mu_{\mathbf{k}}), \tag{9.51}$$

the orthogonality of the Legendre polynomials (Eq. (C.2)) implies that the ratio of the quadrupole to the monopole in linear theory is

$$\frac{P_s^{(2)}(k)}{P_s^{(0)}(k)} = \frac{\frac{4}{3}\beta + \frac{4}{7}\beta^2}{1 + \frac{2}{3}\beta + \frac{1}{5}\beta^2}. \tag{9.52}$$

Figure 9.13 shows Hamilton's efforts to measure the quadrupole-to-monopole ratio in two different redshift surveys. In both cases, nonlinearities are very important

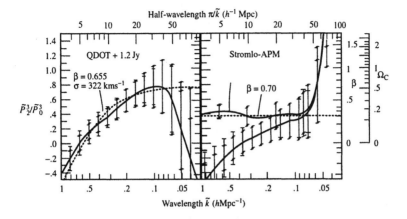

Figure 9.13. The quadrupole-to-monopole ratio for two redshift surveys (Hamilton, 1998). Left panel shows data from two redshift surveys which selected galaxies from the *Infrared Astronomical Satellite* (IRAS): QDOT (Lawrence *et al.*, 1999) picked one out of six galaxies brighter than 0.6 Jansky and the 1.2Jy survey (Strauss *et al.*, 1992) picked all galaxies above that brightness limit. Stromlo-APM survey (Loveday *et al.*, 1996) in right panel contains redshifts for 1 in 20 galaxies seen in APM (Figure 9.1).

and must be handled carefully. In the infrared-selected surveys depicted at left in the figure, Hamilton modeled the nonlinearities with a parameter measuring the small-scale velocity dispersion, σ. Only on scales of order $k \sim 0.1\,h$ Mpc^{-1} does the ratio begin to asymptote to ~ 0.75, implying a value of $\beta \sim 0.7$. The right panel shows a survey of optically selected galaxies. For these, the ratio does not seem to asymptote at all (lower curve) unless nonlinear structures are removed by hand (upper curve). In that case, Hamilton finds a quadrupole-to-monopole ratio closer to 0.4, implying $\beta = 0.3$. Figure 9.14 gives a broader view of the spread in measures of β from redshift and peculiar velocity surveys.

That was the good news. The bad news is that, even if we were to give up hope of measuring β from redshift surveys, we still need to account for redshift space distortions if we want to measure the power spectrum. Blindly measuring the power spectrum by averaging over all directions $\mu_{\mathbf{k}}$ is actually a measure of

$$P_s^{(0)}(k) = \left[1 + \frac{2}{3}\beta + \frac{1}{5}\beta^2\right] P(k), \qquad (9.53)$$

the equality holding only in linear theory. That is, $P_s^{(0)}$ overestimates the power spectrum by as much as a factor of 2. On even moderate scales, as suggested by Figure 9.13, nonlinear effects must be taken into account. The Peacock and Dodds compilation in Figure 7.11 uses a model of the small-scale velocities to do this.

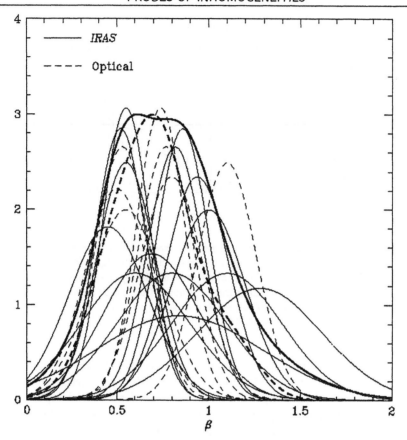

Figure 9.14. Compilation of the likelihood of β from redshift and peculiar velocity surveys (Strauss and Willick, 1995).

9.5 GALAXY CLUSTERS

Until now, we have focused solely on the two-point function: the angular correlation function, the velocity covariance, and the power spectrum. You might have wondered why little has been said about one-point functions, the number density of galaxies for example. To answer this question, first consider an extreme example. How many people are there in the universe? This clearly is an impossible question to answer with the tools we have developed. Putting aside the thorny question of the definition of a "person," we still would have to develop theories of star formation out of the gas in galaxies, then planet formation around stars, then the evolution of life via various biological processes. A prediction of the "person density" in the universe is beyond the scope of ... this book, to say the least.

In a similar, but less extreme, way, a prediction of the galaxy density of the universe from what we have learned about the distribution of matter in the universe involves complicated issues we do not have the tools to address. What fraction of the matter has collapsed into nonlinear structures? How do these nonlinear structures evolve? Do galactic-size nonlinear structures merge? If so, how often? Upon collapse, how do stars form? How are stars distributed? In spiral patterns? Elliptical? One of the exciting developments of the 1990s was the evolution of a number of techniques to answer these questions. In addition to the brute-force approach of numerical simulations, several groups (e.g., Kauffman *et al.*, 1999; Somerville and Primack, 1999; Colberg *et al.*, 2000; Benson *et al.*, 2001; Cooray and Sheth, 2002) developed *semianalytic* techniques which have been remarkably successful at predicting properties and abundances of different galaxy types. Although we will not study these techniques directly here, the one technique we will encounter — the Press–Schechter formalism — forms the basis for much of this work.

Perhaps the fundamental difficulty encountered by one attempting to make predictions about the number density of galaxies is that the galactic scale has already gone nonlinear. Recall from Exercise 7.9 that scales smaller than $\sim 10\,h\,\mathrm{Mpc}^{-1}$ have gone non-linear. What scale in the unperturbed universe encloses the mass of a typical galaxy, $M = 10^{12} M_\odot$? The density in a spherical region of radius R is

$$\rho_{\mathrm{m}} = \frac{M}{4\pi R^3/3}. \tag{9.54}$$

Since $\rho_{\mathrm{m}} = \Omega_m \rho_{\mathrm{cr}}$, we can invert this to find

$$R = 0.951\,h^{-1}\,\mathrm{Mpc}\left(\frac{Mh}{10^{12}\,\Omega_m\,M_\odot}\right)^{1/3}. \tag{9.55}$$

So a galaxy comes from matter within a radius of about 1 Mpc, corresponding to fluctuations on scales of order $k \sim 1\,h\,\mathrm{Mpc}^{-1}$, well into the nonlinear regime.

This answers a question ("Why not try to predict the number density of galaxies?") but begs another: Are there objects, corresponding to scales closer to the linear regime, for which one might be able to make reliable predictions about abundances? If we invert Eq. (9.55) to get the mass enclosed within a sphere of radius R,

$$M = 1.16 \times 10^{15} \Omega_m h^{-1} M_\odot\,\left(\frac{R}{10\,h^{-1}\,\mathrm{Mpc}}\right)^3, \tag{9.56}$$

then we see that clusters of galaxies — with masses up to $10^{15} M_\odot$ — arise from perturbations on just the right scales.

How, then, to predict the abundance of galaxy clusters? The basic insight comes from a paper by Press and Schechter (1974), and the resulting framework is called Press–Schechter theory. To understand their argument, consider the one-dimensional density field in Figure 9.15. The average inhomogeneity is zero, of course. There are regions with relatively large excursions in both the positive and negative direction. Underdensities cannot get smaller than -1 (when the density is zero), but there are some regions in the figure with densities more than three times the average density. It is these rare regions of large overdensity that we are interested in: they *collapse*, accumulate so much excess matter that local gravity becomes more important than the Hubble flow. Particles in this region stop expanding away from each other and are trapped in the local gravitational field.

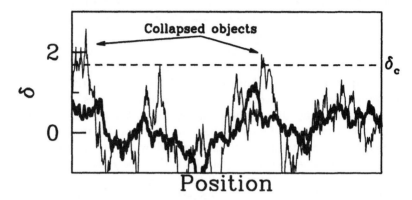

Figure 9.15. Inhomogeneities as a function of 1D position. Dark curve is the same field smoothed on larger scales. Several small-scale fluctuations have collapsed, while the large scale density field does not have any overdensities greater than δ_c, the critical value for collapse.

The Press–Schechter theory predicts the fraction of the volume that has collapsed,

$$f_{\text{coll}}(M(R), z) = \frac{2}{\sqrt{2\pi}\sigma(R, z)} \int_{\delta_c}^{\infty} d\delta \; e^{-\delta^2/2\sigma^2(R,z)}. \qquad (9.57)$$

Here R is the radius over which the density field has been smoothed. This radius is used to compute $\sigma(R, z)$, the rms of the smoothed density field (Exercise 7.9). As you can see from Figure 9.15, the smoothing scale matters. Typically, inhomogeneities on large scales are smaller in magnitude than those on small scales, so small scales collapse first. As time evolves, overdensities grow (proportional to the growth function), so eventually some large-scale inhomogeneities will also collapse. The right-hand side of Eq. (9.57) counts all parts of the Gaussian distribution for which the overdensity is greater than some critical density δ_c. There are several pieces of magic in this formula. First, it assumes that the distribution of inhomogeneities is Gaussian. This is impossible since δ, by definition, never gets smaller than -1. And indeed, it is possible to show that gravity skews an initially Gaussian distribution, producing more underdense regions and a nonnegligible tail of highly overdense regions. Second, the normalization is a bit of a cheat: one would naively not include the factor of 2. Finally, the rms σ in the formula is the *linear* rms, specifically ignoring nonlinear effects. On small scales, there is a huge difference between σ calculated with the linear power spectrum and that with the true nonlinear spectrum. Press–Schechter tells us that the collapsed fraction can be obtained using the linear σ. These peculiarities of the Press–Schechter formalism do not detract from its effectiveness. Numerical simulations (e.g., White, Efstathiou and Frenk, 1993) have shown that it works extremely well. Further, a number of groups (Peacock and

Heavens, 1990; Bond *et al.*, 1991) have justified theoretically some of the aspects of the formula that initially appeared ad-hoc.

To get the collapsed fraction into a form more comparable with observations, first differentiate f_{coll} with respect to M and multiply by a small interval dM. This gives the fraction of the volume collapsed into objects with mass between M and $M + dM$. Multiply this by the average number density of such objects ρ_{m}/M to get the the number density of collapsed objects with mass between M and $M + dM$,

$$dn(M, z) = -\frac{\rho_{\text{m}}}{M}\frac{df_{\text{coll}}(M(R), z)}{dM}dM. \tag{9.58}$$

The minus sign appears here since f_{coll} is a decreasing function of the mass M. Carry out the derivative using the fact that $dM/dR = 3M/R$. Then,

$$\frac{dn(M, z)}{dM} = \sqrt{\frac{2}{\pi}}\frac{\rho_{\text{m}}\delta_c}{3M^2\sigma}e^{-\delta_c^2/2\sigma^2}\left[-\frac{R}{\sigma}\frac{d\sigma}{dR}\right]. \tag{9.59}$$

The term in brackets here, the logarithmic derivative, is close to 1 for most models of interest. The dominant factor in Eq. (9.59), at least for large masses, is the exponential. If σ on a given scale is small, then the number density of collapsed objects on that scale is exponentially suppressed.

Until now, I have sidestepped the question of the numerical value of δ_c, the critical overdensity above which objects collapse. There are two approaches to obtaining δ_c; fortunately, both appear to agree. The first is to rely on a simple model of collapse, a model in which the overdensity is perfectly spherical. One can show that collapse occurs in such a model when $\delta = 1.686$ (for $\rho_{\text{m}} = \rho_{\text{cr}}$). The other way is simply to treat δ_c as a free parameter and calibrate it with numerical simulations (e.g., Eke, Cole and Frenk, 1996). The δ_c obtained this way is close enough to the spherical value that typically one simply adopts $\delta_c = 1.686$.

Measuring the cluster abundance, and therefore testing theories with Eq. (9.59), is a subtle business. One class of difficulties is identifying a cluster. Sophisticated algorthims have been developed to find clusters in a galaxy survey. The second set of difficulties revolves around determining the mass of the cluster. There are several methods of mass determination:

- **X-Ray Temperatures** Hot ionized gas in a cluster emit radiation with a cutoff frequency which is determined by the temperature of the gas. This temperature can be related to the mass of the cluster under certain assumptions.

- **Sunyaev–Zeldovich Distortion** Photons from the CMB pasing through clusters get scattered by the hot gas. This scattering distorts the CMB spectrum as a function of frequency, inducing a decrement at low frequency and excess at high frequency (low-energy photons gain energy from the hot electrons). The shape of the distortion is fixed; the amplitude is another measure of the temperature of the gas, which, again under certain assumptions, can be translated into a measurement of the mass.

- **Weak Lensing** Images of background galaxies are distorted by a foreground cluster. The larger the mass of the cluster, the larger the distortions. Weak lensing is therefore becoming a fabulous tool for measuring masses of clusters without using the temperature.

At least in the first two techniques, the direct measurement is of the cluster temperature. So let's work through a relation between the mass and temperature of a cluster under a simple set of assumptions. Suppose a cluster has virialized so that its kinetic energy is equal to minus half its potential energy. Suppose also that the cluster is spherical with radius $R_{\rm vir}$ with potential (gravitational) energy is equal to $-3GM^2/5R_{\rm vir}$. Then,

$$\frac{1}{2}Mv^2 = \frac{3}{10}\frac{GM^2}{R_{\rm vir}}. \tag{9.60}$$

The overdensity of the cluster $\Delta_{cl} \equiv \rho_{cl}/\rho_{\rm m}$ then allows us to eliminate the radius, since

$$\Delta_{cl} \equiv \frac{M}{4\pi R_{\rm vir}^3 \rho_{\rm m}/3}. \tag{9.61}$$

The temperature is equally apportioned among three directions, so the average velocity squared is $v^2 = 3T/m_p$, where m_p is the proton mass. The temperature can now be expressed in terms of the total mass of the system,

$$T = \frac{m_p}{5}\left[GMH_0\sqrt{\frac{\Delta_c}{2}}\right]^{2/3} \tag{9.62}$$

when $\Omega_m = 1$. Invert to get

$$M = 8.2 \times 10^{13} h^{-1} M_\odot \left(\frac{T}{\rm keV}\right)^{3/2}\sqrt{\frac{178}{\Delta_{cl}}}. \tag{9.63}$$

Here I have normalized Δ_{cl} by its value in the spherical collapse model (with $\rho_{\rm m} = \rho_{\rm cr}$), although again simulations have verified the numerical value.

Figure 9.16 shows the cluster density as a function of temperature for the standard CDM model with $\Omega_m = 1$. One of the most important points made immediately after the COBE detections of anisotropies in 1992 was that this plain-vanilla model predicts too many clusters. Indeed, the abundance of clusters today is often used as an excellent way of normalizing a power spectrum. A typical value for σ_8 from cluster abundances (e.g., Wang and Steinhardt, 1998) is

$$\sigma_8 = 0.5\Omega_m^{-.33-.35\Omega_m} \tag{9.64}$$

with error estimates ranging from just a few percent up to 20%, the latter probably more accurately reflecting uncertainties in the mass determinations.

Another exciting application of the Press–Schechter prediction for cluster abundances is the evolution with redshift. The basic point stems from the exponential

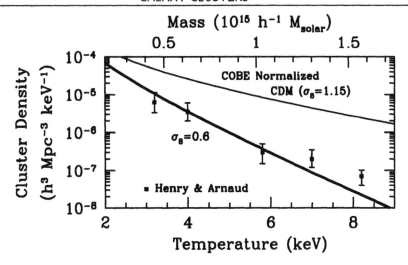

Figure 9.16. The cluster density as a function of temperature. Data from Henry and Arnaud (1991). The two theoretical curves are the Press–Schechter estimates (Eq. (9.59)) of models with $\Omega_m = 1, h = 0.5, \Omega_b = .05$. The only difference between the two is the normalization.

dependence in Eq. (9.59). At high redshift, $\sigma(R, z)$ is necessarily smaller, but how much smaller depends on the underlying cosmology. The growth is fastest in a model with $\Omega_m = 1$ (see Figure 7.12), so for a fixed abundance today, one expects many fewer clusters in such a model. Figure 9.17 illustrates this effect. An $\Omega_m = 1$ model predicts a factor of 700 fewer clusters with masses greater than $3.5 \times 10^{14} M_\odot$ at redshift 0.5 than at redshift zero. By contrast, ΛCDM predicts just a factor of 4 decline. With the observational assault on clusters just getting off the ground, we can expect strong constraints on cosmology to emerge in the coming decade.

SUGGESTED READING

The Large Scale Structure of the Universe (Peebles) has a good description and derivation of the angular correlation function, using Limber's (1953) original derivation. The derivation given in Section 9.1 is based on the Appendix of Kaiser's (1992) work on weak lensing. This derivation has the advantage of being physically intuitive and also generally applicable. It will allow us to easily compute the weak lensing correlation functions we encounter in Chapter 10.

The data discussed in Section 9.1 come from the APM Survey (Maddox *et al.*, 1990). Other recent angular results of note include the Edinburgh/Durham Southern Galaxy Catalogue (EDSGC, Collins, Nichol, and Lumsden, 1992) and the Sloan Digital Sky Survey (York *et al.*, 2000), the first analysis of which is presented in Scranton *et al.* (2002). As I hinted in the text, analysis of these data sets has gotten progressively more sophisticated over time. Baugh and Efstathiou (1993) first inverted the APM angular correlation function and extracted the 3D

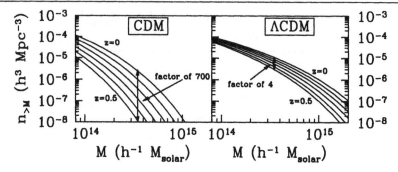

Figure 9.17. The number density of clusters with mass greater than M at different redshifts. *Left panel:* CDM model with $\Omega_m = 1, h = 0.5, \Omega_b = 0.05$ but normalized to give the correct abundance at $z = 0$ ($\sigma_8 = 0.5$). Curves give the abundances in redshift increments of $\delta z = 0.1$ starting from $z = 0$ going out to $z = 0.5$. *Right panel:* ΛCDM model which fits CMB and other data ($\Omega_m = 0.35, h = 0.7, \Omega_b h^2 = 0.02$). Note the relatively slow evolution of clusters with mass $\sim 3 \times 10^{14} M_\odot$ as compared with the critical density model.

power spectrum, confirming that it was quite different from the standard CDM power spectrum on large scales. Dodelson and Gaztanaga (2000) pointed out that the resultant errors on the power spectrum are correlated and that an accurate treatment would also account for the fact that the errors in $w(\theta)$ are also correlated. Eisenstein and Zaldarriaga (2001) and then Efstathiou and Moody (2001) accounted for these correlations leading to the softened conclusions summarized in Figure 9.6. The Eisenstein and Zaldarriaga paper also contains a clear discussion of the relation between $w(\theta)$ and the 2D power spectrum, or the C_l's. They observed that errors on the C_l's are much less correlated, which leads me to believe that C_l's will ultimtely replace $w(\theta)$ as the statistic of choice for angular surveys. Indeed, Huterer, Knox, and Nichol (2001) have analyzed the EDSGC survey with C_l's, and Tegmark *et al.* (2002) obtained C_l's from early SDSS data.

Cosmological Physics (Peacock) is a good resource for peculiar velocities and their effect on redshift surveys. Two important and informative review articles are Strauss and Willick (1995) and Hamilton (1998), the former particularly good for experimental issues involved in determining peculiar velocities and the latter for analyzing galaxy surveys in the presence of redshift space distortions. Another leader in the field, Dekel (1997), has written a good review of the cosmological implications of the peculiar velocity field.

The seminal work on redshift space distortions is by Kaiser (1987), who solved the problem working in Fourier space for linear distortions in the distant-observer, low redshift approximation. Hamilton (1992) found the analogue of this solution in real space. Recently, with large-area, relatively deep surveys coming on line, the generalization for cosmological corrections (i.e., distance is *not* equal to cz/H_0 as z gets large) and all-sky analysis has been carried out by a number of authors. As examples of the work currently going on in the field, Szalay, Matsubara, and Landy (1998) generalized Kaiser's work to large angles while Magira, Ying, and Suto (2000) accounted for cosmological distortions and also nonlinearities and evolution. Peacock and Dodds (1994) analyzed a variety of surveys, arguing that — accounting for redshift space distortions, nonlinearities, and bias properly — the power spectrum has been well measured. Undoubtedly, with the upcoming Sloan Digital Sky Survey, Two Degree Field, and others, more will be learned about the power spectrum in coming years. The semianalytic work referred to on Page 283 is based on the seminal papers of White and Rees (1978) and White and Frenk (1991). Related, but separate form these semianalytic models, is the halo model (reviewed by Cooray and Sheth, 2002) which postulates that all the dark matter is in halos, thereby reducing the clustering problem to (i) the clustering of the halos and (ii) the distribution of matter and galaxies within the halo. Good descriptions of what the halo model is and how it can be used to compare theories with redshift surveys can be found in White (2001); Seljak (2000); and Berlind and Weinberg (2001).

The prediction of the cluster abundance is treated nicely in *Cosmological Physics* (Peacock) and *Cosmological Inflation and Large Scale Structure* (Liddle and Lyth). *Structure Formation in the Universe* (Padmanabhan) has a detailed section on the spherical collapse model, which is the source of the numbers 1.686 and 178 in Section 9.5. In addition to the papers cited in the text, some important cluster normalization papers are Viana and Liddle (1996, 1999) and Pierpaoli, Scott, and White (2001).

EXERCISES

Exercise 1. Suppose the correlation function is defined as

$$\xi(\vec{r}) \equiv \langle \delta(\vec{x})\delta(\vec{x}+\vec{r})\rangle. \tag{9.65}$$

By Fourier expanding each of the δ's and using Eq. (C.20), show that this definition implies that the correlation function is the Fourier transform of the power spectrum.

Exercise 2. Expand the 3D power spectrum in the integral of Eq. (9.9) about $k_3 = 0$. The leading term is the one we considered. Show that the next term is of order $(1/l)^2$, compared with the leading term.

Exercise 3. Rewrite the kernel in Eq. (9.14) as an integral over χ. Show that F is a function of $k\theta$ only.

Exercise 4. Give an order-of-magnitude estimate for the kernel of the angular correlation function.
(a) Consider a shell in Fourier space with radius k and width dk. What fraction of the volume of this shell has $|k_3| < \chi^{-1}$?
(b) Argue that only Fourier modes with $|k_3| < \chi^{-1}$ contribute to the angular correlation function with a weight $\Delta^2(k) = k^3 P(k)/2\pi^2$. Combine this argument

with the fraction computed in part (a) to estimate the kernel relating $w(\theta)$ to $P(k)$. Compare this estimate with Eq. (9.13).

Exercise 5. Decompose the angular correlation function into a sum over spherical harmonics,

$$w(\theta) = \sum_l \frac{2l+1}{4\pi} C_l^{\text{matter}} \mathcal{P}_l\left(\cos\theta\right), \qquad (9.66)$$

where the superscript ^{matter} distinguishes these C_l's from the ones characterizing anisotropies in the CMB, and $\mathcal{P}_l$ here are the Legendre polynomials. Express C_l^{matter} as an integral over the 3D power spectrum. Show that on small scales $C_l^{\text{matter}} = P_2(l)$, where P_2 is the 2D power spectrum introduced in Section 9.1.

Exercise 6. In Section 9.1 we implicitly neglected the evolution of the power spectrum. That is, we assumed that $P(k)$ remains constant with time. Allow $P(k)$ to scale as $(1+z)^\beta$. What is β for linear modes in a flat, matter-dominated universe? Rewrite the kernel in terms of an integral over z, accounting for this evolution.

Exercise 7. Compute (numerically) the linear growth rate f today in an open universe and compare with the approximation $\Omega_m^{0.6}$. What is the fractional error between the approximation and the exact result? Now assume that the universe is flat, with $\Omega_m + \Omega_\Lambda = 1$. Again compare the exact linear growth rate with $\Omega_m^{0.6}$. Show that

$$f = \Omega_m^{0.6} + \frac{\Omega_\Lambda}{70}\left(1 + \frac{\Omega_m}{2}\right) \qquad (9.67)$$

is a better approximation, with no worse than 4% accuracy for $\Omega_m > 0.025$.

Exercise 8. Using CMBFAST, compute the transfer function for standard CDM ($\Omega_m = 1; h = 0.5$) with $\Omega_b = 0.01, 0.05$, and 0.1. Show that the BBKS transfer function is still a reasonable fit as long as

$$\Gamma = \Omega_m h \rightarrow \Omega_m h e^{-2\Omega_b}. \qquad (9.68)$$

Exercise 9. Using CMBFAST or the BBKS transfer function, compute COBE-normalized σ_8 for ΛCDM with $h = 0.7$, $\Omega_\Lambda = 0.7$, $\Omega_m = 0.26$, and $\Omega_b = 0.04$. Locate the model on the bottom left panel of Figure 9.6. What does this imply about the relation between σ_8 (of the mass, which you have just computed) and $(\sigma_8)_g$ (of the galaxies, which APM is sensitive to)?

Exercise 10. Assume that the universe is flat with matter and a cosmological constant. Expand the comoving distance out to a galaxy at redshift z (neglecting peculiar velocities) about $z = 0$. The first-order term in the expansion should give back the redshift space answer. What is the second-order term, the leading correction to redshift space? Express your answer in terms of Ω_m.

Exercise 11. Derive (9.51), using the fact that $\mathcal{P}_4(x) = 35x^4/8 - 15x^2/4 + 3/8$. Show that the definition of the moments in Eq. (9.50) — $P_s^{(l)}(k) = (2l + 1) \int_{-1}^{1} (d\mu_{\mathbf{k}}/2)\mathcal{P}_l(\mu_{\mathbf{k}})P_s(k,\mu_{\mathbf{k}})$ — means that

$$P_s(k,\mu_{\mathbf{k}}) = \sum_l \mathcal{P}_l(\mu_{\mathbf{k}})P_s^{(l)}(k). \qquad (9.69)$$

Exercise 12. In the text we showed how redshift space distortions affect the power spectrum. Show how the redshift space distortions affect the correlation function. Assume linear theory. You will probably need to consult Hamilton (1992), which transforms Kaiser's result to the correlation function in a single (dense) paragraph.

10

WEAK LENSING AND POLARIZATION

The traditional methods of measuring clustering — angular and redshift surveys — are powerful probes of the power spectrum but share a common deficiency. They are measures of the distribution of galaxies, not the distribution of mass. Theories of the early universe can make very accurate predictions about the latter, but not about the former. A very exciting new technology which probes the mass — not the light — distribution is introduced in this chapter. We will see that the inhomogeneities of the matter induce distortions in the observed shapes of distant galaxies due to gravitational lensing. Further, the statistics of these distortions are directly related to the matter power spectrum.

The anisotropies in the CMB are subject to none of the uncertainties or ambiguities which plague the density field.

- We know exactly where the CMB comes from (the surface of last scattering) so there is no analogue of peculiar velocity distortions.
- There is nothing like the mass vs light problem which afflicts the interpretation of galaxy surveys.
- In addition, the mass distribution has gone nonlinear, so a simple comparison of the linear power spectrum derived in Chapter 7 with the data is dangerous. Anisotropies in the CMB are still at the part-in-a-hundred-thousand level, so nonlinearities are for the most part irrelevant.

The C_l's then are easy to interpret and extract information from. Nonetheless, here too we can go beyond what we have already done. Until now we have focused on anisotropies in the temperature field. Compton scattering before decoupling also induced polarization anisotropies. Polarization opens up a new dimension in the study of the CMB. At the very least, it doubles the amount of information contained in the CMB. As we will see in this chapter, the promise of polarization goes well beyond this doubling. Gravity waves — tensor perturbations — produce a particular pattern of polarization that cannot be mimicked by scalar perturbations. Therefore, polarization offers a unique way of searching for gravity waves produced during inflation.

Gravitational lensing and polarization belong in the same chapter primarily because the mathematics describing them is so similar. Both effects can be quantified with a two-by-two symmetric matrix. In lensing, this matrix is the *distortion tensor* encoding information about image distortion. The polarization tensor has a longer history with more famous components Q and U. It is identical mathematically, though. So, the technologies used to study both of these effects are very similar.

10.1 GRAVITATIONAL DISTORTION OF IMAGES

The cosmological gravitational field distorts the paths traveled by light from distant sources to us. This fundamental fact carries with it an enormous amount of cosmological promise. Most important is the idea that light paths respond to *mass*. If we can measure distortions, then, we might be able to infer something about the distribution of mass in the universe. The importance of this inference cannot be overstated: most of what we think we know about this distribution comes from our observations of the galaxy distribution. We hope that, on large scales at least, the two — the mass and the galaxy distribution — are not too different. If we observe the mass distribution directly via distortion of light rays, though, then we need not rely on this hope. We can then directly compare observations with theoretical predictions. For cosmology, therefore, we expect the most important aspect of light ray distortion to be *weak lensing*, wherein the shapes of distant galaxies are distorted (slightly) by intervening foreground mass overdensities. We begin with an overview of image distortion along with a brief discussion of some other applications.

The idea that gravitational fields might distort distant images is as old as general relativity. Indeed, even before Einstein finalized general relativity, he understood the importance of measuring this distortion. Early notebooks of his contain calculations of the magnification of images and of the possibility of a double image of a single source (Renn, Sauer, and Stachel, 1997). And it was detection of gravitational distortion that led to the acceptance of general relativity. In 1919, Eddington led a voyage to the Southern Hemisphere to observe the deflection of starlight during a solar eclipse. The magnitude of this effect (Dyson, Eddington, and Davidson, 1920) was in good agreement with Einstein's new theory.

One of the most spectacular manifestations of gravity bending light paths is strong gravitational lensing. In 1979, Walsh, Carswell, and Weymann observed a multiply imaged quasar, thereby confirming Einstein's early speculations. Light rays leaving the quasar in different directions are focused on the same point (us) by an intervening galaxy. Since then, dozens of multiply imaged quasars have been observed, and we are on the verge of discovering many hundred more in the near future. Exactly what fraction of quasars is lensed is a question that may depend on the background cosmology. In particular, it has been argued that there should be more multiply imaged quasars in a universe with a cosmological constant than in one without (see Exercise 1 and Kochanek, 1996).

There are other examples of gravitational lensing that have an impact on cosmology. Light rays that take different routes to the same endpoint typically arrive at that endpoint at different times. Therefore, two light rays emitted from the same source at the same time which we detect from different directions due to lensing typically arrive at different times. We can measure this time delay by studying sources with variable emission. The delay turns out generally to depend on the Hubble constant, so astronomers have made very accurate measures of H_0 looking for time delays (e.g., Kundic *et al.*, 1997). Another example is microlensing, wherein a lens moves into the line connecting a source to us. When it does, the image is magnified, so that we observe a characteristic variability in the distant source. Microlensing has been used in recent years to find massive compact halo objects (MACHOs) in our galaxy (Alcock *et al.*, 1993). It now appears that MACHOs do not make up all, or even most, of the dark matter in our galaxy. Nonetheless, exactly what and where they are is still a mystery of cosmic significance.

Gravitational Lens in Abell 2218   HST · WFPC2
PF95-14 · ST ScI OPO · April 5, 1995 · W. Couch (UNSW), NASA

Figure 10.1. Foreground galaxies in the cluster Abell 2218 distort the images of background galaxies. Elliptical arcs surround the central region of the cluster at right.

Yet another spectacular manifestation of gravitational lensing is shown in Figure 10.1. The large cluster in the foreground, Abell 2218, distorts the shapes of the background galaxies. This leads to a distinctive pattern of elliptical arcs surrounding the central region of the cluster. Why do the background galaxies appear stretched out elliptically in Figure 10.1? Consider a circular galaxy sitting behind a large mass density with an observer out of the page as in Figure 10.2a. Since the light rays are distorted, we do not expect to see a circular image. Rather, light rays coming from the "bottom" of this source — the ones that pass closer to the central mass region — are bent more than those that do not come as close to the mass. The light rays are bent such that objects at the bottom appear to be *farther away* from

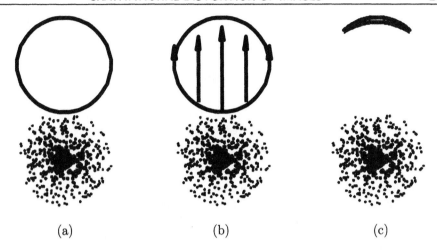

(a)   (b)   (c)

Figure 10.2. (a) Circular galaxy, the source, sits behind a foreground mass distribution represented by points at bottom. The observer is out of the page so that the foreground mass is between the observer and the source. (b) Light rays from source are deflected as they pass by mass distribution. Rays traveling closest to mass get deflected the most. (c) Resulting image is an arc.

the mass. (This is the only subtle part of the argument: rays are bent *toward* the mass distribution, so that as you extrapolate backward, the source appears farther away. See Figure 10.4.) Images will therefore be distorted as in Figure 10.2b. The net effect, therefore, is to turn a circular galaxy into the arc shown in panel c.

A very active field of research uses background galaxies to try to infer the mass distribution of clusters (e.g., Clowe *et al.*, 1998). Most times, the images are not as dramatic as those shown in Figure 10.1. The lack of drama is offset by the huge numbers of background galaxies. By adding up many small distortions, observers have succeeded in obtaining mass estimates for a number of clusters. This idea of statistically averaging small distortions is the hallmark of *weak lensing*. The mass estimates are important information for cosmologists: several cosmological constraints are based on cluster masses and abundances (e.g., Section 9.5, Carlberg *et al.*, 1997; Bahcall *et al.*, 2000).

We will be interested in weak lensing not by a single identifiable lens such as a cluster, but rather by the generic large-scale structure in the universe. Inferring the distribution of the dark matter—i.e., pointing to a spot on the sky and identifying it as an overdense region—is not necessarily the goal. Rather, we should be satisfied if we can measure some simple statistics, for example the correlation function or its Fourier transform, the power spectrum. Indeed, these are the quantities we, as cosmologists, are most interested in anyway. We don't care where the overdense and underdense regions are; we simply want to compare theory with observations. So our main goal here is to relate the observations (which have already begun) of distortions of galaxy images to the underlying mass power spectrum.

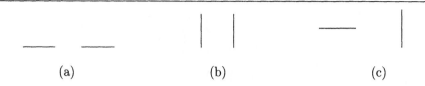

Figure 10.3. Different lensing patterns. Panels (a) and (b) could be produced by a mass distribution (a) above or below the distorted images and (b) in between or on either side of the images. But the alignment in panel (c) could not be produced by lensing.

One final note concerning the correlations of distortions expected from gravitational lensing. We might expect to find the images of two (circular) galaxies to be distorted so that they look like Figure 10.3a if, for example, there is a large overdensity above or beneath this galaxy pair. We might also expect images similar to those in Figure 10.3b if an overdensity exists between them or to either side. However, lensing cannot produce the alignment sketched in panel (c). This fact, which we will shortly prove, is often used as a check against systematic problems afflicting an observation.

10.2 GEODESICS AND SHEAR

We want to solve for the path of a light ray as it leaves a distant source and travels through the inhomogeneous universe. Figure 10.4 shows the geometry and notation, which will be similar to that set up in our discussion of the angular correlation function. The position of the photon at any time is given by $\vec{x}$, with the x_3 component equal to the radial distance χ and the transverse components equal to $\chi\vec{\theta}$. The intensity we observe from a source is

$$I_{\text{obs}}(\vec{\theta}) = I_{\text{true}}(\vec{\theta}_S); \tag{10.1}$$

a source whose image appears at $\vec{\theta}$ is actually at $\vec{\theta}_S$.

To solve for the path of a light ray, we need to use the machinery of general relativity. Recall that in Chapter 4, we used the time component of the geodesic equation to find dp/dt, the rate of change of the magnitude of the momentum. Here we are interested in deflections, so we will need the spatial component,

$$\frac{d^2 x^i}{d\lambda^2} = -\Gamma^i_{\alpha\beta} \frac{dx^\alpha}{d\lambda} \frac{dx^\beta}{d\lambda}; \tag{10.2}$$

in particular, we will need the transverse part. Let's first consider the left side of this equation. We can express the derivatives with respect to affine parameter λ in terms of derivatives with respect to χ using the fact that

$$\frac{d\chi}{d\lambda} = \frac{d\chi}{dt} \frac{dt}{d\lambda}$$

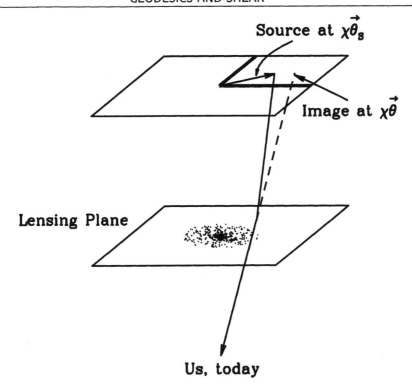

Figure 10.4. A light ray leaving a distance source is distorted as it passes by an intervening overdense region. At all times, the position of the light ray can be characterized by a 2D vector specifying its angular distance from the center of the lens. The ray starts with angular vector $\vec{\theta}_S$, but appears to us to be coming from $\vec{\theta}$.

$$= \frac{-1}{a} \, p(1 - \Psi). \tag{10.3}$$

The first part of this equality $(d\chi/dt = -1/a)$ follows from Eq. (2.42), while the second part comes from Eq. (4.14). The transverse part of $\vec{x}^i$ is equal to $\chi\theta^i$, so the left-hand side of the geodesic equation is

$$\frac{d^2 x^i}{d\lambda^2} = \frac{1}{a} \, p \frac{d}{d\chi} \left[\frac{p}{a} \, \frac{d}{d\chi} \left(\chi\theta^i \right) \right]. \tag{10.4}$$

Here I have dropped the (small) gravitational potential because it multiplies the (small) angle θ^i. We can reduce further by remembering that (at zero order) the momentum p times a remains constant, so removing pa outside the derivative leads to

$$\frac{d^2 x^i}{d\lambda^2} = p^2 \frac{d}{d\chi} \left[\frac{1}{a^2} \, \frac{d}{d\chi} \left(\chi\theta^i \right) \right]. \tag{10.5}$$

Now let's consider the right side of the geodesic equation. Again, changing the derivatives with respect to λ to those with respect to χ leads to

$$\Gamma^i_{\alpha\beta} \frac{dx^\alpha}{d\lambda} \frac{dx^\beta}{d\lambda} = \left(\frac{p}{a}\right)^2 (1-\Psi)^2 \Gamma^i_{\alpha\beta} \frac{dx^\alpha}{d\chi} \frac{dx^\beta}{d\chi}. \tag{10.6}$$

There are three types of terms in the sum over α and β: those with $\alpha = \beta = 0$, those with one index spatial and the other temporal, and finally those in which both α and β are spatial. We have already derived the relevant Christoffel symbols, given in Eq. (5.7). Let's work through the terms one by one:

- When $\alpha = \beta = 0$, we have

$$\Gamma^i_{00} \left(\frac{dt}{d\chi}\right)^2 = \Psi_{,i} = -\Phi_{,i} \tag{10.7}$$

  where the second equality holds since in the late universe there are no anisotropic stresses so $\Phi = -\Psi$.
- When one of the indices is spatial, Γ^i_{0j} is nonzero only when $i = j$. Therefore, the spatial index j must be transverse with $x^j = \chi\theta^j$. Since θ^j is small, we can drop the potential in the Christoffel symbol leading to

$$\Gamma^i_{0j} \frac{dt}{d\chi} \frac{dx^j}{d\chi} = -aH \frac{d}{d\chi} \left[\chi\theta^i\right] \tag{10.8}$$

  with of course an identical term coming from Γ^i_{j0}.
- When both indices are spatial, the Christoffel symbol is proportional to the (small) gravitational potential. When multiplied by the small transverse distance, these terms will be negligible, so we need consider only the term $\Gamma^i_{jk}(dx^j/d\chi)(dx^k/d\chi)$ in which $j = k = 3$ along the radial direction. In that case $x^3 = \chi$, the derivative is trivial, and we have

$$\Gamma^i_{jk} \frac{dx^j}{d\chi} \frac{dx^k}{d\chi} = -\Phi_{,i}. \tag{10.9}$$

Collecting these terms leads to the geodesic equation for transverse motion,

$$\frac{d}{d\chi}\left[\frac{1}{a^2} \frac{d}{d\chi}\left(\chi\theta^i\right)\right] = \frac{2}{a^2}\left[\Phi_{,i} + aH\frac{d}{d\chi}\left[\chi\theta^i\right]\right]. \tag{10.10}$$

The derivative on the left acting on a^{-2} exactly cancels the term proportional to aH on the right, so our final equation for the transverse displacement is

$$\frac{d^2}{d\chi^2}\left(\chi\theta^i\right) = 2\Phi_{,i}. \tag{10.11}$$

This geodesic equation tells us that in a uniform potential, the angular direction $(\chi\theta^i)'$ remains constant, whereas a changing potential perturbs it. The sign is correct: An overdensity centered at $x = y = 0$ has $\Phi > 0$ there, and therefore the derivative of Φ with respect to x ($\Phi_{,i}$ with $i = 1$) is negative for $x > 0$. As such, the force on a light ray passing the overdensity on the positive x-axis is negative, i.e., inward toward the overdensity, as we expect.

Equation (10.11) can be integrated to find the image angle as a function of the source angle. Integrating once gives

$$\frac{d}{d\chi}\left(\chi\theta^i\right) = 2\int_0^\chi d\chi' \Phi_{,i}\left(\vec{x}(\chi')\right) + \text{constant};\tag{10.12}$$

we'll fix the constant momentarily. Integrating again leads to

$$\theta_S^i = \frac{2}{\chi}\int_0^\chi d\chi'' \int_0^{\chi''} d\chi' \Phi_{,i}\left(\vec{x}(\chi')\right) + \text{constant}\tag{10.13}$$

since $\theta^i(\chi) \equiv \theta_S^i$, the value of $\vec{\theta}$ at the source. We now see that the constant is equal to θ^i — the observed angle — since the angle retains its initial value if there is no perturbation. The integral in the χ', χ'' plane is restricted to the shaded region in Figure 10.5 so we can change orders of integration with the χ'' integral ranging

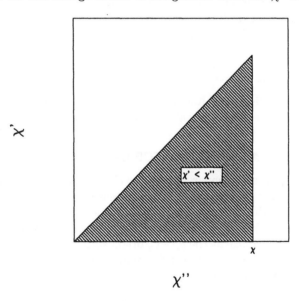

Figure 10.5. Range of integration in the double integral of Eq. (10.13). The shaded region can be expressed as $0 < \chi'' < \chi, 0 < \chi' < \chi''$ or as $\chi' < \chi'' < \chi, 0 < \chi' < \chi$. The latter is more convenient here, since the χ'' integral is then trivial.

from χ' to χ. The χ'' integral is then trivial (since $\Phi_{,i}$ depends only on χ') so

$$\theta_S^i = \theta^i + 2\int_0^\chi d\chi' \Phi_{,i}\left(\vec{x}(\chi')\right)\left(1 - \frac{\chi'}{\chi}\right).\tag{10.14}$$

To describe the shift in the angle experienced by a light ray, it is conventional to define the 2×2 symmetric transformation matrix,

$$A_{ij} \equiv \frac{\partial\theta_S^i}{\partial\theta^j}$$

$$\equiv \begin{pmatrix} 1 - \kappa - \gamma_1 & -\gamma_2 \\ -\gamma_2 & 1 - \kappa + \gamma_1 \end{pmatrix}. \tag{10.15}$$

The parameter κ is called the *convergence*; it describes how an image is magnified. Although this magnification has many important ramifications (e.g., microlensing and multiple images) it is not what is important for the distortions studied in weak lensing. Rather, these distortions are governed by the two components of the *shear*,

$$\gamma_1 = -\frac{A_{11} - A_{22}}{2}$$

$$\gamma_2 = -A_{12}. \tag{10.16}$$

Equation (10.15) says that the components of shear involve derivatives of Eq. (10.14) with respect to angle $\vec{\theta}$.[1] The only dependence on $\vec{\theta}$ is in the argument of the potential, where $\vec{x}(\chi') = \chi'\vec{\theta}$ (for the transverse components). Therefore, the derivative with respect to θ^j can be written as a derivative with respect to x^j (in our notation $,j$) times χ'. Therefore,

$$A_{ij} - \delta_{ij} = \begin{pmatrix} -\kappa - \gamma_1 & -\gamma_2 \\ -\gamma_2 & -\kappa + \gamma_1 \end{pmatrix} = 2 \int_0^\chi d\chi' \Phi_{,ij}\left(\vec{x}(\chi')\right) \chi' \left(1 - \frac{\chi'}{\chi}\right). \tag{10.17}$$

So γ_1 and γ_2 are well-defined functions of the potential. The next section shows how they influence the shapes of galaxy images.

10.3 ELLIPTICITY AS AN ESTIMATOR OF SHEAR

We expect lensing to turn circular images into elliptical ones. To describe this effect, then, we need to come up with quantitative measures of ellipticity, and then see how these are related to the components of shear defined above. The simplest measure of ellipticity starts with the definition of the quadrupole moments of an image. Imagine centering an image at the $\theta_x - \theta_y$ origin such that it has no dipole moment ($\langle \theta_x \rangle = \langle \theta_y \rangle = 0$ where angular brackets are averages over the intensity). Then the quadrupole moments are defined as

$$q_{ij} \equiv \int d^2\theta I_{\text{obs}}(\theta)\theta_i\theta_j. \tag{10.18}$$

A circular image has $q_{xx} = q_{yy}$ and $q_{xy} = 0$. Therefore, two good measures of ellipticity are

$$\epsilon_1 \equiv \frac{q_{xx} - q_{yy}}{q_{xx} + q_{yy}}$$

---

[1]The derivative is formally with respect to the observed angle $\vec{\theta}$, while the right-hand side of Eq. (10.14) depends on the potential at the true position of the light ray. In principle, then, the derivatives which go into the definition of A_{ij} are quite complicated. In practice, though, deflections are sufficiently small that we can ignore the distinction between the final angle $\vec{\theta}$ and the actual angle everywhere along the trajectory. Therefore, on the right-hand side of Eq. (10.14) we evaluate the potential along the undistorted path parameterized by $\vec{\theta}$.

$$\epsilon_2 \equiv \frac{2q_{xy}}{q_{xx} + q_{yy}}. \tag{10.19}$$

Figure 10.6 shows different orientations of elliptical images and the associated

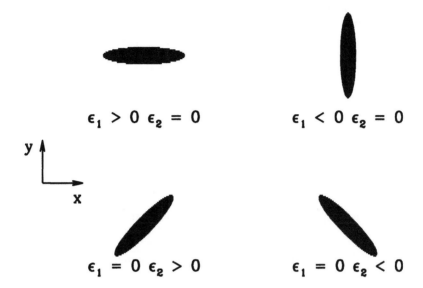

$$\epsilon_1 > 0 \quad \epsilon_2 = 0 \qquad\qquad \epsilon_1 < 0 \quad \epsilon_2 = 0$$

$$\epsilon_1 = 0 \quad \epsilon_2 > 0 \qquad\qquad \epsilon_1 = 0 \quad \epsilon_2 < 0$$

Figure 10.6. Definition of ellipticities ϵ_1 and ϵ_2. Circular images have both ellipticities equal to zero.

values of the ϵ_1 and ϵ_2. With these definitions, we can make more precise the statement at the end of Section 10.1 about correlations of ellipticities. Panel (a) in Figure 10.3 has two galaxies at $\vec{\theta}_1$ and $\vec{\theta}_2$, each with ϵ_1 positive; in panel (b) both galaxies have ϵ_1 negative. In both possible cases, then, the product $\epsilon_1(\vec{\theta}_1)\epsilon_1(\vec{\theta}_2)$ is positive if the x-axis is chosen along the direction connecting the two galaxies. The impossible case is depicted in panel (c) wherein the product is negative. Therefore, we do not expect lensing to produce $\epsilon_1(\vec{\theta}_1)\epsilon_1(\vec{\theta}_2) < 0$.

How are the ellipticities defined in Eq. (10.19) related to the shear defined in Eq. (10.16)? Let's assume that the source is spherical and compute the ellipticity of the image. Focusing on ϵ_1, we have

$$\epsilon_1 = \frac{\int d^2\theta I_{\text{true}}(\vec{\theta}_S) \, [\theta_x \theta_x - \theta_y \theta_y]}{\int d^2\theta I_{\text{true}}(\vec{\theta}_S) \, [\theta_x \theta_x + \theta_y \theta_y]} \tag{10.20}$$

where I have used the equality of Eq. (10.1). The integrals here are over the observed angles $\vec{\theta}$, while the integrands depend in part on the angle from which the photon started at the source, $\vec{\theta}_S$. For small angles, these are related via $\theta_i = (A^{-1})_{ij}\theta_{Sj}$. To do the integrals, then, change dummy variables in the integral to $\vec{\theta}_S$, and write all occurences of θ_i as $(A^{-1}\theta_S)_i$. This leads to

$$\epsilon_1 = \frac{\sum_{ij}\left[(A^{-1})_{xi}(A^{-1})_{xj} - (A^{-1})_{yi}(A^{-1})_{yj}\right]\int d^2\theta_S I_{\text{true}}(\vec{\theta}_S)\theta_{Si}\theta_{Sj}}{\sum_{ij}\left[(A^{-1})_{xi}(A^{-1})_{xj} + (A^{-1})_{yi}(A^{-1})_{yj}\right]\int d^2\theta_S I_{\text{true}}(\vec{\theta}_S)\theta_{Si}\theta_{Sj}}. \tag{10.21}$$

Now the integral over the (true) circular image vanishes unless $i = j$ and so is proportional to δ_{ij}. The proportionality constant is irrelevant since it appears in both the numerator and denominator, so

$$\epsilon_1 = \frac{(A^{-1})_{xi}(A^{-1})_{xi} - (A^{-1})_{yi}(A^{-1})_{yi}}{(A^{-1})_{xi}(A^{-1})_{xi} + (A^{-1})_{yi}(A^{-1})_{yi}}$$

$$= \frac{\left(A_{xx}^{-1}\right)^2 - \left(A_{yy}^{-1}\right)^2}{\left(A_{xx}^{-1}\right)^2 + \left(A_{yy}^{-1}\right)^2 + 2\left(A_{xy}^{-1}\right)^2}. \tag{10.22}$$

It is easy to compute the inverse of the 2×2 matrix A:

$$A^{-1} = \frac{1}{(1-\kappa)^2 - \gamma_1^2 - \gamma_2^2}\begin{pmatrix} 1 - \kappa + \gamma_1 & \gamma_2 \\ \gamma_2 & 1 - \kappa - \gamma_1 \end{pmatrix}, \tag{10.23}$$

so we see that the ellipticity ϵ_1 can be expressed in terms of the shear as

$$\epsilon_1 = \frac{(1 - \kappa + \gamma_1)^2 - (1 - \kappa - \gamma_1)^2}{(1 - \kappa + \gamma_1)^2 + (1 - \kappa - \gamma_1)^2 + 2\gamma_2^2}$$

$$= \frac{4\gamma_1(1 - \kappa)}{2(1 - \kappa)^2 + 2\gamma_1^2 + 2\gamma_2^2}. \tag{10.24}$$

If all the distortions are small, then

$$\epsilon_1 \simeq 2\gamma_1, \tag{10.25}$$

the desired result. A similar equality holds for ϵ_2. By measuring ellipticities of distant galaxies, therefore, we can get an estimate of the shear field, a field which depends manifestly on the underlying gravitational potential via Eq. (10.17).

10.4 WEAK LENSING POWER SPECTRUM

We can now compute the simplest statistics of the shear field, which can be estimated by measuring background galaxy ellipticities. Let's remove the identity from the transformation matrix A,

$$\psi_{ij} \equiv A_{ij} - \delta_{ij}. \tag{10.26}$$

In the absence of inhomogeneities, the apparent angle θ is equal to the source angle θ_S, so $A = I$. Therefore, by removing the identity matrix from A, we have extracted the part describing the distortion of the light ray path due to inhomogeneities. As such, we will refer to ψ as the *distortion* tensor. The last term in (10.17) is ψ_{ij} for a background galaxy (or galaxies) at distance $\chi(z)$ from us. In general, a survey contains a distribution of redshifts. Let's call this distribution $W(\chi)$, just as we did when studying angular correlations in Chapter 9. Again, let's normalize W so that $\int d\chi W(\chi) = 1$. Then, the distortion tensor is

$$\psi_{ij} = 2 \int_0^{\chi_\infty} d\chi W(\chi) \int_0^{\chi} d\chi' \Phi_{,ij}\left(\vec{x}(\chi')\right) \chi' \left(1 - \frac{\chi'}{\chi}\right). \qquad (10.27)$$

We can simplify here by changing orders of integration (almost exactly as depicted in Figure 10.5). Then

$$\psi_{ij}(\vec{\theta}) = \int_0^{\chi_\infty} d\chi \Phi_{,ij}\left(\vec{x}(\chi)\right) g(\chi) \qquad (10.28)$$

where I have dropped the prime and defined

$$g(\chi) \equiv 2\chi \int_\chi^{\chi_\infty} d\chi' \left(1 - \frac{\chi}{\chi'}\right) W(\chi'). \qquad (10.29)$$

On average, each of the components of the distortion tensor is zero: $\langle \psi_{ij} \rangle = 0$. To make our money, therefore, we need to do just what we did for the CMB and galaxy distributions, compute the two-point function, either the angular correlation functions of the different components of ψ_{ij} or their Fourier transforms, the power spectra. To compute these two-point functions, we will be able to essentially copy the results from Section 9.1 as long as we are careful to account for the indices on ψ_{ij}.

To compute the power spectrum of the distortion tensor, $P^{\psi}_{ijkl}(\vec{l})$, let's recall the steps we took when we analyzed the angular correlation function of galaxies (see Table 10.1).

- The distortion tensor in Eq. (10.28) is a function of the 2D vector $\vec{\theta}$ since the argument of the potential is $\vec{x} \simeq \chi(\theta_1, \theta_2, 1)$. As in the case of the galaxy density field, we can Fourier transform ψ_{ij} so that it depends on the 2D vector conjugate to $\vec{\theta}$, $\vec{l}$.
- In the case of the angular galaxy overdensity, we expressed the 2D overdensity as an integral over the 3D overdensity with some weighting function (Eq. (9.3)). Here the situation is identical: g in Eq. (10.29) plays the role of the selection function W there while the 3D field here is not the overdensity δ, but rather $\Phi_{,ij}$.
- Next we found that — in the small-angle limit — the 2D power spectrum is given by an integral over the 3D power spectrum, Eq. (9.10). Here, too, the 2D power spectrum of ψ_{ij} can be expressed as an integral over the 3D power spectrum of the gravitational potential Φ. The only slightly tricky part is computing the 3D power spectrum of $\Phi_{,ij}$. The Fourier transform of $\Phi_{,ij}$ is $-k_i k_j \tilde{\Phi}$ with a variance

$$k_i k_j k_l' k_m' \langle \tilde{\Phi}(\vec{k}) \tilde{\Phi}^*(\vec{k}') \rangle = (2\pi)^3 k_i k_j k_l k_m \delta^3(\vec{k} - \vec{k}') P_\Phi(k). \qquad (10.30)$$

So the 3D power spectrum we need — the one associated with the Fourier transform of $\Phi_{,ij}$ — is $k_i k_j k_l k_m P_\Phi(k)$.

Table 10.1. Similarity between Agular Correlations of Galaxies and Weak Lensing.

	Angular galaxy distribution	Weak lensing
2D observation	$\delta_2(\vec{\theta})$	Distortion tensor $\psi_{ij}(\vec{\theta})$
Weighting function	$W(\chi)$	$g(\chi)$
3D field	δ	$\Phi_{,ij}$
3D power spectrum	$\langle \tilde{\delta}\tilde{\delta}^* \rangle \sim P(k)$	$\langle \tilde{\Phi}_{,ij}\tilde{\Phi}^*_{,lm} \rangle \sim k_i k_j k_l k_m P_\Phi(k)$
2D power spectrum	$P_2(l)$	$P^\psi_{ijlm}(\vec{l})$

I will keep things in terms of P_Φ, but if you are more comfortable with the density power spectrum, you can see from Eq. (7.7) that you need only multiply P_Φ by $9\Omega_m^2 H_0^4 (1+z)^2/(4k^4)$. Applying Eq. (9.10) then leads to

$$\langle \tilde{\psi}_{ij}(\vec{l})\tilde{\psi}_{lm}(\vec{l'}) \rangle = (2\pi)^2 \delta^2(\vec{l}-\vec{l'}) P^\psi_{ijlm}(\vec{l}) \tag{10.31}$$

with the 2D power spectrum

$$P^\psi_{ijlm}(\vec{l}) = \int_0^{\chi_\infty} d\chi \frac{g^2(\chi)}{\chi^2} \frac{l_i l_j l_l l_m}{\chi^4} P_\Phi(l/\chi). \tag{10.32}$$

Equation (10.32) is an expression for the power spectrum of the different components of the distortion tensor. We can turn these into power spectra for the convergence κ and two different components of shear by using Eq. (10.16). Let's work this out for one of the shear components explicitly; the other two are relegated to a problem. Since $\gamma_1 = (\psi_{22} - \psi_{11})/2$, the power spectrum of γ_1 is $1/4$ times $P^\psi_{2222} + P^\psi_{1111} - 2P^\psi_{2211}$. If we decompose the 2D vector $\vec{l}$ into a radial part l and an angle ϕ_l, then $l_1 = l\cos\phi_l$ and $l_2 = l\sin\phi_l$, so

$$P_{\gamma_1}(l, \phi_l) = \left(\sin^4\phi_l + \cos^4\phi_l - 2\sin^2\phi_l\cos^2\phi_l\right) \left[\frac{l^4}{4} \int_0^{\chi_\infty} d\chi \frac{g^2(\chi)}{\chi^6} P_\Phi(l/\chi)\right]. \tag{10.33}$$

Since $\sin^4\phi_l + \cos^4\phi_l + 2\cos^2\phi_l\sin^2\phi_l = 1$, the term in parentheses here is equal to $1 - 4\sin^2\phi_l\cos^2\phi_l = 1 - \sin^2(2\phi_l)$ or $\cos^2(2\phi_l)$. You will show in Exercise 6 that the expression in square brackets is equal to the power spectrum of the convergence,

$$P_\kappa = \frac{l^4}{4} \int_0^{\chi_\infty} d\chi \frac{g^2(\chi)}{\chi^6} P_\Phi(l/\chi). \tag{10.34}$$

Therefore, the power spectrum of γ_1 is

$$P_{\gamma_1}(l, \phi_l) = \cos^2(2\phi_l) \, P_\kappa(l). \tag{10.35}$$

You will show in Exercise 6 that the power spectrum of γ_2 is also proportional to P_κ,

$$P_{\gamma_2}(l, \phi_l) = \sin^2(2\phi_l) \, P_\kappa(l). \tag{10.36}$$

Thus, the power spectra of the two components of shear depend not only on the magnitude of $\vec{l}$ but also on its direction. Figure 10.7 shows the convergence power

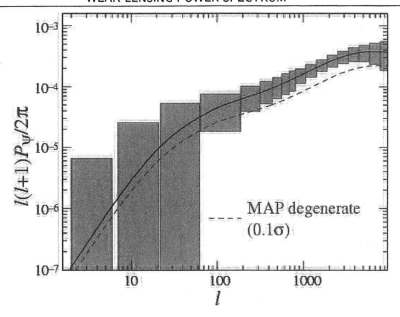

Figure 10.7. The power spectrum of the convergence for two CDM models (Hu and Tegmark, 1999). The models are indistinguishable using CMB data alone, so future weak lensing data (depicted by the error boxes) add invaluable information. The projected error boxes assume a $3°$ survey down to 25th magnitude.

spectrum for two models. Note that, unlike the CMB or even the matter power spectrum, it is essentially featureless.

The power spectra of both shear components are proportional to P_κ, with a prefactor depending on the angle between $\vec{l}$ and a fixed x-axis. You might expect then that a linear combination of the two components depends only on $P_\kappa(l)$ with no dependence on the angle ϕ_l. This is correct. Even more interesting, though, is the possibility that a different linear combination would have a vanishing power spectrum. This also turns out to be correct, and extremely useful. Such a mode has no expected cosmological signal, so any nonzero value is a measure of some systematic effect. By focusing on this "zero" mode, one can identify and eliminate contaminating effects in an experiment. I want to spend some time on this decomposition into two modes — one with signal and one without — not only because of its importance in this case of weak lensing, but also because an exact analogy exists in the case of polarization of the CMB, which we will take up in the next sections. To get ahead of myself, in the case of polarization, the "zero" mode is zero only for scalar perturbations, whereas tensor perturbations do contribute to it. Therefore, we will see in Section 10.9 that this decomposition is a powerful tool with which to detect primordial gravity waves.

Consider then the following two linear combinations of the shear:

$$E(\vec{l}) \equiv \cos(2\phi_l)\gamma_1(\vec{l}) + \sin(2\phi_l)\gamma_2(\vec{l})$$

$$B(\vec{l}) \equiv -\sin(2\phi_l)\gamma_1(\vec{l}) + \cos(2\phi_l)\gamma_2(\vec{l}). \tag{10.37}$$

The power spectrum of each of these modes is easily obtainable in terms of the spectra of γ_1 and γ_2. First the E-mode:

$$P_E = \cos^2(2\phi_l)P_{\gamma_1} + \sin^2(2\phi_l)P_{\gamma_2} + 2\sin(2\phi_l)\cos(2\phi_l)P_{\gamma_1\gamma_2}. \tag{10.38}$$

This expression involves the power spectrum corresponding to $\langle \gamma_1\gamma_2 \rangle$, which is equal to $\cos(2\phi_l)\sin(2\phi_l)$ times the ubiquitous convergence power spectrum. Therefore, P_E is proportional to P_κ with proportionality constant $\cos^4(2\phi_l) + \sin^4(2\phi_l) + 2\cos^2(2\phi_l)\sin^2(2\phi_l) = (\cos^2(2\phi_l) + \sin^2(2\phi_l))^2 = 1$, or

$$P_E = P_\kappa, \tag{10.39}$$

independent of the angle ϕ_l. The calculation for the B-mode is similar:

$$P_B = \sin^2(2\phi_l)P_{\gamma_1} + \cos^2(2\phi_l)P_{\gamma_2} - 2\sin(2\phi_l)\cos(2\phi_l)P_{\gamma_1\gamma_2}$$

$$= 0. \tag{10.40}$$

You can also check that the cross power spectrum $\langle EB \rangle$ vanishes.

The field of weak lensing due to large-scale structure is its infancy. The year 2000 saw the first detections by four independent groups (Van Waerbecke *et al.*, 2000; Bacon, Refregier, and Ellis, 2000; Wittman *et al.*, 2000; Kaiser, Wilson, and Luppino, 2000). They presented the shear correlation function, one example of which is shown in Figure 10.10. We can easily translate the power spectra derived above into angular correlation functions that can be compared with data.

Let's focus on the angular correlation function of γ_1, the Fourier transform of P_{γ_1},

$$w_{\gamma_1}(\vec{\theta}) = \int \frac{d^2l}{(2\pi)^2} e^{i\vec{l}\cdot\vec{\theta}} \cos^2(2\phi_l) \left[\frac{l^4}{4} \int_0^{\chi_\infty} d\chi \frac{g^2(\chi)}{\chi^6} P_\Phi(l/\chi) \right]. \tag{10.41}$$

The variable ϕ_l we are integrating over is the angle between the 2D vector $\vec{l}$ and an arbitrary external x-axis. If we do the angular integral over ϕ_l, then—as you can see from Figure 10.8—the argument of the exponential is quite complicated: $il\theta\cos(\phi_l - \phi)$. Instead, let's integrate over the angle between $\vec{l}$ and $\vec{\theta}$, call it ϕ'. Then,

$$w_{\gamma_1}(\vec{\theta}) = \int_0^\infty \frac{dl}{(2\pi)^2} \frac{l^5}{4} \int_0^{\chi_\infty} d\chi \frac{g^2(\chi)}{\chi^6} P_\Phi(l/\chi) \int_0^{2\pi} d\phi' e^{il\theta\cos\phi'} [\cos(2(\phi' + \phi))]^2. \tag{10.42}$$

The cosine squared in the integrand is equal to

$$\left[\cos(2\phi')\cos(2\phi) - \sin(2\phi')\sin(2\phi) \right]^2$$

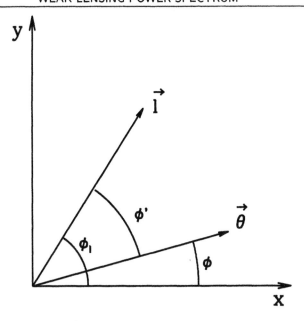

Figure 10.8. Angles made by $\vec{\theta}$ and $\vec{l}$ with respect to an external fixed x-y coordinate system. The angle between $\vec{\theta}$ and $\vec{l}$ is ϕ'.

$$= \cos^2(2\phi')\cos^2(2\phi) - \sin(4\phi')\sin(4\phi)/2 + \sin^2(2\phi')\sin^2(2\phi). \quad (10.43)$$

Thus there are three terms to be integrated over. To do the integral over $\cos^2(2\phi')$, first rewrite it as $(1+\cos(4\phi'))/2$; then recall that the integral of $\cos(n\phi')e^{iz\cos\phi'}$ is equal to $2\pi i^n J_n(z)$ (Eq. (C.21)). Therefore, the integral of $\cos^2(2\phi')$ gives a factor of π times $J_0(l\theta) + J_4(l\theta)$. Using exactly the same arguments, you can see that the integral of $\sin^2(2\phi')$ gives π times $J_0(l\theta) - J_4(l\theta)$. Less obvious is the fact that the integral of $\sin(4\phi')$ vanishes (change integration variable to $\phi'' = \phi' - \pi$ and argue that the integrand is antisymmetric). Therefore,

$$w_{\gamma_1}(\vec{\theta}) = \frac{1}{16\pi}\int_0^\infty dl\; l^5 \int_0^{\chi_\infty} d\chi \frac{g^2(\chi)}{\chi^6} P_\Phi(l/\chi)$$

$$\times \left\{ \cos^2(2\phi)\left[J_0(l\theta) + J_4(l\theta)\right] + \sin^2(2\phi)\left[J_0(l\theta) - J_4(l\theta)\right] \right\}. (10.44)$$

There are many angles floating around, so let me reiterate that $\vec{\theta} = (\theta\cos\phi, \theta\sin\phi)$; that is, ϕ is the angle that $\vec{\theta}$ makes with the x-axis. By changing the l integral into one over 3D wavenumber $k = l/\chi$, we can rewrite this angular correlation function in terms of kernels,

$$w_{\gamma_1}(\vec{\theta}) = \int_0^\infty dk\; k^5 P_\Phi(k) \left[F_+(k\theta)\cos^2(2\phi) + F_-(k\theta)\sin^2(2\phi)\right]. \quad (10.45)$$

Note that here I have assumed that the potential remains constant with time, an assumption which breaks down at late times because of nonlinearities or non-matter domination. The kernels are integrals over radial distance χ modulated by the Bessel functions,

$$F_{\pm}(k\theta) \equiv \frac{1}{16\pi} \int_0^{\chi_\infty} d\chi g^2(\chi) \left[J_0(k\chi\theta) \pm J_4(k\chi\theta) \right]. \qquad (10.46)$$

Figure 10.9 shows these two kernels for background galaxies at redshift $z = 0.9$.

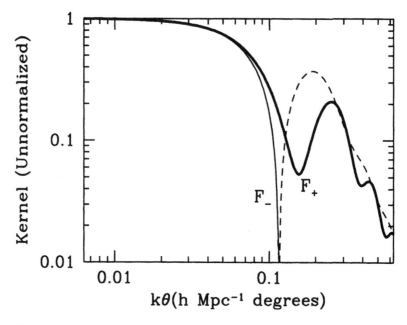

Figure 10.9. The kernels for the shear correlation function assuming all background galaxies are at $z = 0.9$. Dashed region corresponds to negative kernel. If the x-axis is chosen along the line connecting pairs of galaxies, then the $+$ kernel is for $\langle \gamma_1 \gamma_1 \rangle$ and the $-$ for $\langle \gamma_2 \gamma_2 \rangle$. Note that the former is always positive.

If we choose the x-axis to be along the line connecting pairs of galaxies, then we are evaluating the correlation function at $\vec{\theta} = (\theta, 0)$, that is, with $\phi = 0$. In that case, Figure 10.9 shows that w_{γ_1} is always positive, a result we anticipated pictorially in Section 10.1. The correlation function for γ_2, on the other hand, is identical to that in Eq. (10.45) except that $F_{\pm}$ are interchanged. Therefore, w_{γ_2} can, and indeed does, go negative, usually on large angular scales. The final point to take away from the kernels in Figures 10.9 is the rough sense that the shear on a scale of a tenth of degree probes the power spectrum at $k \sim 1\,h\,\text{Mpc}^{-1}$ since this is where the kernel breaks.

Consider then Figure 10.10, which shows results from a survey of three "blank" (i.e., no known clusters of galaxies present) fields over a period of several years. There is a clear detection of ellipticity, presumably due to cosmic shear. The root mean square amplitude of the shear is the square root of the typical numbers on the y-axis, around 0.01. Thus, shear due to large-scale structure has been detected

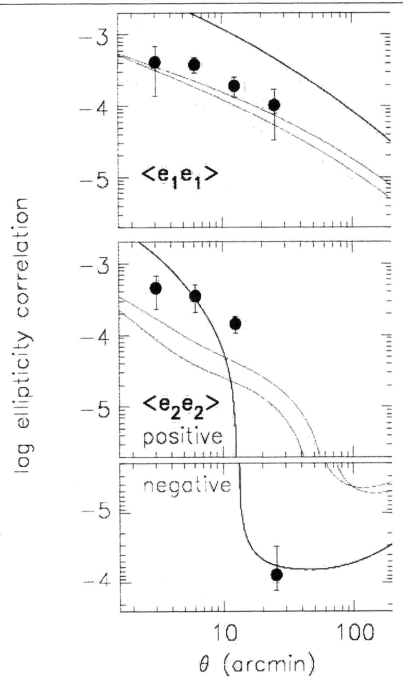

Figure 10.10. Measurement of the shear correlation functions using $145,000$ background galaxies (Wittman *et al.*, 2000). Also shown are a variety of CDM models; topmost in top panel is standard CDM, ruled out here at many sigma. Note that $w_{\gamma_1} = \langle e_1 e_1 \rangle$ remains positive on all angular scales. See color Plate 10.10.

with an amplitudeof order 10^{-2} on angular scales ranging from $1'$ out to about $1°$. Observations planned in the coming years will go far beyond these preliminary results; from these observations we will learn much about the mass distribution of the universe.

10.5 POLARIZATION: THE QUADRUPOLE AND THE Q/U DECOMPOSITION

The radiation in the CMB is expected to be polarized because of Compton scattering at the time of decoupling. A polarization pattern shares a number of mathematical features with the shear induced by gravitational lensing that we have just studied. In addition to these mathematical similarities, they also share similar experimental histories. Whereas anisotropies in the temperature of the CMB and inhomogeneities in the density field were discovered back in the 20th century, weak lensing by large scale structure and polarization of the CMB have just been detected. They are true 21st century phenomena. Therefore, they are both fields rich for study: We are just beginning our observations of them, and they both promise to deliver much cosmological information.

Light traveling in the x-direction corresponds to electric and magnetic fields oscillating in the y-z plane, i.e., transverse to the direction of propagation. If the intensity along the two transverse directions is equal, then the light is unpolarized. Until now, when we have considered the CMB, we have been implicitly studying this case. Now we must account for the possibility that the intensities in the two transverse directions are unequal: that the radiation is polarized.

At first glance, Compton scattering is a perfect mechanism for producing polarized radiation. It allows all transverse radiation to pass through unimpeded, while completely stopping any radiation parallel to the outgoing direction. To see this, consider Figure 10.11 which shows a ray incident from the $+\hat{x}$ direction. This (unpolarized) ray has equal intensity in the $\hat{y}$ and $\hat{z}$ directions. It scatters off an electron at the origin and gets deflected into the $+\hat{z}$ direction.[2] Since the outgoing direction is along the z-axis, none of the (incoming) intensity along the z-axis gets transmitted. By contrast, all of the intensity along the y-axis (which is perpendicular to both the incoming and outgoing directions) is transmitted. The net result is outgoing polarization in the $\hat{y}$ direction.

Obviously, we cannot content ourselves with studying one incoming ray; we must generalize to radiation incident on an electron from all directions. When we do so, we begin to realize that producing polarization will not be quite as easy as it appears from Figure 10.11. Consider first Figure 10.12, which shows a caricature of a much more relevant case: isotropic radiation incident on the electron from all directions. I say "caricature" because I have shown incoming rays from only two directions, the $+\hat{x}$- and $+\hat{y}$-directions. The intensity of the outgoing ray along the

---

[2]Of course radiation gets scattered into all directions with varying probability. Here we consider just one outgoing direction for simplicity. In the next section we account for this probability.

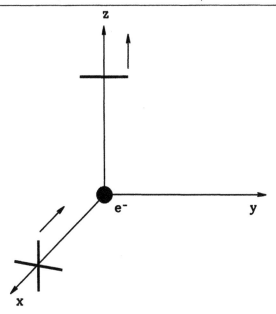

Figure 10.11. Unpolarized radiation moving toward the origin along the x-axis is scattered by an electron into the $+\hat{z}$ direction. Only the $\hat{y}$ component of the radiation remains after scattering. Since there was no incoming $\hat{x}$ polarization, the outgoing radiation is polarized in the $\hat{y}$ direction. (This and the next three figures are adapted from Hu and White, 1997b).

x-axis comes from the radiation incident from the $\hat{y}$ direction, while the outgoing y-intensity comes from the radiation incident from the $\hat{x}$-axis. Since the radiation from both directions has equal intensity (isotropic radiation), though, the outgoing wave is has equal intensity along the $\hat{x}$- and $\hat{y}$-axes: it is unpolarized.

Can anisotropic radiation produce polarization? The simplest example of anisotropy is a dipole pattern, a caricature of which is shown in Figure 10.13. Now the outgoing intensity along the x-axis comes from the $\pm\hat{y}$-incident radiation, which has the average temperature. The outgoing intensity along the y-axis is also neither hot nor cold because it comes from a cold spot (the $-\hat{x}$-direction) and a hot spot (the $+\hat{x}$-direction). The dipole pattern leads therefore only to cancellations and unpolarized outgoing radiation.

To produce polarized radiation, the incoming radiation must have a nonzero quadrupole. Figure 10.14 illustrates the polarization produced by an incoming quadrupole. The hotter (colder) radiation incident from the $\hat{x}$- ($\hat{y}$-) direction produces higher (lower) intensity along the y- (x-) axis for the outgoing wave. Therefore, the intensity of the outgoing wave is greater along the y-axis than along the x-axis: the outgoing radiation is polarized.

The fact that Compton scattering produces polarization only when the incident field has a quadrupole moment has important ramifications for cosmology. We need Compton scattering to produce the polarization, so we need to focus on the epoch

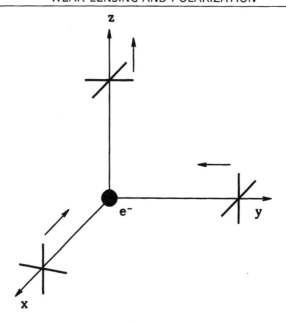

Figure 10.12. Incoming isotropic radiation produces no polarization. Here, since the incoming amplitudes from the $\hat{x}$- and $\hat{y}$-directions are equal, the outgoing intensities along both of these directions are equal, leading to unpolarized radiation.

before electrons and photons have completely decoupled from each other. However, in this epoch electrons and photons are tightly coupled, which we have seen leads to a very small quadrupole. Therefore, we expect polarization from the standard decoupling epoch to be smaller than the anisotropies. Late reionization enhances the polarization at large scales, but does not modify the qualitative conclusion that the polarization signal is expected to be small.

Figure 10.14 depicts polarization in the x-y plane, preferentially in the $\hat{y}$-direction. Alternatively, had the incoming rays been rotated by 45° in the x-y plane, the outgoing polarization would have been along the axis 45° from the x- and y-axes. Polarization therefore can be depicted as a headless vector, with a length corresponding to its magnitude and the orientation of the line describing the axis along which the intensity is greatest. In the 2D plane perpendicular to the direction of propagation, we therefore decompose the intensity into

$$I_{ij} = \begin{pmatrix} T+Q & U \\ U & T-Q \end{pmatrix}. \tag{10.47}$$

The diagonal elements T are the temperature we studied in Chapter 8 (with a uniform part and a perturbation Θ); the two new variables Q and U describe polarization. The pattern in Figure 10.14 has $Q < 0$ and $U = 0$. Note that these definitions of Q and U are identical to the definitions of shear and ellipticity in Section 10.2. Especially relevant is Figure 10.6 where we simply replace ϵ_1 with Q

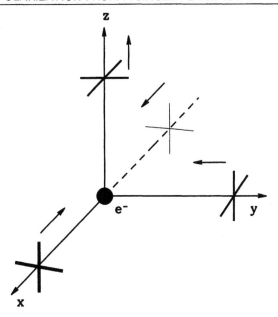

Figure 10.13. Incoming dipole radiation also produces no polarization. (See also color Plate 10.13.) Heavy (thin) lines denote hot (cold) spots. Here the incoming radiation is hotter than average (average is medium thickness) from the $+\hat{x}$-direction, and colder than average from the $-\hat{x}$-direction. The two rays from the $\pm\hat{x}$-directions therefore produce the average intensity for the outgoing ray along the $\hat{y}$-direction. The outgoing intensity along the $\hat{x}$-direction is produced by the ray incident from the $\pm\hat{y}$-directions. Since these have the average intensity, the outgoing intensity is also the average along the $\hat{x}$-direction. The net result is outgoing unpolarized light.

and ϵ_2 with U. A final note: students of electricity and magnetism will no doubt recognize T, Q, and U as three of the four Stokes parameters used to describe polarization. The fourth, V, is nonzero only if polarization is circularly polarized, a phenomenon we do not expect in the early universe, so I have implicitly set $V = 0$ here.

10.6 POLARIZATION FROM A SINGLE PLANE WAVE

The pictures of the previous subsection are important to gain a qualitative understanding of how Compton scattering produces polarization, but they are inefficient tools with which to study the phenomenon quantitatively. The proper tool is the Boltzmann equation. We could proceed now by simply writing down the Boltzmann equation for the Q and U polarization states. In doing so, however, we would lose some of the intuition just gained, so I will take an intermediate tack. We will generalize the discussion of the previous section by summing over all incident rays, not just a handful. This will enable us to make the connection with the distribution Θ

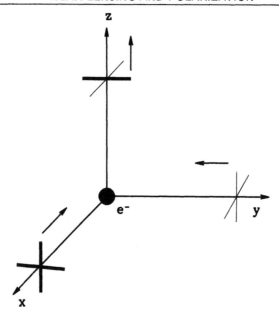

Figure 10.14. Incoming quadrupole radiation produces outgoing polarized light. See also color Plate 10.14.) The outgoing radiation has greater intensity along the y-axis than in the $\hat{x}$-direction. This is a direct result of the hotter radiation incident from the $\hat{x}$-direction.

we have used until now to characterize the photons.

We first need to define the polarization axes in the most general case when the incoming photon arrives from direction $\hat{n}'$. When that direction was $\hat{x}$, as in the previous section, it was clear that polarization was defined as the difference in the intensity along the two perpendicular directions, $\hat{y}$ and $\hat{z}$. In the general case, depicted in Figure 10.15, the direction of the incoming photon is depicted by $\hat{n}'$, and we must integrate over all incoming directions. The two axes perpendicular to this direction are most conveniently taken to be $\hat{\theta}'$ and $\hat{\phi}'$, the standard unit vectors perpendicular to the position vector. These are called $\hat{\epsilon}_1'$ and $\hat{\epsilon}_2'$. We still are interested in the polarization of outgoing photons in the $\hat{z}$-direction, so we can choose the two outgoing polarization axes as $\hat{\epsilon}_1 = \hat{x}$ and $\hat{\epsilon}_2 = \hat{y}$. In short, the incoming polarization vectors are $\hat{\epsilon}_i'$, the outgoing are $\hat{\epsilon}_i$.

The idea that Compton scattering allows the fields transverse to the outgoing direction to pass through unimpeded, while stopping those parallel to the outgoing direction, can be encapsulated by saying that the cross-section for outgoing photons polarized in the $\hat{\epsilon}_i$ direction is proportional to

$$\sum_{j=1}^{2} \left| \hat{\epsilon}_i(\hat{n}) \cdot \hat{\epsilon}_j'(\hat{n}') \right|^2 . \tag{10.48}$$

The Q polarization is the difference between this cross-section for $i = 1$ and $i = 2$, i.e., the difference between the field strength in $\hat{x}$- and $\hat{y}$-directions:

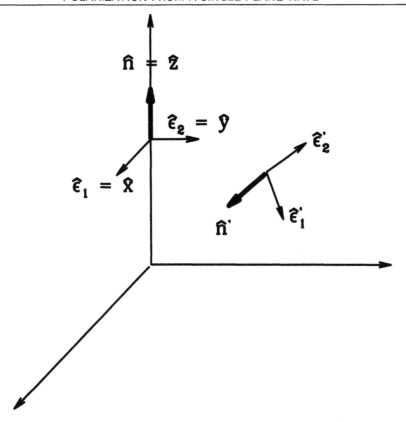

Figure 10.15. Incoming photon from direction $\hat{n}'$ Compton scatters off an electron at the origin producing outgoing photon in direction $\hat{n} = \hat{z}$. The plane perpendicular to the incoming direction is spanned by the two polarization vectors, $\hat{\epsilon}'_1 = \hat{\theta}'$ and $\hat{\epsilon}'_2 = \hat{\phi}'$. The outgoing photon is in the $\hat{z}$ direction, so the polarization vectors are $\hat{\epsilon}_1 = \hat{x}$ and $\hat{\epsilon}_2 = \hat{y}$.

$$\sum_{j=1}^{2} \left| \hat{\epsilon}_1(\hat{n}) \cdot \hat{\epsilon}'_j(\hat{n}') \right|^2 - \sum_{j=1}^{2} \left| \hat{\epsilon}_2(\hat{n}) \cdot \hat{\epsilon}'_j(\hat{n}') \right|^2 = \sum_{j=1}^{2} \left(\left| \hat{x} \cdot \hat{\epsilon}'_j(\hat{n}') \right|^2 - \left| \hat{y} \cdot \hat{\epsilon}'_j(\hat{n}') \right|^2 \right).$$

(10.49)

Integrating over all incoming $\hat{n}'$ directions leads to

$$Q(\hat{z}) = A \int d\Omega_{n'} f(\hat{n}') \sum_{j=1}^{2} \left(\left| \hat{x} \cdot \hat{\epsilon}'_j(\hat{n}') \right|^2 - \left| \hat{y} \cdot \hat{\epsilon}'_j(\hat{n}') \right|^2 \right). \qquad (10.50)$$

Here A is a normalization constant which will not concern us for now, and $f(\hat{n}')$ is the intensity of the radiation incoming from the $\hat{n}'$-direction, and we integrate over all such directions. Note that f depends only on $\hat{n}'$, but not on ϵ'_j: this corresponds to the assumption that the incident radiation is unpolarized.

To take the dot products in Eq. (10.50), we will find it useful to express $\hat{\epsilon}'_1$ and $\hat{\epsilon}'_2$ in terms of their Cartesian coordinates. Since they are equal to $\hat{\theta}'$ and $\hat{\phi}'$, respectively, we have

$$\hat{\epsilon}_1'(\theta', \phi') = (\cos\theta' \cos\phi', \cos\theta' \sin\phi', -\sin\theta')$$

$$\hat{\epsilon}_2'(\theta', \phi') = (-\sin\phi', \cos\phi', 0). \tag{10.51}$$

Now, the dot products become trivial, and we find

$$Q(\hat{z}) = A \int d\Omega_{n'} f(\hat{n}') \left[\cos^2\theta' \cos^2\phi' + \sin^2\phi' - \cos^2\theta' \sin^2\phi' - \cos^2\phi' \right]$$

$$= -A \int d\Omega_{n'} f(\hat{n}') \sin^2\theta' \cos 2\phi'. \tag{10.52}$$

You might recognize the combination of angles here as being propotional to the sum of the spherical harmonics $Y_{2,2} + Y_{2,-2}$ (Eq. (C.10)). Since the spherical harmonics are orthogonal, the integral will pick out the $l = 2, m = \pm 2$ components of the distribution f. That is, nonzero Q will be produced only if the incident radiation has a quadrupole moment. This verifies the argument-by-pictures given in the previous subsection. It is straightforward to derive the corresponding expression for the U-component of polarization (Exercise 10),

$$U(\hat{z}) = -A \int d\Omega_{n'} f(\hat{n}') \sin^2\theta' \sin(2\phi'). \tag{10.53}$$

The combination of sines here is proportional to $Y_{2,2} - Y_{2,-2}$. Again, only an incident quadrupole produces U polarization.

We can now relate the outgoing Q and U to the moments of the incident unpolarized distribution. We'll do this in four steps, in increasing generality.

- First, we'll consider the polariztion induced by a wavevector $\vec{k}$ in the $\hat{x}$-direction.
- Next, we allow $\vec{k}$ to lie anywhere in the $\hat{x}$-$\hat{z}$ plane.
- Then, we consider the most general possible wavevector.
- The first three steps will give us Q and U of the outgoing radiation along the z-axis. We need to generalize this to arbitrary outgoing directions.

The reason that we need to move so slowly is that the photon distribution, $f(\hat{n}')$, takes its cue from the direction of the wavevector. Recall that, in Chapter 4, we wrote the photon distribution as the sum of a zero-order piece — the uniform Planck distribution — and a perturbation, characterized by $\Theta(k, \mu)$ (e.g., Eq. (4.35)). There μ was the dot product of the wave vector $\hat{k}$ and the direction of propagation. Here we have labeled the direction of propagation of the incident photon as $\hat{n}'$, so $\mu = \hat{k} \cdot \hat{n}'$. Thus, $f(\hat{n}')$ in Eq. (10.52) will be an expansion in Legendre polynomials with argument $\hat{k} \cdot \hat{n}'$. This argument is *not* equal to the cosine of θ', since θ' is the angle between the external z-axis and $\hat{n}'$. Relating μ to θ' and ϕ' therefore is not trivial, and we will proceed slowly.

Let's first consider the wave vector $\vec{k}$ to lie in the $\hat{x}$-direction. Then,

$$\mu \equiv \hat{k} \cdot \hat{n}' = \left(\hat{n}' \right)_x$$

$$= \sin \theta' \cos \phi'. \tag{10.54}$$

Recall that we decomposed the perturbation Θ into a sum over Legendre polynomials, so

$$\Theta(k, \hat{k} \cdot \hat{n}') = \sum_l (-i)^l (2l+1) \Theta_l(k) \mathcal{P}_l \left(\hat{k} \cdot \hat{n}' \right)$$

$$\rightarrow -5\Theta_2(k) \mathcal{P}_2 \left(\sin \theta' \cos \phi' \right), \tag{10.55}$$

where the last line follows by substituting our expression for μ (Eq. (10.54)) and considering only the relevant quadrupole part of the sum.

A plane wave with wavevector $\vec{k}$ pointing in the $\hat{x}$-direction therefore has

$$Q(\hat{z}, \vec{k} \parallel \hat{x}) = 5A\Theta_2(k) \int_0^{\pi} d\theta' \sin \theta' \int_0^{2\pi} d\phi' \mathcal{P}_2 \left(\sin \theta' \cos \phi' \right) \sin^2 \theta' \cos 2\phi'. \tag{10.56}$$

Recall that $\mathcal{P}_2(\mu) = (3\mu^2 - 1)/2$. The $-1/2$ part of this gives no contribution to the integral since the ϕ' integral over $\cos(2\phi')$ vanishes. Therefore, we are left with

$$Q(\hat{z}, \vec{k} \parallel \hat{x}) = \frac{15A\Theta_2(k)}{2} \int_0^{\pi} d\theta' \sin^5 \theta' \int_0^{2\pi} d\phi' \cos^2 \phi' \cos 2\phi'. \tag{10.57}$$

The ϕ' integral is $\pi/2$, while the θ' integral — easily done by defining $\mu' = \cos \theta'$ — is $16/15$. So

$$Q(\hat{z}, \vec{k} \parallel \hat{x}) = 4\pi A\Theta_2(k). \tag{10.58}$$

We've now made part of the connection between polarization — represented by Q here — and the formalism of anisotropies — described by Θ in general and Θ_2 specifically for the quadrupole. This expression though applies only in the very simple case that the wavevector points along the x-axis, perpendicular to the line of sight.

Let's generalize this expression to wavevectors pointing in an arbitrary direction in the $\hat{x}$-$\hat{z}$ plane: $\hat{k} = (\sin \theta_k, 0, \cos \theta_k)$. In this case, the factor of $(\hat{k} \cdot \hat{n}')^2$ coming from $\mathcal{P}_2$ is $\sin^2 \theta_k \sin^2 \theta' \cos^2 \phi' + \cos^2 \theta_k \cos^2 \theta'$. The first term is identical to the $\hat{k} \parallel \hat{x}$ case just derived, multiplied by $\sin^2 \theta_k$. The second term introduces no new ϕ' dependence; since the integral over $\cos(2\phi')$ vanishes, it does not contribute. Therefore

$$Q(\hat{z}, \vec{k} \perp \hat{y}) = 4\pi A \sin^2 \theta_k \Theta_2(k). \tag{10.59}$$

In Exercise 10 you will show that there is no U-polarization from this type ($\vec{k}$ in the $\hat{x}$-$\hat{z}$ plane): the polarization is all Q.

For any single plane wave, we can always rotate our coordinate system around the z-axis to ensure that the plane wave lies in the $\hat{x}$-$\hat{z}$ plane, so that Eq. (10.59) applies. When we come to consider the real universe, however, with its super-position of many plane-wave perturbations, we won't have this luxury. Instead, we will need to account for the most general wavevector with orientation $\hat{k} = (\sin\theta_k \cos\phi_k, \sin\theta_k \sin\phi_k, \cos\theta_k)$. For this more general perturbation, you can go through (Exercise 9) exactly the same types of calculations as went into Eq. (10.59) to show that

$$Q(\vec{z}, \vec{k}) = 4\pi A \sin^2\theta_k \cos(2\phi_k)\Theta_2(k)$$

$$U(\vec{z}, \vec{k}) = 4\pi A \sin^2\theta_k \sin(2\phi_k)\Theta_2(k). \tag{10.60}$$

We must make one final generalization. Until now, we have focused solely on the outgoing radiation along the z-axis. Of course, not all outgoing rays will be along the z-axis. (This is what we are looking for: difference in polarization as a function of angle.) To account for arbitrary directions, we need to allow the polar angle θ_k in Eq. (10.60) to be the angle between the observation direction $\hat{n}$ and $\hat{k}$. So $\cos\theta_k \rightarrow \hat{n} \cdot \hat{k}$, and of course $\sin^2\theta_k$ becomes $1 - (\hat{n} \cdot \hat{k})^2$. Therefore, for observations near the z-axis, the outgoing polarization induced by incoming unpolarized incident radiation is

$$Q(\hat{n}, \vec{k}) = 4\pi A \left[1 - \left(\hat{n} \cdot \hat{k} \right)^2 \right] \cos(2\phi_k)\Theta_2(k)$$

$$U(\hat{n}, \vec{k}) = 4\pi A \left[1 - \left(\hat{n} \cdot \hat{k} \right)^2 \right] \sin(2\phi_k)\Theta_2(k). \tag{10.61}$$

These expressions are valid only for directions $\hat{n}$ near the z-axis. This restriction is due to the dependence on the azimuthal angle, ϕ_k. Far from the z-axis, $\cos(2\phi_k)$ and $\sin(2\phi_k)$ give way to much more complicated expressions depending on both $\hat{n}$ and $\hat{k}$. Near the z-axis, though, the relatively simple sine and cosine describe the dependence on azimuthal angle. Thus, we will work in the small angle limit, where all observation directions are close to one another, clustered around the z-axis.

Equations (10.61) allow us to draw polarization patterns around the z-axis for arbitrary $\hat{k}$ modes. Consider the four patterns in Figure 10.16. In each case, the z-axis is out of the page in the center of the panel. For $\vec{k}$ in the $\hat{x}$-$\hat{y}$ plane ($\theta_k = 90°$), Eq. (10.61) says that the strength of the polarization as a function of $\hat{n}_x$ and $\hat{n}_y$ varies as

$$1 - (\hat{n} \cdot \hat{k})^2 = 1 - (\hat{n}_x \hat{k}_x + \hat{n}_y \hat{k}_y)^2. \tag{10.62}$$

That is, deviations from the maximum at $\hat{n}_x = \hat{n}_y = 0$ are small, quadratic in $\hat{n}_x, \hat{n}_y$. The orientation of the polarization in these cases can be either Q (top left panel, $\phi_k = 0$) or U (bottom left panel, $\phi_k = 45°$). For $\vec{k}$ out of the $\hat{x}$-$\hat{y}$ plane, we begin to observe changes in the polarization strength. The two right panels in Figure 10.16 illustrate these changes. Again there can be either Q or U polarization. The most important feature of these patterns is that the polarization strength is always changing in the direction parallel or perpendicular to the sense of the

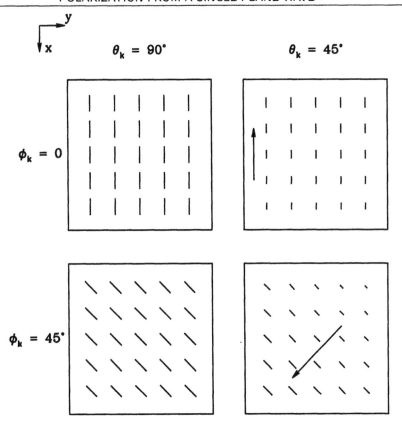

Figure 10.16. Polarization patterns near the z-axis arising from four plane wave perturbations with different $\vec{k}$. E.g., upper left arises from $\vec{k}$ along the x-axis ($\theta_k = 90°, \phi_k = 0$). For $\hat{k}$ in the $\hat{x}$-$\hat{y}$ plane (two left panels with $\theta_k = 90°$), polarization is at a maximum at $\hat{z}$, so little variation is seen. Arrows in right panel show direction in which polarization strength increases. This direction (or the direction perpendicular to it along which polarization remains constant) is aligned with the polarization pattern. This alignment is the hallmark of an E-mode.

polarization. In the top right panel, polarization is aligned with the x-axis, and this is the direction in which the polarization strength is changing. In the bottom right panel, polarization is aligned along $\hat{x}+\hat{y}$, and the change is along the perpendicular direction $\hat{x} - \hat{y}$. We will soon decompose polarization into E and B modes, just as we did the shear pattern in weak lensing. The patterns observed here are all pure E. Indeed, scalar perturbations generate only E modes.

We can also begin to understand the E/B decomposition. The polarization generated by scalar perturbations, the E mode, varies in strength in the same direction as its orientation. This conjures images of an electric field. An electric field from a point source, $\vec{E} = q\hat{r}/r^2$, varies in strength as one moves away from the point source. The electric field is pointed in the same direction: radially away from the source. As one moves in the direction of the field, the strength of the field decreases. In Section 10.9, we will encounter the B mode, and see that — just like a magnetic field — the B mode of polarization varies in strength in a different direction from that in which it is pointing.

10.7 BOLTZMANN SOLUTION

To make quantitative predictions for the polarization expected in the CMB, we must go beyond the treatment of the previous section. There, we sat a single electron at the origin and considered the polarization emerging from incoming radiation with a given distribution. The real problem has lots of electrons coupled to an evolving photon distribution. For this, we need the Boltzmann equation. We wrote down the relevant equations in Chapter 4, although it will take a little bit of work to relate the variable we used there, Θ_P, to Q and U introduced above. The relevant equations ((4.101) and (4.102)) from Chapter 4 are

$$\dot{\Theta}_P + ik\mu\Theta_P = -\dot{\tau}\left[-\Theta_P + \frac{1}{2}\Big(1 - \mathcal{P}_2(\mu)\Big)\Pi\right] \tag{10.63}$$

$$\Pi = \Theta_2 + \Theta_{P2} + \Theta_{P0} \tag{10.64}$$

where $\mu \equiv \hat{k} \cdot \hat{n}$, and Θ_{P0} and Θ_{P2} are the monopole and quadrupole, respectively, of the polarization field.

We are left with the question of the relationship between Θ_P and Q, U. Θ_P is the strength of the polarization, while Q and U together describe both the strength and the orientation. In Chapter 4, we implicitly chose $\hat{k}$ to to lie in the $\hat{x}$-$\hat{z}$ plane in Chapter 4. In that case, we have just seen that the polarization is all Q, so

$$Q(\hat{z}, \vec{k}) = \Theta_P(\hat{z}, \vec{k})$$

$$U(\hat{z}, \vec{k}) = 0 \qquad\qquad\qquad \hat{k} \perp \hat{y}. \tag{10.65}$$

More generally, for arbitrary $\vec{k}$, at least for directions $\hat{n}$ close to $\hat{z}$, we have

$$Q(\vec{k}, \hat{n}) = \Theta_P(\hat{k} \cdot \hat{n}) \cos(2\phi_k)$$

$$U(\vec{k}, \hat{n}) = \Theta_P(\hat{k} \cdot \hat{n}) \sin(2\phi_k). \tag{10.66}$$

Equation (10.66) is a crucial connection between the polarization pattern $Q(\hat{n}), U(\hat{n})$ we are interested in and $\Theta_P(\mu \equiv \hat{k} \cdot \hat{n})$ for which we have Boltzmann equations. Now all we need to do is solve the Boltzmann equations for Θ_P, and then use Eq. (10.66) to construct power spectra for Q and U. We attack the first task in this section and the second in the next.

First, though, to solidify this connection between Θ_P and Q, U, it is instructive to rederive the result of the previous section for an incoming unpolarized wave using the Boltzmann equation. We found there (Eq. (10.59)) that the outgoing polarization (for $\hat{k}$ in the $\hat{x}$-$\hat{z}$ plane) was proportional to $(1 - \mu^2)\Theta_2$ where μ is the cosine of the angle between $\hat{k}$ and $\hat{z}$. Can we get this from the Boltzmann equation? In the absence of any prior polarization, Eq. (10.63) reduces to

$$\dot{\Theta}_P = -\dot{\tau}\frac{1 - \mathcal{P}_2(\mu)}{2}\Pi$$

$$= \frac{-3\dot{\tau}}{2}(1 - \mu^2)\Theta_2. \tag{10.67}$$

Integrating over η, we find that the polarization induced by Compton scattering from incident unpolarized radiation is

$$\Theta_P = \frac{3\tau}{2}(1 - \mu^2)\Theta_2, \tag{10.68}$$

i.e. the optical depth times the quadrupole modulated by the geometric factor $1 - \mu^2$, in agreement with the less formal derivation advanced above. We also see that the strength of the polarization generated is proportional to the optical depth, τ, the integral along the line of sight of the free electron density times the Thomson cross-section.

Now let's solve the Boltzmann equation for the polarization. In analogy to Eq. (8.46), the formal solution to Eq. (10.63) for Θ_P is

$$\Theta_P(\hat{n}, \vec{k}) = \int_0^{\eta_0} d\eta e^{i\vec{k}\cdot\hat{n}(\eta - \eta_0) - \tau(\eta)} S_P(k, \mu, \eta), \tag{10.69}$$

where the source term is

$$S_P(k, \mu, \eta) = -\frac{3}{4}\dot{\tau}\left(1 - \mu^2\right)\Pi. \tag{10.70}$$

Remember that the visibility function is defined as $-\dot{\tau}e^{-\tau}$, so

$$\Theta_P(\hat{n}, \vec{k}) = \frac{3}{4}\left(1 - \mu^2\right)\int_0^{\eta_0} d\eta e^{i\vec{k}\cdot\hat{n}(\eta - \eta_0)}g(\eta)\Pi(k, \eta). \tag{10.71}$$

A reasonable approximation is to assume that we can evaluate the integrand — except for the rapidly changing visibility function — at the time of decoupling (for standard recombination). Then, since the visibility function integrates to unity,

$$\Theta_P(\hat{n}, \vec{k}) \simeq \frac{3\Pi(k, \eta_*)}{4}\left(1 - \mu^2\right)e^{i\vec{k}\cdot\hat{n}(\eta_* - \eta_0)}. \tag{10.72}$$

Neglecting η_* compared with η_0 and rewriting the factors of μ as derivatives leads to

$$\Theta_P(k, \mu) \simeq \frac{3\Pi(k, \eta_*)}{4}\left(1 + \frac{\partial^2}{\partial(k\eta_0)^2}\right)e^{-ik\eta_0\mu}. \tag{10.73}$$

To get the moments Θ_{Pl}, we must multiply Eq. (10.73) by $\mathcal{P}_l(\mu)$ and integrate over all μ as in Eq. (4.99). This gives (Eq. (C.15))

$$\Theta_{Pl}(k) \simeq \frac{3\Pi(k, \eta_*)}{4} \left(1 + \frac{\partial^2}{\partial(k\eta_0)^2}\right) j_l(k\eta_0). \tag{10.74}$$

The sum of the spherical Bessel function and its second derivative can be rewritten using the spherical Bessel equation (C.13) as

$$j_l + j_l'' = -\frac{2j_{l-1}}{k\eta_0} + \frac{2(l+1)}{(k\eta_0)^2} j_l + \frac{l(l+1)}{(k\eta_0)^2} j_l. \tag{10.75}$$

Of the three terms on the right, the last one dominates on small scales. To see this, remember that the spherical Bessel function peaks roughly at $k\eta_0 \sim l$. Physically, this means that anisotropy on an angular scale l is determined by perturbations with wavelength $k^{-1} \sim \eta_0/l$. For our order-of-magnitude estimate, this means that we can take $k\eta_0$ to be of order l in the three terms on the right-hand side. The first is then of order l^{-1}, the second of order l^{-1}, and the last of order $l^2/(k\eta_0)^2 \sim 1$: the last term dominates. Therefore,

$$\Theta_{Pl}(k) \simeq \frac{3\Pi(k, \eta_*)}{4} \frac{l^2}{(k\eta_0)^2} j_l(k\eta_0). \tag{10.76}$$

In the tight coupling limit, we can express Π in terms of the quadrupole, which in turn is related to the dipole. As you can show in Exercise 12, $\Pi = 5\Theta_2/2$. Therefore, the polarization moments today are

$$\Theta_{Pl}(k) \simeq \frac{15\Theta_2(k, \eta_*)}{8} \frac{l^2}{(k\eta_0)^2} j_l(k\eta_0). \tag{10.77}$$

We can go one step further by noting that—in the tightly coupled limit—the quadrupole is proportional to the dipole (Eq. (8.34)). Therefore,

$$\Theta_{Pl}(k) \simeq -\frac{5k\Theta_1(k, \eta_*)}{6\dot{\tau}(\eta_*)} \frac{l^2}{(k\eta_0)^2} j_l(k\eta_0). \tag{10.78}$$

Equation (10.78) is a final expression for the polarization moments today assuming the tightly coupled limit. Three features are worthy of note. First, and most important, the polarization spectrum is seen to be smaller than the anisotropy spectrum by a factor of order $k/\dot{\tau}$ at the time of decoupling. We will quantify this in the next section, but we now understand that it is a direct result of the twin facts that polarization is generated by a quadrupole moment and the quadrupole is suppressed in the early universe due to Compton scattering. Second, we expect there to be oscillations in the polarization power spectrum because $\Theta_{Pl} \propto \Theta_1$, which undergoes acoustic oscillations. More quantitatively, we expect the polarization oscillations, just like the dipole, to be out of phase with the monopole. The peaks and troughs in the temperature anisotropy spectrum, arising primarily from oscillations in the monopole, should then be out of phase with the peaks and troughs in the polarization power spectrum. Finally, there is no analogue here to the integrated Sachs–Wolfe effect which impacts the temperature anisotropy spectrum. Polarization cannot be induced by photons moving through changing gravitational potentials. Therefore, the polarization spectrum today is in some senses a more pristine view of the early universe, uncontaminated by later developments.

10.8 POLARIZATION POWER SPECTRA

Equation (10.78) is an expression for the polarization moments from a single plane wave. In the real universe, we have not just one plane wave, but a superposition of many waves, all with differing amplitudes $\Theta_P(\vec{k}, \hat{n})$. The angular power spectrum from a superposition of plane waves follows from the identical calculation on the temperature anisotropies (Eq. (8.68)):

$$C_{P,l} = \frac{2}{\pi} \int_0^\infty dk \ k^2 \left| \Theta_{Pl}(k) \right|^2 . \tag{10.79}$$

For quite a while, cosmologists computed this power spectrum without reference to Q or U. In 1997, a flurry of papers appeared which derived the power spectra for Q and U. These exploited Eq. (10.66) or large-angle generalizations of it.

Based on our solution for the power spectra of the different components of shear in Section 10.4, we have a sense of what to expect for the power spectra of Q and U. Consider first Figure 10.17. In the small angle limit, Q for example is a function of the 2D vector $\vec{\theta}$, the projection of $\hat{n}$ onto the plane perpendicular to the $\hat{z}$ axis. Thus, we can Fourier transform Q just as we Fourier transformed the shear fields above; its transform will depend on $\vec{l}$, the vector conjugate to $\vec{\theta}$. Based on our experience with weak lensing, we expect the power spectrum of $\tilde{Q}$ to depend not only on the magnitude of $\vec{l}$ but also on its orientation. Looking back at Eq. (10.66), we will not be surprised to find that this power spectrum, $C_{QQ}(\vec{l})$ is proportional to $C_{P,l}$. The proportionality constant is $\cos^2(2\phi_l)$, where ϕ_l is the angle $\vec{l}$ makes with the x-axis. Thus, the factor of $\cos(2\phi_k)$ in Eq. (10.66) becomes $\cos(2\phi_l)$ when we sum over all $\vec{k}$. Similarly, the power spectrum of $\tilde{U}$ is $\sin^2(2\phi_l)C_{P,l}$.

Let's derive this connection between the power spectra of Q and U and that of Θ_P explicitly. We can write the Q polarization as a sum over all plane waves:

$$Q(\vec{\theta}) = \int \frac{d^3k}{(2\pi)^3} \Theta_P(\vec{\theta}, \vec{k}) \cos(2\phi_k) \tag{10.80}$$

The modulating factor $e^{i\vec{x}\cdot\vec{k}}$ is set to 1 here, since we observe from only one position, and we are calling that position $\vec{x} = 0$. To deal with the $\cos(2\phi_k)$ factor, first note that it is equal to $\cos^2 \phi_k - \sin^2 \phi_k$, or in terms of the Cartesian components of $\vec{k}$:

$$\cos(2\phi_k) = \frac{k_x^2 - k_y^2}{k_x^2 + k_y^2}. \tag{10.81}$$

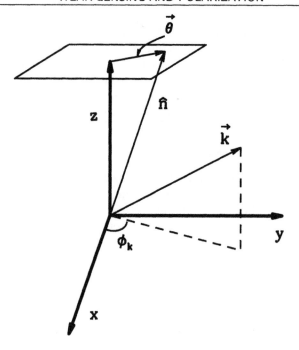

Figure 10.17. Different vectors in polarization. We observe radiation with incoming direction $\hat{n}$, also parameterized by 2D angle θ. Wavevector $\vec{k}$ has an azimuthal angle ϕ_k.

Since Θ_P has the exponential factor $e^{-i\vec{k}\cdot\hat{n}\eta_0}$ (e.g. Eq. (10.69)), and since $\hat{n}_x = \theta_x$ and $\hat{n}_y = \theta_y$, we can rewrite these Cartesian coordinates as derivatives with respect to the $\vec{\theta}$. For example, $k_x \rightarrow [-i\eta_0]^{-1}\partial/\partial\theta_x$. The full $\cos(2\phi_k)$ factor therefore can be written solely as derivatives with respect to $\vec{\theta}$:

$$\cos(2\phi_k) = D_Q(\vec{\theta}) \equiv \left[\frac{\partial^2}{\partial\theta_x^2} + \frac{\partial^2}{\partial\theta_y^2}\right]^{-1}\left(\frac{\partial^2}{\partial\theta_x^2} - \frac{\partial^2}{\partial\theta_y^2}\right). \qquad (10.82)$$

This expression looks formidable, but it is extremely useful for summing up many different k-modes. We can replace Eq. (10.80) with

$$Q(\vec{\theta}) = D_Q(\vec{\theta})\int\frac{d^3k}{(2\pi)^3}\Theta_P(\vec{\theta},\vec{k})$$

$$= D_Q(\vec{\theta})\Theta_P(\vec{\theta}). \qquad (10.83)$$

Both $Q(\vec{\theta})$ and $\Theta_P(\vec{\theta})$ can be written in terms of their Fourier transforms, so that Eq. (10.83) becomes

$$\int\frac{d^2l}{(2\pi)^2}e^{i\vec{l}\cdot\vec{\theta}}Q(\vec{l}) = D_Q(\vec{\theta})\int\frac{d^2l}{(2\pi)^2}e^{i\vec{l}\cdot\vec{\theta}}\Theta_{Pl}. \qquad (10.84)$$

Now D_Q, which is so very complicated in θ-space, becomes very simple, for we know exactly what it looks like when it acts on the exponential $e^{i\vec{l}\cdot\vec{\theta}}$. In that case, it simply becomes $\cos(2\phi_l)$, where ϕ_l is the angle that the 2D vector $\vec{l}$ makes with the x-axis. Therefore, the Fourier transform of $Q(\vec{\theta})$ is

$$\tilde{Q}(\vec{l}) = \tilde{\Theta}_{Pl}\cos(2\phi_l). \tag{10.85}$$

An identical argument says that $U(\vec{l}) = \Theta_{Pl}\sin(2\phi_l)$. Therefore, the power spectra of $\tilde{Q}$ and $\tilde{U}$ are

$$C_{QQ}(\vec{l}) = C_{P,l}\cos^2(2\phi_l)$$

$$C_{UU}(\vec{l}) = C_{P,l}\sin^2(2\phi_l). \tag{10.86}$$

Recall that in the case of weak lensing, we noticed that one could take linear combinations of γ_1 and γ_2 such that the power spectrum of one of the linear combinations vanishes (Eq. (10.40)), while the other is equal to the convergence power spectrum (Eq. (10.39)). Here we can do exactly the same thing. If we define

$$E(\vec{l}) \equiv \tilde{Q}(\vec{l})\cos(2\phi_l) + \tilde{U}(\vec{l})\sin(2\phi_l)$$

$$B(\vec{l}) \equiv -\tilde{Q}(\vec{l})\sin(2\phi_l) + \tilde{U}(\vec{l})\cos(2\phi_l) \tag{10.87}$$

then

$$C_{BB}(\vec{l}) = 0. \tag{10.88}$$

In the small-scale limit, the power in the E-mode is precisely equal to C_P:

$$\lim_{l \gg 1} C_{EE}(\vec{l}) = C_{P,l}. \tag{10.89}$$

Figure 10.18 shows the resultant power spectrum, both the exact numerical result and the approximation of Eq. (10.78) integrated over all modes as dictated by Eq. (10.79). Also shown is the spectrum of temperature anisotropies from Chapter 8. As expected it is higher in amplitude, since polarization is suppressed in the tightly coupled limit. Also as anticipated, the oscillations in the polarization spectrum are out of phase with those in the temperature spectrum. In 2002, the DASI experiment announced the first detection of polarization, a detection shown in Figure 10.18. This is akin to the first detection of shear by large scale structure, the beginning of our journey down promising new paths in cosmology.

The spectrum in Figure 10.18 is shown only on small scales; on larger scales, the treatment of this section needs to be modified (Kamionkowski, Kosowsky, and Stebbins, 1997a; Seljak and Zaldarriaga, 1997). The resultant spectrum has no surprises: it falls off very rapidly on large scales. Since the polarization is proportional to the dipole, which vanishes for large-scale modes, we could have anticipated this result as well. Although I won't go into the technical details of this large-angle

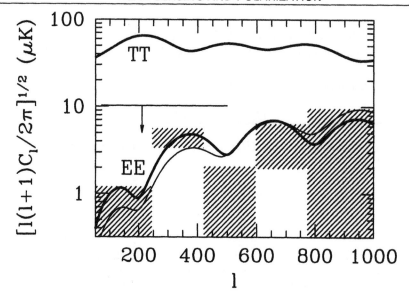

Figure 10.18. Power spectra of temperature and E-mode polarization for the standard CDM model. Thick curves show exact results; thin curve is the tight coupling approximation of Eq. (10.78). Only scalar perturbations have been assumed, so there is no power in the B-mode. Straight line at 10μK is an upper limit from Hedman *et al.*, 2002, while the hatched boxes are the first detection by the DASI experiment (Kovac *et al.*, 2002).

result, the basic idea is that instead of expanding polarization in terms of Legendre polynomials, or ordinary spherical harmonics, one must use tensor spherical harmonics.

One final comment: we have been implicitly assuming until now that the perturbations of interest are scalar. We inserted this assumption early on by writing the plane-wave perturbation as Eq. (10.55). If the perturbations were tensor, the decomposition would have included an azimuthal dependence; recall Eq. (4.115).

10.9 DETECTING GRAVITY WAVES

There is a fundamental difference between the scalar perturbations we have considered in the previous sections and tensor perturbations. A scalar plane-wave perturbation has one direction associated with it: the direction of the wavevector $\vec{k}$. Once this direction is specified, all photon moments depend only on the angle between the incoming photon and the wavevector. Once this angle is specified, there is an azimuthal symmetry about the $\hat{k}$ direction. This rotational symmetry is the reason that only the E mode is produced by scalar perturbations. There are two directions in a polarization field: (i) the direction in which the polarization strength is changing and (ii) the orientation of the polarization. For scalar perturbations, we saw in Figure 10.16 that these directions must be aligned (or perpendicular to each other).

Intuitively, each direction looks to the only vector it knows — $\vec{k}$ — for guidance, and they each arrive at the same end. This alignment is the salient characteristic of the E mode.

The photon distribution from tensor perturbations is not rotationally symmetric about the $\vec{k}$-direction. Gravity waves are pulsations in the metric; these induce an azimuthal dependence to the photon distribution. Recall from Eq. (4.115) that the resultant distribution varies as $\sin(2\phi)$ or $\cos(2\phi)$, where ϕ is the azimuthal angle about the $\vec{k}$-axis. This dependence on ϕ means that there is an additional direction to choose from when the polarization field gets induced. We might expect then that the orientation of the polarization will not necessarily be aligned with the direction of changing polarization strength. That is, we might expect that gravity waves will produce B-mode polarization. This is exactly what we will show in this section.

Before working through the algebra, we need to pause to understand the importance of the B-mode generated by tensor perturbations. Let's start with the difficulty of detecting tensors through the E-mode. Both scalars and tensors contribute to the E-mode, so the only way to disentangle them is to take advantage of differences in their spectra as a function of l. We saw in the case of temperature anisotropies that this is a tricky game, though, for other parameters can change spectra in ways similar to tensors. So even if we had perfect knowledge of the $C_{E,l}$ spectrum (no noise), we would still not necessarily know whether tensors were present. The B-mode is different. There is no contamination from scalar perturbations, so if we observe a B-mode in polarization, we know that it comes from gravity waves. In principle, this realization has unlimited power: no matter how small the tensor signal from inflation (no matter how small $H/m_{\rm Pl}$), we can ultimately detect this signal by searching for a B-mode. In practice, there are contaminants due to nonlinearities, but these are quite small. Estimates (Knox and Song, 2002; Kesden, Cooray, and Kamionkowski, 2002) suggest that the lowest obtainable limit on r, the tensor-to-scalar ratio, is of order 10^{-4}.

Let's compute the polarization pattern from a single plane wave generated by tensor perturbations. This problem is identical to that treated in Section 10.6. To find the outgoing polarization near the z-axis, we need to integrate over the incoming photon distribution. As in Eqs. (10.52) and (10.53), we want

$$\begin{pmatrix} Q \\ U \end{pmatrix} \propto - \int d\Omega' \Theta^T(\Omega') \sin^2\theta' \begin{pmatrix} \cos(2\phi') \\ \sin(2\phi') \end{pmatrix}$$

$$\propto - \int d\Omega' \Theta^T(\Omega') \begin{pmatrix} Y_{2,2}(\Omega') + Y_{2,-2}(\Omega') \\ \frac{1}{i}[Y_{2,2}(\Omega') - Y_{2,-2}(\Omega')] \end{pmatrix}, \qquad (10.90)$$

where I have inserted the photon distribution due to tensor perturbations, Θ^T; recognized the combination of $\sin^2\theta'$ and the azimuthal dependence as $Y_{2,2} \pm Y_{2,-2}$; and neglected the absolute normalization of the polarization.

To complete the calculation, we need to find the angular dependence of Θ^T. This is a bit more difficult than one might expect. Although we know that this angular dependence is $\sin^2\theta' \cos(2\phi')$ (for h_+) or $\sin^2\theta' \sin(2\phi')$ ($h_\times$) for $\vec{k}$ lying along the

z-axis, we need the dependence for a general wavevector $\vec{k}$. One way of finding this dependence is to rotate the coordinate system so that a unit vector pointing in the $\hat{z}$-direction gets rotated so that it points in the $\hat{k}$-direction. The relevant rotation matrix is

$$R = \begin{pmatrix} \cos\theta_k \cos\phi_k & -\sin\phi_k & \sin\theta_k \cos\phi_k \\ \cos\theta_k \sin\phi_k & \cos\phi_k & \sin\theta_k \sin\phi_k \\ -\sin\theta_k & 0 & \cos\theta_k \end{pmatrix}. \tag{10.91}$$

You should verify that R really does take $\hat{z} \to \hat{k}$ and work through a simple derivation of R (Exercise 13). We want to know what R does to Θ^T. To be concrete, let's focus on $h_\times$, so that $\Theta^T \propto \sin^2\theta' \sin(2\phi')$. First, we can reexpress this angular dependence in terms of the unit vector $\hat{n}'$ describing the direction of the incident photon:

$$\sin^2\theta' \sin(2\phi') = 2\sin^2\theta' \sin\phi' \cos\phi'$$

$$= 2\hat{n}'_x \hat{n}'_y. \tag{10.92}$$

Now let's rotate the coordinate system so that the z-axis points in the direction of $\vec{k}$. The anisotropies due to the $h_\times$-mode used to be proportional to $\hat{n}'_x \hat{n}'_y$. In the new coordinate system, they become

$$\Theta^T \propto (R^t \hat{n}')_x (R^t \hat{n}')_y \tag{10.93}$$

where ^{t} denotes transpose.

Now we work through the matrix multiplication and find

$$\Theta^T \propto (\cos\theta_k \cos\phi_k \sin\theta' \cos\phi' + \cos\theta_k \sin\phi_k \sin\theta' \sin\phi' - \sin\theta_k \cos\theta')$$

$$\times (-\sin\theta' \sin\phi_k \cos\phi' + \cos\phi_k \sin\theta' \sin\phi')$$

$$= \sin\theta' \sin(\phi' - \phi_k) \left(\cos\theta_k \sin\theta' \cos(\phi' - \phi_k) - \sin\theta_k \cos\theta' \right)$$

$$= \frac{1}{2} \cos\theta_k \sin^2\theta' \sin\left[2(\phi' - \phi_k) \right] - \sin\theta' \cos\theta' \sin\theta_k \sin(\phi' - \phi_k). \tag{10.94}$$

This last combination can be reexpressed in terms of spherical harmonics: it is a linear combination of $Y_{2,\pm 2}, Y_{2,\pm 1}$, and $Y_{2,0}$. That is, the anisotropy pattern about the wavevector $\vec{k}$ due to gravity waves (the $h_\times$ mode) has a $Y_{2,2} - Y_{2,-2}$ dependence when $\hat{k}$ is along the z-axis. When $\hat{k}$ is general, this dependence gets mixed up among all the $Y_{2,m}$'s. We are interested in the polarization pattern generated by Θ^T; from Eq. (10.90) and the orthogonality property of the spherical harmonics, this means we are interested only in the $Y_{2,\pm 2}$ components of Θ^T. We can now extract these from Eq. (10.94). The last term on the right has a factor of $\sin(\phi' - \phi_k)$, so it is proportional to $Y_{2,\pm 1}$, and we can neglect it. The first term is

$$\frac{1}{2} \cos\theta_k \sin^2\theta' \sin\left[2(\phi' - \phi_k) \right] = -\sqrt{\frac{32\pi}{15}} \frac{\cos\theta_k}{4i} \left[e^{-2i\phi_k} Y_{2,-2}^*(\Omega') - e^{2i\phi_k} Y_{2,2}^*(\Omega') \right]. \tag{10.95}$$

To find Q and U, we dot this into $Y_{2,2} \pm Y_{2,-2}$, so that the integral in Eq. (10.90) leads to

$$\begin{pmatrix} Q(\hat{z}) \\ U(\hat{z}) \end{pmatrix} \propto \Theta_0^T \cos\theta_k \begin{pmatrix} -\sin(2\phi_k) \\ \cos(2\phi_k) \end{pmatrix}. \qquad (10.96)$$

If we move small angles away from the $\hat{z}$-direction, then the azimuthal dependence does not change, and $\cos\theta_k \to \hat{n} \cdot \hat{k}$, so that

$$\begin{pmatrix} Q(\hat{n}) \\ U(\hat{n}) \end{pmatrix} \propto \Theta_0^T \hat{n} \cdot \hat{k} \begin{pmatrix} -\sin(2\phi_k) \\ \cos(2\phi_k) \end{pmatrix}. \qquad (10.97)$$

The polarization pattern described by Eq. (10.97) has a nonzero B-mode. To see this, first consider the definition of B in Eq. (10.87). This definition is in Fourier space, but we remember that using the operator $D_{Q,U}$, we can replace $\vec{l}$ with $\vec{k}$, the wavevector. Therefore, for $\vec{k}$ in the $\hat{x}$-$\hat{z}$ plane ($\phi_k = 0$), the B-mode corresponds to only U polarization. Indeed this is precisely what Eq. (10.97) says is produced by the $h_\times$ mode of gravity waves. So the anisotropies due to gravity waves do produce the B-mode of polarization.

Figure 10.19 shows the polarization patterns due to a single plane wave $h_\times \propto e^{i\vec{k}\cdot\vec{x}}$ for four different wavevectors $\vec{k}$. For example, the top left panel considers $\vec{k}$ lying along the x-axis. In that case, since $\phi_k = 0$, Eq. (10.97) says that the polarization is all U and that the strength scales as

$$\hat{n} \cdot \hat{k} = \hat{n}_x. \qquad (10.98)$$

The strength of the polarization therefore increases as one moves away from the y-axis. The important feature of this pattern, which characterizes the B-mode, is that the strength of the polarization varies in the $\hat{x}$-direction, while the orientation is in the $\hat{x} \pm \hat{y}$ direction. These two directions (varying strength and polarization orientation) are *not* aligned or perpendicular to each other. The other panels show the same feature.

Figure 10.20 shows the anisotropy spectrum in a standard CDM model with an equal amount of tensors and scalars. The T and E spectra are similar to the tensorless case depicted in Figure 10.18. With tensors, the B spectrum is now nonzero, albeit small. Studies suggest that polarization searches will help significantly in the quest to detect small r.

SUGGESTED READING

Gravitational lensing is described in exquisite detail in *Gravitational Lenses* (Schneider, Ehlers, and Falco). An excellent, comprehensive review of weak lensing is in Bartelmann and Schneider (2001). Electromagnetic polarization is a textbook subject, covered in, for example, *Classical Electrodynamics* (Jackson) and *Radiative Processes in Astrophysics* (Rybicki and Lightman). Initial papers on gravitational lensing by large scale structure include Blandford *et al.* (1991), Miralda-Escude

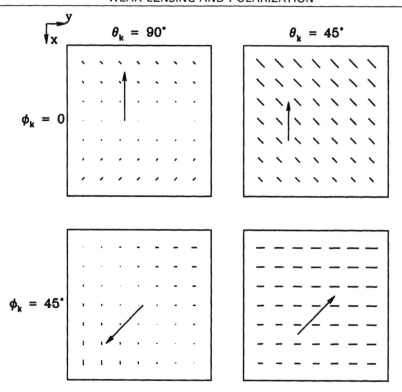

Figure 10.19. Polarization patterns from a single plane wave $h_\times(\vec{k})$ in a plane perpendicular to the z-axis ($\hat{z}$ out of the paper). Patterns from four different $\vec{k}$ are shown. Arrows depict direction of increasing polarization strength. This direction is *not* aligned with the orientation of the polarization.

(1991), and Kaiser (1992). Recent theoretical work connecting lensing observations to cosmological parameters includes Jain and Seljak (1997); Bernardeau, van Waerbecke, and Mellier (1997); and Hu and Tegmark (1999). The onset of the new millenium saw the first detections of lensing by large scale structure in van Waerbecke *et al.* (2000); Wittman *et al.* (2000); Bacon, Refrgier, and Ellis (2000); Kaiser, Wilson and Luppino (2000); and Maoli *et al.* (2001). Active work continues. The future will undoubtedly bring observations of weak lensing on large fields. Two proposal for such observations are the SuperNova Acceleration Probe (SNAP; http://snap.lbl.gov), which presently plans to devote $\sim 20\%$ of its time to weak lensing, and the Large Scale Synaptic Telescope (LSST; http://www.dmtelescope.org).

Polarization of the CMB was studied in the seminal papers of the 1980s by Bond and Efstathiou (1984) and Polnarev (1985). Kosowsky's thesis (1996) is a lucid Boltzmann-esque discussion of this work. The first papers to recognize the importance of the E/B decomposition were Stebbins (1996); Seljak (1997);

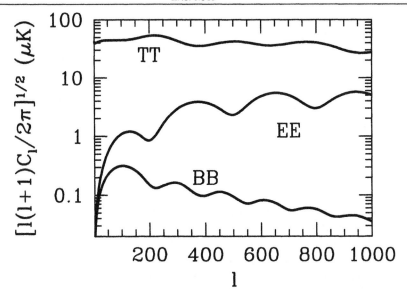

Figure 10.20. Anisotropy spectrum from a standard CDM model with equal amounts of scalar and tensor perturbations ($r = 1$). The T and E spectra come from both scalars and tensors, whereas the B-mode is due solely to tensors.

Kamionkowski, Kosowsky, and Stebbins (1997a,b), and Zaldarriaga and Seljak (1997). I've followed the treatment of Seljak who worked in the small angle limit because the algebra is simpler. The review article of Hu and White (1997b) is perhaps the most accessible introduction into the recent literature on polarization of the CMB, but it is a difficult subject. Even this lucid review with its illuminating pictures requires a lot of effort to understand.

DASI (Kovac *et al.*, 2002) detected polarization at the 5-sigma level. Previous stringent upper limits, which are still valuable on large scales, were obtained by Hedman *et al.* (2000) and Keating *et al.* (2001).

EXERCISES

Exercise 1. The probability that there will be a galaxy massive enough to act as a lens between a quasar at redshift z and us is roughly proportional to the volume between us. Compute

$$V(z) \equiv \int_{x < \chi(z)} d^3x. \qquad (10.99)$$

The integral is trivial, but the dependence on z is not. Numerically compute $V(z)$ in a flat universe with cosmological constant. Plot $V(z)$ vs Ω_Λ for $z = 2, 3, 4$. If the galaxy density does not depend on cosmology, then the expected number of lenses scales simply as this volume. For $z = 3$, what is the ratio of lenses expected in a flat universe with $\Omega_\Lambda = 0.7$ as compared with a flat, matter-dominated universe?

Be warned that Keeton (2002), among others, has made a strong case that the expected lensing frequency does *not* vary this dramatically with cosmology because of differences in the galaxy densities in the different models.

Exercise 2. Compute the magnification of an image in terms of the convergence κ and shear γ_1 and γ_2. Show that, in the limit of weak fields, the magnification μ is related to the convergence via

$$\mu \simeq 1 + 2\kappa. \tag{10.100}$$

Exercise 3. It is often useful to write observable properties of lenses — such as deflection angles and shear — in terms of a *projected potential* ϕ.
(a) Using Eq. (10.14), determine the ϕ such that $\vec{\theta}_S = \vec{\theta} + \nabla\phi$ where ∇ is the gradient with respect to the 2D angular variable $\vec{\theta}$.
(b) Express the transformation matrix defined in Eq. (10.15) in terms of the projected potential.

Exercise 4. When the lens is at a fixed redshift (e.g., a single galaxy or a cluster as opposed to large-scale structure in general) z_L corresponding to comoving distance χ_L from us, show that the projected potential ϕ from the previous problem reduces to

$$\phi(\vec{\theta}; z_L) = 4G\frac{\chi_S - \chi_L}{\chi_S\chi_L} \int d^2R\,\Sigma(\vec{R}) \ln\left|\vec{R} - \chi_L\vec{\theta}\right|. \tag{10.101}$$

Here χ_S is the comoving distance out to the source; $\vec{R}$ is the radius in the plane perpendicular to the line of sight; and $\Sigma(\vec{R})$ is the projected surface density in this plane.

Exercise 5. Compute the observed component of ellipticity ϵ_2 from an intrinsically circular source; express it in terms of the components of the transformation matrix, κ, γ_1, and, most importantly, γ_2.

Exercise 6. (a) Show that the power spectrum of the convergence is given by Eq. (10.35). Show that the power spectrum of γ_2 is given by Eq. (10.36).
(b) Using CMBFAST or the BBKS transfer function, compute numerically P_κ for standard CDM with $\Omega_m = 1, h = 0.5, n = 1$. Assume all background galaxies are at redshift $z = 1$. At what l do you expect your result — based on the linear power spectrum — to lose validity due to nonlinear effects?

Exercise 7. Equation (10.34) gives the power spectrum of the convergence in the small angle limit ($l \gg 1$). The more general expression is (Stebbins, 1996, the extra factor of $(2\pi)^3$ here due to differing power spectrum conventions)

$$C_{\kappa,l} = 4\pi l^2(l+1)^2 \int \frac{d^3k}{(2\pi)^3}\, k^2 I_l^2(k\chi)P_\Phi(k) \tag{10.102}$$

when all background galaxies are at comoving distance χ and

$$I_l(x) \equiv \int_0^1 \frac{dy}{y}(1-y)j_l(xy). \tag{10.103}$$

(a) Verify, either analytically or numerically, that in the small-angle limit Eq. (10.102) reduces to the expression for $P_\kappa(l)$ in Eq. (10.34).
(b) Redo the calculation of the convergence power spectrum for the sCDM model of Exercise 6, this time using the general expression in Eq. (10.102).

Exercise 8. In the text, we computed the angular correlations of galaxies (Chapter 9) and the weak lensing correlation function (this chapter). One can also compute the cross-correlation, which measures how correlated the galaxies are with the mass. One way to measure this is to separate a galaxy sample into foreground and background galaxies and measure the cross-correlation between the two samples. Since they are separated by such large distances, the only possible correlation arises because the background galaxies have been magnified by the foreground mass. This problem allows you to work out the background/foreground correlation function (e.g., Moessner and Jain, 1998). Incidentally, this cross-correlation function can also be measured by the QSO/galaxy correlation function. Suppose the 2D overdensity of foreground galaxies is due solely to intrinsic inhomogenities, so that it is given by Eq. (9.3). Assume that the 2D overdensity of background galaxies arises only from magnification. That is, galaxies that should not be included in the survey because they are intrinsically fainter than the magnitude limit are magnified and so appear brighter, thereby making the cut. If the magnification is μ, then the number of background galaxies in an angular patch is

$$n_b = \bar{n}_b \mu^{2.5s-1}. \tag{10.104}$$

Here $\bar{n}_b$ is the average number of background galaxies, and s is defined as $d\log N(m)/dm$ where $N(m)$ is the number of galaxies at the magnitude limit m. For the present problem, don't worry about where this relation comes from (see Broadhurst, Taylor, and Peacock, 1995, for an explanation).
(a) Express δ_b, the background overdensity, in terms of κ and s using Eq. (10.100).
(b) Find an expression for the convergence $\kappa(\vec{\theta})$ in terms of the mass overdensity. First express it in terms of the relevant components of transformation matrix A of Eq. (10.17), but then eliminate the potential there in favor of the density field δ.
(c) Using these two expressions — Eq. (9.3) and your answer in **(b)** — for the foreground and background overdensities, compute the angular cross-correlation function $w_{bg}(\theta) \equiv \langle \delta_b(\theta)\delta_g(0)\rangle$.

Exercise 9. As the wavevector $\vec{k}$ moves out of the $\hat{x}$-$\hat{z}$ plane, show that the Q-polarization (for outgoing radiation in the $\hat{z}$-direction) changes as $\cos(2\phi_k)$. To do this, first compute $\hat{k}\cdot\hat{n}'$, and then integrate $\mathcal{P}_2(\hat{k}\cdot\hat{n}')$ over solid angle, with the weighting factor $\sin^2\theta'\cos(2\phi')$ derived in Eq. (10.52).

Exercise 10. This problem focuses on the U-component of polarization.

(a) We showed that the Q-component of polarization from unpolarized incident radiation is given by Eq. (10.52), which stems from Eq. (10.50). The Q-component thus depends on the difference between $|\hat{\epsilon}_i \cdot \hat{x}|^2$ and $|\hat{\epsilon}_i \cdot \hat{y}|^2$. For the U-component, $\hat{x}$ and $\hat{y}$ must be replaced by unit vectors rotated $45°$, i.e., $(\hat{x}+\hat{y})/\sqrt{2}$ and $(\hat{x}-\hat{y})/\sqrt{2}$. With this replacement, derive Eq. (10.53).

(b) Show that a plane-wave perturbation with wavevector $\vec{k}$ lying in the $\hat{x}$-$\hat{z}$ plane does not produce any U-polarization in the outgoing $\hat{z}$-direction.

(c) For the most general orientation of the wavevector, $\hat{k} = (\sin\theta\cos\phi, \sin\theta\sin\phi, \cos\theta)$, show that U-polarization is given by Eq. (10.60).

Exercise 11. Draw the polarization patterns near the z-axis arising from a plane-wave scalar perturbation with (a) $\theta_k = \pi/8, \phi_k = \pi/8$; (b) $\theta_k = 3\pi/4, \phi_k = \pi/4$; (c) $\theta_k = 3\pi/4, \phi_k = 0$; and (d) $\theta_k = 3\pi/2, \phi_k = 0$. In each case, show that the sense of polarization is aligned with (or perpendicular to) the direction in which the polarization strength is changing.

Exercise 12. In the tight coupling limit, find an expression for $\Pi \equiv \Theta_2 + \Theta_{P2} + \Theta_{P0}$.

(a) When $\dot{\tau}$ is very large, the terms multiplying it on the right hand side of Eq. (10.63) must cancel. Write down this equality for $\Theta_P(\mu)$ in terms of the moments, Θ_2, Θ_{P2}, and Θ_{P0}.

(b) Expand $\Theta_P(\mu)$ in terms of Legendre polynomials, keeping only the monopole and the quadupole. Then equate the coefficients of $\mathcal{P}_0$ and $\mathcal{P}_2$.

(c) This leads to two equations for three unknowns. Show that solving for the two polarization moments in terms of the temperature quadrupole gives $\Theta_{P0} = 5\Theta_2/4$ and $\Theta_{P2} = \Theta_2/4$.

(d) Use the results of (c) to determine Π in terms of Θ_2.

Exercise 13. This problem concerns the rotation matrix R given in Eq. (10.91).

(a) Act with R on the unit vector $(0,0,1)$ and show that it gets transformed into $\hat{k}$.

(b) Derive R. One way to do this is to first rotate the x-y-z frame about the z-axis by an angle ϕ'. Then, rotate about the y-axis by an angle $-\theta'$. The product of these two rotations is R.

Exercise 14. In the text we considered polarization patterns from a single plane-wave perturbation due to gravity waves. There are actually two such orientations. We considered only $h_\times$. In this problem, consider polarization from h_+. In a frame in which $\hat{k}$ is along the z-axis the anisotropies have a $\sin^2\theta\cos(2\phi)$ dependence.

(a) Find the dependence of Θ^T on angle in the more general frame in which $\hat{k}$ does not lie along the z-axis.

(b) Determine Q and U from this incoming distribution.

(c) Plot the anisotropy pattern near the outgoing $\hat{z}$-direction for the four sets of θ_k, ϕ_k shown in Figure 10.19.

Exercise 15. Find expressions for the cross-correlation spectra between the temperature anisotropy and polarization anisotropy $C_{TQ}(\vec{l})$ and $C_{TU}(\vec{l})$ in terms of Θ_l and Θ_{Pl}. Assume scalar perturbations only. Express $C_{TE,l}$ in terms of these.

11

ANALYSIS

Increasingly, theorists and even busy experimentalists are turning their attention to the fundamental question of how best to analyze a set of data. The main reason for this focus is that the quality and quantity of data have improved dramatically over the past decade. There is every reason to believe that this trend will continue. Anisotropies in the temperature of the CMB have been measured by dozens of experiments already. The satellites MAP and Planck will take these measurements to the next stage, but there is no reason to think this will be the last stage. There are still polarization and very small scale anisotropies to be measured. The power spectrum of matter is probed in a variety of ways; activity here, too, shows no sign of letting up. After the completion of the Sloan Digital Sky Survey and 2DF, the two largest redshift surveys to date, surveyors have begun planning large weak lensing missions and even deeper galaxy surveys. These larger data sets create new challenges in analysis.

A wonderful/disturbing example of these challenges was given recently by Julian Borrill. He used scaling arguments to show that a brute-force algorithm for making a map from the raw data of the Boomerang CMB anisotropy experiment would take 12 years to run on current computers! Already, data sets are far too large for brute-force calculations. And things are rapidly getting more dire. Since typically the number of arithmetic operations scales as the number of pixels cubed, and since MAP and Planck will have of order 10 to 100 times more data than Boomerang, the situation cries out for creative solutions.

Another reason for the recent focus on analysis is one I hope to convey in this chapter: analysis is exciting. The techniques that have been proposed to deal with the complexity of forthcoming data sets are beautiful. The elegance of these techniques is of course enhanced by their importance. But the elegance is there; for this reason alone, it is well worth working through some of these recent advances.

11.1 THE LIKELIHOOD FUNCTION

The basic building block of contemporary analysis is the likelihood function. This is defined as the probability that a given experiment would get the data it did given a theory. This seemingly simple definition is exceedingly powerful. Once we have the likelihood function, with a caveat or two, we can determine the parameters of the theory (best estimate is the place in parameter space where the likelihood function is largest) along with errors (determined by the width of the likelihood function). We start with a simple example and move on to the likelihood function for the CMB and then a galaxy survey.

11.1.1 Simple Example

Suppose you want to weigh somebody. Since you are a scientist, you know that, in addition to the measurement, you should also report an uncertainty. So you set up 100 different scales and record the person's weight on each of these different scales. Given these 100 numbers, what value should you report for the weight and the uncertainty in the weight? We all know the answer to this question, so let's introduce the formalism of the likelihood function in this simple context.

The likelihood function is the probability of getting the hundred numbers given a theory. Our theory will be that each measurement is the sum of a constant signal (the person's weight) w and noise, with the noise drawn from a Gaussian distribution with mean zero and variance σ_w^2. Thus our "theory" has two free parameters, w and σ_w. If only one data point d was taken, the probability of getting d given the theory would be

$$P[d|w, \sigma_w] \equiv \mathcal{L}(d; w, \sigma_w) = \frac{1}{\sqrt{2\pi\sigma_w^2}} \exp\left\{-\frac{(d-w)^2}{2\sigma_w^2}\right\}. \qquad (11.1)$$

Here and throughout, $P[x|y]$ denotes the probability of x given y. Equation (11.1) simply restates the assumptions that $d - w$ is equal to noise and that the noise is drawn from a Gaussian distribution with standard deviation σ_w. In the limit that σ_w becomes very small, this function becomes sharply peaked at $d = w$. Since we are making $N_{\rm m} = 100$ independent measurements, the likelihood function is the product of all the individual likelihood functions. That is,

$$\mathcal{L} = \frac{1}{(2\pi\sigma_w^2)^{N_{\rm m}/2}} \exp\left\{-\frac{\sum_{i=1}^{N_{\rm m}}(d_i - w)^2}{2\sigma_w^2}\right\}. \qquad (11.2)$$

Notice that, although the data are distributed as a Gaussian, the likelihood function is not Gaussian in all the theoretical parameters (it is in w but is not in σ_w).

We are interested in the value of the theoretical parameters w and σ_w. Thus, we don't want $P[\{d_i\}|w, \sigma_w]$, which is the likelihood function we have computed. Rather, we want $P[w, \sigma_w|\{d_i\}]$. To obtain the latter from the former we can use a simple relation from elementary probability theory,

$$P[B \cap A] = P[B|A]P[A]$$

$$= P[A|B]P[B]. \tag{11.3}$$

In this context $A = \{d_i\}$ and $B = \{w, \sigma_w\}$, so the equality between the two lines of Eq. (11.3) means that

$$P[w, \sigma_w|\{d_i\}] = \frac{P[\{d_i\}|w, \sigma_w] \ P[w, \sigma_w]}{P[\{d_i\}]}. \tag{11.4}$$

The denominator can be rewritten by realizing that when we integrate the probability $P[w, \sigma_w|d]$ over all values of the parameters w, σ_w, we must get 1. So the denominator is equal to the integral of the numerator over w, σ_w. As a result, the denominator does not depend on the parameters w and σ_w, so it does not affect the place in parameter space where the likelihood function peaks or the width of the likelihood function. For the most part, then, we are free to ignore it.

To get the probability we want we need the likelihood function — the first term in the numerator — and also the *prior* probability $P[w, \sigma_w]$. If we possess prior information about these quantities, we might use this information here. If we want to be conservative, and assume nothing, we put in a uniform prior for the parameters. Then,

$$P[w, \sigma_w|\{d_i\}] \propto \mathcal{L}, \tag{11.5}$$

the proportionality constant being independent of the parameters and therefore of little interest. Many people find this idea of using prior information disturbing. Indeed, even the conservative choice of a uniform prior is not as innocent as it sounds. If we had taken the parameter to be σ_w^2 instead of σ_w and we had assumed a prior uniform in σ_w^2 (i.e., that equal intervals of σ_w^2 are equally likely), we would get a different answer for the final probability (try it!). Nonetheless, the dependence on the prior is a problem only in cases where the data are not very discriminatory. If the data do have discriminatory power, then the likelihood function $P[\{d_i\}|w, \sigma_w]$ will be sharply peaked and any reasonable prior will not affect the final results.

We can now find best-fit values for our parameters w and σ_w. Simply find the place in parameter space where the likelihood function is largest. In this simple example, we can proceed analytically by differentiating $\mathcal{L}$ with respect to each of the parameters. First consider the derivative with respect to w.

$$\frac{\partial \mathcal{L}}{\partial w} = \frac{\sum_{j=1}^{N_m}(d_j - w)}{\sigma_w^2 (2\pi\sigma_w^2)^{N_m/2}} \exp\left\{-\frac{\sum_{i=1}^{N_m}(d_i - w)^2}{2\sigma_w^2}\right\}. \tag{11.6}$$

For this derivative to be zero, we set the prefactor

$$\sum_{i=1}^{N_m}(d_i - w) = 0 \tag{11.7}$$

or equivalently, the likelihood is a maximum when

$$w = \bar{w} = \frac{1}{N_m}\sum_{i=1}^{N_m} d_i, \tag{11.8}$$

the expected answer. Similarly, we can find what the most probable value of σ_w^2 is by computing

$$\frac{\partial \mathcal{L}}{\partial \sigma_w^2} = \mathcal{L} \left[-\frac{N_{\mathrm{m}}}{2\sigma_w^2} + \frac{\sum_{i=1}^{N_{\mathrm{m}}} (d_i - w)^2}{2\sigma_w^4} \right] \tag{11.9}$$

and setting it equal to zero. Solving for the variance σ_w^2, we find a most probable value of

$$\sigma_w^2 = \frac{\sum_{i=1}^{N_{\mathrm{m}}} (d_i - \bar{w})^2}{N_{\mathrm{m}}}, \tag{11.10}$$

again the expected result.

We have found the best-fit values of our theoretical parameters. What is the error in these best-fit values? The error is just proportional to the width of the likelihood function. A simple way to approximate the width is to assume that $\mathcal{L}$ is Gaussian in the parameters, or equivalently that $\ln \mathcal{L}$ is quadratic in the parameters. We know in general that the variance of a Gaussian distribution is twice the inverse of the coefficient of the quadratic term, so we can simply identify the variance (the square of the error) by computing this coefficient. Let's work this out explicitly for w:

$$\ln \mathcal{L}(w) = \ln \mathcal{L}(\bar{w}) + \frac{1}{2} \left. \frac{\partial^2 \ln \mathcal{L}}{\partial w^2} \right|_{w=\bar{w}} (w - \bar{w})^2$$

$$= \ln \mathcal{L}(\bar{w}) - \frac{N_{\mathrm{m}}}{2\sigma_w^2} (w - \bar{w})^2. \tag{11.11}$$

Thus the width of the likelihood function at its maximum is $\sigma_w / N_{\mathrm{m}}^{1/2}$. This is the one-sigma error in our determination of w. This too is familiar: as more measurements are taken, the noise gets beaten down by a factor of 1 over the square root of the number of independent measurements. It is important to reiterate that the uncertainty on our estimate of the weight is *not* equal to σ_w.

Two numbers then sum up all N_{m} measurements: our best guess for the person's weight — in this case $\bar{w}$ given by Eq. (11.8) — and the error on this estimate, here equal to $\sigma_w / N_{\mathrm{m}}^{1/2}$. Therefore, we can compress all 100 measurements into just two by rewriting the likelihood function as

$$\mathcal{L} = \frac{1}{\sqrt{2\pi C_N}} \exp \left\{ \frac{-(w - \bar{w})^2}{2C_N} \right\} \tag{11.12}$$

where the variance due to noise is now

$$C_N = \frac{\sigma_w^2}{N_{\mathrm{m}}}. \tag{11.13}$$

This form of the likelihood has precisely the same maximum and width as does the form with all N_{m} data points. Thus it has compressed all the information in the likelihood function into two numbers, $\bar{w}$ and C_N.

Moving away from the weight metaphor, we can apply the above to the CMB. Instead of a true weight, the signal s is the true CMB temperature at a given point on the sky. The many measurements of the signal correspond to many measurements of the temperature at that point. The signal at that spot is a constant, and the data are the sum of the constant signal plus noise (atmospheric, instrumental, or both). The compression of all the different measurements into one data point with an associated error as in Eq. (11.12) is called map-making. We will take this up in more detail in Section 11.5. Now, though, we must move beyond the likelihood of Eq. (11.12). For we know that no theory predicts a value for the temperature at a particular position on the sky; i.e., no theory predicts s. Rather all theories predict a distribution of temperatures, from which s at a given pixel is drawn. We must now incorporate this distribution into the likelihood function.

11.1.2 CMB Likelihood

Let's convert the notation of the previous subsection to the CMB. The true temperature anisotropy in a given spot on the sky s replaces w, while the data point $\bar{w}$ (really the average of many measurements) becomes the estimated value of this temperature anisotropy, call it Δ.[1] The variance of this estimator C_N, which represents the spread of the measurements, is also given. How can this set of data (Δ, C_N) be compared with theory? The simplest theories, such as inflation, predict that the signal in a given spot on the sky is drawn from a Gaussian distribution. So the probability that the sky temperature falls in a range between s and $s + ds$ is

$$P(s)ds = \frac{1}{\sqrt{2\pi C_S}} \exp\left\{\frac{-s^2}{2C_S}\right\} ds. \qquad (11.14)$$

Here C_S is the variance expected due to the signal alone, independent of any noise. This variance is directly related to the C_l's in a manner we will explore in Section 11.2.

In order to get the likelihood function, we have to convolve the probability distribution of Eq. (11.14) with the likelihood function of Eq. (11.12). Schematically

$$P[\Delta|C_S] = \sum_s P[\Delta|s] \, P[s|C_S]. \qquad (11.15)$$

More concretely, the likelihood function is an integral over all possible values of the true anisotropy:

$$\mathcal{L} = \int_{-\infty}^{\infty} \frac{ds}{\sqrt{2\pi C_S}} \exp\left\{\frac{-s^2}{2C_S}\right\} \frac{1}{\sqrt{2\pi C_N}} \exp\left\{\frac{-(\Delta - s)^2}{2C_N}\right\}. \qquad (11.16)$$

The argument of the exponential here is quadratic in s so it is straightforward to carry out the integration over s. Let us rewrite the argument of the exponential as

---

[1] In Section 11.5 we explore ways to go from the raw data, the timestream, to the pixelized map represented by Δ. For now, we assume that this step has already been taken.

$$-\frac{s^2 C}{2 C_S C_N} + \Delta s / C_N - \frac{\Delta^2}{2 C_N} = -\frac{C}{2 C_S C_N}\left[s - \frac{C_S \Delta}{C}\right]^2 + \frac{C_S \Delta^2}{2 C C_N} - \frac{\Delta^2}{2 C_N} \quad (11.17)$$

where the full covariance matrix is defined as

$$C \equiv C_S + C_N. \quad (11.18)$$

Changing variables in the integral over s to $x \equiv s - C_S \Delta / C$ leads to

$$\mathcal{L} = \frac{1}{\sqrt{2\pi C_N}} \exp\left\{\frac{C_S \Delta^2}{2 C C_N} - \frac{\Delta^2}{2 C_N}\right\} \int_{-\infty}^{\infty} \frac{dx}{\sqrt{2\pi C_S}} \exp\left\{-\frac{C x^2}{2 C_S C_N}\right\}$$

$$= \sqrt{\frac{1}{2\pi C}} \exp\left\{\frac{-\Delta^2}{2 C}\right\}. \quad (11.19)$$

This is our final expression for the likelihood function for a one pixel experiment. This form is exactly what one expects: the measured temperature should be distributed like a Gaussian with a variance given by the sum of the variances due to noise and signal.

We can easily generalize Eq. (11.19) to the more realistic case with a measurement of N_p pixels. Then the likelihood function is

$$\mathcal{L} = \frac{1}{(2\pi)^{N_p/2} (\det C)^{1/2}} \exp\left\{-\frac{1}{2} \Delta C^{-1} \Delta\right\} \quad (11.20)$$

where now Δ is the data vector consisting of all N_p measurements and C is the full covariance matrix. In general, the noise covariance matrix can often be close to diagonal, but the theoretical covariance matrix is not diagonal. Thus, the hard computational part of evaluating the likelihood function is taking the determinant and the inverse of the $N_p \times N_p$ matrix C. The passage from one theoretical parameter (C_S) to a full matrix of parameters creates complications besides the computational. If there was only parameter, observers could quote results in the form of one number. Now that all correlations need to be included, one needs to allow for many different theoretical parameters, in principle all $N_p(N_p + 1)/2$ elements of the (symmetric) covariance matrix C_S. In practice of course this is not done. First of all, the covariance in all theories depends on the angular distance between two points, so elements of the matrix corresponding to two sets of points separated by the same distance are identical. Equivalently, a given theory is associated with a full set of C_l's; as we will see shortly, these can be used to construct all the elements of C_S. The second simplification is that most experiments have not been sensitive to individual C_l's but rather to the average power over a range of l, i.e., in a given *band*. So, analysts typically fit for *bandpowers*, a fitting which requires even fewer parameters to be determined.

As mentioned the matrix $C = C_S + C_N$ is typically not diagonal. However, we can get some nice insight into the likelihood function by considering the special case when C is diagonal and proportional to the identity matrix (all diagonal elements the same). In that case,

$$\mathcal{L} \propto \frac{1}{(C_S + C_N)^{N_p/2}} \exp \left\{ -\frac{1}{2} \frac{\sum_{i=1}^{N_p} \Delta_i^2}{C_S + C_N} \right\}. \tag{11.21}$$

We can easily find the value of C_S which maximizes the likelihood function in this case. Differentiating with respect to C_S leads to

$$\frac{\partial \mathcal{L}}{\partial C_S} = \mathcal{L} \times \left[\frac{-(N_p/2)}{(C_S + C_N)} + \frac{1}{2} \frac{\sum_{i=1}^{N_p} \Delta_i^2}{(C_S + C_N)^2} \right]. \tag{11.22}$$

If we set this to zero, we find that the likelihood function is maximized at

$$C_S = \frac{1}{N_p} \sum_{i=1}^{N_p} \Delta_i^2 - C_N. \tag{11.23}$$

Thus a useful rule of thumb for estimating the signal in a CMB experiment is to calculate the variance of the data points (the first term on the right in Eq. (11.23)) and compare it with the average noise per pixel (the second term on the right). If the data has larger variance than the noise, the theoretical signal is simply the difference between the two.

We can also calculate the error on this determination of C_S. As we saw in Section 11.1.1, the error is related to the second derivative of the log of the likelihood function:

$$\sigma_{C_S} = \left(\frac{-\partial^2 \ln \mathcal{L}}{\partial C_S^2} \right)^{-1/2}. \tag{11.24}$$

In this case, it is easy to calculate the derivative. Differentiating Eq. (11.22) once more leads to

$$\frac{\partial^2 \ln \mathcal{L}}{\partial C_S^2} = \frac{(N_p/2)}{(C_S + C_N)^2} - \frac{\sum_{i=1}^{N_p} \Delta_i^2}{(C_S + C_N)^3}. \tag{11.25}$$

At the peak of the likelihood we can replace the $\sum_{i=1}^{N_p} \Delta_i^2$ by $N_p[C_S + C_N]$, so

$$\sigma_{C_S} = \sqrt{\frac{2}{N_p}} (C_S + C_N). \tag{11.26}$$

Equation (11.26) is a simplified version of a very handy, useful formula which can be used to assess how accurately a given experiment will determine parameters. This simplified version gives the errors on our one theoretical parameter, C_S. The more general formula gives the corresponding errors when the free parameters are the C_l's themselves. In that case,

$$\sigma_{C_l} = \sqrt{\frac{2}{(2l+1)f_s}} (C_l + C_{N,l}). \tag{11.27}$$

The only change moving from Eq. (11.26) to Eq. (11.27) is that the number of pixels – or equivalently the number of independent measurements – has been replaced by

$(2l+1)f_s$, where f_s is the fraction of the sky covered. This makes perfect sense, for in the full sky limit, one can measure at best $2l + 1$ a_{lm}'s; that is, one can sample the distribution characterized by C_l only $2l + 1$ times. In fact, this is a fundamental limit on the accuracy with which we can measure the C_l's. Even if there is no noise $(C_N = 0)$, there remains a fundamental uncertainty in the theoretical parameters (either C_S or C_l) due to the fact that we only have one sky on which to take measurements. This limit, which we have already encountered in Chapter 8, is called *sample variance*, or in the limit of an all-sky survey *cosmic variance*.

11.1.3 Galaxy Surveys

At first, one might think that analysis of galaxy surveys would be completely different from CMB analysis. There *are* a number of differences. The galaxy distribution is fundamentally 3D, while the CMB anisotropies are a function of angular position only. Also, CMB experiments measure a continuous field, the temperature field, a function of position. Galaxy surveys count discrete objects (galaxies). A survey is simply a list of positions of these objects. Another difference is that the CMB temperatures are drawn from a Gaussian distribution, whereas the galaxies supposedly trace the underlying mass distribution, which — at least on small scales — has already "gone nonlinear." Nonlinearities inevitably produce non-Gaussianity, even if the primordial distribution is Gaussian.

Despite these, and other, differences, analysts have in recent years come to realize that many of the same techniques can be applied to data from both the CMB and galaxy surveys. To solidify the CMB–galaxy survey connection, we need to formalize the concept of a pixel. In the case of the CMB, the notion of a pixel is so natural that I didn't even bother to define it above. For galaxy surveys, following the treatment of Tegmark *et al.* (1998), we can define the data in pixel i as

$$\Delta_i \equiv \int d^3x \ \psi_i(\vec{x}) \left[\frac{n(\vec{x}) - \bar{n}(\vec{x})}{\bar{n}(\vec{x})} \right]. \tag{11.28}$$

Here $n(\vec{x})$ is the galaxy density at $\vec{x}$ and $\bar{n}$ is the expected number of galaxies at $\vec{x}$, i.e., the number there would be if the distribution was uniform. The weighting function ψ_i, which determines the pixelization, will be discussed shortly, but first let's understand operationally how to determine n and $\bar{n}$ from a survey. A simple way is to divide the volume into small sub-volumes, each of which is much smaller than the total survey, but large enough to contain many (e.g., greater than 10) galaxies. The density of a given sub-volume is then the number of galaxies in it divided by its volume. For a uniform survey, the average density $\bar{n}$ would just be the total number of galaxies divided by the total volume.[2]

There are two popular choices for ψ_i, choices which determine the pixelization. First is "counts-in-cells," wherein

---

[2] A caveat: if the pixels as defined by ψ_i are overlapping, more care must be taken in computing $\bar{n}$ for then the same galaxy could appear in more than one pixel.

$$\psi_i^{\text{CIC}}(\vec{x}) = \begin{cases} \bar{n}(\vec{x}) & \text{if } \vec{x} \text{ is in the } i\text{th sub-volume} \\ 0 & \text{otherwise} \end{cases} \tag{11.29}$$

In this case, the sub-volumes themselves are the pixels, and Δ_i is the over(under)-density in the ith sub-volume. Another useful pixelization scheme is a set of Fourier pixels, which emerge from choosing

$$\psi_i^{\text{Fourier}}(\vec{x}) = \frac{e^{i\vec{k}_i \cdot \vec{x}}}{V} \begin{cases} 1 & \vec{x} \text{ inside the survey volume} \\ 0 & \vec{x} \text{ outside survey volume} \end{cases} \tag{11.30}$$

Here V is the volume of the survey. In this case, the pixels are not spatial, but rather live in the Fourier domain. Still, even in this case, Δ_i is the fractional overdensity in the pixel.

No matter which pixelization is chosen, one cannot hope to write down a simple expression for the likelihood function, the probability of getting a set of $\{\Delta_i\}$ given a theory. The theory of galaxy formation is simply too complicated. Indeed, even if one assumes that the galaxy density perfectly traces the mass overdensity, the complications from gravity alone make the likelihood function non-Gaussian. Nonetheless, progress can still be made by noting that the expectation value of Δ_i is zero by construction, with a covariance matrix[3]

$$\langle \Delta_i \Delta_j^* \rangle = (C_S)_{ij} + (C_N)_{ij} \tag{11.31}$$

exactly like the CMB case. We will discuss the signal covariance matrix in detail in Section 11.2. The noise covariance matrix is actually easier than the corresponding CMB matrix, which depends on the atmosphere, pointing, instrumental noise, scan strategy, and other experimental details. In a galaxy survey, even if there was no signal, the expected value of the square of the density, $\langle n^2(\vec{x}) \rangle$, would still differ from $\bar{n}^2$ simply because there are only a finite number of galaxies in a given sub-volume. Thus, even in the absence of any intrinsic clustering of the galaxies $\langle \Delta_i \Delta_j \rangle$ would be nonzero because of *Poisson noise*. You can show in Exercise 3 that the covariance matrix due to Poisson noise is

$$(C_N)_{ij} = \int d^3x \frac{\psi_i(\vec{x}) \psi_j^*(\vec{x})}{\bar{n}(\vec{x})}. \tag{11.32}$$

Armed with this noise covariance matrix and the signal covariance matrix we explore next, galaxy survey analysts can use many of the same techniques as the CMB analysts.

11.2 SIGNAL COVARIANCE MATRIX

Until now, we have sidestepped the question of how the expected variance in a given experiment is related to the underlying power spectrum. That is, we have learned that the predictions of a given theory are a set of C_l's and $P(k)$. If we want to relate theory to experiment — and we do! — we need to know how to turn this set of predictions into a covariance matrix C_S.

---

[3]The angular brackets here denote an average over the distribution from which Δ_i is drawn.

11.2.1 CMB Window Functions

For simplicity, let us first consider the diagonal element of the covariance matrix:

$$C_{S,ii} \equiv \langle s_i s_i \rangle \qquad \text{(no sum over } i) \qquad (11.33)$$

where the average $\langle \ldots \rangle$ is over many realizations of the theoretical distribution and the subscript i labels the pixel. The temperature difference reported in each pixel can be expressed as

$$s_i = \int d\hat{n} \Theta(\hat{n}) B_i(\hat{n}) \qquad (11.34)$$

where B_i is the beam pattern at the ith pixel and Θ is the underlying temperature. As an example, the beam pattern from the MSAM experiment is shown in Figure 11.1. It is typical of the patterns produced by many CMB experiments: the difference of the temperature in two (or more) regions of the sky, and in each region the temperature is sampled by a beam which is roughly Gaussian.

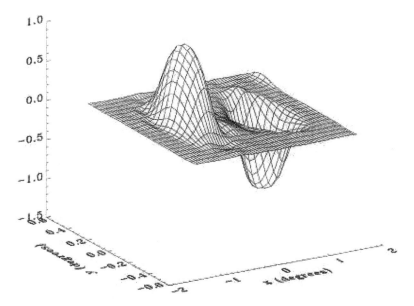

Figure 11.1. The beam pattern for the MSAM experiment (see Wilson *et al.*, 2000, for a summary). The anisotropy reported at a given pixel is roughly the difference between the temperature at ± 0.5 degrees from the center of the pixel. The beamwidth is also on the order of half a degree.

To find C_S we square Eq. (11.34) after expressing the temperature field as an expansion over spherical harmonics as in Eq. (8.60):

$$\frac{C_{S,ii}}{T^2} = \int d\hat{n} \int d\hat{n}' B_i(\hat{n}) B_i(\hat{n}') \sum_{lm} Y_{lm}(\hat{n}) \sum_{l'm'} Y^*_{l'm'}(\hat{n}') \langle a_{lm} a^*_{l'm'} \rangle. \qquad (11.35)$$

Using Eq. (8.63), we find that the sums over $l'm'$ collapse to give

$$\frac{C_{S,ii}}{T^2} = \int d\hat{n} \int d\hat{n}' B_i(\hat{n}) B_i(\hat{n}') \sum_l C_l \sum_m Y_{lm}(\hat{n}) Y_{lm}^*(\hat{n}'). \qquad (11.36)$$

But $\sum_m Y_{lm}(\hat{n}) Y_{lm}^*(\hat{n}') = (2l+1) P_l(\hat{n} \cdot \hat{n}')/4\pi$, so

$$\frac{C_{S,ii}}{T^2} = \sum_l \frac{2l+1}{4\pi} C_l W_{l,ii} \qquad (11.37)$$

where the window function is defined as

$$W_{l,ii} \equiv \int d\hat{n} \int d\hat{n}' B_i(\hat{n}) B_i(\hat{n}') P_l(\hat{n} \cdot \hat{n}'). \qquad (11.38)$$

Until now, we have been thinking of $\hat{n}, \hat{n}'$ as three-dimensional unit vectors. If $\hat{n}'$ and $\hat{n}$ are sufficiently close to each other, though, we can use the flat space approximation. The three-dimensional unit vectors can be safely approximated as two-dimensional vectors $\vec{x}, \vec{x}'$ in the transverse directions. The distance between $\vec{x}$ and $\vec{x}'$ (measured in radians) is then equal to the angle between $\hat{n}$ and $\hat{n}'$. In this limit, the argument of the Legendre polynomial in Eq. (11.38) becomes

$$\vec{n} \cdot \vec{n}' = \cos\left(|\vec{x} - \vec{x}'|\right). \qquad (11.39)$$

The diagonal window function is therefore

$$W_{l,ii} = \int d^2 x \int d^2 x' B_i(\vec{x}) B_i(\vec{x}') P_l\left(\cos\left(|\vec{x} - \vec{x}'|\right)\right). \qquad (11.40)$$

A useful property of Legendre polynomials is that they become equal to the zero-order Bessel function in the limit of large l (the small-angle limit we are working under here). So,

$$P_l\left(\cos\left(|\vec{x} - \vec{x}'|\right)\right) \to J_0\left(l|\vec{x} - \vec{x}'|\right)$$

$$= \frac{1}{2\pi} \int_0^{2\pi} d\phi\, e^{-il|\vec{x} - \vec{x}'|\cos\phi}, \qquad (11.41)$$

where the last line is an integral representation of the Bessel function (Eq. (C.21)). We can simplify further by promoting l to a 2D vector with direction chosen so that the angle between $\vec{l}$ and $\vec{x} - \vec{x}'$ is equal to ϕ. Then, the argument of the exponential simplifies to $-i\vec{l} \cdot (\vec{x} - \vec{x}')$. This form is so useful because the $\vec{x}$ integral for example is now

$$\int d^2 x B_i(\vec{x}) e^{-i\vec{l} \cdot \vec{x}} \equiv \tilde{B}_i(\vec{l}), \qquad (11.42)$$

where $\tilde{B}_i$ is the Fourier transform of the beam pattern. The $\vec{x}'$ integral is the complex conjugate of this, so the window function simplifies to

$$W_{l,ii} = \frac{1}{2\pi} \int_0^{2\pi} d\phi\, \left|\tilde{B}_i(\vec{l})\right|^2. \qquad (11.43)$$

Thus calculating the window function reduces to a two-step process:

- Calculate the 2D Fourier transform of the beam pattern.
- Find the angular average of the square of this transform.

The window function is a function of the experiment only and indeed contains information about the beam size and chopping angle of the experiment. However, it is not the whole story. A complete evaluation of the likelihood function entails calculating all of the elements of the covariance matrix C_S. A given off-diagonal element of the matrix is given by Eqs. (11.37) and (11.38), with one of the indices i changed to j. The matrix is symmetric, so it is characterized by $N_p(N_p + 1)/2$ elements where N_p is the number of pixels. Thus, for an N_p-pixel experiment there are really $N_p(N_p + 1)/2$ window functions!

11.2.2 Examples of CMB Window Functions

Gaussian Beam. Let us take a break from formalism and calculate a simple (diagonal) window function. Consider a Gaussian beam; this is a good approximation to many CMB experiments. The beam pattern for the ith pixel is

$$B_i(\vec{x}) = \frac{1}{2\pi\sigma^2} \exp\left(-\frac{(\vec{x} - \vec{x}_i)^2}{2\sigma^2}\right). \tag{11.44}$$

We may choose $\vec{x}_i$ to be zero for the window function computation. The Fourier transform of the beam is also a Gaussian,

$$\tilde{B}_i(\vec{l}) = \frac{1}{2\pi\sigma^2} \int d^2x \, e^{-i\vec{l}\cdot\vec{x}} \exp\left(-\frac{x^2}{2\sigma^2}\right)$$

$$= e^{-l^2\sigma^2/2}. \tag{11.45}$$

In this simple case, $\tilde{B}$ does not depend on the direction of $\vec{l}$, so there is no need to take the angular average. The window function is then simply the square of the Fourier transform,

$$W_{l,ii} = e^{-l^2\sigma^2}. \tag{11.46}$$

The window function falls off sharply at large l. Large l corresponds to small angular scales. Structure on scales smaller than the beam size is inevitably washed away and undetectable. Figure 11.2 illustrates the series of steps, from beam function to Fourier transform to window function.

There are two subtleties associated with the Gaussian window function. First, one must avoid the temptation to set σ equal to the number which is often quoted in papers, the full width half maximum (FWHM). The latter is twice the value of x for which $B(\vec{x})$ drops to half of its maximum. So $\sigma = \text{FWHM}/(\sqrt{8\ln(2)}) = 0.4245$ FWHM. The second subtlety has to do with normalization. It is crucial to determine how observers have normalized their output. The prefactor in Eq. (11.44) ensures that if the temperature field was uniform, the reported temperture would be equal to the underlying one. In this simple case, the choice is obvious; in general factors of 2 can easily be lost.

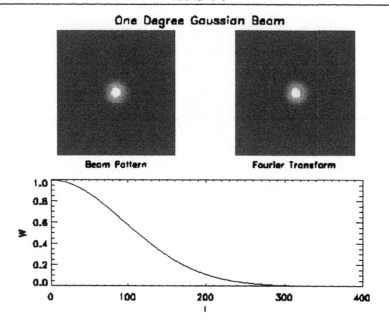

Figure 11.2. A one-degree Gaussian beam is shown in the upper left panel and its Fourier transform, also a Gaussian, in the upper right. The resulting window function is shown in the bottom panel.

Differencing a Gaussian Beam. As another straightforward example, let us consider an experiment which takes the difference between the temperatures at two adjacent points on the sky. For simplicity, let us first assume that the Gaussian beam is infinitely small, so it can be approximated as a Dirac delta function. Then,

$$B(x,y) = \delta(y)\left[\delta(x - x_0) - \delta(x + x_0)\right], \tag{11.47}$$

where the *chopping angle*, or the distance between the plus and minus position, is $2x_0$. The Fourier transform of this is straightforward:

$$\tilde{B}(\vec{l}) = 2i\sin(l_x x_0). \tag{11.48}$$

The window function is the angular average over all $\vec{l}$ directions. Choosing the angle between $\vec{l}$ and the $x-$ axis to be ϕ, we have

$$W_l = \frac{4}{2\pi}\int_0^{2\pi} d\phi \; \sin^2\left[lx_0\cos\phi\right]$$

$$= \frac{1}{\pi}\int_0^{2\pi} d\phi \left(1 - \cos\left[2lx_0\cos\phi\right]\right)$$

$$= 2 \left(1 - P_l \big[\cos(2x_0) \big] \right), \tag{11.49}$$

where the last line follows from Eq. (11.41).

Until now, we have neglected the finite width of the beam. However, this turns out to be very simple to rectify. A realistic beam will be the convolution of the chop described by Eq. (11.47) with a finite beam size:

$$B(x,y) = \frac{1}{2\pi\sigma^2} \int dx' dy' \exp\left\{ - \frac{(x-x')^2 + (y-y')^2}{2\sigma^2} \right\}$$

$$\times \, \delta(y') \left[\delta(x' - x_0) - \delta(x' + x_0) \right]. \tag{11.50}$$

Recall, though, that the Fourier transform of the convolution of two functions is simply equal to the product of the two Fourier transforms. The angular averaging over this product is unaffected since the Gaussian has no angular dependence. Therefore, the final window function is

$$W_l = e^{-l^2 \sigma^2} \left(1 - P_l \big[\cos(2x_0) \big] \right). \tag{11.51}$$

This window function is shown in Figs. 11.3 and 11.4 along with the beam pattern

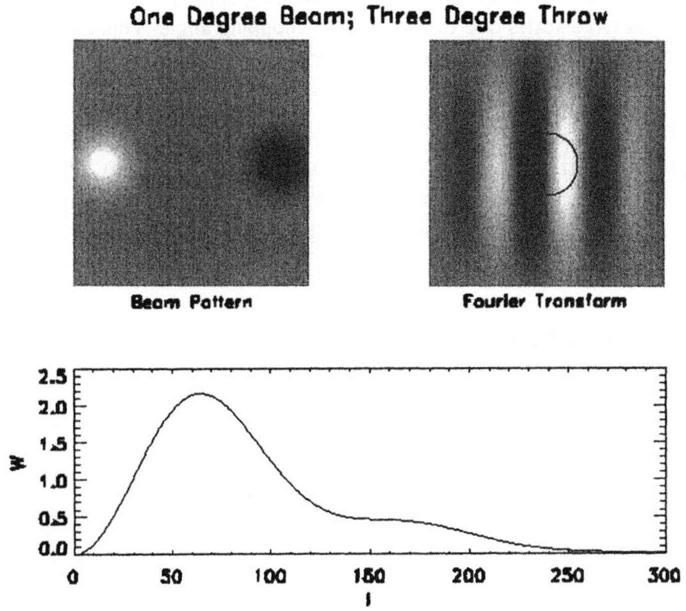

One Degree Beam; Three Degree Throw

Beam Pattern   Fourier Transform

Figure 11.3. Differencing a Gaussian beam. Upper left panel shows the beam pattern and the upper right its Fourier transform. The circle in the upper right corresponds to $l = 50$. The bottom panel shows the window function.

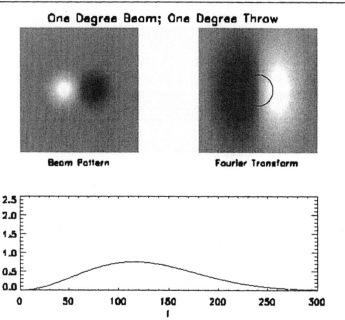

Figure 11.4. Same as Figure 11.3 except the chopping angle is now much smaller. Note the lack of support at large scales due to the reduced chopping angle. This can also be seen in the Fourier transform which vanishes near the center of the circle.

for two different chopping angles. Note that for l much smaller than $1/x_0$, the Fourier transform vanishes. So, unlike the undifferenced beam, there is no support for small l. That is, chopping removes information about structure on large scales. As the chopping angle gets larger, more and more information is obtained about large scales.

11.2.3 Window Functions for Galaxy Surveys

We now consider the signal covariance matrix for galaxy surveys. By setting the term in square brackets in Eq. (11.28) to $\delta(\vec{x})$, we see that the signal covariance matrix is equal to

$$(C_S)_{ij} = \langle \Delta_i \Delta_j \rangle \Big|_{\text{no noise}} = \int d^3x \, d^3x' \, \psi_i(\vec{x}) \psi_j(\vec{x}') \xi(\vec{x} - \vec{x}'). \qquad (11.52)$$

The correlation function ξ appears here because it is equal to the expectation value of the product of two overdensities, Eq. (9.7). Since ξ is the Fourier transform of the power spectrum, we see that the signal covariance matrix for a galaxy survey is

$$(C_S)_{ij} = \int d^3x \, d^3x' \int \frac{d^3k}{(2\pi)^3} \frac{d^3k'}{(2\pi)^3} \frac{d^3k''}{(2\pi)^3} \tilde{\psi}_i(\vec{k}) \tilde{\psi}_j^*(\vec{k}') P(k'') e^{i[\vec{k}+\vec{k}'']\cdot\vec{x} - i[\vec{k}'+\vec{k}'']\cdot\vec{x}'}$$

$$= \int \frac{d^3k}{(2\pi)^3} P(k) \tilde{\psi}_i(\vec{k}) \tilde{\psi}_j^*(\vec{k}). \tag{11.53}$$

The second equality follows simply after integrating over $\vec{x}$ and $\vec{x}'$ to get 3D Dirac delta functions and using these to perform the integrals over $\vec{k}'$ and $\vec{k}''$. It is convenient to define the window function as the angular part of this integral

$$W_{ij}(k) \equiv \int \frac{d\Omega_k}{4\pi} \tilde{\psi}_i(\vec{k}) \tilde{\psi}_j^*(\vec{k}) \tag{11.54}$$

so that

$$(C_S)_{ij} = \int_0^\infty \frac{dk}{k} \left[\frac{k^3 P(k)}{2\pi^2} \right] W_{ij}(k). \tag{11.55}$$

Notice that the window function for galaxy surveys has the identical form as that for CMB experiments. In both cases, it is the angular average of the square of the Fourier transform of the weighting function (either B or ψ). Also, you should recognize the quantity in square brackets in Eq. (11.55) as $\Delta^2(k)$, the contribution to the variance per $\ln(k)$. Let's turn to some examples of window functions of galaxy surveys.

Volume-Limited Survey. Consider a survey which observes all galaxies within a radius R from us. If we use Fourier pixelization (Eq. (11.30)), then the Fourier transform of the weighting function is

$$\tilde{\psi}_i(\vec{k}) = \int_{|\vec{x}|<R} \frac{d^3x}{V} e^{-i\vec{k}\cdot\vec{x}} e^{i\vec{k}_i\cdot\vec{x}}. \tag{11.56}$$

We will shortly carry out this integral, square, and then average over all angles to get the diagonal window function of Eq. (11.54). First, though, let's ask what we expect qualitatively. Equation (11.56) is the Fourier transform of the survey volume as a function of $\vec{k} - \vec{k}_i$. The survey volume is a sphere of radius R. In general, when a function is confined to a region $x < R$, the Fourier transform is confined to $k < 1/R$. Here then, $\tilde{\psi}$ will be nonzero only when $|\vec{k} - \vec{k}_i|$ is less than $1/R$. The window function therefore should peak at $k = k_i$ and have a width of order $1/R$.

More quantitatively, the integral in Eq. (11.56) is

$$\tilde{\psi}_i(\vec{k}) = \frac{4\pi}{V|\vec{k} - \vec{k}_i|} \int_0^R dx \, x \, \sin\left(|\vec{k} - \vec{k}_i|x\right)$$

$$= \frac{4\pi}{V\left(|\vec{k} - \vec{k}_i|\right)^3} \left[-|\vec{k} - \vec{k}_i|R\cos\left(|\vec{k} - \vec{k}_i|R\right) + \sin\left(|\vec{k} - \vec{k}_i|R\right) \right] \tag{11.57}$$

The diagonal window function is the angular average of the square of this. Defining $y \equiv |\vec{k} - \vec{k}_i|R$, this angular average is

$$W_{ii}(k) = \frac{(4\pi R^3)^2}{V^2} \int_{-1}^1 \frac{d\mu}{2} \int_0^{2\pi} \frac{d\phi}{2\pi} \frac{(\sin y - y\cos y)^2}{y^6}$$

$$= \frac{8\pi^2 R^6}{V^2} \int_{-1}^{1} \frac{d\mu}{y^6} \, (\sin y - y \cos y)^2 \tag{11.58}$$

where μ is the cosine of the angle between $\vec{k}$ and $\vec{k}_i$. Integrating over y instead of μ leads to

$$W_{ii}(k) = \frac{9}{2kk_i R^2} \int_{|k-k_i|R}^{(k+k_i)R} \frac{dy}{y} \, j_1^2(y) \tag{11.59}$$

since the volume of the survey $V = 4\pi R^3/3$. This window function is shown in Figure 11.5 for several different values of $k_i R$. Notice that modes with wavelength much smaller than the size of the survey $k_i R \gg 1$ do indeed have window functions sharply peaked at $k = k_i$, with a width of order $1/R$. The largest wavelength modes, however, pick up contributions from all scales (e.g., the $k_i R = 3$ curve in Figure 11.5). Not surprisingly, surveys do not do a good job measuring the power on wavelengths comparable to their sizes.

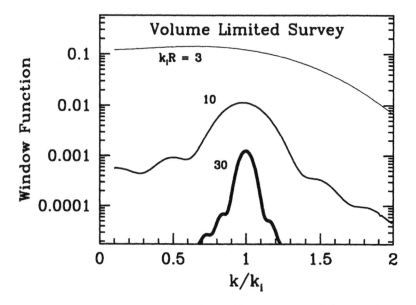

Figure 11.5. The window function for a volume-limited survey. Narrow window functions enable one to determine the power spectrum at the wavenumber of interest more accurately. Modes with wavelengths comparable to the size of the survey (e.g., here $k_i R = 3$) have broad window functions.

The height of the window function is also important, for it determines the amplitude of the signal covariance matrix. When considering modes with wavelengths k_i^{-1} much smaller than R, you will show in Exercise 7 that

$$(C_S)_{ii} \simeq \frac{P(k_i)}{V}. \tag{11.60}$$

It is instructive to compare this to the corresponding element of the noise matrix, as given in Eq. (11.32). For the diagonal elements, $|\psi_i|^2$ in the integrand is $1/V^2$ as long as $\vec{x}$ is in the survey volume. Thus,

$$(C_N)_{ii} = \frac{1}{\bar{n}V}, \qquad (11.61)$$

equal to the inverse of the total number of galaxies in the survey. The ratio of the diagonal elements of the signal and noise covariance matrices is therefore

$$\frac{(C_S)_{ii}}{(C_N)_{ii}} \simeq P(k_i)\bar{n}. \qquad (11.62)$$

Harking back to Eq. (11.27), we identify this ratio as the ratio of cosmic variance to Poisson noise. A rough estimate is that $\bar{n} \sim 1\,h^3\,\mathrm{Mpc}^{-3}$, so cosmic variance dominates as long as the power spectrum is larger than $1\,h^{-3}\,\mathrm{Mpc}^3$. Looking back to Figure 7.11, we see that on large scales, this is always satisfied. On small scales, eventually the power spectrum does drop beneath $1\,h^{-3}\,\mathrm{Mpc}^3$, the linear power spectrum of standard CDM at $k \sim 10\,h\,\mathrm{Mpc}^{-1}$. On very small scales, therefore, Poisson noise becomes important.

Pencil-Beam Survey. Now consider a survey which is very deep, but also very narrow, with the general shape of a pencil (Figure 11.6). The Fourier transform of the

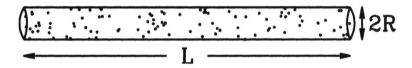

Figure 11.6. A pencil-beam survey with depth L much larger than the typical size of the narrow dimensions, R. The z-axis in the text is taken to be aligned with the long dimension.

weighting function in this case is

$$\tilde{\psi}_i(\vec{k}) = \int \frac{d^3x}{\pi R^2 L} e^{i(\vec{k}_i - \vec{k})\cdot\vec{x}} \Theta(z + L/2)\Theta(L/2 - z)\Theta(R^2 - x^2 - y^2) \qquad (11.63)$$

where Θ is a step function, equal to 1 if its argument is positive and zero otherwise. It is easiest to carry out this integral in cylindrical coordinates, wherein $x^2 + y^2 = r^2$. If we define $\vec{q} \equiv \vec{k}_i - \vec{k}$, then

$$\tilde{\psi}_i(\vec{k}) = \frac{1}{\pi R^2 L} \int_{-L/2}^{L/2} dz\, e^{iq_z z} \int_0^R dr\, r \int_0^{2\pi} d\theta\, e^{iq_r r \cos\theta}. \qquad (11.64)$$

The azimuthal integral can be done using Eq. (C.17), the integral over z using Eq. (C.15), so that

$$\tilde{\psi}_i(\vec{k}) = \frac{2}{R^2} j_0(q_z L/2) \int_0^R dr\, r J_0(q_r r). \qquad (11.65)$$

Finally the integral over r is $RJ_1(q_r R)/q_r$, which you can see by differentiating the integral with respect to $q_r R$ and then using Eq. (C.22). Therefore, the Fourier transform of the weighting function is

$$\tilde{\psi}_i(\vec{k}) = \frac{2}{(q_r R)} j_0(q_z L/2) J_1(q_r R). \qquad (11.66)$$

Equation (11.66) indicates that the Fourier transform of the weighting function is anisotropic. Indeed, even before deriving the various flavors of Bessel functions, we should have expected this Fourier transform to be compact along the q_z direction and broad in the transverse plane, i.e., shaped like a disk. This flows from our intuition that the Fourier transform of a function localized within a radius R should be localized within a region $1/R$. Indeed, in the z direction, $j_0(q_z L/2)$ falls off once q_z gets larger than $2/L$. The same holds for $J_1(q_r R)$: it becomes small for $q_r > 1/R$. The ringing associated with these Bessel functions is a manifestation of the fact that the Fourier transform of a top-hat function oscillates for large wavenumbers (e.g., Exercise 6).

To get the window function for a pencil-beam survey, we need to average Eq. (11.66) over all directions $\hat{k}$. This average will differ for different $\vec{k}_i$. Let's choose $\vec{k}_i$ to point in the $\hat{z}$-direction as one concrete example. The averaging will pick up contributions only when $\vec{q} \equiv \vec{k}_i - \vec{k}$ has z-component smaller than L^{-1} and transverse component smaller than R^{-1}. Since the transverse component of $\vec{k}_i$ is zero in this example, many $\vec{k}$ will contribute, as long as their transverse component is smaller than R^{-1}. Therefore, we expect the window function to get contributions from many wavenumbers, not to be sharply peaked around $|\vec{k}_i|$. A similar argument holds for other directions $\hat{k}_i$. Figure 11.7 shows the window function for a pencil-beam survey. As expected, it is broader than that for a symmetric, volume-limited survey: a given scale k_i picks up contributions from smaller scales $k > k_i$.

11.2.4 Summary

We have determined the signal covariance matrix for CMB experiments which measure a filtered version of the temperature in a given set of pixels and for galaxy surveys which measure the overdensity in a given set of pixels. Not surprisingly, the fundamental relation between the covariance matrix and and the underlying power spectrum is very similar in both cases: the connection is provided by a window function determined by the experimental/observational specifications. It is interesting to point out that we have encountered *natural* window functions in Chapters 9 and 10. The angular correlation function of Section 9.1 is simply the signal covariance matrix of measurements of the 2D galaxy distribution. Recall from Eq. (9.13) that this too is an integral of the 3D power spectrum convolved with a window function (we called it a *kernel* back then). The same is true for the peculiar velocity (Eq. (9.29)) and the shear field that can be measured with weak lensing (Eq. (10.32)). In all of those cases, the window function is determined partly by the

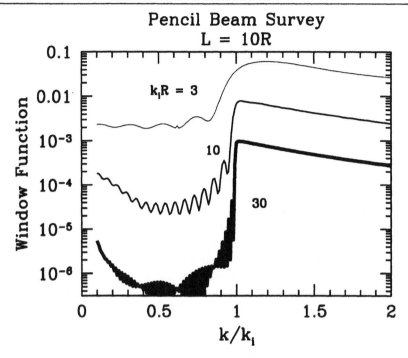

Figure 11.7. The window function for a pencil-beam survey with length 10 times larger than width. These window functions are for $\vec{k}_i$ pointing in the $\hat{z}$-direction, along the long dimension of the survey.

observational strategy (e.g., how deep one goes), but also by the intrinsic nature of the measurement itself. For example, the 2D galaxy distribution intrinsically is a projection of the 3D distribution. So it is seductive to think of CMB anisotropy experiments and 3D galaxy surveys as more pristine measurements of the power spectrum. Analysts are hard to seduce, though. They recognize that, mathematically, the different sets of measurements can all be analyzed in the same fashion. So the likelihood function can be used on virtually all cosmological observations. It is a very powerful tool.

It is also simple. Were it not for the size of modern cosmological data sets, we would be done. At least in the Gaussian case (CMB or even galaxy surveys on large scales), the likelihood function is given by Eq. (11.20). The data points are simply the pixelized temperatures or overdensities, while the covariance matrix is the sum of the noise and signal covariances. The noise covariance matrix is usually estimated from the data itself, while the signal covariance matrix is computed by convolving the theory (i.e., the C_l's or the power spectrum) with the window function. In principle, then, one could compute this likelihood function at many points in parameter space, find its maximum (this constitutes the set of best-fit parameters), and the contour delineating the region in which, say, 95% of the volume lies. This contour would then be the 95% confidence region of the parameters. Many experiments of the previous decade, especially CMB experiments, have been analyzed in this brute-force fashion. Times are changing, though, and the brute-force approach has already become impractical.

11.3 ESTIMATING THE LIKELIHOOD FUNCTION

To illustrate the need for new techniques of likelihood computation, let us consider a concrete example: the data set from the Boomerang anisotropy experiment (Netterfield *et al.*, 2001). There are 57,000 pixels on the sky covered in this set. The theory and noise covariance matrices are both nondiagonal and both $57,000 \times 57,000$ dimensional matrices. Inverting these beasts with present computers is possible, although slow.[4] If we needed to invert only once, this might be acceptable. But, we need to evaluate the likelihood function at many points in parameter space to find its maximum and the region at which it falls to, say, 5% of its maximum. This would be barely manageable if the parameter space was one-dimensional. A one-dimensional fit, though, would lose most of the information contained in the map. The data are actually sensitive to the power on many different scales. Therefore, the parameter space — the amplitude of the power on these many different scales — is multidimensional, "multi-" here of order 20. The likelihood function should in principle be computed about 10 times in each dimension, for a total of 10^{20} computations. Since each inversion takes several hours, this is not feasible. All of these estimates are for the Boomerang experiment. The MAP satellite will have 10 times as many pixels and be sensitive to a wider range of scales. Planck will be more sensitive still. Thus, we need new techniques, shortcuts, for evaluating the likelihood function and finding its maximum and its width.

11.3.1 Karhunen–Loève Techniques

The first technique was discovered many years ago and reinvented by a number of people over the past few years to deal with both CMB data (Bunn, 1995; Bond, 1995) and with data from galaxy surveys (Vogeley and Szalay, 1996). It is a method for speeding up the computation of the likelihood function. The fundamental idea is simple: any experiment, no matter how good, will have many *modes* which are useless, fundamentally contaminated by noise.[5] If it was obvious which modes were most noisy, then we could greatly simplify the likelihood calculation by not using those modes. If only 10% of the modes carried useful information — and this figure is roughly what is found in many present-day experiments — then the data set would be reduced by a factor of 10. The covariance matrices would now be $(N/10) \times (N/10)$ and the inversion (which scales as N^3) would speed up by a factor of 1000. It is a nice, simple idea. The only problem is finding the modes which are useful.

---

[4]Note that the problems outlined here assume that we are handed the map of 57,000 pixels. Mapmaking is actually the most difficult computational part of the analysis!

[5]A mode here is defined as a linear combination of the data points.

If the signal and noise covariance matrices were diagonal, then it would be simple to ascertain which modes had high signal-to-noise. The pixels with diagonal elements $C_S > C_N$ would have signal-to-noise greater than 1; the others would be low-signal modes. The problem is to identify the low modes in the more realistic case where the covariance matrices are not diagonal. This is precisely what Karhunen–Loève does. To illustrate the technique, let us first write it down formally and then work out a simple example explicitly.

We assume there are N_p data points, Δ_i. Each data point is presumed to be the sum of both signal s_i and noise n_i. Each of these are assumed to be uncorrelated (the noise knows nothing about the signal and vice versa). Thus the full covariance matrix is

$$\langle \Delta_i \Delta_j \rangle \equiv C_{ij} = C_{S,ij} + C_{N,ij}. \tag{11.67}$$

The Karhunen–Loève method takes advantage of the fact that instead of computing the likelihood function using Δ_i and its covariance matrix C, we could instead use rotated data

$$\Delta_i' \equiv R_{ij} \Delta_j \tag{11.68}$$

where R is a real matrix. The covariance matrix associated with Δ' will be

$$C_{ij}' = \langle (R\Delta)_i (R\Delta)_j \rangle$$

$$= R_{ii'} R_{jj'} C_{i'j'} \tag{11.69}$$

In matrix notation this is simply

$$C' = RCR^T \tag{11.70}$$

where R^T is the transpose of R.

The Karhunen–Loève method consists of three such rotations.

1. R_1 : Diagonalize C_N
2. R_2 : Set $C_N' = I$
3. R_3 : Diagonalize C_S'

The first step is always possible since C_N is a real, symmetric matrix. Once C_N has been diagonalized, it is trivial to perform step 2: simply choose R_2 to be diagonal with elements equal to 1 over the square root of the diagonal elements that emerge from step 1. Finally, step 3 is straightforward since again C_S is a real symmetric matrix. Let's evaluate the new theory and noise covariance matrices. The theory matrix is

$$C_S' = R_3 R_2 R_1 C_S R_1^T R_2^T R_3^T. \tag{11.71}$$

Note that, since R_2 is diagonal it is equal to its transpose. The matrix C_S' is a diagonal matrix. Now consider C_N'. After step 2, it was simply the identity matrix. So we need consider only the effect of step 3. In fact, since R_3 is unitary, it has no effect ($R_3 I R_3^T = I$). Thus, C_N' is still equal to the identity matrix. This has profound implications. It means that the elements of (the diagonal matrix) C_S' are a measure of the signal-to-noise squared of the modes! The data points

$$\Delta_i' = (R_3 R_2 R_1)_{ij} \Delta_j \qquad (11.72)$$

then have diagonal covariance matrix

$$\langle \Delta_i' \Delta_j' \rangle = \begin{cases} 1 + C_{S,ii}' & i = j \\ 0 & i \neq j \end{cases}. \qquad (11.73)$$

These modes can be ordered according to their signal-to-noise values. Modes with large C_S' can be kept; those with C_S' significantly smaller than 1 can be eliminated from the analysis.

Let us work through a simple example to see how Karhunen–Loève picks out the highest signal-to-noise modes. The example is a simple two-pixel experiment with diagonal noise:

$$C_N = \begin{pmatrix} \sigma_n^2 & 0 \\ 0 & \sigma_n^2 \end{pmatrix}. \qquad (11.74)$$

The signal covariance matrix does have correlations between the two pixels so

$$C_S = \sigma_s^2 \begin{pmatrix} 1 & \epsilon \\ \epsilon & 1 \end{pmatrix} \qquad (11.75)$$

where σ_s is the expected rms in the pixel and $-1 < \epsilon < 1$ measures how correlated the signal is between the two pixels. Steps 1 and 2 of the Karhunen–Loève method are particularly simple since C_N is diagonal. Thus,

$$R_1 = I \quad ; \quad R_2 = \frac{1}{\sigma_n} I. \qquad (11.76)$$

To complete step 3, we need to diagonalize

$$R_2 R_1 C_S R_1^T R_2 = \frac{\sigma_s^2}{\sigma_n^2} \begin{pmatrix} 1 & \epsilon \\ \epsilon & 1 \end{pmatrix}. \qquad (11.77)$$

To diagonalize the matrix in Eq. (11.77), we must solve

$$\frac{\sigma_s^2}{\sigma_n^2} \begin{pmatrix} \cos\theta & \sin\theta \\ -\sin\theta & \cos\theta \end{pmatrix} \begin{pmatrix} 1 & \epsilon \\ \epsilon & 1 \end{pmatrix} \begin{pmatrix} \cos\theta & -\sin\theta \\ \sin\theta & \cos\theta \end{pmatrix} = \begin{pmatrix} C_{S,11}' & 0 \\ 0 & C_{S,22}' \end{pmatrix} \qquad (11.78)$$

for the rotation angle θ. Carrying out the multiplication on the left side leads to

$$\frac{\sigma_s^2}{\sigma_n^2} \begin{pmatrix} 1 + \epsilon\sin(2\theta) & \epsilon\cos(2\theta) \\ \epsilon\cos(2\theta) & 1 - \epsilon\sin(2\theta) \end{pmatrix} = \begin{pmatrix} C_{S,11}' & 0 \\ 0 & C_{S,22}' \end{pmatrix}. \qquad (11.79)$$

Equality in the off-diagonal elements holds if $\theta = \pi/4$, so the new theory covariance matrix is

$$C_S' = \frac{\sigma_s^2}{\sigma_n^2} \begin{pmatrix} 1 + \epsilon & 0 \\ 0 & 1 - \epsilon \end{pmatrix}. \qquad (11.80)$$

The rotation matrix which diagonalized the theory matrix is the first one on the left in Eq. (11.78) with $\theta = \pi/4$, so

$$R_3 = \frac{1}{\sqrt{2}} \begin{pmatrix} 1 & 1 \\ -1 & 1 \end{pmatrix}. \tag{11.81}$$

The new modes are $\Delta' = R_3 R_2 \Delta$; explicitly, they are

$$\Delta_1' = \frac{(\Delta_1 + \Delta_2)}{\sqrt{2}\sigma_n}$$

$$\Delta_2' = \frac{(-\Delta_1 + \Delta_2)}{\sqrt{2}\sigma_n}. \tag{11.82}$$

The new modes (Eq. (11.82)) and their covariance matrix (Eq. (11.80)) are easy to understand if we consider the special cases of $\epsilon = 0$ and $\epsilon = +1$. If $\epsilon = 0$ the two modes have the same signal-to-noise, σ_s/σ_n. If the expected signal is large, these modes both carry information; if not, the noise swamps the signal. In either case, each mode — the sum and the difference — is equally (un)important. If $\epsilon = +1$ then the theory predicts the same signal in each pixel. In that case, the difference mode (Δ_2') is worthless, since only noise contributes to it. We see this from the fact that $(C_S')_{22}$ goes to zero as $\epsilon \to 1$. Its signal-to-noise is zero. The other mode — the sum of the two pixels — has signal-to-noise of $\sqrt{2}\sigma_s/\sigma_n$ since the two measurements beat down the noise by a factor of $\sqrt{2}$. This would of course have emerged from the full 2×2 likelihood analysis. But, using both modes in the analysis is a waste of time, a waste which is detected and obliterated by the Karhunen–Loève method.

Bunn (1995) and Bond (1995) independently analyzed the COBE data by looking at Karhunen–Loève modes. Figure 11.8 shows several such modes: clearly the ones with the highest signal-to-noise are the large-scale modes, indicating that COBE was sensitive to large-angle anisotropy. The smallest signal-to-noise modes are the small-scale modes which COBE did not have the resolution to measure.

Vogeley and Szalay (1996) first applied this technique to galaxy surveys. In this context, the Karhunen–Loève method has another useful feature. Recall from Section 11.2.3 that on large scales, the signal covariance matrix is larger than Poisson noise, while on small scales the reverse is true. When we order the modes, then, large-scale modes will have the largest ratio of signal-to-noise. Thus the Karhunen–Loève basis will preferentially pick out large-scale modes. This is extremely useful because we are often most interested in eliminating small-scale modes — which are afflicted by nonlinearities and bias — from an analysis. The Karhunen–Loève method does this automatically! An example is shown in Figure 11.9. For example, the first eigenmode roughly weights pixels only on the basis of their distance from us. It essentially takes the difference between the number of galaxies close to us and the number at moderate distances. The second mode takes a different component of the dipole, the difference between the number of galaxies on the left and the number on the right. Modes with lower weight take successively higher moments of the galaxy distribution.

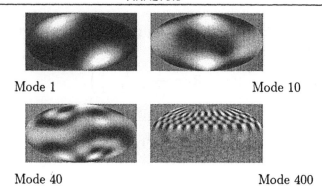

Mode 1   Mode 10

Mode 40   Mode 400

Figure 11.8. Modes of the COBE anisotropy experiment (from Bunn, 1995) ordered according to their signal-to-noise. The mode with the largest signal-to-noise (upper left) is sensitive predominantly to the quadrupole, while the mode with the smallest signal (bottom right) is sensitive to much smaller scale structure.

There are several drawbacks to the Karhunen–Loève method. First, we need to assume a C_S to begin with, in order to identify the modes that are worthless. Although this might seem like a big problem — the modes that are thrown out for one choice of C_S could conceivably be important for another — people who have studied the issue assure us that it is not. They claim that the choice of the important modes is relatively insensitive to the input, assumed C_S. Another drawback is computational. Once the important modes are chosen, C_S' needs to be recalculated at many points in parameter space. In most of parameter space it will not be diagonal at all (it is only diagonal at the special point, the C_S that was chosen as the input spectrum initially). Thus at every point in parameter space, we still need to invert nondiagonal covariance matrices. This drawback is of course partially offset by the fact that — by virtue of the much smaller size of the matrices — the computation is now much faster. Nonetheless in many instances this is not enough to make the full computation managable. We must find still other ways of reducing the computational burden.

Before turning to one such way, let me mention one more use of the Karhunen–Loève method. It is an extremely useful consistency check. This is perhaps best illustrated with an example. The Python experiment (Coble *et al.*, 1999) measured anisotropy over a large (for that time) area, with very low signal-to-noise. In principle, this is a good idea because the large area beats down cosmic variance. In practice, though, it presents challenges because it is difficult to check the consistency of the data. One typically wants to break up the data into several subsets and make sure that each subset sees the same signal. With Python, this was very difficult because noise dominated each subset. One way to check for consistency then is to work with Karhunen–Loève modes. In this basis, each data point d_i' should be drawn from a Gaussian distribution with variance equal to $1 + C_{S,ii}'$. The histogram of $d_i'/(1 + C_{S,ii}')^{1/2}$ should then look Gaussian. Figure 11.10 shows this histogram

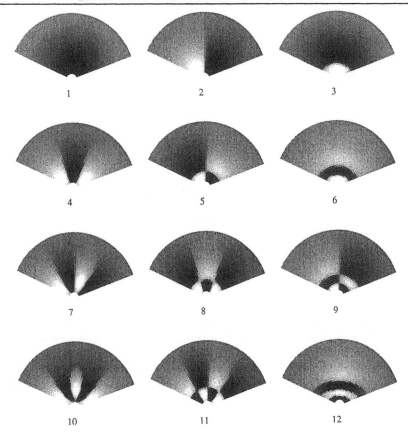

Figure 11.9. The 12 modes with highest signal-to-noise for the CFA2 survey (from Vogeley and Szalay, 1996). Modes pick out the large-scale structure of the galaxy distribution. In each case, we are at the bottom of the slice, and the top region is farthest from us.

using a preliminary noise matrix. There were 650 measurements, but about 70 of these were eliminated for reasons that needn't concern us, so ignore the central spike at $d = 0$. If the remaining 580 data points were distributed as a Gaussian we would expect about two of them (0.3%) to have absolute value greater than 3, and none of them to have absolute value greater than 4. In fact, Figure 11.10 shows that nine modes are more than 4-sigma away from zero. The distribution is definitely not Gaussian with this noise model.

This analysis led the Python team to question the model they had constructed for the noise covariance. (Adjacent points were more correlated than they had allowed for.) Redoing the analysis with a new noise matrix led to the results in Figure 11.11. This technique for testing data quality has been used in a number of other venues, often identifying signs of trouble.

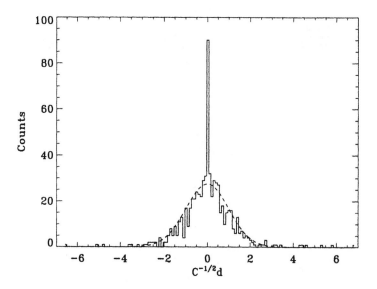

Figure 11.10. Histogram of data from one modulation of the Python CMB anisotropy measurement with a preliminary noise matrix. The data are in Karhunen–Loéve basis in which the covariance matrix is diagonal, so $C^{-1/2}d$ should be distributed as a Gaussian with variance equal to unity. The central spike should be ignored as 70 of the modes have been set to zero. The best fit Gaussian is the solid line. The counts are lower than the best-fit Gaussian in the central region, but above it in the tails.

11.3.2 Optimal Quadratic Estimator

One simple way of speeding up the likelihood calculation is to employ one of many successful *root-finding* algorithms. We are searching for the place where the likelihood function is a maximum, so we want to find

$$\left.\frac{\partial \mathcal{L}}{\partial \lambda}\right|_{\lambda=\bar{\lambda}} = 0 \qquad (11.83)$$

where for simplicity I've assumed that the likelihood function depends on only one parameter λ (we'll generalize this shortly) and $\bar{\lambda}$ is its value at the maximum of the likelihood.

An efficient way to find the root is to consider the derivative of the likelihood function evaluated at some trial point $\lambda = \lambda^{(0)}$. Expand this derivative around $\lambda^{(0)}$ in a Taylor expansion:

$$\mathcal{L}_{,\lambda}(\bar{\lambda}) = \mathcal{L}_{,\lambda}(\lambda^{(0)}) + \mathcal{L}_{,\lambda\lambda}(\lambda^{(0)})\left(\bar{\lambda} - \lambda^{(0)}\right) + \dots \qquad (11.84)$$

where I have introduced the notation of writing partial derivatives as subscripted commas:

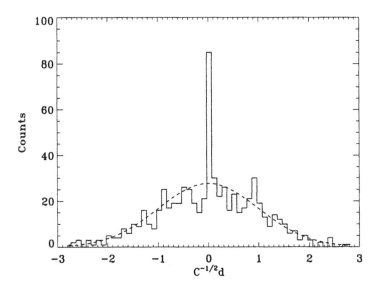

Figure 11.11. Same as Figure 11.10, but this time with an improved noise matrix. Note that there are no data points with absolute value greater than $3C^{1/2}$. The counts are consistent with a Gaussian distribution.

$$\mathcal{L}_{,\lambda} \equiv \frac{\partial \mathcal{L}}{\partial \lambda} \quad ; \quad \mathcal{L}_{,\lambda\lambda} \equiv \frac{\partial^2 \mathcal{L}}{\partial \lambda \partial \lambda}. \tag{11.85}$$

Since $\mathcal{L}$ is maximized at $\bar{\lambda}$, its derivative vanishes there; hence the right-hand side of Eq. (11.84) must also vanish. Setting it to zero leads to a simple expression for $\bar{\lambda}$:

$$\bar{\lambda} \simeq \lambda^{(0)} - \frac{\mathcal{L}_{,\lambda}(\lambda^{(0)})}{\mathcal{L}_{,\lambda\lambda}(\lambda^{(0)})} \tag{11.86}$$

where the $\simeq$ sign acknowledges that we have neglected higher order terms in the Taylor expansion in Eq. (11.84).

The solution in Eq. (11.86) assumes that the likelihood function is a quadratic function of the parameter λ. In fact, it is nothing of the sort: even in the simplest cases the likelihood function is not a quadratic function. For example, far from its maximum, $\mathcal{L}$ typically becomes exponentially small. A much better approximation therefore is that $\mathcal{L}$ is a Gaussian function, so that $\ln(\mathcal{L})$ is quadratic in λ. We can repeat the derivation above since the place where $\mathcal{L}$ is maximized is also the place where $\ln \mathcal{L}$ is a maximum. The estimator for λ is now

$$\hat{\lambda} = \lambda^{(0)} - \frac{(\ln \mathcal{L})_{,\lambda}(\lambda^{(0)})}{(\ln \mathcal{L})_{,\lambda\lambda}(\lambda^{(0)})}. \tag{11.87}$$

Figure 11.12 illustrates the first iteration of this algorithm, called the Newton–Raphson method.

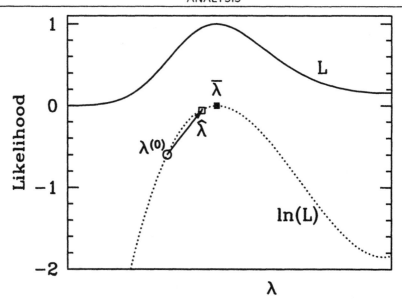

Figure 11.12. A sample likelihood function and its derivatives. The root finding alogrithm starts from the point $\lambda^{(0)}$ (open circle) and moves to $\hat{\lambda}$ (open square) which is quite close to the true maximum of the likelihood, $\bar{\lambda}$ (filled square). The Newton–Raphson technique does this by evaluating the first and second derivatives of $\ln(L)$ at the trial point $\lambda^{(0)}$. The method would not work as well if the derivatives of L were used, because L is not even approximately quadratic away from $\bar{\lambda}$.

To find an estimate for the best-fit value of λ, we need to calculate the derivatives in Eq. (11.87). In the case of the CMB, $\mathcal{L}$ is given explicitly by Eq. (11.20). Thus we need to differentiate the log of Eq. (11.20):

$$(\ln \mathcal{L})_{,\lambda} = \frac{\partial}{\partial \lambda}\left[-\frac{1}{2}\ln(\det C) - \frac{1}{2}\Delta C^{-1}\Delta\right]. \tag{11.88}$$

The covariance matrix here C depends on the theoretical parameter, λ. We can use the identity $\ln \det(C) = \mathrm{Tr}\ln(\mathrm{C})$ and the fact that $C^{-1}_{,\lambda} = -C^{-1}C_{,\lambda}C^{-1}$ to get

$$(\ln \mathcal{L})_{,\lambda} = \frac{1}{2}\Delta C^{-1}C_{,\lambda}C^{-1}\Delta - \frac{1}{2}\mathrm{Tr}[\mathrm{C}^{-1}\mathrm{C}_{,\lambda}]. \tag{11.89}$$

Here, the trace of $C^{-1}C_{,\lambda}$ is the sum of $(C^{-1})_{ij}\partial C_{ji}/\partial \lambda$ over all pixels i and j. Getting the second derivative requires more of the same types of steps. We find

$$(\ln \mathcal{L})_{,\lambda\lambda} = -\Delta C^{-1}C_{,\lambda}C^{-1}C_{,\lambda}C^{-1}\Delta + \frac{1}{2}\mathrm{Tr}[\mathrm{C}^{-1}\mathrm{C}_{,\lambda}\mathrm{C}^{-1}\mathrm{C}_{,\lambda}]$$

$$+ \frac{1}{2}\left(\Delta C^{-1}C_{,\lambda\lambda}C^{-1}\Delta - \mathrm{Tr}[\mathrm{C}^{-1}\mathrm{C}_{,\lambda\lambda}]\right). \tag{11.90}$$

Equation (11.90) gives the second derivative of $\ln \mathcal{L}$ with respect to the parameter λ. By definition, this is minus the *curvature* of the likelihood function:

$$\mathcal{F} \equiv -\frac{\partial^2 \ln \mathcal{L}}{\partial \lambda^2}. \tag{11.91}$$

The curvature is particularly important when evaluated at the maximum of the likelihood function, for there it measures how rapidly the likelihood falls away from the maximum (since the first derivative vanishes). If the curvature is small, then the likelihood changes slowly and the data are not very constraining: the resulting uncertainities on the parameter will be large. Conversely, large curvature translates into small uncertainties.[6]

We could take the ratio of Eqs. (11.89) and (11.90) to get an estimate for the maximum of the likelihood. This is *not* what is usually done. Rather, one typically sets $\Delta\Delta \rightarrow \langle \Delta\Delta \rangle = C$ in the second derivative. Upon doing this, the last line of Eq. (11.90) vanishes, and we are left with

$$\hat{\lambda} = \lambda^{(0)} + F_{\lambda\lambda}^{-1} \frac{\Delta C^{-1} C_{,\lambda} C^{-1} \Delta - \text{Tr}[C^{-1} C_{,\lambda}]}{2} \tag{11.92}$$

where F is defined as

$$F_{\lambda\lambda} \equiv \langle \mathcal{F} \rangle$$

$$= \frac{1}{2} \text{Tr}[C_{,\lambda} C^{-1} C_{,\lambda} C^{-1}]. \tag{11.93}$$

That is, F is the average of the curvature over many realizations of signal and noise, both of which in this case are assumed to be drawn from Gaussian distributions. Here C and its derivatives are evaluated at the input point, $C(\lambda^{(0)})$.

There are several important features about Eq. (11.92), our estimator for the value of λ which maximizes the likelihood function. As the title of this subsection promised, it is a *quadratic* estimator: it is of the general form $A\Delta^2 + B$. The only hard part was determining the coefficients which lead to the best algorithm for finding the root. In the spirit of the Newton–Raphson method, Eq. (11.92) is best used iteratively. One assumes an input spectrum, uses it to determine a new input parameter ($\hat{\lambda}$ in Eq. (11.92)), then uses the new input parameter to find a new best-fit value, and so on until the process converges. In practice, analysts have found that very few iterations are needed until convergence. Nonetheless, we must be somewhat wary of quadratic estimators. It is possible that they will lead us to local maxima in parameter space. Finally, the foregoing discussion assumes there

---

[6]The correspondence between the curvature matrix and parameter uncertainties is often more quantitative than this. See *Numerical Recipes* (Press *et al.*, 1992), Chapter 15, for a detailed discussion. The bottom line, though, is that, in many cases, the inverse of the curvature matrix is a very good approximation to the error matrix on the parameters. E.g., in the one parameter case we are presently considering, the 1-σ, 68% confidence level on λ is approximately equal to the inverse square root of the curvature matrix.

is only one free parameter, but Eqs. (11.92) and (11.93) are easily generalizable to the more relevant case when many parameters are allowed to vary. If we have many parameters $\lambda_\alpha = \lambda_1, \lambda_2, \ldots$, then the quadratic estimator for each is

$$\hat{\lambda}_\alpha = \lambda_\alpha^{(0)} + F_{\alpha\beta}^{-1} \frac{\Delta C^{-1} C_{,\beta} C^{-1} \Delta - \text{Tr}[C^{-1} C_{,\beta}]}{2} \tag{11.94}$$

where the *Fisher matrix* is defined as

$$F_{\alpha\beta} \equiv \langle -\frac{\partial^2 (\ln \mathcal{L})}{\partial \lambda_\alpha \partial \lambda_\beta} \rangle$$

$$= \frac{1}{2} \text{Tr}[C_{,\alpha} C^{-1} C_{,\beta} C^{-1}]. \tag{11.95}$$

Putting aside the details which led to Eq. (11.94), we can appreciate that the result is remarkable. We can now hope to find the values of the parameters which maximize the likelihood function without blindly covering the whole parameter space. A very small number of matrix manipulations suffice to determine these best-fit values. This is a huge advantage, too good to pass up considering the alternative.

We still need to find a way to evaluate the errors on the parameters. Had we evaluated $\mathcal{L}$ everywhere, we could easily identify the region in parameter space ruled out at, say, the 95% level. How do we identify such regions using the quadratic estimator? To answer this question we need to remove ourselves from the derivation above and simply notice that Eq. (11.94) is an estimator for the true best-fit values of the parameters. If we think of it in this way — rather than as the result of a root-finding algorithm — we can study its distribution. Since the distributions for both signal and noise are known (they are assumed Gaussian with covariance matrices C_S and C_N, respectively), we can calculate the estimator's expectation value and variance.

First, let's consider its expectation value:

$$\langle \hat{\lambda}_\alpha \rangle = \lambda_\alpha^{(0)} + F_{\alpha\beta}^{-1} \frac{\langle \Delta C^{-1} C_{,\beta} C^{-1} \Delta \rangle - \text{Tr}[C^{-1} C_{,\beta}]}{2}. \tag{11.96}$$

Here the covariance matrix, its derivatives, and the Fisher matrix have all been evaluated at the trial point $\lambda_\alpha = \lambda_\alpha^{(0)}$. The expectation value $\langle \Delta\Delta \rangle$ on the other hand is equal to the *true* covariance matrix, $C(\bar{\lambda}_\alpha)$. We can expand $C(\bar{\lambda}_\alpha)$ about $\lambda_\alpha^{(0)}$:

$$C(\bar{\lambda}_\alpha) \simeq C + C_{,\alpha'} \left(\bar{\lambda}_{\alpha'} - \lambda_{\alpha'}^{(0)} \right). \tag{11.97}$$

Therefore the expectation value of the quadratic estimator is

$$\langle \hat{\lambda}_\alpha \rangle = \lambda_\alpha^{(0)} + \frac{1}{2} F_{\alpha\beta}^{-1} \left\{ \text{Tr}\left[C^{-1} C_{,\beta} C^{-1} C \right] \right.$$

$$+\mathrm{Tr}\left[C^{-1}C_{,\beta}C^{-1}C_{,\alpha'}\right]\left(\bar{\lambda}_{\alpha'} - \lambda_{\alpha'}^{(0)}\right) - \mathrm{Tr}[C^{-1}C_{,\beta}]\Bigg\}. \qquad (11.98)$$

The first term in brackets cancels the third. The remaining trace is twice the Fisher matrix, so upon multiplying by F^{-1}, we are left with

$$\langle\hat{\lambda}_{\alpha}\rangle = \bar{\lambda}_{\alpha}. \qquad (11.99)$$

So the quadratic estimator we have been considering is *unbiased*: the expectation values of the set of $\hat{\lambda}_{\alpha}$ are equal to the true parameters $\bar{\lambda}_{\alpha}$, no matter what set of parameters are assumed at the outset.

We are also interested in the variance of the estimator:

$$\langle(\hat{\lambda}_{\alpha} - \bar{\lambda}_{\alpha})(\hat{\lambda}_{\beta} - \bar{\lambda}_{\beta})\rangle = \left(F^{-1}\right)_{\alpha\beta}. \qquad (11.100)$$

This equality, which I will leave as an exercise, holds if we are truly at the maximum of the likelihood function and if the data points really are distributed as a Gaussian. If these conditions hold, then the expected errors on the parameters are equal to the square root of the diagonal elements of F^{-1}. This is a magic limit, for there is a theorem, the *Cramer–Rao inequality*, that states that no method can measure the parameters with errors smaller than this (e.g., Kendall and Stuart, 1969). This makes sense since the errors from a full likelihood computation could not possibly be smaller than the width of the likelihood. This width in turn is determined by the curvature, and the Fisher matrix is simply the ensemble average of the curvature. Equation (11.100) tells us that on average, the quadratic estimator of Eq. (11.94) will reach this optimal limit.

Given any point in parameter space, we can calculate the associated Fisher matrix. Thus a simple way to assign error bars to the parameters determined via the quadratic estimator is to use the Fisher matrix evaluated at that point in parameter space. Bond, Jaffe, and Knox (1998), among others, have shown that this prescription works well: i.e., it agrees with a more complete tracing out of the likelihood contours.

Equation (11.100) is useful for other reasons as well. As is apparent from Eq. (11.95), the Fisher matrix — and hence the expected errors on any set of parameters — can be evaluated *without any data*. It will serve us well in Section 11.4.3 when we set out to determine how well upcoming experiments will be able to determine parameters.

We have derived the quadratic estimator in a way which might lead you to believe that it is restricted to the CMB. Namely, our derivation assumed that the likelihood function is Gaussian, true for the CMB but not for galaxy surveys. Even without the assumption of a Gaussian likelihood function, though, the quadratic estimator of Eq. (11.94) can be applied to galaxy surveys. Like any quadratic estimator, it has a mean and a variance. We have just seen that, for Gaussian distributions, it has the lowest variance possible. On large scales, where the galaxy distribution is Gaussian, therefore, it is extremely relevant. Even on small scales, where nonlinearities add to the variance, it is often competitive with other, more traditional estimators.

11.4 THE FISHER MATRIX: LIMITS AND APPLICATIONS

The Fisher matrix plays a key role in describing the ability of a given experiment to constrain parameters. It is difficult to gain much insight, though, from the definition in Eq. (11.95). Fortunately, in the case of full sky coverage, the Fisher matrix can be computed analytically. This analytic computation can then be extended — via a plausibility argument — to the more realistic case of partial sky coverage. This calculation is presented next for both the CMB and galaxy surveys. The most popular use of the Fisher matrix is as a tool for forecasting. How well do we expect a given experiment (even a hypothetical one) to determine cosmological parameters? The Fisher matrix is ideally suited for this task, and we will see some startling expectations from upcoming experiments.

11.4.1 CMB

The trace in Eq. (11.95) is a sum of the diagonal elements of the matrix $[C_{,\alpha}C^{-1}C_{,\beta}C^{-1}]_{ij}$ where i,j index the pixels used in the map. There are two decisions that need to be made before the Fisher matrix can be computed. First, what pixelization scheme should we use, and second what parameters λ_α are we interested in? For a full-sky CMB experiment, we choose as our parameters the C_l's themselves. That is, we take each individual C_l as a free parameter and ask how well an experiment can determine it. To avoid confusion (both the covariance matrix and the C_l's are C's), let's call each parameter λ_l instead of C_l, at least while working through the algebra. This answers the second question. The best way to deal with the first question — how to pixelize — in the case of the CMB is to use the a_{lm}'s. That is, instead of using the pixelized temperatures $\Theta(\hat{n})$, use

$$a_{lm} = \int d\Omega Y_{lm}^*(\hat{n})\Theta(\hat{n}) \tag{11.101}$$

as the data values. Each *pixel* then is labeled by l and m, so a given row (or column) in the covariance matrix corresponds to a fixed valued of l and m. Explicitly, since we start with the quadrupole,

$$C = \begin{pmatrix} C_{l=2,m=-2;l'=2,m'=-2} & C_{2,-2;2,-1} & \cdots & C_{2,-2;2,2} & C_{2,-2;3,-3} & \cdots \\ C_{2,-1;2,-2} & C_{2,-1;2,-1} & \cdots & C_{2,-1;2,2} & C_{2,-1;3,-3} & \cdots \\ & & \vdots & & & \\ C_{3,-3;2,-2} & C_{3,-3;2,-1} & \cdots & C_{3,-3;2,2} & C_{3,-3;3,-3} & \cdots \\ & & \vdots & & & \end{pmatrix}. \tag{11.102}$$

As usual the covariance matrix is the sum of the signal and noise covariance matrices. From Eq. (8.63), the signal covariance matrix would be $\delta_{ll'}\delta_{mm'}\lambda_l$ (remember that we're using λ_l instead of C_l) if the window function were unity. Let's assume that the experiment measures the anisotropy with a beam size σ. Then the signal covariance matrix must be multiplied by $e^{-l^2\sigma^2}$. The noise covariance matrix is a little trickier. You will show in Exercise 11 that, in the case of uncorrelated, uniform noise, it is $\delta_{ll'}\delta_{mm'}w^{-1}$. Here w is the *weight* defined as

$$w = \left[(\Delta\Omega)\sigma_n^2\right]^{-1} \qquad (11.103)$$

where $\Delta\Omega$ is the size in radians of the real space pixels and σ_n is the noise per pixel. Putting these two together, we have

$$C_{lm;l'm'} = \delta_{ll'}\delta_{mm'}\left[\lambda_l e^{-l^2\sigma^2} + w^{-1}\right]. \qquad (11.104)$$

With these simple assumptions, we can take the inverse of the covariance matrix C and also find its derivative with respect to the parameters, the λ_l's. The inverse of the covariance matrix is

$$\left(C^{-1}\right)_{lm;l'm'} = \delta_{ll'}\delta_{mm'}\left[\lambda_l e^{-l^2\sigma^2} + w^{-1}\right]^{-1} \qquad (11.105)$$

while the derivative of the covariance matrix with respect to the parameter λ_α is

$$C_{lm;l'm',\alpha} = \delta_{ll'}\delta_{mm'}\delta_{l\alpha}e^{-l^2\sigma^2}. \qquad (11.106)$$

We can now construct the Fisher matrix; the only difficult task will be keeping track of indices. Very explicitly,

$$F_{\alpha\alpha'} = \frac{1}{2}C_{lm;l'm',\alpha}C^{-1}_{l'm';l''m''}C_{l''m'';l'''m''',\alpha'}C^{-1}_{l'''m''';lm}$$

$$= \frac{1}{2}\left(\delta_{ll'}\delta_{mm'}\delta_{l\alpha}e^{-l^2\sigma^2}\right)\left(\frac{\delta_{l'l''}\delta_{m'm''}}{\lambda_{l'}e^{-l'^2\sigma^2} + w^{-1}}\right)\left(\delta_{l''l'''}\delta_{m''m'''}\delta_{l''\alpha'}e^{-l''^2\sigma^2}\right)$$

$$\times \left(\frac{\delta_{l'''l}\delta_{m'''m}}{\lambda_l e^{-l^2\sigma^2} + w^{-1}}\right) \qquad (11.107)$$

with the implicit sum over $ll'l''l'''mm'm''m'''$. Consider first the Kronecker deltas with subscripts m, m', m'', and m'''. There are four of these; summing over all the subscripts besides m contracts these four to

$$\sum_{m'm''m'''} \delta_{mm'}\delta_{m'm''}\delta_{m''m'''}\delta_{m'''m} = \delta_{mm}. \qquad (11.108)$$

Then, summing over all m leads to a factor of $2l + 1$. The remaining sums over l and its cousins leads to a simple factor of $\delta_{\alpha\alpha'}$. Therefore in the all-sky limit, the Fisher matrix for a CMB experiment is

$$F_{ll'} = \frac{2l + 1}{2}\delta_{ll'}e^{-2l^2\sigma^2}\left[C_l e^{-l^2\sigma^2} + w^{-1}\right]^{-2} \qquad (11.109)$$

where I have gone back to C_l here since the covariance matrix does not appear.

In an all-sky survey, therefore, the Fisher matrix for the C_l's is diagonal. There are no correlations between adjacent C_l's. The errors on a given C_l expected from an all-sky experiment can be read off from Eq. (11.109). The errors are equal to $\sqrt{F^{-1}}$, so

$$\delta C_l = \sqrt{\frac{2}{2l+1}} \left[C_l + w^{-1} e^{l^2 \sigma^2} \right]. \tag{11.110}$$

As anticipated in Eq. (11.26), there are thus two sources of error: (i) cosmic variance, proportional to the signal itself C_l and (ii) noise — atmospheric or instrumental — as encoded in the weight w and the smoothing determined by the beam width σ. The factor of $2l + 1$ in the denominator also traces back to Eq. (11.26); it is the number of independent samples used to estimate a given C_l.

No experiment will ever cover the entire sky, since the CMB cannot be observed in the plane of our galaxy. Even MAP and Planck, two satellites designed to map CMB anisotropy from space, will therefore cover a fraction of the sky $f_{\text{sky}} < 1$. Recalling that the factor of $2l + 1$ in the denominator of Eq. (11.26) counts the number of samples, we could guess that this factor must be multiplied by f_{sky}. This leads to

$$\delta C_l = \sqrt{\frac{2}{(2l+1)f_{\text{sky}}}} \left[C_l + w^{-1} e^{l^2 \sigma^2} \right]. \tag{11.111}$$

This formula enables one to project the errors obtainable for any given experiment. The three characteristics of the experiment which determine the error on C_l are the sky coverage; the weight; and the beam width.

11.4.2 Galaxy Surveys

The analogue of the all-sky CMB experiment is a volume-limited galaxy survey as the volume gets arbitrarily large. This limit applies to all modes with wavelengths k^{-1} much smaller than the typical size of the survey. We have already computed the signal and noise covariance matrices in this limit, Eqs. (11.60) and (11.61), so the covariance matrix for Fourier pixels is

$$C_{\vec{k}_i, \vec{k}_j} = \frac{\delta_{ij}}{V} \left(P(k_i) + \frac{1}{\bar{n}} \right). \tag{11.112}$$

To compute the Fisher matrix, we will need the inverse of this and its derivative with respect to whatever parameters we choose. The inverse is simple,

$$C_{\vec{k}_i, \vec{k}_j}^{-1} = \frac{\delta_{ij} V}{P(k_i) + \frac{1}{\bar{n}}}. \tag{11.113}$$

For our parameters, we will choose the amplitude of the power spectrum in a set of narrow k-bins, each with width Δk. The power in the bin with $k_\alpha < k < k_\alpha + \Delta k$

will be denoted P_α. The derivative of the covariance matrix with respect to P_α is therefore

$$C_{ij,\alpha} \equiv \frac{\partial C_{\vec{k}_i,\vec{k}_j}}{\partial P_\alpha} = \frac{\delta_{ij}}{V} d_{i\alpha} \tag{11.114}$$

with

$$d_{i\alpha} \equiv \begin{cases} 1 & k_\alpha < |\vec{k}_i| < k_\alpha + \Delta k \\ 0 & \text{otherwise} \end{cases}. \tag{11.115}$$

The Fisher matrix is now

$$\begin{aligned} F_{\alpha\alpha'} &= \frac{1}{2} C_{ij,\alpha} C_{jj'}^{-1} C_{j'i',\alpha'} C_{i'i}^{-1} \\ &= \frac{1}{2} \delta_{ij} d_{i\alpha} \frac{\delta_{jj'}}{P(k_j) + \frac{1}{\bar{n}}} \delta_{j'i'} d_{i'\alpha'} \frac{\delta_{i'i}}{P(k_i) + \frac{1}{\bar{n}}}. \end{aligned} \tag{11.116}$$

The sums over j, j', i' are straightforward, leaving

$$F_{\alpha\alpha'} = \frac{1}{2} \frac{\sum_i d_{i\alpha} d_{i\alpha'}}{\left(P_\alpha + \frac{1}{\bar{n}}\right)\left(P_{\alpha'} + \frac{1}{\bar{n}}\right)}. \tag{11.117}$$

As long as the k-bins do not overlap, k_i cannot be in two different bins so the product $d_{i\alpha} d_{i\alpha'}$ requires $\alpha = \alpha'$. The sum then is over all vectors $\vec{k}_i$ in a spherical shell with radius k_α and width Δk. This sum is $4\pi k_\alpha^2 \Delta k V$. Therefore, the Fisher matrix is diagonal, with elements

$$F_{\alpha\alpha'} = \delta_{\alpha\alpha'} \frac{4\pi k_\alpha^2 \Delta k V}{2\left(P_\alpha + \frac{1}{\bar{n}}\right)^2}. \tag{11.118}$$

The error on the power spectrum in this limit is the inverse square root,

$$\delta P_\alpha = \sqrt{\frac{2}{4\pi k_\alpha^2 \Delta k V}} \left(P_\alpha + \frac{1}{\bar{n}}\right). \tag{11.119}$$

This is identical in form to the errors on the C_l's. The denominator in the prefactor counts the number of modes in a given estimate; the first term is cosmic variance; and the last is the noise, in this case Poisson noise.

11.4.3 Forecasting

One of the great promises of upcoming cosmological experiments is that they will determine many of the presently unknown cosmological parameters. How does one predict the expected uncertainties in cosmological parameters from future experiments? The answer is surprisingly simple. Let's consider a CMB experiment as an example. Start with the following:

- A set of C_l's that are assumed to describe the true universe

- The uncertainty on C_l from a given experiment, δC_l, assumed to be given by Eq. (11.111)
- The set of cosmological parameters, $\{\lambda_\alpha\}$, for which we want to forecast errors

The observed C_l's in this universe will be close to the true C_l's; indeed, if we form

$$\chi^2\left(\{\lambda_\alpha\}\right) = \sum_l \frac{\left(C_l(\{\lambda_\alpha\}) - C_l^{\mathrm{obs}}\right)^2}{(\delta C_l)^2}, \tag{11.120}$$

then we expect this χ^2 to reach a minimum at the point in parameter space where $\lambda_\alpha = \bar\lambda_\alpha$, the actual values of the parameters. Of course, we do not now know what those values are, but even without that information, we can ask how quickly $\chi^2\left(\{\lambda_\alpha\}\right)$ changes as λ_α moves away from $\bar\lambda_\alpha$. If it increases rapidly, then the errors on the parameters will be very small; if the χ^2 changes little, then there will be large errors on the parameters.

To quantify this, we can expand χ^2 about its minimum at $\bar\lambda_\alpha$. Let's first do this in the case of one parameter; the generalization to many parameters will be straightforward. In the one-parameter case,

$$\chi^2(\lambda) = \chi^2(\bar\lambda) + \mathcal{F}(\lambda - \bar\lambda)^2. \tag{11.121}$$

The linear term in Eq. (11.121) vanishes since χ^2 is a minimum at $\bar\lambda$. The coefficient of the quadratic term is

$$\mathcal{F} = \frac{1}{2}\frac{\partial^2 \chi^2}{\partial \lambda^2}\bigg|_{\lambda=\bar\lambda}. \tag{11.122}$$

The curvature here, $\mathcal{F}$, measures how rapidly the χ^2 changes away from its minimum. As such, the error on λ is simply $1/\sqrt{\mathcal{F}}$. So all we need to do in order to determine the expected errors on a parameter is calculate $\mathcal{F}$. Note that $\mathcal{F}$ is the curvature of the likelihood function only if the likelihood function is equal to $e^{-\chi^2/2}$, that is, if the errors on the C_l's are Gaussian distributed. In fact, they are not, so $\mathcal{F}$ as given by Eq. (11.122) is not really the curvature, $-\partial^2 \ln\mathcal{L}/\partial\lambda^2$. Nonetheless, the distribution is close enough to Gaussian that the error estimates that arise are expected to be quite accurate.

The second derivative of χ^2 contains two terms:

$$\mathcal{F} = \sum_l \frac{1}{(\delta C_l)^2}\left[\left(\frac{\partial C_l}{\partial \lambda}\right)^2 + (C_l - C_l^{\mathrm{obs}})\frac{\partial^2 C_l}{\partial \lambda^2}\right]. \tag{11.123}$$

The second term in the sum over l is traditionally neglected. The idea (as elucidated in *Numerical Recipes*) is that the difference $C_l - C_l^{\mathrm{obs}}$ will sometimes be negative, sometimes positive. On average, there will be much cancellation, so the first term will dominate. Thus, the general practice is to take

$$\mathcal{F} \to \sum_l \frac{1}{(\delta C_l)^2}\frac{\partial C_l}{\partial \lambda}\frac{\partial C_l}{\partial \lambda}. \tag{11.124}$$

Equivalently, you can think of the dropping of the second term as taking an average over the whole distribution, thereby replacing the curvature matrix with the Fisher matrix (although again keeping in mind that this is not the true curvature or Fisher matrix since the C_l's are not distributed as Gaussians). The generalization of this to many parameters is simply

$$F_{\alpha\beta} = \sum_l \frac{1}{(\delta C_l)^2} \frac{\partial C_l}{\partial \lambda_\alpha} \frac{\partial C_l}{\partial \lambda_\beta}.$$ (11.125)

In order to predict how accurately parameters will be known, then, we simply need to know the experiment's specifications (to determine δC_l) and the derivatives of the C_l's around their (assumed) true values.

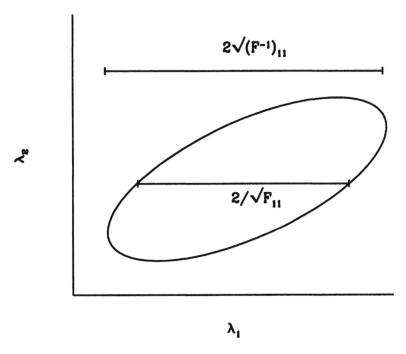

Figure 11.13. Error ellipse in a 2D parameter space. If no prior information is known about λ_2, then the error on λ_1 is $\sqrt{(F^{-1})_{11}}$. If, however, λ_2 is fixed, the error on λ_1, $(F_{11})^{-1/2}$ is smaller.

Assuming a Gaussian distribution, the one-sigma uncertainty on a one parameter fit is $1/\sqrt{F}$. How about if more than one parameter is allowed to vary? Figure 11.13 illustrates the situation in a two-dimensional setting. If the parameter λ_2 is assumed known, then the error on λ_1 is still $1/\sqrt{F_{11}}$. However, if λ_2 is allowed to vary, the error on λ_1 is now $\sqrt{(F^{-1})_{11}}$. It is instructive to prove this explicitly by noting that we are assuming that the joint probability for the two parameters is

$$P(\lambda_1, \lambda_2) \propto \exp\left\{-\frac{1}{2}\lambda_i F_{ij}\lambda_j\right\} \tag{11.126}$$

where I have assumed that the distribution peaks at $\lambda_i = 0$ for simplicity. Allowing λ_2 to vary is equivalent to integrating this probability distribution over all possible values of λ_2. This is referred to as *marginalizing* over λ_2. Then,

$$P(\lambda_1) = \int d\lambda_2 P(\lambda_1, \lambda_2)$$

$$\propto \exp\left\{-\frac{\lambda_1^2}{2}\left(\frac{F_{11}F_{22} - F_{12}F_{21}}{F_{22}}\right)\right\} \tag{11.127}$$

where the second line comes from carrying out the λ_2 integration explicitly. The term in parentheses in the exponential — $[F_{11}F_{22} - F_{12}F_{21}]/F_{22}$ — is simply equal to $1/(F^{-1})_{11}$. So the one-sigma error is indeed given by $\sqrt{(F^{-1})_{11}}$.

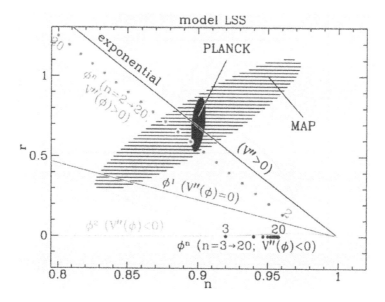

Figure 11.14. Expected 95% uncertainty on the inflationary parameters n and r from MAP and Planck (from Dodelson, Kinney, and Kolb, 1997). (See color Plate 11.14.) Three other parameters (normalization, Ω_B, and h) have been marginalized over. Every inflationary model gives a unique prediction somewhere in this plane; many such predictions are plotted.

Figure 11.14 shows the expected uncertainties from MAP and Planck for two inflationary parameters, those which determine the primordial spectrum. Note that I have fixed the "true" model to be one in which the spectral index $n = 0.9$ and the tensor-to-scalar ratio $r = 0.7$. Different fiducial models often lead to quite different error bars. The ellipse in Figure 11.14 has been drawn after marginalizing over

Table 11.1. Marginalized Errors for ΛCDM for various experiments.

	Map	Planck	Map+SDSS	Planck+SDSS
h	0.22	0.13	0.029	0.022
Ω_m	0.24	0.14	0.036	0.027
Ω_Λ	0.19	0.11	0.042	0.024
$\ln(\Omega_b h^2)$	0.060	0.010	0.050	0.010
m_ν (eV)	0.58	0.26	0.33	0.21
Y_P	0.020	0.013	0.020	0.013
n	0.048	0.008	0.040	0.008
r	0.18	0.012	0.16	0.012
τ	0.022	0.004	0.021	0.004

three other cosmological parameters: normalization, baryon density, and Hubble constant. To do this, one starts with the five-dimensional Fisher matrix, inverts, and then considers only the 2×2 part of the inverse. This 2×2 part defines the ellipses drawn in the figure.

Also plotted in Figure 11.14 are the predictions from a wide variety of inflationary models. The figure argues persuasively that we will indeed be able to distinguish among different inflationary models with upcoming CMB experiments. This is a remarkable finding: we will learn about physics at 10^{15} GeV or higher using CMB data. If more parameters are allowed to vary, then the errors on any one parameter naturally get larger. However, this can be offset by using other observations, most notably those from large-scale structure. Table 11.1 presents the marginalized errors on a number of parameters expected from the MAP, Planck, and Sloan Digital Sky Survey experiments (Eisenstein, Hu, and Tegmark, 1999). Here the errors include the measurement of polarization in MAP and more importantly Planck.

11.5 MAPMAKING AND INVERSION

Until now, we have discussed ways of analyzing a *map*, a collection of pixels with data Δ and noise covariance matrix C_N. How does one make such a map? How do we go from the timestream of data to a pixelized set of spatial Δ's? Most work on this issue has focused on the temperature maps in the CMB, so I'll use this as an example. We will see that mapmaking is essentially an inversion problem. So the mapmaking techniques discussed here are applicable to a broad range of problems in physics and astronomy.

Let's first state the problem. An experiment amounts to a *timestream* of data, d_t. Each number in the timestream corresponds to data taken at a given point on the sky. The data are assumed to be composed of signal plus noise:

$$d_t = P_{ti} s_i + \eta_t. \tag{11.128}$$

Here subscript t denotes a given element in the timestream; i denotes spatial pixel; P is the pointing matrix; s is the underlying temporally constant, but spatially varying signal; and η is the temporal noise. The pointing matrix encodes information about where the receiver is pointing. That is, it associates with each temporal measurement t a given pixel i. It is an $N_t \times N_p$ matrix where N_t is the number of temporal measurements and N_p is the number of spatial pixels. The pointing matrix has a special form: every row has only one nonzero entry equal to 1, corresponding to the pixel on the sky being observed at the time denoted by the row. Each column, however, typically has many nonzero entries corresponding to all the times a given spot has been observed. The noise η is assumed to be Gaussian with a covariance matrix N. There are techniques to determine N directly from the data, but to simplify the discussion, we will assume that N is known.

What is the best way to make a map from this timestream? One forms a χ^2,

$$\chi^2 \equiv \sum_{tt'ij} \left(d_t - P_{ti}s_i \right) N_{tt'}^{-1} \left(d_{t'} - P_{t'j}s_j \right), \tag{11.129}$$

and minimizes with respect to the signal s. Indeed, if the noise is Gaussian, then the likelihood function is proportional to $e^{-\chi^2/2}$ and minimizing the χ^2 is equivalent to maximizing $\mathcal{L}$. Taking the derivative with respect to s_i leads to

$$\frac{\partial \chi^2}{\partial s_i} = -2 \sum_{tt'j} P_{ti} N_{tt'}^{-1} \left(d_{t'} - P_{t'j}s_j \right). \tag{11.130}$$

Set the derivative equal to zero:

$$\sum_{tt'j} P_{ti} N_{tt'}^{-1} P_{t'j} s_j = \sum_{tt'j} P_{ti} N_{tt'}^{-1} d_{t'}. \tag{11.131}$$

The terms multiplying s_j on the left are an $N_p \times N_p$ matrix,

$$\left(C_N^{-1} \right)_{ij} \equiv \sum_{tt'} P_{ti} N_{tt'}^{-1} P_{t'j}. \tag{11.132}$$

Multiply both sides by the inverse of this (C_N itself) to find that the χ^2 is minimized when s is equal to

$$\Delta_i = C_{N,ij} P_{ti} N_{tt'}^{-1} d_{t'}. \tag{11.133}$$

In matrix notation,

$$\Delta = C_N P^T N^{-1} d \tag{11.134}$$

where ^{T} denotes transpose. The noise matrix of this map is equal to

$$C_N = \left(P^T N^{-1} P \right)^{-1}, \tag{11.135}$$

a fact which you can verify by taking $\langle \Delta\Delta \rangle$.

A simple limit of Eq. (11.134) emerges when the timestream noise is diagonal and uniform (this is unrealistic). In that case, the elements of C_N become

$$C_{N,ij} \to N \left(\sum_t P_{it}^T P_{tj} \right)^{-1} \tag{11.136}$$

where now N is simply a number, the diagonal element of the timestream noise. Recall that for a given t, P_{ti} is nonzero for only one pixel i. So the product $P_{it}^T P_{tj}$ vanishes unless $i = j$ and the receiver at time t was pointing at pixel i. Thus the sum over t counts the number of times the receiver sampled pixel i; call this number π_i. In this artificial case of uniform, uncorrelated noise, therefore, the noise covariance matrix for the map is diagonal with elements N/π_i. This makes sense: as a given pixel is sampled more times, the standard deviation goes down as $\pi_i^{-1/2}$. The map now becomes

$$\Delta_i \to \frac{1}{\pi_i} \sum_t P_{it}^T d_t. \tag{11.137}$$

That is, one simply averages all the data points corresponding to the given pixel (exactly like Eq. (11.8)).

Figure 11.15 shows a more realistic implementation of Eq. (11.134), from the long-duration balloon flight of Boomerang (Netterfield *et al.*, 2001), launched on December 29, 1998. The map covers a region a region of about 700 square degrees with 7-arcminute pixels. Therefore, the map required $N_p \simeq 50,000$, while the timestream contained of order 2×10^8 data points. Some tricks were needed to avoid direct inversion and multiplication of the large matrices P and N. Nonetheless, the basis for the Boomerang map, indeed for all CMB maps, is Eq. (11.134).

The raw data from which the map is made need not be the timestream. Instead, the raw data could consist of a series of modulations, e.g., the difference between the temperature at two points. Reconstructing a map from a set of modulations sounds like a much different problem than doing the same from the timestream. In fact, it is mathematically identical: the data d is the sum of a signal and noise. The signal can be thought of as a matrix acting on the underlying temperature field. This matrix does not have the exact form as the pointing matrix (i.e., only one nonzero element in each row), but it is a matrix nonetheless, and all the operations in Eqs. (11.134) and (11.135) can be carried out. There is a big advantage in using a map constructed in this fashion as opposed to the modulated data themselves. Ultimately, the main use of the data will be to estimate parameters in a likelihood analysis. As we have seen, one must construct the signal covariance matrix in order to carry out such an analysis. The signal covariance matrix for a map is extremely simple — the window function is simply $P_l(\cos\theta_{ij})$ — whereas that for modulated data is quite difficult to obtain (recall Section 11.2).

Equation (11.134) and the corresponding noise matrix in Eq. (11.135) are even more general than this. They apply to any problem in which the data are a sum of some processed signal and noise, i.e., to an extraordinarily wide range of problems in physics and astronomy. Consider just two examples. First, the angular correlation

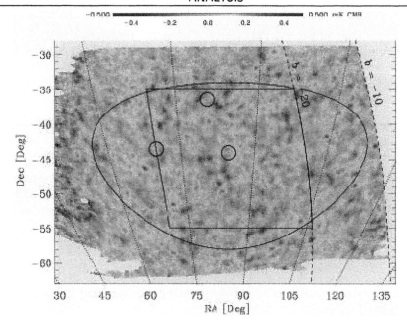

Figure 11.15. A map of the CMB temperature from observations by Boomerang (Netterfield *et al.*, 2001), a long-duration balloon flight at the South Pole. (See also color Plate 11.15.) Hot and cold spots have amplitudes as large as $500\mu K$. Circles shows quasars identified in these radio observations. The large elliptical region delineates data analyzed to obtain bandpowers. The rectangular region is an earlier data set.

function is a sum of the 3D power spectrum processed by a kernel plus noise. One can apply Eq. (11.134) directly to obtain the 3D power spectrum, simply replacing the pointing matrix with the kernel. In the next section, we will see another example, the extraction of the CMB signal from data contaminated by foregrounds.

11.6 SYSTEMATICS

A *systematic* error is one which remains even after averaging over many data samples. Systematic errors are the most worrisome aspect of most cosmological observations. Knowing this, observers typically take many precautions against them, submitting raw data to a wide variety of consistency checks. Many of these tests are the result of common sense and intuition. Nothing much formal can be said about them. Here I want to focus on several ... systematic ways of dealing with such effects.

11.6.1 Foregrounds

One of the biggest obstacles to observing the anisotropies in the CMB are *foregrounds*, other sources of radiation which also emit at microwave frequencies. The list of foregrounds is long and includes anything in space that might come between us and the radiation left over from the Big Bang. There is dust, synchrotron radiation, and free-free or bremsstrahlung emission, all emanating from our galaxy (but extending to regions of the sky far from the galactic plane). There are also extragalactic sources of radiation, point sources and clusters of galaxies. All of these have the potential to contaminate an experiment searching only for a cosmic signal. The magnitude of this problem is hinted at in the nomenclature. Usually in science, a possible source of systematic error is called a *background*. In CMB physics, we cannot call these things backgrounds: the "B" in CMB precludes that possibility. We must acknowledge that the cosmic signal is coming from farther away than any possible contaminant and we must deal with the real possibility that the cosmic signal will be smaller than some of these contaminants.

The problem of foregrounds has in the past few years been demonstrated to be manageable. There are a number of reasons for this good fortune (it is good fortune: if we were living deeper in the galaxy foreground amplitudes would be considerably larger). First observers have been very successful at finding the coolest portions of the sky and using only these regions. Second, foreground amplitudes have proven to be smaller than the cosmic signal in a fairly wide part of frequency space. Figure 11.16 shows the intensity of several galactic foregrounds and the CMB. The amplitudes of each of these components varies across the sky, but the relative amplitudes shown in Figure 11.16 are fairly typical. At very high frequencies, dust dominates, and at very low frequencies synchrotron and bremsstrahlung become important. But, in the range from 30 to 200 GHz, the CMB anisotropies often have the largest intensities.

The final reason foregrounds can be managed — and the one I want to focus on here — can also be gleaned from Figure 11.16. The spectral shapes of the foregrounds are all different, different from one another and from the blackbody shape of the CMB anisotropy. This raises the possibility that detections at different frequencies can be used to extract the CMB signal from the foregrounds. As analysts, we must find the optimal way to perform this extraction. Given measurements at several different frequencies, what is the best algorithm for finding the cosmic signal? How effective do we expect this extraction to be?

First we need to set up some notation. Instead of intensity (or brightness) B, which has units of ergs cm^{-2} sec^{-1} Hz^{-1} , it is convenient to introduce the brightness temperature or *antenna temperature*, defined as

$$T_{\text{ant}} \equiv \frac{B}{2\nu^2}$$

$$= 2\pi\nu f, \tag{11.138}$$

which has dimensions of kelvins (recall that we are working in units with $k_B = \hbar = c = 1$). The frequency ν is related to the momentum we used in earlier chapters via $p = 2\pi\nu$ and f is the familiar phase space density.

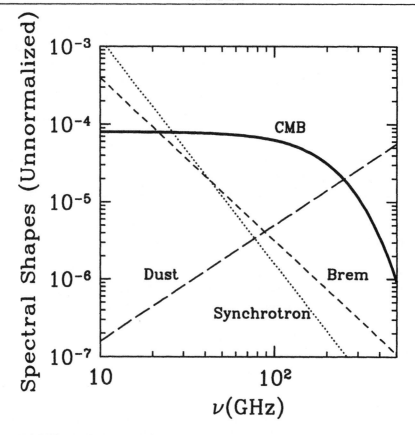

Figure 11.16. Spectral shapes of the dominant galactic foregrounds and the CMB blackbody temperature anisotropy.

For the CMB we know that f is given by Eq. (4.35), so the antenna temperature is

$$\frac{T^{\mathrm{cmb}}_{\mathrm{ant}}}{T} = \frac{x}{e^x - 1} + \Theta \frac{x^2 e^x}{(e^x - 1)^2} \tag{11.139}$$

where

$$x \equiv \frac{p}{T} = \frac{2\pi\nu}{T}. \tag{11.140}$$

The first term in Eq. (11.139) is the monopole, which does not interest us. The second contains information about the shape of the spectrum of the CMB anisotropies. It is useful then to neglect the first term and write

$$\frac{T^{\mathrm{cmb}}_{\mathrm{ant}}(\nu)}{T} = \Theta W^{\mathrm{cmb}} \left(2\pi\nu/T\right) \tag{11.141}$$

where W^α will be the *shape* vector for different components. That for the CMB is

$$W^{\mathrm{cmb}} = \frac{x^2 e^x}{(e^x - 1)^2}. \tag{11.142}$$

Every component can be written in this fashion: the product of an amplitude (in this case Θ), which is frequency independent, and the shape vector, which carries no information about the amplitude. For a blackbody distribution like the CMB, the amplitude has a name: Θ is called the *thermodynamic temperature*. Note that the CMB shape vector goes to 1 in the limit of small frequencies, the Rayleigh–Jeans limit. At high frequencies, it falls off exponentially.

I have called W a shape *vector* implying that it has a number of components. These components are the different frequency channels at which a given experiment takes data. We will label these with a subscript c while allowing for many possible foreground components:

$$\vec{W}^\alpha = W_c^\alpha \qquad c = 1, \ldots, N_{\mathrm{ch}} \quad ; \quad \alpha = 0, \ldots, N_{\mathrm{foregrounds}}. \tag{11.143}$$

Here the CMB component is associated with index $\alpha = 0$. Thus, the data from a given experiment in a given pixel (we focus on only one spatial pixel) on the sky is in the form of a set of antenna temperatures, d_c, at all the different frequency channels. Our model is that this data set is the sum of the contributions from the CMB, foregrounds, and Gaussian noise:

$$d_c = \sum_{\alpha=0}^{N_{\mathrm{foregrounds}}} W_c^\alpha \Theta^\alpha + n_c. \tag{11.144}$$

We assume that we know the covariance matrix of the noise N and all the spectral shapes[7] W_c^α. The question is, how do we best determine the CMB anisotropy?

This problem has the exact same form as does the general inversion problem of Section 11.5. We want to obtain estimates of the amplitude Θ^α. We can immediately write down the minimum variance estimator for Θ^α:

$$\Delta^\alpha = (C_N)_{\alpha\beta} W_c^\beta N_{cd}^{-1} d_d \tag{11.145}$$

with covariance matrix

$$\left(C_N^{-1}\right)_{\alpha\beta} = W_c^\alpha (N^{-1})_{cd} W_d^\beta. \tag{11.146}$$

Let's work out a simple example to bring these formulae, which are so ubiquitous, to life.

---

[7]This assumption is true for the CMB, and very nearly true for synchrotron and bremsstrahlung which have shapes which are fairly constant over the sky. The assumption is worst for dust. A number of groups have explored the consequences of incorrect assumptions about the shapes or allowed for some freedom in the shapes.

Consider an experiment taking measurements at two frequencies, both in the Rayleigh–Jeans regime. The shape vector for the CMB then has two components, both equal to 1:

$$\vec{W}^0 = (1,1). \tag{11.147}$$

Let's also assume that the noise is uncorrelated from one frequency channel to the next and is uniform with diagonal element, σ_n^2. First let's consider the case of zero foregrounds. In that case, the covariance matrix is just a single number, the inverse of

$$WN^{-1}W \rightarrow (1 \quad 1) \begin{pmatrix} 1/\sigma_n^2 & 0 \\ 0 & 1/\sigma_n^2 \end{pmatrix} \begin{pmatrix} 1 \\ 1 \end{pmatrix}$$

$$= \frac{2}{\sigma_n^2}. \tag{11.148}$$

The inverse square root of this is the noise, $\sigma_n/\sqrt{2}$. The two channels reduce the noise by a factor of the square root of 2 (if there were three channels in our example, the factor would be $\sqrt{3}$, etc.). The mimimum variance estimator is given by Eq. (11.145),

$$\Delta \rightarrow \frac{\sigma_n^2}{2}(1,1) \begin{pmatrix} 1/\sigma_n^2 & 0 \\ 0 & 1/\sigma_n^2 \end{pmatrix} \begin{pmatrix} d_1 \\ d_2 \end{pmatrix}$$

$$= \frac{d_1 + d_2}{2}. \tag{11.149}$$

We simply average the two data points.

Now suppose there is one foreground to worry about, say synchrotron emission, with shape vector

$$\vec{W}^1 = (1,1/2). \tag{11.150}$$

Typically, the intensity of synchrotron emission scales as ν^{-1} (see, e.g., Rybicki and Lightman, 1979), so its antenna temperature falls off as ν^{-3}. Thus a shape vector $(1,1/2)$ follows from observing at, e.g., $\nu_1 = 20$ GHz and $\nu_2 = 25$ GHz.

Now the covariance matrix C_N is a two by two matrix. Then Eq. (11.146) becomes

$$C_N^{-1} = W_c^\alpha (N^{-1})_{cd} W_d^\beta = \begin{pmatrix} 1 & 1 \\ 1 & 1/2 \end{pmatrix} \begin{pmatrix} 1/\sigma_n^2 & 0 \\ 0 & 1/\sigma_n^2 \end{pmatrix} \begin{pmatrix} 1 & 1 \\ 1 & 1/2 \end{pmatrix}$$

$$= \frac{1}{\sigma_n^2} \begin{pmatrix} 2 & 3/2 \\ 3/2 & 5/4 \end{pmatrix}. \tag{11.151}$$

Already at this stage, we can glean some important information about the experiment. Recall from the discussion in Section 11.4 that the inverses of the diagonal elements of C_N^{-1} are the unmarginalized variances, the errors if all other parameters are known. In this case, there is one parameter besides the amplitude of the CMB, the amplitude of the foreground. If we assume that is known, then the error on the CMB temperature will be the inverse square root of the $_{00}$ component of the matrix in Eq. (11.151). This is $\sigma_n/\sqrt{2}$, in agreement with the calculation above. To find the errors if we know nothing about the foreground amplitudes, take the inverse of this to find the covariance matrix,

$$C_N = 4\sigma_n^2 \begin{pmatrix} 5/4 & -3/2 \\ -3/2 & 2 \end{pmatrix}. \tag{11.152}$$

The $_{00}$ component of this gives the marginalized variance, $5\sigma_n^2$. The ratio of the marginalized error to the unmarginalized error is a measure of how severely the unknown foregrounds degrade our ability to determine the CMB temperature. It is called the *foreground degradation factor*, or simply the FDF. In this case, the FDF is equal to $\sqrt{10}$.

In this example, we can now determine the minimum variance estimator for the CMB temperature. Following Eq. (11.145), we write

$$\Delta^0 = 4\sigma_n^2 \, (5/4, -3/2) \begin{pmatrix} 1 & 1 \\ 1 & 1/2 \end{pmatrix} \begin{pmatrix} 1/\sigma_n^2 & 0 \\ 0 & 1/\sigma_n^2 \end{pmatrix} \begin{pmatrix} d_1 \\ d_2 \end{pmatrix}$$

$$= -d_1 + 2d_2. \tag{11.153}$$

This should be no surprise. The best estimator for the CMB temperature is completely insensitive to the amplitude of the foreground component. For, if the foreground really does have shape vector $(1, 1/2)$, the linear combination $-d_1 + 2d_2$ from the foreground is equal to zero.

In real life, one must find the minimum variance estimators at many different spatial pixels. The formula of Eq. (11.145) remains identical in this more general case: one simply adds an index for spatial pixel. It is often most convenient to work with the a_{lm}'s instead of the temperature as a function of angular coordinates. Then, the minimum variance estimator for a_{lm} often weights the different frequency channels differently depending upon how the noise in each channels scales with l. An example is shown in Figure 11.17, based on the five frequencies of the MAP experiment. The figure shows how the best estimator for a_{lm} weighs the five different frequencies in the absence of foregrounds. At low l, all the channels have similar noise, so the best estimator is just the average of the five. The beam size is frequency dependent, however, largest at the lowest frequencies. Therefore, at high l, the lowest channels have no signal. Only the highest frequency channel can be used. Indeed, one sees that the minimum variance estimator gradually drops a channel at a time as l gets larger.

Often prior information on the foregrounds exists, in the form of an estimate of the power spectrum of each foreground component, i.e., its C_l. This prior information can be incorporated into the minimum variance estimate; see, e.g., Exercise 15. Figure 11.18 shows the results of accounting for foregrounds in the MAP experiment. The difference between the weighting scheme in this figure and that in Figure

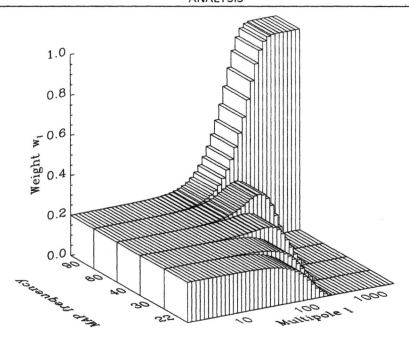

Figure 11.17. The minimum variance linear combination of the frequency channels of MAP in the *absence* of foregrounds (from Tegmark *et al.*, 2000). The noise at low l is identical in all channels, so the minimum variance estimator weights them equally. At high l, the highest frequency channel has the lowest noise, so the best estimator uses only that channel.

11.17 is striking, especially at low l. No longer does one weight all the different channels identically. Rather, a complicated set of weights must be used to project out the foreground contamination.

Note that foregrounds do indeed fit the definition of a systematic effect. If one neglected synchrotron emission in the previous example, the naive estimate of the CMB temperature $(d_1 + d_2)$ would be wrong no matter how small the noise. You might argue that, over the whole sky, the average "wrongness" would cancel out, since there are an equal number of hot and cold spots of foregrounds. The power spectrum, though, the key quantity of interest, *would* be contaminated: it would be the sum of C_l^{CMB} and C_l^{synch}, again even if there was no noise.

11.6.2 Mode Subtraction

A common problem in cosmological observations is that a particular mode is contaminated by some external source. An example might be a region of space very close to our galaxy, where foregrounds are particularly important. In a galaxy survey, it might be a dusty region. Dust tends to absorb high-frequency light, and so redden an object's image. This leads to less flux in blue bands, often pushing a galaxy outside the flux limits of a survey. This push can be crucial, because typ-

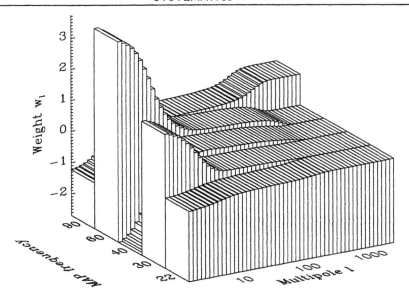

Figure 11.18. The minimum variance linear combination of the frequency channels of MAP in the *presence* of foregrounds (from Tegmark *et al.*, 2000). Compare with Figure 11.17 to see that, especially at low l, foreground contamination dictates a more complicated linear combination of the channels.

ically most of a survey's galaxies lie close to the flux limits. Yet another example is a CMB experiment which has no sensitivity to the average temperature over a given set of pixels because of atmospheric contamination. Another is a galaxy survey in which a given stripe is contaminated because a CCD went bad at the time of observation. And there are many more examples.

One way of dealing with such contamination is to subtract it off. This is commonly done in galaxy surveys by applying the *reddening correction* due to dust from an external dust map (the best one at present is by Schlegel, Finkbeiner, and Davis, 1998).

There is another way of dealing with mode contamination, one which is rapidly growing in importance, as precision cosmology becomes a reality. This technique is based on the twin observations that (i) often there is quite a bit of uncertainty in the amplitude of the contamination and (ii) a given experiment often measures many, many modes. Since the second fact is true, we can often do without the offending mode entirely without losing too much information. Since the first fact is true, we often *should* do without the offending mode, for it may lead us to an incorrect place in parameter space.

How can we eliminate a contaminated mode from an experiment? Let's start with the case where the contaminated mode is a single spatial pixel. In that case, one simple way to ensure that the pixel carries no weight in the likelihood analysis is to add to the covariance matrix a huge number in the diagonal element corresponding to the offending pixel. Then, no matter the value observed in this pixel, the likelihood function will not be affected. This simple idea — adding large noise to a contaminated mode — can be extended to more complicated modes, those not localized to one spatial pixel.

As an example, consider a two-pixel CMB experiment, in which the atmosphere contributes identically to both pixels. Thus, the average temperature of the CMB cannot be determined. The contaminant is assumed to be 100% correlated in the two pixels, so we add to the noise covariance matrix

$$C_{\text{con}} = \kappa \begin{pmatrix} 1 & 1 \\ 1 & 1 \end{pmatrix}, \tag{11.154}$$

where κ is a very large number. Suppose the noise — in the absence of this constraint — is uniform and uncorrelated. Then the new noise covariance matrix is

$$C_N = \sigma_n^2 \begin{pmatrix} 1 & 0 \\ 0 & 1 \end{pmatrix} + \kappa \begin{pmatrix} 1 & 1 \\ 1 & 1 \end{pmatrix}. \tag{11.155}$$

In the likelihood function, we add the noise covariance matrix to the signal covariance matrix. Again, for simplicity assume that the pixels are far enough away from each other so that the signal is uncorrelated. Then, the likelihood function depends on the full covariance matrix

$$C = \begin{pmatrix} \sigma_n^2 + \sigma_s^2 + \kappa & \kappa \\ \kappa & \sigma_n^2 + \sigma_s^2 + \kappa \end{pmatrix}. \tag{11.156}$$

In particular, the likelihood function depends on the determinant and inverse of this matrix (recall Eq. (11.20)). In the limit that κ is very large, the determinant becomes $2\kappa(\sigma_n^2 + \sigma_s^2)$ and the inverse is

$$C^{-1} \to \frac{1}{2(\sigma_n^2 + \sigma_s^2)} \begin{pmatrix} 1 & -1 \\ -1 & 1 \end{pmatrix}. \tag{11.157}$$

Therefore, in this limit, the likelihood function becomes,

$$\mathcal{L} \to \frac{1}{2\pi\sqrt{2\kappa(\sigma_n^2 + \sigma_s^2)}} \exp\left\{ -\frac{(\Delta_1 - \Delta_2)^2}{2(\sigma_n^2 + \sigma_s^2)} \right\}. \tag{11.158}$$

That is, apart from an irrelevant normalization constant, the likelihood function is a Gaussian in $\Delta_1 - \Delta_2$ (the difference between the observed temperatures in the two pixels) with variance equal to $\sigma_n^2 + \sigma_s^2$. Thus adding the constraint matrix of Eq. (11.154) is our way of telling the likelihood function to ignore the average temperature.

In this simple example, we could have written down Eq. (11.158) from the outset, but with more complicated modes, the constraint formalism is extremely powerful. What is the generalization of Eq. (11.154) for an arbitrary contamination? Suppose the contaminated mode is of the form m_i where index i labels the pixels. Thus, in the average example, m_i would be equal to $(1,1)$. One adds to the noise covariance matrix the outer product of this vector times a large number κ:

$$(C_{\text{con}})_{ij} = \kappa m_i m_j. \tag{11.159}$$

This is precisely what we did above for the average, but this expression is now completely general and allows for elimination of any contaminated mode. These matrices, often called *constraint* matrices, have been used extensively in recent CMB analyses, most notably in the interferometric experiment, DASI.

SUGGESTED READING

I am not a professional statistician and this chapter no doubt glosses over some important concepts in statistics. Nonetheless, I believe this chapter does do justice to the recent flurry of activity in cosmological analysis. Readers interested in more general statistics treatments might consult *The Advanced Theory of Statistics* (Kendall and Stuart). An essential reference is *Numerical Recipes* (Press *et al.*) for all numerical work, and Chapter 15 especially for some of the analysis issues discussed here. A couple of nice early papers on the CMB are Readhead *et al.* (1989) and Bond *et al.* (1991). The former takes the likelihood function further than we did here. For example, it deals with frequentist tests and defines such terms as significance and power, which are extremely important in statistics. The Bond *et al.* paper is a concise introduction to Bayesian analysis of a CMB experiment. Among the ideas explained clearly there are CMB window functions; the likelihood function; dealing with an unknown average; and, to top it off, the idea that CMB experiments will probe the baryon density.

The discussion in Section 11.1.3 is based on a similar treatment in Tegmark *et al.* (1998), which cemented the connection between CMB analyses and galaxy surveys. It also deals with pixelization schemes other than just the two in Section 11.1.3. Window functions for the CMB are discussed in many places. A nice recent treatment is given by Souradeep and Ratra (2001).

Karhunen and Loéve decompositions were introduced to the cosmology community by Bunn (1995) and Bond (1995), who used them to analyze COBE, and by Vogeley and Szalay (1996) in the context of galaxy surveys. The optimal quadratic estimator was introduced by Tegmark (1997) and Bond, Jaffe, and Knox (1998), the latter using the Newton–Raphson motivation I stressed in the text. The former focused on the *minimum variance* aspect, which you can prove in Exercise 10. Earlier, Feldman, Kaiser, and Peacock (1994) had computed an optimal estimator for galaxy surveys which turns out to be the small-scale limit of the optimal quadratic estimator. The Fisher matrix was introduced by Fisher (1935). Knox (1995) computed the Fisher matrix in the all-sky CMB case, Tegmark *et al.* (1998) for galaxy surveys. Jungman *et al.* (1996) used the Fisher matrix (although they didn't call it that) to give the first forecast of parameter determination. There have been many follow-up papers improving and tweaking various parts of the forecast. Some of the improvements are discussed by Eisenstein, Hu, and Tegmark (1999). The curvature matrix and the covariance matrix of errors on parameters are covered in detail in *Numerical Recipes.*

Mapmaking and indeed many of the issues in CMB analysis are reviewed by Bond *et al.* (1999). Foregrounds have been discussed by many authors. The text borrows most heavily from Dodelson (1997) and Tegmark *et al.* (2000). Other works of note include Tegmark and Efstathiou (1996) and Bouchet and Gispert (1999). Mode subtraction along the lines discussed in Section 11.6.2 was introduced by Bond, Jaffe, and Knox (1998). A nice example of its use is in Halverson *et al.* (2002).

EXERCISES

Exercise 1. In the simple example of Section 11.1.1, show that a prior uniform in σ_n^2 gives a final probability distribution in σ_n different from the one in Eq. (11.5).

Exercise 2. In the simple example of Section 11.1.1, we found the error on the signal s. What is the error on the other theoretical parameter of the model σ_n?

Exercise 3. Derive the expression for the covariance matrix due to Poisson sampling, Eq. (11.32).
(a) Divide the survey region into small sub-volumes. Assume that the number of galaxies in a given sub-volume is drawn from a Poisson distribution with mean $\bar{n}$ (assume $\bar{n}$ is constant in all sub-volumes for simplicity),

$$P(n) = \frac{(\bar{n})^n e^{-\bar{n}}}{n!}. \tag{11.160}$$

Determine the expectation values $\langle n \rangle$ and $\langle n^2 \rangle$ for this distribution.
(b) Rewrite Eq. (11.28) as

$$\Delta_i = v \sum_\alpha \psi_i(\vec{x}_\alpha) \left[\frac{n(\vec{x}_\alpha) - \bar{n}}{\bar{n}} \right] \tag{11.161}$$

where α indexes each sub-volume of size v. Using the results of **(a)**, and assuming that there is no intrinsic clustering, determine $\langle \Delta_i \Delta_j \rangle$. Show that it is given by Eq. (11.32). You'll have to change the sums back into integrals.

Exercise 4. Determine the noise covariance matrix in a galaxy survey using counts-in-cells. Assume the cells are spherical with radius R, and find $(C_N)_{ij}$ as a function of the separation between two cell centers; call it r_{ij}. Assume that the survey is volume limited, that is, that $\bar{n}$ is constant everywhere within a volume V.

Exercise 5. Do a full likelihood analysis of the University of California at Santa Barbara's 1990-1 CMB experiment carried out at the South Pole (Gaier *et al.*, 1992).
(a) Determine the window function. The chopping angle was 2.1° and the beam width (FWHM) 1.35°. The anisotropy was measured at nine positions, each separated by 2.1° on the sky. Neglect off-diagonal elements.
(b) Fit a flat band power (i.e., take $\mathcal{C} \equiv l(l+1)C_l/2\pi$ constant) to the following data:

Position	$\delta T(\mu K)$	$\sigma_n(\mu K)$
1	-30.5	25.9
2	-3.2	26.5
3	29.2	26.1
4	-10.8	26.3
5	-8.7	28.8
6	23.1	26.4
7	4.7	26.5
8	-24.7	26.6
9	20.3	25.8

Assume that the noise and the signal are both uncorrelated from pixel to pixel.
(c) The likelihood function you obtain should peak at band power equal to zero.
Find the 95% CL upper limit on the band power, defined by the point for which

$$\int_0^{C_U} d\mathcal{C}\mathcal{L}(\mathcal{C}) = 0.95 \int_0^\infty d\mathcal{C}\mathcal{L}(\mathcal{C}). \tag{11.162}$$

This upper limit was reported around the same time as the COBE detection of
anisotropies. Compare the two results.

Exercise 6. Find the Fourier transform of the top-hat function $f(x) = \Theta(x + R)\Theta(R - x)$ where Θ is the step function, equal to 1 when its argument is positive
and zero otherwise.

Exercise 7. Find the diagonal elements of the covariance matrix C_S for a volume
limited survey for modes $k_i \gg R^{-1}$. Show that they are given by Eq. (11.60).

Exercise 8. Compute the off-diagonal window function for the two types of galaxy
surveys mentioned in Section 11.2.3, a volume-limited survey and a pencil-beam
survey. Take both $\vec{k}_i$ and $\vec{k}_j$ parallel to the z-axis (which in the pencil beam survey
is aligned with the long distance L). Plot the window function at $k = k_i$ as a
function of k_j. At the point $k_i = k_j$, you should recapture the corresponding points
in Figures 11.5 and 11.7.

Exercise 9. Prove Eq. (11.100). Assume there is only one parameter λ and that
the likelihood function is Gaussian in the overdensities Δ. Further assume that you
have iterated enough times so that the input parameter $\lambda^{(0)}$ is equal to the true
value $\bar{\lambda}$.

Exercise 10. Derive the optimal quadratic estimator of Eq. (11.92) by minimizing the variance of a general quadratic estimator subject to the constraint that
its expectation value is unbiased. Hint: Use a Lagrange multiplier to enforce the
constraint.

Exercise 11. Consider an all-sky CMB experiment with spatial pixels of area $\Delta\Omega$.
Assume that the experiment measures the temperature in each pixel with Gaussian
noise σ_n. The noise is thus assumed to be uniform (the same everywhere on the

sky) and uncorrelated (from one pixel to the next). Determine the noise covariance matrix for a_{lm}. If the pixel size is cut in half (for the same experiment), each pixel will get less observing time by a factor of 2. The noise will then go up for each pixel by a factor of $\sqrt{2}$. Show that these two changes (smaller pixels; more noise per pixel) leave the noise covariance matrix for a_{lm} unchanged.

Exercise 12. Estimate the expected errors on C_l for the following experiments: (i) COBE, (ii) Boomerang, and (iii) MAP, and (iv) Planck.

Exercise 13. The full-sky limits for the Fisher matrix derived in Section 11.4 can be used to find the optimal quadratic estimator. The results should not surprise you.
(a) Given a set of a_{lm} from a full-sky CMB experiment with uniform weight w, find the optimal quadratic estimator for C_l.
(b) Given a set of pixelized Fourier overdensities from a 3D galaxy survey, find the optimal quadratic estimator for $P(k)$.
(c) From your answers to (a) and (b), discuss qualitatively the effectiveness of the optimal quadratic estimator. When do you expect it to perform differently from the naive quadratic estimator, $\sum_m |a_{lm}|^2/(2l+1)$ for C_l and $\sum_{\vec{k}} |\delta_{\vec{k}}|^2/(4\pi k^2 \Delta k)$ for $P(k)$ (where Δk is the width of the k-bin)?

Exercise 14. Show that the noise covariance matrix of a map using the estimator in Eq. (11.134) is C_N, as given in Eq. (11.135)

Exercise 15. Suppose one had prior information about a foreground, in the form of an assumed power spectrum, C_l^α, where α labels the foreground component.
(a) Find the miminum variance estimator for the temperature and the associated covariance matrix.
(b) Consider the example of Section 11.6.1 with two frequencies in the Rayleigh–Jeans regime and one foreground with shape vector $\vec{W}^1 = (1, 1/2)$. What is the best estimator of the CMB temperature if the foreground has assumed mean equal to zero and variance equal to that of the noise (σ_n^2)? What is the new error on the determination of the CMB temperature? Compare both of these to the case treated in the text when no information about the foreground amplitude was assumed.

APPENDIX A

SOLUTIONS TO SELECTED PROBLEMS

The problems at the end of each chapter have a broad range of difficulty. Some are simply repeating calculations in the text in a slightly different context; others are fairly elementary applications of basic formulae; while some are quite difficult, culled from recent papers. Here are some selected solutions. The solutions are heavily weighted to the first several chapters, especially Chapter 2, because it is important to be comfortable with the background cosmology before proceeding to tackle perturbations.

CHAPTER 1

Exercise 1 The ratio

$$\frac{\rho_\Lambda}{3H^2/(8\pi G)} = (\rho_\Lambda/\rho_c)_0 \left(\frac{H_0}{H}\right)^2 \tag{A.1}$$

where subscript 0 means evaluate today, where it is assumed to be 0.7. Again, by assumption, the universe is forever radiation dominated (clearly not true today, but a good approximation early on), so $H/H_0 = a^{-2}$. The temperature also scales as a^{-1}, so $H/H_0 = (T/T_0)^2$ with $T_0 = 2.7K = 2.3 \times 10^{-4}$ eV. So,

$$\frac{\rho_\Lambda}{3H^2/(8\pi G)} = 0.7 \left(\frac{T_0}{T}\right)^4. \tag{A.2}$$

At the Planck scale, $T_0/T = 2.3 \times 10^{-4}/1.22 \times 10^{28}$, so

$$\frac{\rho_\Lambda}{3H^2/(8\pi G)} = 9 \times 10^{-128}. \tag{A.3}$$

This is the so-called fine-tuning problem: for the cosmological constant to be important today, it had to have been fine-tuned to an absurdly small value at early times. It's a deep problem.

Exercise 2 We need to do the integral

$$t_0 = \frac{1}{H_0} \int_0^1 \frac{da}{a} \left[\Omega_\Lambda + \frac{1 - \Omega_\Lambda}{a^3} \right]^{-1/2} \tag{A.4}$$

for $\Omega_\Lambda = 0.7$ and 0. The latter case can be done analytically:

$$\int_0^1 \frac{da}{a} a^{3/2} = \frac{2}{3}. \tag{A.5}$$

So $t_0 = 2/3H_0 = 0.67 \times 10^{10} h^{-1}$ yrs. When Ω_Λ is not zero, the integral needs to be done numerically. I find

$$\int_0^1 \frac{da}{a} \left[0.7 + \frac{0.3}{a^3} \right]^{-1/2} = 0.96. \tag{A.6}$$

So for fixed Hubble constant, a cosmological constant universe is older than a matter-dominated one, older by a factor of $0.96/0.67 = 1.43$. For $h = 0.7$, a cosmological constant universe has an age of 14 billion years, in accord with other observations of the age of the universe.

Exercise 4 An inverse wavelength is ν/c, so replacing ν everywhere in Eq. (1.8) by c/λ leads to

$$I_\nu = \frac{4\pi\hbar c}{\lambda^3} \frac{1}{\exp\{2\pi\hbar c/\lambda k_B T\} - 1}. \tag{A.7}$$

This is energy per Hz; we want energy per cm^{-1}, so we need to multiply by c, leaving

$$I_{1/\lambda} = \frac{4\pi\hbar c^2}{\lambda^3} \frac{1}{\exp\{2\pi\hbar c/\lambda k_B T\} - 1}. \tag{A.8}$$

Plugging in numbers leads to

$$I_{1/\lambda} = 1.2 \times 10^{-5} \text{erg sec}^{-1} \text{ cm}^{-1}\text{sr}^{-1} \left(\frac{\text{cm}}{\lambda}\right)^3 \frac{1}{\exp\{0.53\text{cm}/\lambda\} - 1}. \tag{A.9}$$

A quick check verifies that this agrees with Figure 1.10.

To find the peak, differentiate I with respect to $1/\lambda$ and set equal to zero. This leaves

$$\lambda = \frac{1}{3} \frac{(2\pi\hbar c/k_B T)}{1 - \exp\{-2\pi\hbar c/\lambda k_B T\}}. \tag{A.10}$$

So $1/\lambda_{\text{peak}}$ is $3/.53\text{cm}^{-1}$. The exact coefficient, accounting for the exponential is 2.82, so $1/\lambda_{\text{peak}} = 5.3\text{cm}^{-1}$, exactly where it occurs in Figure 1.10.

CHAPTER 2

Exercise 1

   (a) To get from kelvin to eV, use $k_B = \text{eV}/(11605K)$. So $2.725\text{K} \rightarrow k_B 2.725\text{K} = (2.725/11605)$ eV. Or 2.348×10^{-4} eV.

(b) Since $T_0 = 2.348 \times 10^{-4}$ eV,

$$\rho_\gamma = \frac{\pi^2 T_0^4}{15} = 2.000 \times 10^{-15} \text{eV}^4. \tag{A.11}$$

To get this in g cm^{-3}, first divide by $(\hbar c)^3 = (1.973 \times 10^{-5} \text{eV cm})^3$ to get 0.2604eV cm^{-3}. Then to change from eV to grams, remember that the mass of the proton is either 1.673×10^{-24} g or 0.9383×10^9 eV, so 1 eV $= 1.783 \times 10^{-33}$ g. Therefore, $\rho_\gamma = 4.643 \times 10^{-34}$ g cm^{-3}.

(c) We have parametrized $H_0 = 100\, h$ km sec^{-1} Mpc^{-1}, or using the fact that one Mpc is equal to 3.1×10^{19} km, $H_0 = 3.23\, h \times 10^{-18}$ sec^{-1}. To get this into inverse cm, divide by the speed of light, $c = 3 \times 10^{10}$ cm sec^{-1}; then $H_0 = 1.1\, h \times 10^{-28}$ cm. Or $H_0^{-1} = 9.3\, h^{-1} \times 10^{27}$ cm.

(d) To get the Planck mass (1.2×10^{28} eV) into kelvins, multiply by $k_B^{-1} = 11605\text{K/eV}$; then $m_{\text{Pl}} = 1.4 \times 10^{32}$ K. To get it into inverse cm, divide by $\hbar c = 1.97 \times 10^{-5}$ eV cm to get $m_{\text{Pl}} = 6.1 \times 10^{32}$ cm^{-1}. To get this is units of time, multiply by the speed of light to get $m_{\text{Pl}} = 6.1 \times 10^{32} \times 3 \times 10^{10}$ cm sec^{-1}, or $m_{\text{Pl}} = 1.8 \times 10^{43}$ sec^{-1}.

Exercise 7

Start with

$$\Gamma^0_{\mu\nu} = \frac{g^{0\alpha}}{2} \left[\frac{\partial g_{\alpha\mu}}{\partial x^\nu} + \frac{\partial g_{\alpha\nu}}{\partial x^\mu} - \frac{\partial g_{\mu\nu}}{\partial x^\alpha} \right] \tag{A.12}$$

where μ, ν range from 0 t0 2, 0 being the time index, 1 corresponding to θ, and 2 to ϕ. Since the metric is diagonal, $g^{0\alpha}$ is nonzero only when $\alpha = 0$ in which case it is -1. So

$$\Gamma^0_{\mu\nu} = \frac{-1}{2} \left[\frac{\partial g_{0\mu}}{\partial x^\nu} + \frac{\partial g_{0\nu}}{\partial x^\mu} - \frac{\partial g_{\mu\nu}}{\partial t} \right]. \tag{A.13}$$

All of these terms vanish: the first two since g_{00} is a constant, and the last because none of the metric elements depend on $x^0 = t$. So $\Gamma^0_{\mu\nu} = 0$ for all μ, ν.

Next consider

$$\Gamma^\theta_{\mu\nu} = \frac{g^{\theta\alpha}}{2} \left[\frac{\partial g_{\alpha\mu}}{\partial x^\nu} + \frac{\partial g_{\alpha\nu}}{\partial x^\mu} - \frac{\partial g_{\mu\nu}}{\partial x^\alpha} \right]. \tag{A.14}$$

Again since the metric is diagonal, and $g^{\theta\theta} = 1/r^2$, this reduces to

$$\Gamma^\theta_{\mu\nu} = \frac{1}{2r^2} \left[\frac{\partial g_{\theta\mu}}{\partial x^\nu} + \frac{\partial g_{\theta\nu}}{\partial x^\mu} - \frac{\partial g_{\mu\nu}}{\partial \theta} \right]. \tag{A.15}$$

Only the $g_{\phi\phi}$ component depends on one of our variables, so only it is nonzero when differentiated. Therefore, the first two terms vanish and the last is nonzero only when $\mu = \nu = \phi$, in which case it is

$$\Gamma^\theta_{\phi\phi} = \frac{1}{2r^2} \left[-r^2 \frac{\partial \sin^2 \theta}{\partial \theta} \right] = -\sin\theta \cos\theta. \tag{A.16}$$

Finally, when the upper index is ϕ, we have

$$\Gamma^\phi_{\mu\nu} = \frac{1}{2r^2 \sin\theta} \left[\frac{\partial g_{\phi\mu}}{\partial x^\nu} + \frac{\partial g_{\phi\nu}}{\partial x^\mu} - \frac{\partial g_{\mu\nu}}{\partial \phi} \right]. \tag{A.17}$$

The last term vanishes since none of the metric elements depend on ϕ; the first two are nonzero only if one of the indices μ, ν is equal to ϕ and the other is θ, so

$$\Gamma^\phi_{\phi\theta} = \Gamma^\phi_{\theta\phi} = \frac{\cos\theta}{\sin\theta}. \tag{A.18}$$

The geodesic equation is

$$\frac{d^2 x^\mu}{d\lambda^2} = -\Gamma^\mu_{\alpha\beta} P^\alpha P^\beta \tag{A.19}$$

with

$$P^\mu \equiv \frac{dx^\mu}{d\lambda}. \tag{A.20}$$

Let's apply this to the $\mu = \theta$ component. The left-hand side is

$$\frac{d^2\theta}{d\lambda^2} = \frac{d}{\lambda}\frac{dt}{d\lambda}\dot\theta = E^2 \ddot\theta \tag{A.21}$$

since $E = dt/d\lambda$ is constant. The Christoffel symbol on the right-hand side $\Gamma^\theta_{\alpha\beta}$ is nonzero only when $\alpha = \beta = \phi$ in which case it is $-\sin\theta\cos\theta$. So,

$$\ddot\theta - \sin\theta\cos\theta \, (\dot\phi)^2 = 0. \tag{A.22}$$

For the second equation, consider the ϕ component of the geodesic equation,

$$\frac{d^2\phi}{d\lambda^2} = -\Gamma^\phi_{\alpha\beta} P^\alpha P^\beta. \tag{A.23}$$

Again the left-hand side is simply $E^2 \ddot\phi$. The right-hand side gets nonzero contributions when $\alpha = \theta, \beta = \phi$ or an identical term when $\alpha = \phi, \beta = \theta$. Therefore,

$$\ddot\phi + 2\frac{\cos\theta}{\sin\theta}\dot\theta\dot\phi = 0. \tag{A.24}$$

Incidentally this is equivalent to

$$\frac{d}{dt}\left(\dot\phi \sin^2\theta\right) = 0 \tag{A.25}$$

and the conserved quantity in parentheses is the angular momentum.

The Ricci scalar is

$$\mathcal{R} = g^{\mu\nu} R_{\mu\nu} = -R_{00} + \frac{1}{r^2} R_{\theta\theta} + \frac{1}{r^2 \sin^2\theta} R_{\phi\phi}. \tag{A.26}$$

The time-time component vanishes since all Γ's with time components are zero. We need to compute the two spatial components. First, consider

$$R_{\theta\theta} = \frac{\partial \Gamma^{\alpha}_{\ \theta\theta}}{\partial x^{\alpha}} - \frac{\partial \Gamma^{\alpha}_{\ \theta\alpha}}{\partial \theta} + \Gamma^{\alpha}_{\ \beta\alpha}\Gamma^{\beta}_{\ \theta\theta} - \Gamma^{\alpha}_{\ \beta\theta}\Gamma^{\beta}_{\ \theta\alpha}. \tag{A.27}$$

The first and third terms vanish since the Christoffel symbol with two lower θ's vanishes. For the same reason, the index α in the second term must be equal to ϕ, and both β and α in the last term must equal ϕ:

$$R_{\theta\theta} = -\frac{\partial(\cos\theta/\sin\theta)}{\partial \theta} - \left(\frac{\cos\theta}{\sin\theta}\right)^2. \tag{A.28}$$

Carrying out the derivative then gives

$$R_{\theta\theta} = \left[1 + \frac{\cos^2\theta}{\sin^2\theta}\right] - \left(\frac{\cos\theta}{\sin\theta}\right)^2 = 1. \tag{A.29}$$

The other spatial component is

$$R_{\phi\phi} = \frac{\partial \Gamma^{\alpha}_{\ \phi\phi}}{\partial x^{\alpha}} - \frac{\partial \Gamma^{\alpha}_{\ \phi\alpha}}{\partial \phi} + \Gamma^{\alpha}_{\ \beta\alpha}\Gamma^{\beta}_{\ \phi\phi} - \Gamma^{\alpha}_{\ \beta\phi}\Gamma^{\beta}_{\ \phi\alpha}. \tag{A.30}$$

The Christoffel symbol in the first term is nonzero only if $\alpha = \theta$, while the one in the second term is always zero. In the third term β must be equal to θ to make the second Christoffel symbol be nonzero, and then $\alpha = \phi$. In the last term β can be θ and $\alpha = \phi$ or vice versa, so

$$R_{\phi\phi} = \frac{\partial \Gamma^{\theta}_{\ \phi\phi}}{\partial \theta} + \Gamma^{\phi}_{\ \theta\phi}\Gamma^{\theta}_{\ \phi\phi} - \Gamma^{\phi}_{\ \theta\phi}\Gamma^{\theta}_{\ \phi\phi} - \Gamma^{\theta}_{\ \phi\phi}\Gamma^{\beta}_{\ \phi\theta}. \tag{A.31}$$

The middle two terms cancel leaving

$$R_{\phi\phi} = -\frac{\partial(\sin\theta\cos\theta)}{\partial \theta} + \sin\theta\cos\theta\frac{\cos\theta}{\sin\theta}. \tag{A.32}$$

Carrying out the derivative gives

$$R_{\phi\phi} = -\cos^2\theta + \sin^2\theta + \cos^2\theta = \sin^2\theta. \tag{A.33}$$

Summing up, we get

$$\mathcal{R} = \frac{1}{2r^2}. \tag{A.34}$$

The Ricci scalar is therefore a measure of the curvature of the space.

Exercise 9 Accumulating the various Γ's leads to

$$\frac{d^2x^i}{d\lambda^2} = -2\frac{\dot{a}}{a}\frac{dt}{d\lambda}\frac{dx^i}{d\lambda}. \tag{A.35}$$

Change to differentiation with respect to η using the facts that $dt/d\lambda = E$ and $d\eta/d\lambda = E/a$. Then the geodesic equation becomes

$$\frac{E}{a}\frac{d}{d\eta}\left(\frac{E}{a}\frac{dx^i}{d\eta}\right) = -2\frac{\dot{a}}{a}\frac{E^2}{a}\frac{dx^i}{d\eta}. \tag{A.36}$$

Since $E/a \propto a^{-2}$, when the derivative on the left acts on E/a, the resulting term (proportional to $dx^i/d\eta$) exactly cancels the term on the right, leaving the result of Eq. (2.99).

Exercise 10 The age integral is

$$t(a) = \int_0^a \frac{da'}{a'H(a')}. \tag{A.37}$$

Since we are assuming only matter and radiation, we can take

$$H(a) = H_0 \sqrt{\rho/\rho_{cr}} = \sqrt{\frac{1}{a^3} + \frac{\Omega_r}{a^4}} \tag{A.38}$$

where the $1/a^3$ term is from matter with density equal to the critical density. When the density in matter is equal to critical, $\Omega_r = a_{eq} = 4.15 \times 10^{-5}h^{-2}$. Therefore, the age integral is

$$t = \frac{1}{H_0} \int_0^a \frac{da'a'}{\sqrt{a' + a_{eq}}}. \tag{A.39}$$

Integrate by parts to get

$$H_0 t = 2a\sqrt{a + a_{eq}} - 2 \int_0^a da' \sqrt{a' + a_{eq}}. \tag{A.40}$$

Carrying out the last integral leads to

$$H_0 t = 2a\sqrt{a + a_{eq}} - \frac{4}{3}\left\{[a + a_{eq}]^{3/2} - a_{eq}^{3/2}\right\}. \tag{A.41}$$

At very early times, such as when the temperature was 0.1 MeV, a is much smaller than a_{eq}, so

$$t \to \frac{a^2}{2H_0\sqrt{a_{eq}}} \quad ; \quad a \ll a_{eq}. \tag{A.42}$$

This limit is easiest to see directly in the integral of Eq. (A.39), but you can also get it by Taylor expanding Eq. (A.41). When the temperature is 0.1 MeV, the scale factor is $2.35 \times 10^{-4}\,\mathrm{eV}/0.1\,\mathrm{MeV} = 2.35 \times 10^{-9}$, the temperature today divided by 0.1 MeV. Plugging in numbers leads to

$$t\,(0.1\,\mathrm{MeV}) = 4.28 \times 10^{-16} \times 9.78 \times 10^9\,\mathrm{yr} = 130\,\mathrm{sec}. \tag{A.43}$$

At $T = 1/4$ eV, $a = 9.4 \times 10^{-4}$, significantly larger than $a_{eq} = 8.5 \times 10^{-5}$ with $h = 0.7$, so $H_0 t \to (2/3)a^{3/2}$. So,

$$t\,(1/4\,\mathrm{eV}) = 270,000\,\mathrm{yr}. \tag{A.44}$$

Exercise 12 The angle subtended is the physical distance divided by the angular diameter distance

$$\theta(z) = \frac{5\,\mathrm{kpc}(1+z)}{\chi(z)}. \tag{A.45}$$

In a flat, matter-dominated universe, χ is given by Eq. (2.43). When $z = 0.1$ (1), the term in brackets in Eq. (2.43) is equal to 0.0465 (0.293). The comoving distance out to z is, therefore,

$$\chi = \begin{cases} 280h^{-1}\,\text{Mpc} & z = 0.1 \\ 1760h^{-1}\,\text{Mpc} & z = 1 \end{cases}. \tag{A.46}$$

Carrying out the division and converting radians to arcsec (1 radian equals 2.06×10^5 arcsec) leads to

$$\theta = \begin{cases} 4.0h'' & z = 0.1 \\ 1.2h'' & z = 1 \end{cases}. \tag{A.47}$$

In a universe with $\Omega_\Lambda = 0.7, \Omega_m = 0.3$, χ must be computed numerically. At $z = 1$, I find χ to be larger than in the flat, matter-dominated case by a factor of 1.3, so the angular size will be smaller by this factor, down to $0.9h''$. At $z = 0.1$ the difference in comoving distances is only 5%, so the angular size goes down to $3.8h''$ in the cosmological constant case.

Exercise 13 Rewriting Eq. (1.8) in terms of momentum $p = h\nu/c = 2\pi\hbar\nu/c$ and recognizing the denominator there as $1/f$ leads to

$$I_\nu = f\frac{4\pi p^3}{(2\pi)^3} \tag{A.48}$$

with $\hbar = c = 1$. So the energy density is the integral of this over all frequencies, with a factor of 4π to count photons from all directions (i.e., I_ν is per steradian):

$$\rho_\gamma = 4\pi \int_0^\infty d\nu\, I_\nu. \tag{A.49}$$

This can be converted into an integral over momentum, with $d\nu = dp/(2\pi)$:

$$\rho_\gamma = 2 \int_0^\infty dp\, I_\nu. \tag{A.50}$$

Exercise 15 We want to compute $\rho = -T_0^0$. Setting $\mu = \nu = 0$ leads to

$$T^0{}_0 = -g_i \int \frac{dP_1 dP_2 dP_3}{(2\pi)^3}(-\det[g_{\mu\nu}])^{-1/2} P^0 f_i. \tag{A.51}$$

The matrix $g_{\mu\nu}$ is diagonal, so the determinant is simply the product of the diagonal elements, $-a^6$. By definition, $p^2 = g^{ij}P_iP_j = a^{-2}\delta_{ij}P_iP_j$. So $p_i \equiv \hat{p}_i p = P_i/a$ with $\hat{p}_i$ a unit vector pointing in the direction of the momentum. Therefore, $d^3P = a^3 d^3p$ and the factors of a precisely cancel those coming from the determinant. We're left with

$$T_0^0 = -g_i \int \frac{d^3p}{(2\pi)^3} P^0 f_i. \tag{A.52}$$

The four vector P_μ squared is equal to $-m^2$, the mass of the particle, so $g_{00}(P^0)^2 = -m^2 - g_{ij}P^iP^j = -m^2 - p^2$. Since $g_{00} = -1$, $P^0 = \sqrt{p^2 + m^2}$, in accord with Eq. (2.59).

Exercise 17 The energy density of a massless boson is $g\pi^2 T^4/30$, while that of a fermion is 7/8 times this. So,

$$s = \frac{2\pi^2}{45}\left[\sum_{i=\text{bosons}} g_i T_i^3 + \frac{7}{8}\sum_{i=\text{fermions}} g_i T_i^3\right] \tag{A.53}$$

accounting for the possibility that different species have different temperatures.

CHAPTER 3

Exercise 1 The number density of a species with degeneracy $g = 2$ is

$$n = 2 \int \frac{d^3p}{(2\pi)^3} f(p). \tag{A.54}$$

For the distributions we will consider, the phase space density f depends only on the magnitude of the momentum, so the angular part of the integral can be performed leading to the a factor of 4π; therefore,

$$n = \frac{1}{\pi^2} \int_0^\infty dp \, p^2 f(p). \tag{A.55}$$

First let's consider the high m/T limit. In this case, the limit of the Boltzmann distribution is $\exp[-(m + p^2/2m)/T]$. I claim, though, that this is precisely the limit of both the Fermi–Dirac and Bose–Einstein distributions:

$$\frac{1}{e^{E/T} \pm 1} \rightarrow e^{-E/T} \tag{A.56}$$

since $E \simeq m \gg T$ so that the exponential in the denominator dwarfs the 1. Therefore the low-temperature limit of all three distributions is

$$n^{\text{low T}} = \frac{e^{-m/T}}{\pi^2} \int_0^\infty dp \, p^2 e^{-p^2/2mT}. \tag{A.57}$$

To do the integral, define a dimensionless parameter $x \equiv p/\sqrt{2mT}$. In terms of the variable $dpp^2 = [2mT]^{3/2} dx x^2$, so

$$n^{\text{low T}} = \frac{e^{-m/T}}{\pi^2} [2mT]^{3/2} \int_0^\infty dx \, x^2 e^{-x^2}. \tag{A.58}$$

But the integral is equal to $\sqrt{\pi}/2$, so we have

$$n^{\text{low T}} = 2 e^{-m/T} \left(\frac{mT}{2\pi} \right)^{3/2}. \tag{A.59}$$

The high-temperature Boltzmann limit is

$$n^{\text{Hi T, Boltz}} = \frac{1}{\pi^2} \int_0^\infty dp \, p^2 e^{-p/T}. \tag{A.60}$$

Defining the dummy variable $x \equiv p/T$ leads to

$$n^{\text{Hi T, Boltz}} = \frac{1}{\pi^2} T^3 \int_0^\infty dx \, x^2 e^{-x}. \tag{A.61}$$

The x integral is equal to 2. So,

$$n^{\text{Hi T, Boltz}} = \frac{2T^3}{\pi^2}. \tag{A.62}$$

The Bose–Einstein and Fermi–Dirac integrals similarly are

$$n^{\text{Hi T, BE/FD}} = \frac{T^3}{\pi^2} \int_0^\infty \frac{dx\, x^2}{e^x \mp 1}. \tag{A.63}$$

The integrals can be written in terms of the Riemann zeta function, via Eq. (C.27). So the integral in Eq. (A.63) with the minus sign — the Bose–Einstein distribution — is $\zeta(3)\Gamma(3) = 2\zeta(3)$. The integral with the plus sign — the Fermi–Dirac distribution — is $3\zeta(3)\Gamma(3)/4 = 3\zeta(3)/2$, so

$$n^{\text{Hi T}} = \frac{\zeta(3)T^3}{\pi^2} \begin{cases} 2 & \text{Bose–Einstein} \\ 3/2 & \text{Fermi–Dirac} \end{cases}.$$

By the way, $\zeta(3) \simeq 1.202$, so there are more bosons than fermions for the same temperature, and these bracket the Boltzman amount. All of course are proportional to T^3.

Exercise 6 The photon number density is $411\,\text{cm}^{-3}$, while the baryon number density is $n_b = \rho_b/m_p = \rho_{\text{cr}}\Omega_b/m_p$. Plugging in numbers gives

$$n_b = \Omega_b \frac{1.879h^2 \times 10^{-29}\text{g cm}^{-3}}{1.673 \times 10^{-24}\text{g}} = 1.12 \times 10^{-5}\Omega_b h^2\,\text{cm}^{-3}. \tag{A.64}$$

So η_b, the ratio of the baryon to the photon number density, is indeed given by Eq. (3.11).

Exercise 11 To find this ratio, we compute the entropy density $(\mathcal{P} + \rho)/T$ at the two times. In both cases, only relativistic particles contribute to the entropy density significantly so that Eq. (A.53) holds. At high temperatures, the following particles contribute to the energy density: quarks ($g_* = 5 \times 3 \times 2$ for the five least massive types — up, down, strange, charm, bottom — each with three colors and two spin states); anti-quarks ($g_* = 30$ again); leptons ($g_* = 6 \times 2$ for the six types — $e, \nu_e, \mu, \nu_\mu, \tau, \nu_\tau$ — each with two spin states); anti-leptons ($g_* = 12$ again); photons (2); and gluons ($g_* = 8 \times 2$ for eight possible colors each with two spin states). This totals up to

$$g_* = 2 + 16 + \frac{7}{8}(30 + 30 + 12 + 12) = 91.5. \tag{A.65}$$

The sixth quark, the top quark, does not contribute because it is too heavy to be around at these temperatures $m_t \simeq 175$ GeV. Today entropy comes only from photons and neutrinos. The former contribute 2 to g_*; the latter contribute $(7/8) \times 3 \times 2 \times (4/11)^{4/3} = 1.36$, so today $g_* = 3.36$. Since the product sa^3 remains constant, we have

$$\left[g_*(aT)^3\right]\Big|_{T=10\text{ Gev}} = \left[g_*(aT)^3\right]\Big|_{T_0}. \tag{A.66}$$

Therefore,

$$\frac{(aT)^3\big|_{T=10\text{ Gev}}}{(a_0T_0)^3} = \frac{3.36}{91.5} = \frac{1}{27}. \tag{A.67}$$

CHAPTER 4

Exercise 1 First integrate Eq. (4.6) over all momentum. This gives

$$\frac{\partial n}{\partial t} + \frac{\partial (nv)}{\partial x} = 0, \tag{A.68}$$

the $\partial f/\partial p$ term vanishing after integrating by parts and noticing that $f = 0$ at $p = \pm\infty$ (there are no particles with infinite momentum). This is the continuity equation. To get the Euler equation, first multiply by p/m and then integrate over all momentum. This gives

$$\frac{\partial (nv)}{\partial t} + \frac{\partial}{\partial x} \int_{-\infty}^{\infty} \frac{dp}{2\pi} \frac{p^2}{m^2} + \frac{kx}{m} n = 0 \tag{A.69}$$

where the last term follows from an integration by parts. The integral over p^2 yields two terms, one a *bulk velocity* term, v^2, and the second a pressure term, P. Using the continuity equation reduces this to

$$\dot{v} + v\frac{\partial v}{\partial x} + \frac{1}{n}\frac{\partial P}{\partial x} + \frac{kx}{m} = 0. \tag{A.70}$$

Exercise 4 From Eq. (3.3), the electron distribution function peaks at zero momentum, with a maximum value of $e^{(\mu-m_e)/T}$. To relate the chemical potential to the density, recall that $n = e^{\mu/T}n^{(0)}$, so in the low-temperature limit (Eq. (3.6)):

$$e^{\mu/T} = \frac{n_e}{2}\left(\frac{2\pi}{m_eT}\right)^{3/2} e^{m_e/T}. \tag{A.71}$$

So the maximum value of f_e is $(n_e/2)(2\pi/m_eT)^{3/2}$. Divide Eq. (3.44) by the Thomson cross-section to get $n_e = 1.12 \times 10^{-5}\Omega_B h^2 \mathrm{cm}^{-3}$ today including both ionized and captured electrons. Taking the electron temperature to be equal to the photon temperature today gives $2\pi/m_eT = 2.04 \times 10^{-11}\mathrm{cm}^2$. Putting back in the factors of a leads to

$$f_e^{\mathrm{MAX}} = 10^{-21}\Omega_B h^2 a^{-3/2}. \tag{A.72}$$

This expression holds only up to $T \leq m_e$, corresponding to $a \simeq 4.6 \times 10^{-10}$. So, as long as the temperature is well below the electron mass, f_e is very small.

Exercise 7 The difference between the amplitude we used in the derivation in Section 4.3 and the more accurate one given in the problem is $2\pi\sigma_T m_e^2[3\cos(\hat{p}\cdot\hat{p}') - 1]$. The combination is square brackets is twice the second Legendre polynomial. Rewrite using the addition formula of spherical harmonics; then the difference becomes

$$2\pi\sigma_T m_e^2 \frac{8\pi}{5}\sum_{m=-2}^{2} Y_{2m}(\hat{p})Y_{2m}^*(\hat{p}'). \tag{A.73}$$

This is the quantity we need to insert into the multiple integral in Eq. (4.49) in place of $\mathcal{M}^2$. When we do this, only the $m = 0$ term will contribute since all other

$Y_{2m}(\hat{p}')$ have an azimuthal dependence which integrates to zero. Therefore, the new collision term due to anisotropic Compton scattering is

$$\delta C[f(\vec{p})] = \frac{\pi^2 n_e \sigma_T}{p} \mathcal{P}_2(\mu) \int \frac{d^3 p'}{(2\pi)^3 p'} \mathcal{P}_2(\hat{\gamma}' \cdot \hat{k})$$

$$\times \left\{ \delta(p - p') + (\vec{p} - \vec{p}') \cdot \vec{v} \frac{\partial \delta(p - p')}{\partial p'} \right\} \{ f(\vec{p}') - f(\vec{p}) \}, \quad \text{(A.74)}$$

where I have used the fact that $Y_{20} = -\sqrt{5} \mathcal{P}_2 / \sqrt{4\pi}$. The only term which survives the angular integral is the one proportional to $\delta(p - p') f(\vec{p}')$, leaving

$$\delta C[f(\vec{p})] = -\frac{n_e \sigma_T}{2p} \mathcal{P}_2(\mu) \int_0^\infty dp' p' \delta(p - p') p' \frac{\partial f^{(0)}}{\partial p'}$$

$$\times \int_{-1}^{1} \frac{d\mu}{2} \mathcal{P}_2(\mu) \Theta(\mu). \quad \text{(A.75)}$$

The angular integral gives $-\Theta_2$. Then integrating over the Dirac δ-function yields

$$\delta C[f(\vec{p})] = +p \frac{\partial f^{(0)}}{\partial p} \frac{n_e \sigma_T}{2p} \mathcal{P}_2(\mu) \Theta_2. \quad \text{(A.76)}$$

This adds a factor of $-\mathcal{P}_2 \Theta_2 / 2$ inside the square brackets of Eq. (4.54) and explains the corresponding factor in Eq. (4.100).

CHAPTER 5

Exercise 4 In Fourier space,

$$\epsilon_{ijk}(\hat{k}_k \hat{k}_l - \delta_{kl}/3) G_{,jl}^L = -k^2 \epsilon_{ijk}(\hat{k}_k \hat{k}_j - \hat{k}_j \hat{k}_k/3) G^L$$

$$= -2k^2/3 \epsilon_{ijk} \hat{k}_j \hat{k}_k G^L = 0 \quad \text{(A.77)}$$

since ϵ_{ijk} is antisymmetric under interchange of j and k while $\hat{k}_j \hat{k}_k$ is symmetric. The combination is also traceless since $\delta_{ij}(\hat{k}_i \hat{k}_j - \delta_{ij}/3) = 0$.

Exercise 7 (a) By definition,

$$\Gamma^i_{jk} = \frac{g^{ii'}}{2} [g_{i'j,k} + g_{i'k,j} - g_{jk,i'}]. \quad \text{(A.78)}$$

All derivatives here are spatial, and the only spatially varying part of the metric is the first-order piece $\mathcal{H}$. Therefore, we can again use the zero-order $g^{ii'} = \delta_{ii'}/a^2$, leaving Eq. (5.43).

(b) The product $\Gamma^\alpha_{\beta j} \Gamma^\beta_{i\alpha}$ vanishes when both indices α and β are zero (because $\Gamma^0_{0i} = 0$) and when both indices are spatial (because then each Christoffel symbol is first order). Therefore, this product is

$$\Gamma^\alpha_{\beta j} \Gamma^\beta_{i\alpha} = \Gamma^0_{kj} \Gamma^k_{i0} + \Gamma^k_{0j} \Gamma^0_{ik}$$

$$= \Gamma^0_{kj}\Gamma^k_{i0} + (i \leftrightarrow j). \tag{A.79}$$

But

$$\Gamma^0_{kj}\Gamma^k_{i0} = \frac{1}{2}\left(2Hg_{jk} + a^2\mathcal{H}_{jk,0}\right)\left(H\delta_{ik} + \frac{1}{2}\mathcal{H}_{ik,0}\right)$$

$$= (H)^2 g_{ij} + a\frac{da}{dt}\mathcal{H}_{ij,0}. \tag{A.80}$$

We must remember to add back in the same set of terms with i and j interchanged. This just introduces a factor of 2, so

$$\Gamma^\alpha_{\beta j}\Gamma^\beta_{i\alpha} = 2\,(H)^2 g_{ij} + 2a\frac{da}{dt}\mathcal{H}_{ij,0}. \tag{A.81}$$

CHAPTER 6

Exercise 5 There are 411 photons per cm^{-3} today; the Hubble volume is $(4\pi/3)[3000\,h^{-1}\,\mathrm{Mpc}]^3 = 3.3 \times 10^{84}\,h^{-3}\,\mathrm{cm}^3$. So the total number of photons is $1.4 \times 10^{87}\,h^{-3}$. This number remains roughly constant throughout the matter and radiation eras since the number density scales as T^3, the physical volume as a^3, and the temperature as a^{-1}. So another problem of the classical cosmology is: Why is the entropy of the universe so large?

Inflation solves this problem. At first the solution seems obvious: inflation makes the scale factor grow exponentially fast, thereby increasing the product aT and hence the entropy. In fact, the solution is not quite that simple because during inflation, the exponential expansion is adiabatic: the temperature still falls as a^{-1}. So near the end of inflation the temperature has dropped rapidly enough so that if the entropy was initially of order unity, it remained of order unity.

The production of entropy actually takes place at the end of inflation during the reheating process. Even though the temperature at the end of inflation is extremely small, the energy density (which is almost completely in the scalar field) is not. When the energy in the scalar field transforms into radiation, the temperature of the radiation shoots up from its very low value of T to $\rho^{1/4} \gg T$. Thus, the reheating process is responsible for the large entropy we see today. Another way to say this is to point out that inflation is a very ordered state: the universe supercools while the field is trapped in a false vacuum. The transition to the true vacuum is a transition to the very disordered state of equilibrium.

Exercise 11 (a) With this substitution, the equation becomes

$$\frac{d^2\tilde{v}}{d\eta^2} + \frac{2}{\eta}\frac{d\tilde{v}}{d\eta} + \left(k^2 - \frac{2}{\eta^2}\right)\tilde{v}. \tag{A.82}$$

Defining $x \equiv k\eta$, we see that $\tilde{v}$ satisfies the spherical Bessel equation of order 1 (Eq. (C.13)).

(b) The two general solutions are $j_1(x)$ and $y_1(x)$. The general solution is therefore $Aj_1 + By_1$. Writing these out explicitly leads to

$$v = \eta \tilde{v} = \eta \left(A \frac{\sin x - x \cos x}{x^2} - B \frac{\cos x + x \sin x}{x^2} \right)$$

$$= \frac{1}{2k^2 \eta} \left(e^{ik\eta} \left[-iA - Ak\eta - B + iBk\eta \right] \right.$$

$$\left. + e^{-ik\eta} \left[iA - Ak\eta - B - iBk\eta \right] \right). \tag{A.83}$$

When $k\eta$ is very large and negative, we want $v \to e^{-ik\eta}/\sqrt{2k}$, so the coefficient of $e^{+ik\eta}$ in this limit, proportional to $-A + iB$, must vanish. Thus, $A = iB$. The coefficient of $e^{-ik\eta}$ is

$$\frac{1}{2k^2 \eta} \left[-2Ak\eta \right] = \frac{-A}{k}. \tag{A.84}$$

This must equal $(2k)^{-1/2}$, so $A = -(k/2)^{1/2}$. Therefore the correct solution is

$$v = \frac{-1}{\sqrt{2k}k\eta} \left(e^{-ik\eta} \left[i - k\eta \right] \right) \tag{A.85}$$

in agreement with Eq. (6.57).

Exercise 13 The two components of Einstein's equations are

$$k^2 \Psi + 3aH \left(\dot{\Psi} + aH\Psi \right) = 4\pi G a^2 \delta T^0_0$$

$$ik_i (\dot{\Psi} + aH\Psi) = -4\pi G a \delta T^0_i. \tag{A.86}$$

Here I have simply copied the results from Chapter 5, replacing Φ with $-\Psi$. Multiply the second of these by $3iaHk_i/k^2$, and then add the two equations to get

$$k^2 \Psi = 4\pi G a^2 \left[\delta T^0_0 - \frac{3Hk_i \delta T^0_i}{k^2} \right]. \tag{A.87}$$

On large scales, the left-hand side is negligible, so the terms in brackets on the right must sum to zero, giving Eq. (6.85).

CHAPTER 7

Exercise 4

(c) To do the integral, introduce a new dummy variable $x \equiv \sqrt{1+y}$. Then Eq. (7.31) becomes

$$\Phi = \frac{3\Phi(0)}{2} \frac{\sqrt{1+y}}{y^3} \int_1^{\sqrt{1+y}} dx \frac{(x^2-1)^2(3x^2+1)}{x^2}. \tag{A.88}$$

Now integrate by parts using the fact that the integral of $1/x^2$ is equal to $-1/x$. The surface term is proportional to the numerator and so vanishes at the lower limit, when $x = 1$. Therefore,

$$\Phi = \frac{3\Phi(0)}{2} \frac{\sqrt{1+y}}{y^3} \left[-\frac{y^2(4+3y)}{\sqrt{1+y}} + \int_1^{\sqrt{1+y}} dx \left(18x^4 - 20x^2 + 2\right) \right]$$

$$= \frac{3\Phi(0)}{2} \frac{\sqrt{1+y}}{y^3} \left[-\frac{y^2(4+3y)}{\sqrt{1+y}} + \left(\frac{18}{5}x^5 - \frac{20}{3}x^3 + 2x\right)|_1^{\sqrt{1+y}} \right]. \tag{A.89}$$

Evaluating the terms in parentheses at the upper and lower limits leads to Eq. (7.32).

Exercise 9

$$\sigma_R^2 = \left\langle \left[\int d^3x \delta(x) W_R(x) \right]^2 \right\rangle$$

$$= \left\langle \left[\int \frac{d^3k}{(2\pi)^3} \tilde{\delta}(\vec{k}) \tilde{W}_R^*(\vec{k}) \right]^2 \right\rangle \tag{A.90}$$

where $\tilde{\ }$ denotes Fourier transform, and I have used the fact that since $W_R(x)$ is real, $\tilde{W}_R(\vec{k}) = \tilde{W}_R^*(-\vec{k})$. Also I have evaluated δ_R at the origin; and the angular brackets denote the average, now over all realizations of $\tilde{\delta}(\vec{k})$. Squaring and using the fact that

$$\langle \delta(\vec{k})\delta(\vec{k}') \rangle = (2\pi)^3 \delta^3(\vec{k} + \vec{k}') P(k) \tag{A.91}$$

leads to

$$\sigma_R^2 = \int \frac{d^3k}{(2\pi)^3} P(k) \left| \tilde{W}_R(\vec{k}) \right|^2. \tag{A.92}$$

It remains only to compute the Fourier transform of the top-hat window function,

$$\tilde{W}_R(\vec{k}) = \int d^3x W_R(\vec{x}) e^{-i\vec{k}\cdot\vec{x}}$$

$$= \frac{2\pi}{V_R} \int_0^R dx x^2 \int_{-1}^1 d\mu e^{ikx\mu}. \tag{A.93}$$

Note that I have normalized the window function so that the integral over it is unity; hence the factor of $V_R = 4\pi R^3/3$. Carrying out the remaining angular and radial integrals leads to

$$\tilde{W}_R(k) = \frac{3}{kR^3} \int_0^R dx x \sin(kx)$$

$$= \frac{3}{k^3 R^3} \left[-kR\cos(kR) + \sin(kR) \right]. \tag{A.94}$$

By way of solving Exercise 10, note that

$$\Delta^2(k) = \frac{4\pi}{(2\pi)^3} k^3 P(k). \tag{A.95}$$

CHAPTER 8

Exercise 2 Assume a solution of the form $x = e^{i\omega t}$. The damping equation then becomes a quadratic equation for ω:

$$\omega^2 - \frac{ib}{m}\omega - \frac{k}{m} = 0. \tag{A.96}$$

Solving with $k/m > \gamma^2 \equiv (b/2m)^2$ leads to

$$\omega = i\gamma \pm \omega_1. \tag{A.97}$$

The frequency is now $\omega_1 \equiv [k/m - \gamma^2]^{1/2}$, smaller than in the undamped case. The amplitude is also damped by $e^{-\gamma t}$.

Exercise 9 Use the addition theorem of spherical harmonics (C.12) to write

$$\mathcal{P}_{l'}(\hat{\gamma} \cdot \hat{k}) = \frac{4\pi}{2l+1} \sum_{m'} Y_{l'm'}^*(\hat{\gamma}) Y_{l'm'}(\hat{k}). \tag{A.98}$$

Then the angular integral becomes an integral over the product of two spherical harmonics, which — because of orthogonality — is equal to 1 if $l' = l$ and $m' = m$ and zero otherwise. This leads directly to the desired reult.

Exercise 12 I get the result show in Figure A.1. The integral of the cross-term is significantly smaller than that of either of the squares, so there is no interefence between the monopole and dipole.

Exercise 17 The generalization of Eq. (8.67) to tensors gives

$$C_l^T = \sum_{l'l''} (-i)^{l'+l''} (2l'+1)(2l''+1) \int \frac{d^3k}{(2\pi)^3} \Theta_{l'}^T(k)\Theta_{l''}^{T,*}(k) I_{lml'}(k) I_{lml''}^*(k) \tag{A.99}$$

where I have defined

$$I_{lml'}(k) \equiv \sqrt{\frac{8\pi}{15}} \int d\Omega \mathcal{P}_{l'}(\hat{k} \cdot \hat{\gamma}) Y_{lm}(\Omega) \left[Y_{22}(\Omega) + Y_{2-2}(\Omega) \right]. \tag{A.100}$$

The factor of $[8\pi/15]^{1/2}[Y_{22} + Y_{2-2}]$ is the combination $\sin^2\theta\cos(2\phi)$ which appears in Eq. (4.115), so this expression is valid only for the $+$ mode. However, the $\times$ mode gives exactly the same result.

The integral $I_{lml'}$ is not trivial. By rewriting the Legendre polynomial as $[4\pi/(2l'+1)]^{1/2}Y_{l'0}/i^{l'}$, we can turn $I_{ll'}$ into an integral over the product of three

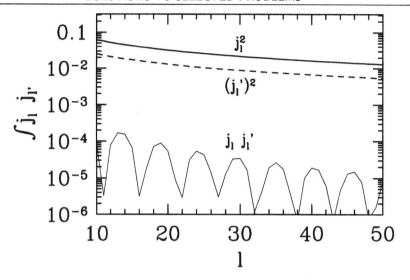

Figure A.1. The integrals of products of spherical Bessel functions.

spherical harmonics. Such integrals are intensively studied in quantum mechanics and can be expressed in terms of the Wigner 3-j symbols. By the way, my favorite reference for these things — especially useful for this integral — is *Quantum Mechanics* (Landau and Lifshitz), like all the other texts in their Course of Theoretical Physics a wonderful investment. The integral is then

$$I_{lml'} = \sqrt{\frac{32\pi^2}{15(2l'+1)}} \frac{1}{i^{l'}} \langle lm|Y_{22} + Y_{2-2}|l'0\rangle \qquad (\text{A.101})$$

which vanishes unless $m = 2$ or $m = -2$. When m takes on one of these two values, the matrix element is

$$\langle l2|Y_{22} + Y_{2-2}|l'0\rangle = i^{l'-l} \begin{pmatrix} l & 2 & l' \\ 0 & 0 & 0 \end{pmatrix} \left[\frac{5(2l'+1)(2l+1)}{4\pi}\right]^{1/2} \begin{pmatrix} l & 2 & l' \\ -2 & 2 & 0 \end{pmatrix}.$$
$$(\text{A.102})$$

The first 3-j symbol here, the one with the bottom row all zero, vanishes unless the sum of the elements in the top row $l + l' + 2$ is even. And of course l' cannot differ from l by more than 2 since the combination of $Y_{22} Y_{l'0}$ leads to angular momenta ranging from $l' - 2$ to $l' + 2$. So the only time the matrix element is nonzero is when $l' = l - 2, l, l + 2$. Using Table 9 in Section 106 of *Quantum Mechanics* leads to the final result:

$$I_{lml'} = \sqrt{\frac{8\pi}{3}} \sqrt{2l+1}\, i^{-l}\, (\delta_{m,2} + \delta_{m,-2}) [c_{-2}\delta_{l',l-2} + c_0\delta_{l',l} + c_2\delta_{l',l+2}] \quad (\text{A.103})$$

where here $\delta_{m,2}$ (and all other δ's) is the Kronecker delta, equal to 1 if $m = 2$ and zero otherwise. The coefficients are

$$c_{-2} = \frac{\sqrt{6}}{4} \frac{[(l-1)l(l+1)(l+2)]^{1/2}}{(2l-3)(2l-1)(2l+1)}$$

$$c_0 = \frac{-2\sqrt{6}}{4} \frac{[(l-1)l(l+1)(l+2)]^{1/2}}{(2l-1)(2l+1)(2l+3)}$$

$$c_2 = \frac{\sqrt{6}}{4} \frac{[(l-1)l(l+1)(l+2)]^{1/2}}{(2l+1)(2l+3)(2l+5)}. \tag{A.104}$$

The result in Eq. (8.93) then follows.

Exercise 18

(a) On large scales, we can take the matter-dominated solution for h, so

$$\Theta_{l,i} = \frac{-1}{2} \int_{\eta_*}^{\eta_0} d\eta \; j_l[k(\eta_0 - \eta)] \frac{d}{d\eta} \left[\frac{3 j_1(k\eta)}{k\eta} \right] P_h^{1/2}. \tag{A.105}$$

Here I have used the fact that the initial amplitude of the gravity waves is $P_h^{1/2}$ with the time dependence given in the square brackets. Plug this into Eq. (8.93) to get

$$C_l^T = 2 \frac{9(l-1)l(l+1)(l+2)}{4\pi} \int_0^\infty dk \; k^2 P_h(k) \left| \int_0^{\eta_0} d(k\eta) \frac{j_2(k\eta)}{k\eta} \right.$$

$$\left. \times \left[\frac{j_{l-2}(k[\eta_0-\eta])}{(2l-1)(2l+1)} + 2\frac{j_l(k[\eta_0-\eta])}{(2l-1)(2l+3)} + \frac{j_{l+2}(k[\eta_0-\eta])}{(2l+1)(2l+3)} \right] \right|^2, \tag{A.106}$$

where I have set the lower limit on the time integral to zero since $\eta_* \ll \eta_0$. Also, I have used the identity $(j_1/x)' = -j_2/x$. The factor of 2 out in front comes from the sum over the $+$ and $\times$ components. Using Eq. (6.100) for P_h (in the slow-roll approximation $\epsilon = 0$ and $\nu = 3/2$) and defining new integration variables $y \equiv k\eta_0$ and $x \equiv k\eta$ leads to

$$C_l^T = 36 \left(\frac{H_{\text{inf}}}{m_{\text{Pl}}} \right)^2 (l-1)l(l+1)(l+2) \int_0^\infty \frac{dy}{y} \left| \int_0^y dx \frac{j_2(x)}{x} \right.$$

$$\left. \times \left[\frac{j_{l-2}(y-x)}{(2l-1)(2l+1)} + 2\frac{j_l(y-x)}{(2l-1)(2l+3)} + \frac{j_{l+2}(y-x)}{(2l+1)(2l+3)} \right] \right|^2. \tag{A.107}$$

Here H_{inf} denotes the Hubble rate during inflation, or more precisely the Hubble rate when the modes in question crossed the horizon (when $k\eta = -1$ early on). This expression does well on the low multipoles. To get even better results stick in the transfer function of Eq. (5.88).

(b) For the $l = 2$ mode, the double integral in Eq. (A.107) is equal to 2.139×10^{-4}, so $C_2^T = 0.185(H/m_{\text{Pl}})^2$. The scalar C_2 is equal to $\pi \delta_H^2/12$. Using Eq. (6.100) for δ_H leads to

$$r = 13.86\epsilon. \tag{A.108}$$

(c) Combining with Eq. (6.104), we expect

$$r = -6.93 n_T. \tag{A.109}$$

For many models, the inflationary parameter $\delta = -\epsilon$, so

$$n - 1 = -n_T. \tag{A.110}$$

CHAPTER 9

Exercise 2 Expand the power spectrum about $k_3 = 0$:

$$P\left(\sqrt{k_3^2 + (H_0\kappa/\chi)^2}\right) = P(H_0\kappa/\chi) + \frac{1}{2k}\frac{dP}{dk}|_{k_3=0}k_3^2 + \ldots. \tag{A.111}$$

For a smooth power spectrum dP/dk is of order P/k, so the coefficient of k_3^2 is of order P/k^2. For us, $k = H_0\kappa/\chi$, so this coefficient is of order $P\chi^2/(H_0\kappa)^2$. We can write k_3^2 as $-H_0^2\partial^2/\partial\chi^2$ acting on the exponential of Eq. (9.9). Assuming the selection function is relatively smooth, this is of order H_0^2/χ^2. So the first correction to the leading term is of order $1/\kappa^2$, which is small as long as the angular scales probed are not too large.

Exercise 3 Define the dummy variable $\chi \equiv H_0\kappa/k$. Then

$$F = H_0 \int \frac{d\chi}{2\pi\chi} J_0(k\theta\chi/H_0)W^2(\chi), \tag{A.112}$$

an expression which clearly depends only on the combination $k\theta$.

Exercise 5 To express C_l^{matter} in terms of w, multiply both sides of Eq. (9.66) by $\mathcal{P}_{l'}(\cos\theta)$ and integrate over $\mu \equiv \cos\theta$. This gives

$$C_l^{\text{matter}} = 2\pi \int_{-1}^{1} d\cos\theta \mathcal{P}_l(\cos\theta)w(\theta). \tag{A.113}$$

Express w as an integral over the 2D power spectrum as in the first line of Eq. (9.13). Then,

$$C_l^{\text{matter}} = \int_0^{\infty} dl'\, l' P_2(l') \int_{-1}^{1} d\cos\theta \mathcal{P}_l(\cos\theta) J_0(l'\theta). \tag{A.114}$$

Note the difference in P's: the first P_2 here is the 2D power spectrum, the second $\mathcal{P}_l$ is the Legendre polynomial. In the limit that l' is large, the Bessel function becomes

$$J_0(l'\theta) \to \mathcal{P}_{l'}(\cos\theta). \tag{A.115}$$

Therefore, the integral over θ vanishes unless $l = l'$, in which case it is equal to $2/(2l+1)$. The integral over l' is identical to a sum over l' at large l' since $dl' \to 1$. The factor of $2/(2l+1)$ in the denominator cancels the factor of l' in the numerator, leaving the desired equality between the 2D power spectrum and C_l^{matter}.

CHAPTER 10

Exercise 2 The total intensity received at the detector is the angular integral of $I_{\text{obs}}(\theta)$ over θ. The total intensity emitted is the angular integral of $I_{\text{true}}(\theta_S)$ over θ_S. The magnification μ is the ratio of the two:

$$\mu \equiv \frac{\int d^2\theta I_{\text{obs}}(\theta)}{\int d^2\theta_S I_{\text{true}}(\theta_S)}. \tag{A.116}$$

Change variables in the denominator to θ, leading to a factor of $\det(A)$ where A is defined in Eq. (10.15). Recall now that $I_{\text{true}}(\theta_S) = I_{\text{obs}}(\theta)$, so except for the determinant, the numerator and denominator cancel. This leaves

$$\mu = \frac{1}{\det(A)} = \frac{1}{(1-\kappa)^2 - (\gamma_1^2 + \gamma_2^2)}. \tag{A.117}$$

If all the perturbations are small, then the magnification depends only on κ:

$$\mu \simeq 1 + 2\kappa. \tag{A.118}$$

Exercise 3 (a) Reading off from Eq. (10.14), we see immediately that

$$\phi = 2 \int_0^{\chi_S} d\chi \, \frac{\chi_S - \chi}{\chi_S \chi} \, \Phi(\chi\vec{\theta}, \chi), \tag{A.119}$$

where I have let $\chi \to \chi_S$ in Eq. (10.14) and replaced the dummy variable χ' there with χ. The only subtlety here is the extra factor of χ in the denominator. This comes from changing the derivative with respect to position (the comma in Eq. (10.14)) to an angular derivative.

Exercise 4 Recall that, in the Newtonian limit, the gravitational potential can be written in terms of the mass density:

$$\Phi(\vec{x}) = -G \int \frac{d^3 x'}{|\vec{x} - \vec{x}'|} \rho(\vec{x}'). \tag{A.120}$$

We will do this integral in cylindrical coordinates, so that $\vec{x}' = (\vec{R}, \chi')$. Thus,

$$\phi \simeq -2G \frac{\chi_S - \chi_L}{\chi_S \chi_L} \int d^2 R \int d\chi' \rho(\vec{R}, \chi') \int_0^{\chi_S} \frac{d\chi}{\sqrt{(\vec{R} - \chi_L\vec{\theta})^2 + (\chi - \chi')^2}} \tag{A.121}$$

where I have set $\chi = \chi_L$ in the slowly varying factors out front. The innermost integral can be done analytically: it is equal to

$$2\ln \left| x + \sqrt{(\vec{R} - \chi_L\vec{\theta})^2 + x^2} \right| \Bigg|_0^\infty$$

where I have set the upper limit to infinity because there is no contribution to the relevant part of the projected potential from large x. In fact, the only part which depends on $\vec{\theta}$ (and hence is relevant when derivatives are taken) comes from the lower limit: $-2\ln|\vec{R} - \chi_L\vec{\theta}|$. The integral over χ' then becomes the surface density leaving the desired result.

CHAPTER 11

Exercise 4 The noise matrix is

$$(C_N)_{ij} = \frac{1}{\bar{n}} \int d^3x \psi_i(\vec{x}) \psi_j(\vec{x}) \tag{A.122}$$

in this case of constant $\bar{n}$. Let's consider first the diagonal elements of the matrix. For these, both ψ_i and ψ_j require x to be within a radius R of the center of the ith cell, so

$$(C_N)_{ii} = \bar{n} \int_{x<R} d^3x = \frac{4\pi\bar{n}R^3}{3} \tag{A.123}$$

(no sum over i intended). For cells separated by more than $2R$, the integral vanishes since $\vec{x}$ cannot be within a distance R of both cell centers. For distances less than $2R$ there is some overlap and the integral becomes

$$\bar{n} \int_{x<R} d^3x \Theta\left(R - |\vec{x} - \vec{r}|\right) = 2\pi\bar{n} \int_0^R dx x^2 \int_{-1}^1 d\mu \Theta\left(R - \sqrt{x^2 + r^2 - 2xr\mu}\right) \tag{A.124}$$

where $\vec{r}$ is the difference between the positions of the two cell centers and Θ is the step function equal to 1 if its argument is positive and zero otherwise. The μ integral therefore goes runs from $(x^2 + r^2 - R^2)/2xr$ up to 1. If this lower limit is greater than 1, then the μ integral vanishes; otherwise it is unity. The only contribution then comes when the lower limit is less than 1, which happens when x lies between $r \pm R$. The integral is therefore

$$\int_{r-R}^R dx x^2 \left[1 - \frac{x^2 + r^2 - R^2}{2xr}\right] = \frac{1}{2r} \int_{r-R}^R dx x \left[2xr - (x^2 + r^2 - R^2)\right]. \tag{A.125}$$

The x integral here is then tedious but completely straightforward since the integrand is simply powers of x. I find that

$$(C_N)_{ij} = \frac{\pi\bar{n}R^3}{12} \left(2 - \frac{r_{ij}}{R}\right)^2 \left(4 + \frac{r_{ij}}{R}\right). \tag{A.126}$$

Exercise 7 We need to compute the integral of Eq. (11.55). Since the window function is sharply peaked for small scale modes, we can set k everywhere to k_i. Then inserting our explicit expression for the window function in a volume-limited survey (Eq. (11.59)), we are left with

$$(C_S)_{ii} \simeq \frac{9P(k_i)}{4\pi^2 R^2} \int_0^\infty dk \int_{|k-k_i|R}^{(k+k_i)R} \frac{dy}{y} j_1^2(y). \tag{A.127}$$

The best way to do the integrals here is to switch orders of integration. Consider the Figure A.2, which shows the region of integration. The region below the horizontal

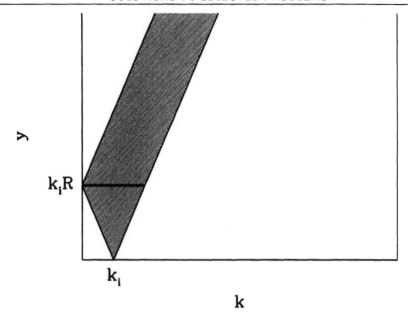

Figure A.2. Region of integration for the double integral in Eq. (A.128). The region below the horizontal line constitutes the first term, above the second term.

line corresponds to $y < k_i R$ and $k_i - y/R < k < k_i + y/R$. In the region above k is bounded by $y/R \pm k_i$. Therefore,

$$\int_0^\infty dk \int_{|k-k_i|R}^{(k+k_i)R} \frac{dy}{y}\, j_1^2(y) = \int_0^{k_i R} \frac{dy}{y}\, j_1^2(y) \int_{k_i-y/R}^{k_i+y/R} dk + \int_{k_i R}^\infty \frac{dy}{y}\, j_1^2(y) \int_{y/R-k_i}^{k_i+y/R} dk$$

$$= \frac{2}{R} \int_0^{k_i R} dy\, j_1^2(y) + 2k_i \int_{k_i R}^\infty \frac{dy}{y}\, j_1^2(y). \qquad (A.128)$$

In the limit $k_i R \gg 1$, the first term here is much larger than the second (since $j_1^2(y)$ goes as $1/y^2$ for large y). In the first integral, we may replace the upper limit by infinity, again since $k_i R$ is large. The resulting integral is (Eq. (C.17)) $\pi\Gamma(3/2)/4\Gamma(5/2) = \pi/6$. Multiplying this by $2/R$ and then by $9/4\pi^2 R^2$ leads to a factor of $1/V$.

Exercise 9 We want to compute the variance

$$\left\langle \left(\hat\lambda - \bar\lambda\right)^2 \right\rangle = \left\langle \left(F_{\lambda\lambda}^{-1} \frac{\Delta C^{-1} C_{,\lambda} C^{-1}\Delta - \mathrm{Tr}[C^{-1}C_{,\lambda}]}{2}\right)^2 \right\rangle \qquad (A.129)$$

where the estimator $\hat\lambda$ is given by Eq. (11.92), and I have assumed that $\lambda^{(0)} = \bar\lambda$, i.e., we are at the true maximum of the likelihood function. Upon squaring there are terms with no Δ; those with two Δ's; and those with four. The ones with two Δ's can be simply evaluated by using

$$\langle \Delta_i \Delta_j \rangle = C_{ij} \tag{A.130}$$

where the indices label the pixels. Since the distribution is assumed Gaussian, the expectation value of four Δ's is

$$\langle \Delta_i \Delta_j \Delta_k \Delta_l \rangle = C_{ij} C_{kl} + C_{ik} C_{jl} + C_{il} C_{jk}. \tag{A.131}$$

Putting in these expectation values leads to

$$\langle \left(\hat{\lambda} - \bar{\lambda} \right)^2 \rangle = \frac{F_{\lambda\lambda}^{-2}}{4} \left\{ \left[C^{-1} C_{,\lambda} C^{-1} \right]_{ij} \left[C^{-1} C_{,\lambda} C^{-1} \right]_{kl} (C_{ij} C_{kl} + C_{ik} C_{jl} + C_{il} C_{jk}) \right.$$

$$\left. - \left(\mathrm{Tr}[C^{-1} C_{,\lambda}] \right)^2 \right\}. \tag{A.132}$$

The $C_{ij} C_{kl}$ terms lead to $(\mathrm{Tr}[C^{-1} C_{,\lambda}])^2$, cancelling the similar term on the last line, so

$$\langle \left(\hat{\lambda} - \bar{\lambda} \right)^2 \rangle = \frac{F_{\lambda\lambda}^{-2}}{4} \left[C^{-1} C_{,\lambda} C^{-1} \right]_{ij} \left[C^{-1} C_{,\lambda} C^{-1} \right]_{kl} (C_{ik} C_{jl} + C_{il} C_{jk}). \tag{A.133}$$

Both terms here contribute identically (giving one factor of 2). The matrix multiplication simplifies since all matrices are symmetric. For example,

$$\left[C^{-1} C_{,\lambda} C^{-1} \right]_{ij} \left[C^{-1} C_{,\lambda} C^{-1} \right]_{kl} C_{ik} C_{jl} = \mathrm{Tr} \left[C^{-1} C_{,\lambda} C^{-1} C_{,\lambda} \right] \tag{A.134}$$

and we recognize the right-hand side as $2F_{\lambda\lambda}$ (another factor of 2). Therefore,

$$\langle \left(\hat{\lambda} - \bar{\lambda} \right)^2 \rangle = F_{\lambda\lambda}^{-1}. \tag{A.135}$$

It is important to keep in mind that this equality holds only if the overdensities are distributed as Gaussians and if we truly have reached the point in parameter space which is the true maximum.

Exercise 15 (a) Use a likelihood approach. The likelihood function for the parameters, here the amplitudes of the different components Θ^α, is proportional to $e^{-\chi^2/2}$ with

$$\chi^2 = (d - W\Theta) N^{-1} (d - W\Theta) + \sum_{\alpha=1}^{N_{\text{foregrounds}}} (\Theta^\alpha)^2 / C^\alpha. \tag{A.136}$$

The first term is identical to that generated by Eq. (11.144). The second accounts for the prior, that Θ^α has mean zero and variance C^α. Maximizing the likelihood corresponds to minimizing the χ^2, with respect to the parameters Θ. Since the χ^2 is quadratic, the minimization leads to a linear equation for Δ^α, the minimum variance estimator of Θ^α:

$$\Delta = \left(WN^{-1}W + C^{-1}\right)^{-1} WN^{-1}d. \tag{A.137}$$

The new covariance matrix is the first term on the right,

$$C_N = \left(WN^{-1}W + C^{-1}\right)^{-1}. \tag{A.138}$$

Here the matrix C is diagonal with $_{00}$ element equal to zero, and the other diagonal elements equal to the assumed power spectra of the foregrounds.

(b) If there is only one foreground with shape vector $W^1 = (1, 1/2)$, and if this foreground has assumed power equal to the noise, then the new inverse covariance matrix goes from that in Eq. (11.151) to the same matrix with $1/\sigma_n^2$ added to the $_{11}$ component. Thus,

$$C_N^{-1} = \frac{1}{\sigma_n^2} \begin{pmatrix} 2 & 3/2 \\ 3/2 & 9/4 \end{pmatrix}. \tag{A.139}$$

Note that the $_{00}$ component of this is unchanged, as it must be since it is the inverse covariance if all foregrounds are known. The inverse of this gives the new covariance matrix,

$$C_N = \frac{4\sigma_n^2}{9} \begin{pmatrix} 9/4 & -3/2 \\ -3/2 & 2 \end{pmatrix}. \tag{A.140}$$

Immediately, we see that the noise in the presence of foregrounds is σ_n. This is a factor of $\sqrt{5}$ smaller than if we had no prior knowledge of the foreground amplitude. It is only a factor of $\sqrt{2}$ larger than the case without foregrounds; thus the new FDF in this case is $\sqrt{2}$. The minimum variance estimator is

$$\Delta^0 = \frac{4\sigma_n^2}{9} (9/4, -3/2) \begin{pmatrix} 1 & 1 \\ 1 & 1/2 \end{pmatrix} \begin{pmatrix} 1/\sigma_n^2 & 0 \\ 0 & 1/\sigma_n^2 \end{pmatrix} \begin{pmatrix} d_1 \\ d_2 \end{pmatrix}$$

$$= \frac{d_1 + 2d_2}{3}. \tag{A.141}$$

APPENDIX B

NUMBERS

Numbers in parentheses denote one standard deviation uncertainties in last digits (e.g., the Rydberg $\epsilon_0 = 13.60569172 \pm 5.3 \times 10^{-7}$ eV). The vast majority of these numbers, at least the physical constants, come from the Particle Data Group (Groom *et al.*, 2001).

B.1 PHYSICAL CONSTANTS

Fine structure constant	α	$=$	$1/137.03599976(50)$
Rydberg	ϵ_0	$=$	$m_e c^2 \alpha^2 / 2$
		$=$	$13.60569172(53)\,\text{eV}$
Thomson cross-section	σ_T	$=$	$8\pi \alpha^2 \hbar^2 / 3 m_e^2 c^2$
		$=$	$0.665245854(15) \times 10^{-24}\,\text{cm}^2$
Neutron lifetime	τ_n	$=$	$885.7(0.8)\,\text{sec}$
Speed of light	c	$=$	$2.99792458 \times 10^{10}\,\text{cm sec}^{-1}$
Fermi constant	G_F	$=$	$1.16639(1) \times 10^{-5}\,\text{GeV}^{-2}(\hbar c)^3$

Newton's constant	G	$=$	$6.673(10) \times 10^{-8} \, \mathrm{cm^3 g^{-1} sec^{-2}}$
		$=$	$\hbar c^5 m_{\mathrm{Pl}}^{-2}$
Reduced Planck's constant	$\hbar$	$=$	$6.58211889(26) \times 10^{-16} \, \mathrm{eV\ sec}$
		$=$	$1.973269602(77) \times 10^{-5} \, \mathrm{eV\ cm}/c$
Boltzmann constant	k_B	$=$	$8.617342(15) \times 10^{-5}], \mathrm{eV\ K^{-1}}$
Electron mass	m_e	$=$	$0.510998902(21) \, \mathrm{MeV}/c^2$
Neutron mass	m_n	$=$	$939.565330(38) \, \mathrm{MeV}/c^2$
Proton mass	m_p	$=$	$1.67262158(13) \times 10^{-24} \, \mathrm{g}$
		$=$	$938.271998(38) \, \mathrm{MeV}/c^2$
Planck mass	m_{Pl}	$=$	$1.221 \times 10^{19} \, \mathrm{GeV}/c^2$
		$=$	$1.094 \times 10^{-38} \, M_\odot$
Neutron–proton mass difference	$\mathcal{Q}$	$=$	$1.2933 \, \mathrm{MeV}/c^2$

B.2 COSMOLOGICAL CONSTANTS

Cosmic microwave background	ρ_γ	$=$	$\pi^2 k_B^4 T^4 / 15(\hbar c)^3$
energy density		$=$	$2.47 \times 10^{-5} h^{-2} (T/T_0)^4 \rho_{\mathrm{cr}}$
Critical density	ρ_{cr}	$=$	$1.879 \, h^2 \times 10^{-29} \, \mathrm{g\ cm^{-3}}$
		$=$	$2.775 \, h^2 \times 10^{11} \, M_\odot \mathrm{Mpc^{-3}}$
		$=$	$8.098 \, h^2 \times 10^{-11} \, \mathrm{eV^4}/(\hbar c)^3$

Massive neutrino density	$\Omega_\nu h^2$	$=$	$(m_\nu/94\,\text{eV})$
Massless neutrino density (N generations)	$\Omega_\nu h^2$	$=$	$1.68 \times 10^{-5}(N/3)$
Scale factor at equality	a_{eq}	$=$	$4.15 \times 10^{-5}(\Omega_m h^2)^{-1}$
Wavenumber at equality	k_{eq}	$=$	$0.073\,\Omega_m h^2\,\text{Mpc}^{-1}$
Hubble constant	H_0	$=$	$100h\,\text{km}\,\text{sec}^{-1}\,\text{Mpc}^{-1}$
		$=$	$2.133h \times 10^{-42}\,\text{GeV}/\hbar$
		$=$	$1.023h \times 10^{-10}\,\text{year}^{-1}$
Solar mass	$M_\odot$	$=$	$1.989 \times 10^{33}\,\text{g}$
		$=$	$1.116 \times 10^{57}\,\text{GeV}/c^2$
Parsec	pc	$=$	$3.0856 \times 10^{18}\,\text{cm}$
Cosmic microwave background temperature today	T_0	$=$	$2.725(2)\,\text{K}$
		$=$	$2.348 \times 10^{-4}\,\text{eV}/k_B$

APPENDIX C

SPECIAL FUNCTIONS

Here is a very brief summary of special functions, focusing primarily on properties relevant to the calcuations in the text. For a more complete treatment, see, e.g., *Handbook of Mathematical Functions* (Abramowitz and Stegun).

C.1 LEGENDRE POLYNOMIALS

The Legendre polynomial $\mathcal{P}_l(\mu)$ is an lth-order polynomial of μ. For $-1 \leq \mu \leq 1$, $\mathcal{P}_l$ has l zeroes in this interval. Some special values are

$$\mathcal{P}_0(\mu) = 1 \quad ; \quad \mathcal{P}_1(\mu) = \mu \quad ; \quad \mathcal{P}_2(\mu) = \frac{3\mu^2 - 1}{2}. \tag{C.1}$$

The property observed in these first few, that $\mathcal{P}_l$ is an even function of μ for l even and an odd function for l odd, holds for all l. They are orthogonal so that

$$\int_{-1}^{1} d\mu \mathcal{P}_l(\mu) \mathcal{P}_{l'}(\mu) = \delta_{ll'} \frac{2}{2l + 1}. \tag{C.2}$$

To generate the higher order ones starting from the low ones, one can use the recurrence relation

$$(l + 1)\mathcal{P}_{l+1}(\mu) = (2l + 1)\mu\mathcal{P}_l(\mu) - l\mathcal{P}_{l-1}(\mu). \tag{C.3}$$

This relation is useful for expressing the Boltzmann equations in terms of moments.

C.2 SPHERICAL HARMONICS

Spherical harmonics are eigenfunctions of the angular part of the Laplacian,

$$\left[\frac{1}{\sin\theta} \frac{\partial}{\partial\theta} \left(\sin\theta \frac{\partial}{\partial\theta} \right) + \frac{1}{\sin^2\theta} \frac{\partial^2}{\partial\phi^2} \right] Y_{lm}(\theta, \phi) = -l(l + 1)Y_{lm}(\theta, \phi). \tag{C.4}$$

In the text, we decomposed the CMB temperature into spherical harmonics (Eq. (8.60)); this decomposition is the analogue of a Fourier decomposition in flat

418

space. The CMB temperature is defined on the sphere, i.e., is a function of θ, ϕ, while the 3D galaxy density, for example, is a function of all three spatial coordinates so is expanded in Fourier coefficients. Some special values are

$$Y_{00}(\theta, \phi) = \frac{1}{\sqrt{4\pi}} \tag{C.5}$$

$$Y_{10}(\theta, \phi) = i\sqrt{\frac{3}{4\pi}} \cos(\theta) \tag{C.6}$$

$$Y_{1,\pm 1}(\theta, \phi) = \mp i\sqrt{\frac{3}{8\pi}} \sin(\theta) e^{\pm i\phi} \tag{C.7}$$

$$Y_{20}(\theta, \phi) = \sqrt{\frac{5}{16\pi}} \left(1 - 3\cos^2\theta\right) \tag{C.8}$$

$$Y_{2,\pm 1}(\theta, \phi) = \pm i\sqrt{\frac{15}{8\pi}} \cos\theta \sin\theta e^{\pm i\phi} \tag{C.9}$$

$$Y_{2,\pm 2}(\theta, \phi) = -\sqrt{\frac{15}{32\pi}} \sin^2\theta e^{\pm 2i\phi}. \tag{C.10}$$

These functions are orthogonal, with normalization

$$\int d\Omega Y_{lm}^*(\Omega) Y_{l'm'}(\Omega) = \delta_{ll'}\delta_{mm'}. \tag{C.11}$$

Another useful expression is the Legendre polynomial in terms of a sum of products of the spherical harmonics:

$$\mathcal{P}_l(\hat{x} \cdot \hat{x}') = \frac{4\pi}{2l+1} \sum_{m=-l}^{l} Y_{lm}(\hat{x}) Y_{lm}^*(\hat{x}'). \tag{C.12}$$

C.3 SPHERICAL BESSEL FUNCTIONS

Spherical Bessel functions are crucial in the study of the CMB and large-scale structure in part because they project the inhomogeneities at last scattering onto the sky today. They satisfy the differential equation

$$\frac{d^2 j_l}{dx^2} + \frac{2}{x}\frac{dj_l}{dx} + \left[1 - \frac{l(l+1)}{x^2}\right] j_l = 0. \tag{C.13}$$

The lowest several are

$$j_0(x) = \frac{\sin(x)}{x} \qquad ; \qquad j_1(x) = \frac{\sin x - x\cos x}{x^2}. \tag{C.14}$$

The key integral relating Legendre polynomials to spherical Bessel functions is

$$\frac{1}{2} \int_{-1}^{1} d\mu P_l(\mu) e^{iz\mu} = \frac{j_l(z)}{(-i)^l}.$$ (C.15)

The inverted version of this leads to a useful expansion for Fourier basis functions:

$$e^{i\vec{k}\cdot\vec{x}} = \sum_{l=0}^{\infty} i^l (2l+1) j_l(kx) P_l(\hat{k}\cdot\hat{x}).$$ (C.16)

Another important integral for the Sachs-Wolfe effect is

$$\int_0^\infty dx\, x^{n-2} j_l^2(x) = 2^{n-4}\pi \frac{\Gamma\left(l+\frac{n}{2}-\frac{1}{2}\right)\Gamma(3-n)}{\Gamma\left(l+\frac{5}{2}-\frac{n}{2}\right)\Gamma^2\left(2-\frac{n}{2}\right)}.$$ (C.17)

Another important relation which eliminates derivatives is

$$\frac{dj_l}{dx} = j_{l-1} - \frac{l+1}{x} j_l.$$ (C.18)

C.4 FOURIER TRANSFORMS

Our Fourier convention is

$$f(\vec{x}) = \int \frac{d^3k}{(2\pi)^3} e^{i\vec{k}\cdot\vec{x}} \, \tilde{f}(\vec{k})$$

$$\tilde{f}(\vec{k}) = \int d^3x\, e^{-i\vec{k}\cdot\vec{x}} \, f(\vec{x}).$$ (C.19)

The power spectrum is then the Fourier transform of the correlation function, with

$$\langle \tilde{\delta}(\vec{k}) \tilde{\delta}(\vec{k}') \rangle = (2\pi)^3 \delta^3(\vec{k} - \vec{k}') P(k).$$ (C.20)

C.5 MISCELLANEOUS

We just need a couple of relations involving ordinary Bessel functions,

$$J_n(x) = \frac{i^{-n}}{\pi} \int_0^\pi d\theta\, e^{ix\cos\theta} \cos(n\theta)$$ (C.21)

and

$$\frac{d}{dx}\left[x J_1(x) \right] = x J_0(x).$$ (C.22)

The Γ function for integers is simply related to factorials:

$$\Gamma(n+1) = n!.$$ (C.23)

More generally

$$\Gamma(x+1) = x\Gamma(x)$$ (C.24)

even if x is not an integer. The Sachs–Wolfe integral (Eq. (C.17)) for a Harrison–Zel'dovich–Peebles spectrum ($n = 1$) depends on

$$\Gamma(3/2) = \frac{\sqrt{\pi}}{2}. \tag{C.25}$$

The Riemann zeta function is useful for evaluating integrals in statistical mechanics. In particular,

$$\zeta(s) = \frac{1}{\Gamma(s)} \int_0^\infty dx \frac{x^{s-1}}{e^x - 1} = \frac{1}{(1 - 2^{1-s})\Gamma(s)} \int_0^\infty dx \frac{x^{s-1}}{e^x + 1}. \tag{C.26}$$

Some low integer Riemann zeta functions are

$$\zeta(2) = \frac{\pi^2}{6} \quad ; \quad \zeta(3) = 1.202 \quad ; \quad \zeta(4) = \frac{\pi^4}{90}. \tag{C.27}$$

APPENDIX D

SYMBOLS

Symbol	Explanation	First page listed
$\dot{\ }$	Derivative with respect to time (before Chapter 4) or conformal time (afterwards)	
$\alpha^{(2)}$	Recombination rate of hydrogen	71
β	Ionization rate of hydrogen	71
$\Gamma^\mu_{\alpha\beta}$	Christoffel symbol	30
$\gamma_{1,2}$	Two components of shear	300
Γ	Parameter determining the power spectrum	205
δ_b	Baryon overdensity	106
$test_i$	Estimated anisotropy in pixel i	340
$\Delta^2(k)$	Dimensionless power on scale k	185
δ	Dark matter overdensity	104
δ	Slow-roll parameter (Chapter 6 only)	155
$\delta^D(\vec{k} - \vec{k}')$	Dirac delta function in D dimensions	16
$\delta\phi$	Perturbation to the scalar field driving inflation	152
δT^μ_ν	Perturbation to energy–momentum tensor	163
δ_{ij}	Kronecker delta = $0(i \neq j)$ or $1(i = j)$	27
δ_H	Amplitude of primordial perturbations at horizon	171
ϵ	Slow-roll parameter	155
$\hat{\epsilon}$	Polarization unit vector	97
$\epsilon_{1,2}$	Two components of ellipticity	301
ϵ_0	Ionization energy of hydrogen, 13.6 eV	70
η	Conformal time	34
η_*	Conformal time at recombination	218
η_{eq}	Conformal time at matter–radiation equality	213
η_b	Baryon-to-entropy ratio	62
η_{prim}	Conformal time at the end of inflation	149
$\eta_{\mu\nu}$	Minkowski metric	26

Symbol	Explanation	First page listed
Θ	Perturbation to photon distribution	93
Θ_l	Legendre moment of photon perturbation	110
Θ_P	Polarization perturbation	111
Θ_r	Perturbation to radiation $= \rho_\gamma \Theta + \rho_\nu \mathcal{N}$	135
Θ^T	Photon perturbation due to tensor perturbations	116
κ	Convergence	300
Λ	Cosmological constant	10
μ	Cosine of the angle between $\hat{k}$ and $\hat{p}$	101
$\xi(r)$	3D correlation function	264
ξ^0, ξ	Generators of coordinate transformations	133
ρ_b	Baryon energy density	41
ρ_{cr}	Critical energy density	3
ρ_{de}	Dark energy density	50
ρ_{dm}	Dark matter energy density	123
ρ_{m}	Matter energy density	38
ρ_γ	Energy density of photons	40
ρ_ν	Energy density of neutrinos	46
ρ_r	Energy density of all radiation	38
σ_T	Thomson cross-section	72
$\tau(\eta)$	Optical depth of photons back to conformal time η	101
$\dot{\tau}$	Scattering rate	101
τ_n	Neutron lifetime	67
Φ	Scalar perturbation to metric	87
Φ_{p}	Primordial value of Φ set during inflation	183
$\phi^{(0)}$	Zero-order value of the field driving inflation	152
$\chi(z)$	Comoving distance out to redshift z	34
χ_∞	Comoving distance to redshift infinity	263
Ψ	Scalar perturbation to metric	87
ψ_{ij}	2×2 distortion tensor	302
Ω_i	Energy density in ith species over ρ_{cr}	10
Ω_k	Ratio of curvature density to critical density	35
A_{ij}	2×2 transformation matrix	300
a	Scale factor of the universe	2
a_*	Scale factor at recombination	186
a_{eq}	Scale factor at matter–radiation equality	51
a_{late}	Scale factor after which perturbations evolve as D_1	183
B	B-mode of polarization or weak lensing	306
B_D	Binding energy of deuterium	65
C	Full covariance matrix	341
$\mathcal{C}$	Band power	389
C_l^{matter}	Angular power spectrum for matter	290

Symbol	Explanation	First page listed
c_s	Sound speed	82
C_N	Covariance matrix due to the noise	339
C_S	Covariance matrix due to the signal	340
D_1	Growth function	183
d_A	Angular diameter distance	35
d_L	Luminosity distance	36
E	E-mode of polarization or weak lensing	306
$F_{\alpha\beta}$	Fisher matrix	366
$\mathcal{F}$	Curvature matrix	365
f	Distribution function, often referring to photons	38
$f_{\rm dm}$	Distribution function of dark matter	102
f_e	Distribution function of electrons	95
$f^{(0)}$	Zero-order distribution function of photons	93
$g(\eta)$	Visibility function	236
g_*	Effective relativistic degrees of freedom	67
$g_{\mu\nu}$	Metric	25
g_i	Number of spin states of species i	38
G	Newton's constant	3
$G_{\mu\nu}$	Einstein tensor	32
h	Parameter for Hubble constant	5
$\tilde{h}$	Variable tracing tensor perturbations	158
$h_\times, h_+$	Tensor perturbations to metric	116
$\mathcal{H}$	3D matrix describing tensor perturbations	126
H	Hubble rate of expansion	3
H_0	Hubble rate today	3
k	Wavenumber	101
$k_i = k^i$	Wavevector	101
$k_{\rm eq}$	Wavenumber crossing horizon at $a_{\rm eq}$	194
$k_{\rm nl}$	Wavenumber of nonlinearity	185
$k_{\rm p}$	Location of acoustic peaks	229
$\mathcal{L}$	Likelihood function	337
$\mathcal{M}$	Particle physics amplitude for a process	59
m_e	Electron mass	70
m_n	Neutron mass	64
m_ν	Neutrino mass	46
$m_{\rm Pl}$	Planck mass	53
m_p	Proton mass	64
N_p	Number of pixels in an experiment	341
$n_{\rm b}$	Baryon number density	62
$n_{\rm dm}$	Dark matter number density	103
$n_{\rm dm}^{(0)}$	Zero-order dark matter number density	104

Symbol	Explanation	First page listed
$n^{(0)}$	Equilibrium number density	61
$\mathcal{N}$	Perturbation to neutrino distribution function	111
$\mathcal{P}_l$	Legendre polynomial of order l	112
$\mathcal{P}$	Pressure	37
P^α	4D comoving energy–momentum vector	31
p	Proper momentum	56
$P(k)$	Power spectrum of matter	16
$P_\Phi(k)$	Gravitational potential power spectrum	167
$\hat{p}^i = \hat{p}_i$	Unit direction vector	90
$\mathcal{Q}$	Proton–neutron mass difference	65
Q	Stokes parameter	312
r	Tensor/scalar ratio	248
r_s	Sound horizon	228
$R_{\mu\nu}$	Ricci tensor	32
$\mathcal{R}$	Ricci scalar $= g^{\mu\nu}R_{\mu\nu}$	32
R	Baryon-to-photon ratio, $3\rho_b/4\rho_\gamma$	82
s	Entropy density	40
t	Age of the universe	2
T_{ant}	Antenna temperature	379
T	Zero-order photon temperature	4
$T_{\mu\nu}$	Stress–energy tensor	32
U	Stokes parameter	312
$\vec{v}_{\text{b}} = \hat{k}v_{\text{b}}$	Velocity of baryons	96
$\vec{v} = \hat{k}v$	Velocity of dark matter	103
v_{H}	Velocity due to Hubble expansion	261
v_{pec}	Peculiar velocity	261
w	Pressure to energy-density ratio	50
$w(\theta)$	Angular correlation function	266
X_e	Free electron fraction	70
X_n	Neutron abundance	66
$X_{n,\text{EQ}}$	Equilibrium neutron abundance	66
Y_p	Mass fraction of ^4He	69
y	Scale factor normalized to 1 at a_{eq}	190
y_H	y when mode crosses horizon	202
Y_{EQ}	Equilibrium abundance of dark matter particles	74
z	Redshift	7
z_*	Redshift at recombination	51
z_{eq}	Redshift at matter–radiation equality	51

BIBLIOGRAPHY

Papers typically appear first as preprints on the archive at `http://arXiv.org`. Those which have not been published in paper journals by 2002 are listed below with their *astro-ph* number. For example, the Stebbins paper referenced with `astro-ph/9609149` can be accessed at `http://arXiv.org/abs/astro-ph/?9609149`. These articles are often even easier to retrieve than those in the standard journals.

M. Abramowitz and I. Stegun, *Handbook of Mathematical Functions, with Formulas, Graphs, and Mathematical Tables* (Dover, 1974)

R. Abusaidi *et al.*, *Physical Review Letters* **84**, 5699 (2000)

A. Albrecht and A. S. Stebbins, *Physical Review Letters* **68**, 2121 (1992)

A. Albrecht and P. J. Steinhardt, *Physical Review Letters* **48**, 1220 (1982)

C. Alcock *et al.*, *Nature* **365**, 621 (1993)

R. A. Alpher, J. W. Follin, and R. C. Herman, *Physical Review* **92**, 1347 (1953)

R. A. Alpher and R. Herman, *Physics Today*, 24 (August 1988)

F. Atrio-Barandela and A. G. Doroshkevich, *Astrophysical Journal* **420**, 26 (1994)

D. Bacon, A. Refrgier, and R. S. Ellis, *Monthly Notices of Royal Astronomical Society* **318**, 625 (2000)

J. N. Bahcall, *Neutrino Astrophysics* (Cambridge University Press, Cambridge, 1989)

N. A. Bahcall, L. M. Lubin, and V. Dorman, *Astrophysical Journal* **447**, L81 (1995)

N. Bahcall *et al.*, *Astrophysical Journal* **541**, 1 (2000)

J. M. Bardeen, *Physical Review* **D22**, 1882 (1980)

J. M. Bardeen, J. R. Bond, N. Kaiser, and A. S. Szalay, *Astrophysical Journal* **304**, 15 (1986)

J. M. Bardeen, P. J. Steinhardt, and M. S. Turner, *Physical Review* **D28**, 679 (1983)

M. Bartelmann and P. Schneider, *Physics Reports* **340**, 291 (2001)

L. Baudis *et al.*, *Physical Review* **D63**, 022001 (2000)

C. M. Baugh and G. Efstathiou, *Monthly Notices of Royal Astronomical Society* **265**, 145 (1993)

R. H. Becker *et al.*, *Astronomical Journal* **122**, 2850 (2001)

C. L. Bennett *et al.*, *Astrophysical Journal* **464**, L1 (1996)

A. J. Benson *et al.*, *Monthly Notices of Royal Astronomical Society* **327**, 1041 (2001)

A. A. Berlind and D. H. Weinberg, astro-ph/0109001

R. Bernabei *et al.*, *Physics Letters* **B450**, 448 (1999)

F. Bernardeau, L. V. Waerbecke, and Y. Mellier, *Astronomy and Astrophysics* **322**, 1 (1997)

J. Bernstein, *Three Degrees Above Zero: Bell Labs in the Information Age* (Scribner, 1984)

J. Bernstein, *Kinetic Theory in the Expanding Universe* (Cambridge University Press, Cambridge, 1988)

J. Bernstein, L. S. Brown, and G. Feinberg, *Reviews of Modern Physics* **61**, 25 (1989)

R. D. Blandford, A. B. Saust, T. G. Brainerd, and J. Villumsen, *Monthly Notices of Royal Astronomical Society* **251**, 600 (1991)

G. R. Blumenthal, S. M. Faber, J. R. Primack, and M. J. Rees, *Nature* **311**, 57 (1984)

J. D. Bjorken and S. D. Drell, *Relativistic Quantum Mechanics* (McGraw-Hill, New York, 1965)

J. R. Bond, *Physical Review Letters* **74**, 4369 (1995)

J. R. Bond, S. Cole, G. Efstathiou, and N. Kaiser, *Astrophysical Journal* **379**, 440 (1991)

J. R. Bond *et al.*, *Computing in Science and Engineering* March-April, 21, (1999)

J. R. Bond and G. Efstathiou, *Astrophysical Journal* **285**, L45 (1984)

J. R. Bond, G. Efstathiou, P. M. Lubin, and P. R. Meinhold, *Physical Review Letters* **66**, 2179 (1991)

J. R. Bond, G. Efstathiou, and J. Silk, *Physical Review Letters* **45**, 1980 (1980)

J. R. Bond, A. H. Jaffe, and L. Knox, *Physical Review* **D57**, 2117 (1998)

J. R. Bond and A. Szalay, *Astrophysical Journal* **274**, 443 (1983)

J. R. Bond *et al.*, *Physical Review Letters* **72**, 13 (1994)

A. Bottino, F. Donato, N. Fornengo, and S. Scopel, *Physical Review* **D63**, 125003 (2001)

F. Bouchet and R. Gispert, *New Astronomy* **4**, 443 (1999)

R. Brandenberger, R. Kahn, and W. H. Press, *Physical Review* **D28**, 1809 (1983)

T. J. Broadhurst, A. N. Taylor, and J. A. Peacock, *Astrophysical Journal* **438**, 49 (1995)

E. F. Bunn, Ph.D. Thesis, U.C. Berkeley, 1995

E. F. Bunn and M. White, *Astrophysical Journal* **480**, 6 (1997)

S. Burles and D. Tytler, *Astrophysical Journal* **499**, 699 (1998)

S. Burles, K. M. Nollett, and M. S. Turner, astro-ph/9903300

R. R. Caldwell, R. Dave, and P. J. Steinhardt, *Physical Review Letters* **80**, 1582 (1998)

R. Carlberg *et al.*, *Astrophysical Journal* **478**, 462 (1997)

D. Clowe *et al.*, *Astrophysical Journal* **497**, L61 (1998)

K. A. Coble, Ph.D. Thesis, U. Chicago, 1999, astro-ph/9911414

K. A. Coble, S. Dodelson, and J. A. Frieman, *Physical Review* **D55**, 1851 (1997)

K. A. Coble *et al.*, *Astrophysical Journal* **519**, L5 (1999)

J. M. Colberg *et al.*, *Monthly Notices of Royal Astronomical Society* **319**, 209 (2000)

M. Colless *et al.*, *Monthly Notices of Royal Astronomical Society* **328**, 1039 (2001)

C. A. Collins, R. C. Nichol, and S. L. Lumsden, *Monthly Notices of Royal Astronomical Society* **154**, 295 (1992)

A. Cooray and R. Sheth, `astro-ph/0206508`

E. Corbelli and P. Salucci, *Monthly Notices of Royal Astronomical Society* **311**, 441 (2000)

R. Cowsik and J. McClelland, *Physical Review Letters* **29**, 669 (1972)

R. Crittenden *et al.*, *Physical Review Letters* **71**, 324 (1993)

P. de Bernardis *et al.*, *Nature* **404**, 955 (2000)

A. Dekel, `astro-ph/9705033`, in Proceedings of the 3rd ESO-VLT Workshop on "Galaxy Scaling Relations: Origins, Evolution and Applications," ed. L. da Costa (Springer)

S. Dodelson, *Astrophysical Journal* **482**, 577 (1997)

S. Dodelson, E. I. Gates, and A. S. Stebbins, *Astrophysical Journal* **467**, 10 (1996)

S. Dodelson and E. Gaztanaga, *Monthly Notices of Royal Astronomical Society* **312**, 774 (2000)

S. Dodelson and J. M. Jubas, *Astrophysical Journal* **439**, 503 (1995)

S. Dodelson, W. H. Kinney, and E. W. Kolb, *Physical Review* **D56**, 3207 (1997)

S. Dodelson *et al.*, *Astrophysical Journal* **572**, 140 (2002)

A. G. Doroshkevich, Ya. B. Zel'dovich, and R. A. Sunyaev, *Soviet Astronomy* **22**, 523 (1978)

F. W. Dyson, A. S. Eddington, and C. Davidson, *Philosophical Transactions of the Royal Society* **220A**, 291 (1920)

J. Edsjo and P. Gondolo, *Physical Review* **D56**, 1879 (1997)

G. Efstathiou, in *Physics of the Early Universe,* ed. J. A. Peacock, A. F. Heavens, and A. T. Davies (Edinburgh University Press, Edinburgh, 1990)

G. Efstathiou and S. J. Moody, *Monthly Notices of Royal Astronomical Society* **325**, 1603 (2001)

D. J. Eisenstein, W. Hu, and M. Tegmark, *Astrophysical Journal* **518**, 2 (1999)

D. J. Eisenstein and M. Zaldarriaga, *Astrophysical Journal* **546**, 2 (2001)

V. R. Eke, S. Cole, and C. S. Frenk, *Monthly Notices of Royal Astronomical Society* **282**, 263 (1996)

O. Elgaroy *et al.*, `astro-ph/0204152`

G. F. R. Ellis and R. M. Williams, *Flat and Curved Space-Times* (Clarendon Press, Oxford, 1990)

J. Ellis, T. Falk, K. A. Olive, and M. Schmitt, *Physics Letters* **B413**, 355 (1997)

X. Fan *et al.*, *Astronomical Journal* **123**, 1247 (2002)

H. Feldman, N. Kaiser, and J. A.Peacock, *Astrophysical Journal* **426**, 23 (1994)

T. Ferris, *The Whole Shebang* (Touchstone Books, 1998)

R. A. Fisher, *Journal of the Royal Statistical Society* **98**, 39 (1935)

D. J. Fixsen *et al.*, *Astrophysical Journal* **473**, 576 (1996)

W. Freedman, *Scientific American* (March 1998)

W. Freedman *et al.*, *Astrophysical Journal* **553**, 47 (2001)

W. Freudling *et al.*, `astro-ph/9904118`, *Astrophysical Journal* **523**, 1 (1999)

Y. Fukuda *et al.*, *Physical Review Letters* **81**, 1562 (1998)

M. Fukugita, C. J. Hogan, and P. J. E. Peebles, *Astrophysical Journal* **503**, 518 (1998)

T. Gaier *et al.*, *Astrophysical Journal* **398**, L1 (1992)

G. Gerstein and Ya. B. Zel'dovich, *Zhurnal Eksperimental'noi i Teoreticheskoi Fiziki, Pis'ma v Redakts* **4**, 174 (1966)

G. Gibbons, S.Hawking, and T. Vachaspati eds., *The Formation and Evolution of Cosmic Strings* (Cambridge University Press, Cambridge 1990)

Gradshteyn and Ryzhik, *Table of Integrals, Series, and Products* (Academic Press, San Diego 1994)

L. Grego *et al.*, *Astrophysical Journal* **552**, 2 (2001)

D. E. Groom *et al.*, *The European Physical Journal*, **C15**, 1 (2000) and 2001 off-year partial update for the 2002 edition available on the PDG WWW pages (URL: http://pdg.lbl.gov/)

J. E. Gunn *et al.*, *Astrophysical Journal* **223**, 1015 (1978)

A. Guth, *Physical Review* **D23**, 347 (1981)

A. Guth and E. J. Weinberg, *Nuclear Physics* **B212**, 321 (1983)

A. Guth, *The Inflationary Universe* (Perseus, 1998)

A. Guth and S.-Y. Pi, *Physical Review Letters* **49**, 1110 (1982)

A. Guth and S.-Y. Pi, *Physical Review* **D32**, 1899 (1985)

N. W. Halverson *et al.*, *Astrophysical Journal* **568**, L28 (2002)

A. J. S. Hamilton, *Astrophysical Journal* **385**, L5 (1992)

A. J. S. Hamilton, `astro-ph/9708102`, in "The Evolving Universe" ed. D. Hamilton, Kluwer Academic, p. 185-275 (1998)

A. J. S. Hamilton and M. Tegmark, `astro-ph/0008392`

S. Hanany *et al.*, *Astrophysical Journal* **545**, L5 (2000)

S. Hawking, *Physics Letters* **B115**, 295 (1982)

S. W. Hawking, I. G. Moss, and J. M. Stewart, *Physical Review* **D26**, 2681 (1982)

M. P. Haynes *et al.*, *Astronomical Journal* **117**, 2039 (1999)

M. M. Hedman *et al.*, *Astrophysical Journal* **573**, L73 (2002)

J. P. Henry and K. A. Arnaud, *Astrophysical Journal* **372**, 410 (1991)

C. Hogan, *The Little Book of the Big Bang* (Copernicus, 1998)

D. W. Hogg, `astro-ph/9905116`

S. Hollands and R. M. Wald, `gr-qc/0205058`

W. Hu and D. J. Eisenstein, *Astrophysical Journal* **498**, 497 (1998)

W. Hu, M. Fukugita, M. Zaldarriaga, and M. Tegmark, `astro-ph/0006436`

W. Hu and N. Sugiyama, *Astrophysical Journal* **444**, 489 (1995)

W. Hu and N. Sugiyama, *Astrophysical Journal* **471**, 542 (1996)

W. Hu and M. Tegmark, *Astrophysical Journal* **514**, L65 (1999)

W. Hu and M. White, *Astrophysical Journal* **479**, 568 (1997a)

W. Hu and M. White, *New Astronomy* **2**, 323 (1997b)

E. Hubble, *Proceedings of the National Academy of Sciences* **15** (1929)

D. Huterer, L. Knox, and R. C. Nichol, *Astrophysical Journal* **555**, 547 (2001)

Y. I. Izotov and T. X. Thuan, *Astrophysical Journal* **500**, 188 (1998)

J. D. Jackson, *Classical Electrodynamics*, (Wiley, New York, 1998)

B. Jain and U. Seljak, *Astrophysical Journal* **484**, 560 (1997)

H. E. Jorgenson, E. Kotok, P. Naselsky, and I. Novikov, *Astronomy and Astrophysics* **294**, 639 (1995)

G. Jungman, M. Kamionkowski, and K. Griest, *Physics Reports* **267**, 195 (1996)

G. Jungman, M. Kamionkowski, A. Kosowsky, and D. N. Spergel, *Physical Review Letters* **76**, 1007 (1995)

N. Kaiser, *Monthly Notices of Royal Astronomical Society* **227**, 1 (1987)

N. Kaiser, *Astrophysical Journal* **388**, 272 (1992)

N. Kaiser, G. Wilson, and G. Luppino, `astro-ph/0003338`

M. Kamionkowski, A. Kosowsky, and A. Stebbins, *Physical Review Letters* **78**, 2058 (1997a)

M. Kamionkowski, A. Kosowsky, and A. Stebbins, *Physical Review* **D55**, 7368 (1997b)

M. Kamionkowski, D. N. Spergel, and N. Sugiyama, *Astrophysical Journal* **426**, L57 (1994)

G. L. Kane, C. F. Kolda, L. Roszkowski, and J. D. Wells, *Physical Review* **D49**, 6173 (1994)

G. Kauffmann, J. M. Colberg, A. Diaferio, and S. D. M. White, *Monthly Notices of Royal Astronomical Society* **303**, 188 (1999)

B. G. Keating, *Astrophysical Journal* **560**, L1 (2001)

C. R. Keeton, *Astrophysical Journal* **575**, L1 (2002)

M. G. Kendall and A. Stuart, *The Advanced Theory of Statistics*, Vol. II (Van Nostrand, New York, 1969)

M. Kesden, A. Cooray, and M. Kamionkowski, *Physical Review Letters* **89**, 011304 (2002)

L. Knox, *Physical Review* **D52**, 4307 (1995)

L. Knox and Y.-S. Song, *Physical Review Letters* **89**, 011303 (2002)

C. S. Kochanek, *Astrophysical Journal* **466**, 638 (1996)

H. Kodama and M. Sasaki, *Prog. Theor. Phys. Suppl.* **78**, 1 (1984)

E. W. Kolb, *Blind Watchers of the Sky* (Perseus, 1996)

E. W. Kolb and M. S. Turner, *The Early Universe* (Addison-Wesley, Redwood City, CA 1990)

A. Kosowsky, *Annals of Physics* **246**, 49 (1996)

J. Kovac *et al.*, `astro-ph/0209478`

L. M. Krauss and B. Chaboyer, `astro-ph/0111597`

T. Kundic *et al.*, *Astrophysical Journal* **482**, 75 (1997)

L. D. Landau and E. M. Lifshitz, *Quantum Mechanics* (Pergamon, Oxford 1977)

A. Lawrence *et al.*, *Monthly Notices of Royal Astronomical Society* **308**, 897 (1999)

A. T. Lee *et al.*, *Astrophysical Journal* **561**, L1 (2001)

A. R. Liddle and D. H. Lyth, *Physics Reports* **231**, 1 (1993)

A. R. Liddle and D. H. Lyth, *Cosmological Inflation and Large Scale Structure* (Cambridge University Press, Cambridge, 2000)

J. E. Lidsey et al, *Reviews of Modern Physics* **69**, 373 (1997)

E. M. Lifshitz, *Journal of Physics (Moscow)* **10**, 116 (1946)

D. N. Limber, *Astrophysical Journal* **117**, 134 (1953)

A. Linde, *Physics Letters* **B108**, 389 (1982)

A. Linde, *Particle Physics and Inflationary Cosmology* (Harwood Academic, Chur, Switzerland 1990)

J. Loveday, B. A. Peterson, S. J. Maddox, and G. Efstathiou, *Astrophysical Journal Supplement* **107**, 201 (1996)

D. H. Lyth and A. Riotto, *Physics Reports* **314**, 1 (1999)

C.-P. Ma and E. Bertschinger, *Astrophysical Journal* **455**, 7 (1995)

S. J. Maddox, G. Efstathiou, W. J. Sutherland, and J. Loveday, *Monthly Notices of Royal Astronomical Society* **242**, 43p (1990)

H. Magira, Y. P. Ying, and Y. Suto, *Astrophysical Journal* **528**, 30 (2000)

R. Maoli *et al.*, astro-ph/0011251

G. Marx and A. Szalay in *Neutrino '72,* ed. A. Frenkel and G. Marx (OMKDT-Technoinform, Budapest, 1972)

J. C. Mather *et al.*, *Astrophysical Journal* **420**, 439 (1994)

J. C. Mather *et al.*, *Astrophysical Journal* **512**, 511 (1999)

A. D. Miller *et al.*, *Astrophysical Journal* **524**, L1 (1999)

J. Miralda-Escude, *Astrophysical Journal* **380**, 1 (1991)

C. Misner, K. S. Thorne, and J. A. Wheeler, *Gravitation* (Freeman, San Francisco 1973)

R. Moessner and B. Jain, *Monthly Notices of Royal Astronomical Society* **294**, L18 (1998)

R. N. Mohapatra and P. B. Pal, *Massive Neutrinos in Physics and Astrophysics* (World Scientific, 1991)

V. F. Mukhanov, H. A. Feldman , and R. H. Brandenberger, *Physics Reports* **215**, 203 (1992)

P. Nath and R. Arnowitt, *Physics Letters* **B289**, 368 (1992)

C. B. Netterfield *et al.*, *Astrophysical Journal* **571**, 604 (2002)

K. A. Olive, *Nucl. Phys. Proc. Suppl.* **80**, 79 (2000)

K. A. Olive, G. Steigman, and T. P. Walker, *Physics Reports* **333**, 389 (2000)

J. M. O'Meara *et al.*, *Astrophysical Journal* **552**, 718 (2001)

T. Padmanabhan, *Structure Formation in the Universe* (Cambridge Universe, Cambridge 1993)

B. E. J. Pagel, E. A. Simonson, R. J. Terlevich, and M. Edmunds, *Monthly Notices of Royal Astronomical Society* **255**, 325 (1992)

R. B. Partridge, *3K: The Cosmic Microwave Background* (Cambridge University, Cambridge 1995)

J. A. Peacock, *Cosmological Physics* (Cambridge University Press, Cambridge, 1999)

J. A. Peacock and S. J. Dodds, *Monthly Notices of Royal Astronomical Society* **267**, 1020 (1994)

J. A. Peacock and A. F. Heavens, *Monthly Notices of Royal Astronomical Society* **243**, 133 (1990)

P. J. E. Peebles, *Astrophysical Journal* **146**, 542 (1966)

P. J. E. Peebles, *Astrophysical Journal* **153**, 1 (1968)

P. J. E. Peebles, *Astrophysical Journal* **263**, L1 (1982)

P. J .E. Peebles, *The Large-Scale Structure of the Universe* (Princeton University Press, Princeton, 1980)

P. J. E. Peebles and J. T. Yu, *Astrophysical Journal* **162**, 815 (1970)

U.-L. Pen and D. N. Spergel, *Physical Review* **D51**, 4099 (1995)

W. J. Percival *et al.*, astro-ph/0105252

L. Perivolaropoulos and T. Vachaspati, *Astrophysical Journal* **423**, 77 (1994)

S. Perlmutter *et al.*, *Astrophysical Journal* **517**, 565 (1999)

M. Persic and P. Salucci, *Monthly Notices of Royal Astronomical Society* **258**, 14 (1992)

E. Pierpaoli, D. Scott, and M. White, *Monthly Notices of Royal Astronomical Society* **325**, 77 (2001)

M. H. Pinsonneault, G. Steigman, T. P. Walker, and V. K. Narayanan, astro-ph/0105439

A. G. Polnarev, *Soviet Astronomy* **29**, 607 (1985)

W. H. Press and P. Schechter, *Astrophysical Journal* **187**, 425 (1974)

W. H. Press *et al.*, *Numerical Recipes in Fortran*, 2nd ed. (Cambridge University Press, Cambridge, 1992)

C. Pryke *et al.*, astro-ph/0104490

M. Rauch *et al.*, *Astrophysical Journal* **489**, 7 (1997)

A. C. S. Readhead *et al.*, *Astrophysical Journal* **346**, 566 (1989)

A. G. Riess *et al.*, *Astronomical Journal* **116**, 1009 (1998)

J. Renn, T. Sauer, and J. Stachel, *Science* **275** 184 (1997)

L. Roszkowski, *Physics Letters* **B309**, 329 (1993)

V. Rubin, *Scientific American*, June (1983).

G. B. Rybicki and A. P. Lightman, *Radiative Processes in Astrophysics* (Wiley, New York, 1979)

R. K. Sachs and A. M. Wolfe, *Astrophysical Journal* **147**, 73 (1967)

A. Sandage, *Astrophysical Journal* **152**, L149 (1968)

W. Saunders *et al.*, *Monthly Notices of Royal Astronomical Society* **317**, 55 (2000)

R. J. Scherrer and M. S. Turner, *Physical Review* **D33**, 1585 (1986)

D. J. Schlegel, D. P. Finkbeiner, and M. Davis, *Astrophysical Journal* **500**, 525 (1998)

P. Schneider, J. Ehlers, and E. E. Falco, *Gravitational Lenses* (Springer-Verlag, Heidelberg, 1992)

D. N. Schramm and M. S. Turner, *Reviews of Modern Physics* **70**, 303 (1998)

B. F. Schutz, *A First Course in General Relativity* (Cambridge University Press, Cambridge, 1990)

R. Scranton *et al.* (The SDSS Collaboration), *Astrophysical Journal* **579**, 48 (2002)

S. Seager, D. Sasselov, and D. Scott, *Astrophysical Journal* **128**, 407 (2000)

W. Rindler, *Essential Relativity* (Springer, 1977)

U. Seljak, *Astrophysical Journal* **435**, L87 (1994)

U. Seljak, *Astrophysical Journal* **482**, 6 (1997)

U. Seljak, `astro-ph/0001493`

U. Seljak and M. Zaldarriaga, *Astrophysical Journal* **469**, 437 (1996)

U. Seljak and M. Zaldarriaga, *Physical Review Letters* **78**, 2054 (1997)

R. K. Sheth and G. Tormen, *Monthly Notices of Royal Astronomical Society* **308**, 119 (1999)

J. Silk, *Astrophysical Journal* **151**, 459 (1968)

J. Silk, *The Big Bang* (W. H. Freeman & Co, 2001)

E. Skillman and R. C. Kennicutt, *Astrophysical Journal* **411**, 655 (1993)

E Skillman *et al.*, *Astrophysical Journal* **431**, 172 (1994)

G. Smoot *et al.*, *Astrophysical Journal* **396**, L1 (1992)

R. S. Somerville and J. R. Primack, *Monthly Notices of Royal Astronomical Society* **310**, 1087 (1999)

T. Souradeep and B. Ratra, `astro-ph/0105270`

A. A. Starobinsky, *Physics Letters* **B117**, 175 (1982)

A. Stebbins, `astro-ph/9609149`

E. Stewart and D. Lyth, *Physics Letters* **B302**, 171 (1993)

M. Strauss *et al.*, *Astrophysical Journal Supplement* **83**, 29 (1992)

M. A. Strauss and J. A. Willick, *Physics Reports* **261**, 271 (1995)

A. S. Szalay, T. Matsubara, and S. D. Landy, *Astrophysical Journal* **498**, L1 (1998)

M. Tegmark, *Astrophysical Journal* **46**, L35 (1996)

M. Tegmark and G. Efstathiou, *Monthly Notices of Royal Astronomical Society* **281**, 1297 (1996)

M. Tegmark, *Physical Review* **D55**, 5895 (1997)

M. Tegmark, A. Taylor, and A. Heavens, *Astrophysical Journal* **480**, 22 (1997)

M. Tegmark *et al.*, *Astrophysical Journal* **499**, 555 (1998)

M. Tegmark, D. J.Eisenstein, W. Hu, and A. de Oliveira-Costa, *Astrophysical Journal* **530**, 133 (2000)

M. Tegmark *et al.*, *Astrophysical Journal* **571**, 191 (2002)

K. S. Thorne, *Black Holes and Time Warps* (Norton, New York 1994)

M. S. Turner, M. White, and J. E. Lidsey, *Physical Review* **D48**, 4613 (1993)

L. van Waerbecke *et al.*, *Astronomy and Astrophysics* **358**, 30 (2000)

P. T. P. Viana and A. R. Liddle, *Monthly Notices of Royal Astronomical Society* **281**, 323 (1996)

P. T. P. Viana and A. R. Liddle, *Monthly Notices of Royal Astronomical Society* **303**, 535 (1999)

A. Vilenkin and E. P. S. Shellard, *Cosmic Strings and Other Topological Defects* (Cambridge University Press, Cambridge 2000)

M. S. Vogeley and A. S. Szalay, *Astrophysical Journal* **465**, 34 (1996)

R. V. Wagoner, *Astrophysical Journal* **179**, 343 (1973)

R. V. Wagoner, W. A. Fowler, and F. Hoyle, *Astrophysical Journal* **148**, 3 (1967)

R. M. Wald, *General Relativity* (University of Chicago, Chicago 1984)

D. Walsh, R. F. Carswell, and R. J. Weymann, *Nature* **279**, 381 (1979)

L. Wang and P. J. Steinhardt, *Astrophysical Journal* **508**, 483 (1998)

S. Weinberg, *Gravitation and Cosmology* (Wiley and Sons, New York 1972)

S. Weinberg, *The First Three Minutes* (Perseus Books 1993)

M. White, *Monthly Notices of Royal Astronomical Society* **321**, 1 (2001)

S. D. M. White, G. Efstathiou, and C. S. Frenk, *Monthly Notices of Royal Astronomical Society* **262**, 1023 (1993)

S. D. M. White and C. S. Frenk, *Astrophysical Journal* **379**, 52 (1991)

S. D. M. White, C. S. Frenk, and M. Davis, *Astrophysical Journal* **274**, L1 (1983)

S. D. M. White, J. F. Navarro, A. E. Evrard, and C. S. Frenk, *Nature* **366**, 429 (1993)

S. D. M. White and M. Rees, *Monthly Notices of Royal Astronomical Society* **183**, 341 (1978)

G. Wilson *et al.*, *Astrophysical Journal* **532**, 57 (2000)

M. L. Wilson and J. Silk, *Astrophysical Journal* **243**, 14 (1981)

D. M. Wittman *et al.*, *Nature* **405**, 143 (2000)

J. Yang, M. S. Turner, G. Steigman, D. N. Schramm, and K. A. Olive, *Astrophysical Journal* **281**, 493 (1984)

D. G. York *et al.*(The SDSS Collaboration), *Astronomical Journal* **120**, 1579 (2000)

M. Zaldarriaga and D. Harari, *Physical Review* **D52**, 3276 (1995)

M. Zaldarriaga and U. Seljak, *Physical Review* **D55**, 1830 (1997)

F. Zwicky, *Physica Acta* **6**, 124 (1933)

INDEX

ABOUT THE AUTHOR

Scott Dodelson is Head of the Theoretical Astrophysics Group at Fermilab and Associate Professor in the Department of Astronomy and Astrophysics at the University of Chicago. He received his Ph.D. from Columbia University and was a research fellow at Harvard before coming to Fermilab and Chicago. He is the author of more than seventy papers on cosmology, most of which focused on the cosmic microwave background and the large scale structure of the universe. Dodelson is a theoretical cosmologist, but has worked with several experiments, including the Sloan Digital Sky Survey and the Python and MSAM anisotropy experiments.

Printed and bound by CPI Group (UK) Ltd, Croydon, CR0 4YY

03/10/2024

01040313-0015